Health Informatics

Practical Guide for Healthcare and Information Technology Professionals

Fifth Edition

ROBERT E. HOYT MD FACP

Editor

ANN K. YOSHIHASHI MD FACE

Associate Editor

NORA J. BAILEY MSP

Associate Editor

Health Informatics

Practical Guide for Healthcare and Information Technology Professionals

Disclaimer

Every effort has been made to make this book as accurate as possible but no warranty is implied. The information provided is on an "as is" basis. The authors and the publisher shall have neither liability nor responsibility to any person or entity with respect to any loss or damages arising from the information contained in this book. The views expressed in this book are those of the author and do not necessarily reflect the official policy or position of the University of West Florida, Department of the Navy, Department of Defense, nor the U.S. Government.

Editors

Robert E. Hoyt MD, FACP

Director Medical Informatics

School of Allied Health and Life Sciences

University of West Florida

Pensacola, FL

Assistant Professor of Medicine

Adjunct Assistant Professor of Family Medicine

The Uniformed Services University of the Health Sciences

Bethesda, MD

Ann K. Yoshihashi MD, FACE

Guest Lecturer, School of Allied Health and Life Sciences

Medical Informatics

University of West Florida

Medical Analyst

Navy Medicine Support Command

Pensacola, FL

Nora J. Bailey, MSP

Program Director

Office of Economic Development and Engagement

University of West Florida

Pensacola, FL

Contributors

Kenneth G. Adler MD, MMM
Medical Director of Information Technology
Arizona Community Physicians
Tucson, AZ

Elmer V. Bernstam MD, MSE
Professor of Biomedical Informatics and Internal Medicine
School of Biomedical Informatics and Medical School
The University of Texas Health Science Center
Houston, TX

Trevor Cohen MD, MBChB, PhD
Assistant Professor of Biomedical Informatics
School of Biomedical Informatics
The University of Texas Health Science Center
Houston, TX

Robert W. Cruz
Director
Software Engineering
Computer Programs and Systems, Inc
Mobile, AL

LaJuana Ehlers, MEd, RT(R)(M)
Assistant Professor
Radiologic Sciences
Mesa State College
Health Science Department
Grand Junction, CO

Reynald Fleury MPH
Global Strategic Marketing Manager -Oncology
GE Healthcare
Amersham, United Kingdom

M. Chris Gibbons MD, MPH
Associate Director
Johns Hopkins Urban Health Institute
Assistant Professor of Medicine, Public Health and Health Informatics
Johns Hopkins Medical Institutions
Baltimore, MD

Ronald W. Gimbel PhD
Interim Chairman
Biomedical Informatics
Uniformed Services University
Bethesda, MD

Jorge Herskovic MD, PhD
Assistant Professor
School of Biomedical Informatics
The University of Texas Health Science Center
Houston, TX

Brent Hutfless MS, CISSP, GSLC
IT Security Manager
Austal USA
Mobile, AL

Todd Johnson PhD
Professor
Division of Biomedical Informatics
University of Kentucky
Lexington, KY

Ken Masters PhD
Assistant Professor of Medical Informatics
Medical Education Unit
College of Medicine & Health Sciences
Sultan Qaboos University
Sultanate of Oman
Editor in Chief, *The Internet Journal of Medical Education*

Justice Mbizo Dr PH
Assistant Professor
Masters of Public Health Program
School of Allied Health and Life Sciences
University of West Florida
Pensacola, FL

M. Hassan Murad MD, MPH
Associate Professor of Medicine
Mayo Clinic College of Medicine
Rochester, MN

Indra Neil Sarkar, PhD, MLIS
Assistant Professor
Departments of Microbiology & Molecular Genetics and Computer Science
Director of Biomedical Informatics
University of Vermont
Burlington, VT

John Sharp MSSA, PMP, FHIMSS
Manager
Research Informatics
Quantitative Health Sciences
Cleveland Clinic
Cleveland, OH

Brandy G. Ziesemer, RHIA, CCS
Health Information Manager and Associate Professor
Lake-Sumter Community College
Leesburg, FL

Preface to the Fifth Edition

Perhaps the most obvious change in the fifth edition of our textbook is the title change. We feel that health informatics is the most accurate term to describe the contents of the book. The American Medical Informatics Association (AMIA) no longer favors the term medical informatics, instead preferring the phrase health informatics. AMIA states that health informatics is "applied research and practice in clinical and public health informatics." We also changed the formatting of the book and graphics to enhance its readability.

Following the publication of the fourth edition of this textbook we witnessed the increased adoption of electronic health records in the United States as a result of legislation such as the HITECH Act. It will likely take longer to measure the impact of other programs related to this Act such as the Statewide Health Information Exchange Cooperative Agreement and Beacon Community programs. It is important for the average reader to understand the implications of such sweeping legislation on the implementation of HIT into the practice of medicine. The fifth edition emphasizes the importance of Meaningful Use of electronic health records in multiple chapters.

In this edition we have re-written all chapters to reflect the changes brought about by federal programs and the exceedingly rapid pace of change in technology in general. Two new chapters have been added; a chapter on healthcare data, information and knowledge and another on ethical issues related to health informatics. We removed a final chapter on "Emerging Trends"; replacing it with a "Future Trends" section in every chapter. Case examples have been added to most chapters to highlight interesting initiatives. Additionally, we added several national and two international co-authors with the goal to include more HIT projects outside the United States and try to reach a broader audience.

The authors have made every attempt to provide the most up-to-date information about health informatics based on the continuous review of medical and lay literature. Our goal is to present the most recent changes and the most interesting concepts. We are dedicated to presenting the issues fairly and objectively and have attempted to present both sides of any controversy. We consider our approach to be most consistent with applied informatics and not theoretical informatics. Specifically, we have sought to discuss actual hardware, software, concepts and initiatives we believe are important. Our book is intended to be an introduction to the field of health informatics that will entice individuals to go further in their education. In our experience, individuals from very diverse backgrounds are interested in this relatively new field. This introductory textbook should give readers, especially those new to healthcare or technology, a better understanding of this burgeoning field. Approximately 1500 medical literature references and web links are included in this book that help direct readers to additional information.

While we are vendor agnostic we are not opposed to presenting interesting hardware and software, including open source, we think will be of interest to our readers. One of the goals of this book is to promote and disseminate innovations that might help healthcare workers as well as technology developers. The fact that we mention specific hardware or software or web-based applications does not mean we endorse the vendor; instead, it is our attempt to highlight an interesting concept that might lead others in a new direction.

We appreciate feedback regarding how to make this book as user friendly, accurate, up-to-date and educational as possible. Please note that all proceeds will be donated to support the advancement of health informatics education.

Robert E. Hoyt MD FACP
Ann K. Yoshihashi MD FACE
Nora J. Bailey MSP

Table of Contents

Table of Figures

Table of Tables

1

Overview of Health Informatics

ROBERT E. HOYT

ELMER V. BERNSTAM

Learning Objectives

After reading this chapter the reader should be able to:

- State the definition and origin of health informatics

- Identify the forces behind health informatics

- Describe the key players involved in health informatics

- State the potential impact of the ARRA and HITECH Act on health informatics in the United States

- List the barriers to health information technology (HIT) adoption

- Describe educational and career opportunities in health informatics

Introduction

"During the past few decades the volume of medical knowledge has increased so rapidly that we are witnessing an unprecedented growth in the number of medical specialties and subspecialties. Bringing this new knowledge to the aid of our patients in an economical and equitable fashion has stressed our system of medical care to the point where it is now declared to be in a crisis. All these difficulties arise from the present, nearly unmanageable volume of medical knowledge and the limitations under which humans can process information."

- Marsden S. Blois, *Information and Medicine: The Nature of Medical Descriptions*, 1984

Health informatics began as a new field of study in the 1950s-1960s time frame but only recently gained recognition as an important component of many aspects of healthcare. Its emergence is partly due to the multiple challenges facing the practice of medicine today. As the 1984 quote above indicates, the growth in the volume of medical knowledge and patient information that has occurred due to better understanding of human health has resulted in more treatments and interventions that produce more information. Likewise, the increase in specialization has also created the need to share and coordinate patient information. Furthermore, clinicians need to be able to access medical information expeditiously, regardless of location or time of day. Technology has the potential to help with each of those areas. With the advent of the internet, high speed computers, voice recognition, wireless and mobile technology, healthcare professionals today have many more tools available at their disposal. However, in general, technology is advancing faster than healthcare professionals can assimilate it into their practice of medicine. One could also argue that there is a critical limitation of current information technology that manages data and not information. Thus, there is a mismatch between what we need (i.e. something to help us manage meaningful data = information) and what we have (ineffective ways to manage information). Additionally, given the volume of data and rapidly changing technologies, there is a great need for ongoing Informatics education of all healthcare workers.

In this chapter we will present an overview of health informatics with emphasis on the factors that helped create and sustain this new field and the key players involved.

Data, Information, Knowledge, Wisdom Hierarchy

Informatics is the science of information and the blending of people, biomedicine and technology. Individuals who practice informatics are known as informaticians or informaticists, such as, a nurse informaticist. There is an information hierarchy that is important in the information sciences, as depicted in the pyramid in Figure 1.1. Notice that there is much more data than information, knowledge or wisdom. As data are consumed and analyzed the amount of knowledge and wisdom produced is much smaller. The following are definitions to better understand the hierarchy:

- Data are symbols or observations reflecting differences in the world. Data are the plural of datum (singular). Thus, a datum is the lowest level of abstraction, such as a number in a database (e.g. 5), or packets sent across a network (e.g. 10010100). Note that there is no meaning associated with data; the 5 could represent five fingers, five minutes or have no real meaning at all. Modern computers process data accurately and rapidly.

- Information is meaningful data or facts from which conclusions can be drawn by humans or computers. For example, *five fingers* has meaning in that it is the number of fingers on a normal human hand. Modern computers do not process information, they process data. This is a fundamental problem and challenge in informatics.

- Knowledge is information that is justifiably considered to be true. For example, a rising prostate specific antigen (PSA) level suggests an increased likelihood of prostate cancer.

- Wisdom is the critical use of knowledge to make intelligent decisions and to work through situations of signal versus noise. For example, a rising PSA could mean prostate infection and not cancer.

Figure 1.1: Information hierarchy

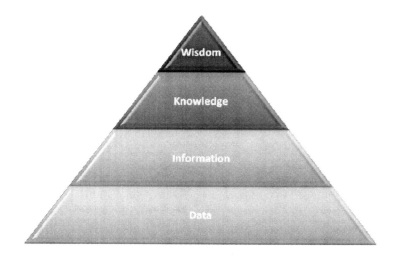

Health information technology provides the tools to generate information from data that humans (clinicians and researchers) can turn into knowledge and wisdom.[1-2] Thus, enabling and improving human decision making with usable information is a central concern of informaticians. This concept is discussed in much more detail in Chapter 2 on healthcare data, information and knowledge.

Another important concept to understand about data is that there are different levels of data (Figure 1.2). Paper forms would be considered level 1 with serious limitations, in regards to sharing, storing and analyzing. Level 2 data could be scanned-in documents. Level 3 data are entered into a computer and are data that are structured and retrievable, but not computable between different computers. Level 4 data are computable data. That means the data are electronic, capable of being stored in data fields and computable because it is in a format that disparate computers can share (interoperable) and interpret (analyzable).

Figure 1.2: Levels of data (Courtesy Government Accounting Office)

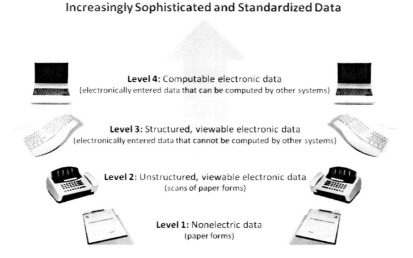

Therefore, the information sciences tend to promote data in formats that can be rapidly transmitted, shared and analyzed. Paper records and reports do not allow this, without a great deal of manual labor. The advent of electronic health records, health information exchanges (HIEs) and multiple hospital electronic information systems provided the ability and the need to collate and analyze large amounts of data to improve health and financial decisions. Enterprise systems have been developed that: integrate disparate information (clinical, financial and administrative); archive data; provide the ability to data mine using business intelligence and analytic tools. Figure 1.3 demonstrates a typical enterprise data system.

Figure 1.3: Enterprise data warehouse and data mining

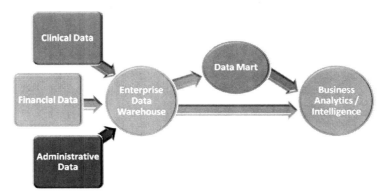

Informatics Definitions

Health informatics is the field of information science concerned with management of healthcare data and information through the application of computers and other technologies. In reality, it is more about applying information in the healthcare field than it is about technology per se. That is one of the many reasons it is different than a pure information technology (IT) position in a healthcare organization. Technology merely facilitates the collection, storage, transmission and analysis of data. This field also includes data standards (such as HL7) and controlled medical vocabularies (such as SNOMED) we will cover in the chapter on data standards.

The definition of health informatics is dynamic because the field is relatively new and rapidly changing. The following are four definitions frequently cited:

- "science of information, where information is defined as data with meaning. Biomedical informatics is the science of information applied to, or studied in the context of biomedicine. Some, but not all of this information is also knowledge"[3]

- "scientific field that deals with resources, devices and formalized methods for optimizing the storage, retrieval and management of biomedical information for problem solving and decision making"[4]

- "application of computers, communications and information technology and systems to all fields of medicine - medical care, medical education and medical research"[5]

- "understanding, skills and tools that enable the sharing and use of information to deliver healthcare and promote health"[6]

Health informatics is also known as *clinical informatics or medical informatics* and *biomedical informatics* in some circles. If the information science deals primarily with actual applications and programs and not theory, it can be referred to as applied informatics. *Bioinformatics*, on the other hand involves the integration of biology and technology and can be defined as the:

- "analysis of biological information using computers and statistical techniques; the science of developing and utilizing computer databases and algorithms to accelerate and enhance biological research."[7]

Some prefer the broader term *biomedical informatics* because it encompasses bioinformatics as well as medical, dental, nursing, public health, pharmacy, medical imaging and veterinary informatics.[8] As we move closer to integrating human genetics into the day-to-day practice of medicine this more global definition may gain traction. We have chosen to use health informatics throughout the book for consistency. The Americian Medical Informatics Association no longer uses the term medical informatics and they include clinical informatics and public health informatics under health informatics, which in turn falls under biomedical informatics.

Health information technology (HIT or healthIT) is defined as the application of computers and technology in healthcare settings.

Health information management (HIM) traditionally focused on the paper medical record and coding. With the advent of the electronic health record HIM specialists now have to deal with a new set of issues, such as privacy and multiple new concepts such as voice recognition.

For a discussion of the definition, concepts and implications (e.g. distinguishing from other related fields) of this field, we refer you to a 2010 article by Bernstam, Smith and Johnson and a 2009 article by Hersh.[3,9]

Background

Given the fact that most businesses incorporate technology into their enterprise fabric, one could argue that it was just a matter of time before the tectonic forces of medicine and technology collided. As more medical information was published and more healthcare data became available as a result of computerization, the need to automate, collect and analyze data escalated. Also, as new technologies such as electronic health records appeared, ancillary technologies such as disease registries, voice recognition and picture archiving and communication systems arose to augment functionality. In turn, these new technologies prompted the need for expertise in health information technology that spawned new specialties and careers.

Health informatics emphasizes *information brokerage*; the sharing of a variety of information back and forth between people and healthcare entities. Examples of medical information that needs to be shared include: lab results, x-ray results, vaccination status, medication allergy status, consultant's notes and hospital discharge summaries. Medical informaticians harness the power of information technology to expedite the transfer and analysis of data, leading to improved efficiencies and knowledge. The field also interfaces with other fields such as the health sciences, computer sciences, biomedical engineering, biology, library sciences and public health, to mention a few. Informatics training, therefore, must be expansive and in addition to the topics covered in the chapters of this book must include IT knowledge about networks and systems, usability, process re-engineering, workflow analysis and redesign quality improvement, project management, leadership, teamwork, implementation and training.

Health information technology (HIT) facilitates the processing, tranmsmission and analysis of information and HIT interacts with many important functions in healthcare organizations and serves as a common thread. (Figure 1.4). This is one of the reasons the Joint Commission created the management of information management standard for hospital certification.[10]

Figure 1.4: Information, information technology and healthcare functions

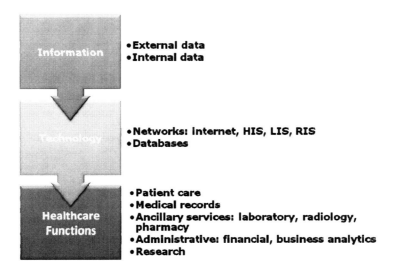

Many aspects of health informatics noted in Figure 1.4 are interconnected. To accomplish data collection and analysis there are hospital information systems (HISs) that collect financial, administrative and clinical information and subsystems such as the laboratory (LISs) and radiology information systems (RISs). As an example, a healthcare organization is concerned that too many of its diabetics are not well controlled and believes it would benefit by offering a diabetic web portal. With a portal, diabetics can upload blood sugars and blood pressures to a central web site so that diabetic educators and/or clinicians can analyze the results and make recommendations. They also have the option to upload physiologic parameters via their smart phone. The following technologies and issues are involved with just this one initiative and we will cover each in other chapters:

- The web-based portal involves consumer (patient) informatics and telemedicine

- Use of a smart phone is an important type of mobile technology

- Management of diabetes requires online medical resources, evidence based medicine, clinical practice guidelines, disease managment and an electronic health record with a disease registry

- If the use of the diabetic web portal improves diabetic control, clinicians may be eligible for improved reimbursement, known as pay-for-performance, a quality improvement strategy

There are multiple forces driving the adoption of health information technology, but the major ones are the need to: increase the efficiency of healthcare (i.e. decrease medical costs and improve physician productivity), improve the quality (patient outcomes) of healthcare, resulting in improved patient safety. Over the past 40 years, there has been increasing recognition that wide variation in practice cannot be justified. For example, patients in some areas of the United States are undergoing more invasive procedures than similar patients in other areas. Thus, there has been a movement to standardize the care of common and expensive conditions, such as coronary artery disease, congestive heart failure and diabetes. Computerized clinical practice guidelines are one way to provide advice at the point of care and we will discuss this in more detail in the chapter on evidence based medicine.

In this book we will discuss the driving forces motivating informatics and their inter-relationships. In addition to the motivation to deliver more efficient, safer and less costly healthcare, there is the natural diffusion of technology which also exerts an influence. In other words, as technologies such as wireless and voice recognition become more common place, easier to use and less expensive, they will have an inevitable impact on the practice of medicine. Technological innovations appear at a startling pace as stated by Moore's Law.

"The number of transistors on a computer chip doubles every 1.5 years"[11]

- Gordon Moore, co-founder Intel Corporation 1965

Moore's Law describes the exponential growth of transistors in computers. Technology will continue to evolve at a rapid rate but it is important to realize that it often advances in an asynchronous manner. For example, laptop computers have advanced greatly with excellent processor speed and memory but their utility is limited by a battery life of roughly 4-6 hours. This is a significant limitation given the fact that most nurses now work eight to 12 hour shifts, so short battery life is one factor that currently limits the utility of laptop computers in healthcare. This may be overcome with tablet computers or a new battery design.

The electronic health record (EHR), covered in Chapter 3, could be considered the centerpiece of health informatics with its potential to improve patient safety, medical quality, productivity and data retrieval. EHRs will likely become the focal point of all patient encounters in the future. Multiple resources that are currently standalone programs are being incorporated into the EHR, e.g. electronic prescribing, physician/patient education, genetic profiles and artificial intelligence. It is anticipated that EHR use will eventually be shown to improve patient outcomes like morbidity and mortality as a result of decision support tools that decrease medication errors and standardize care with embedded clinical guidelines. However, at present, because EHRs do not adequately support clinicians' information needs and workflow, they do little to improve patient care and in some cases have been shown to reduce the quality of care.[12] Informaticians will play a major role in helping to reverse this trend. It will not be enough to simply store electronic data; it must be shared among disparate partners. We will address health information exchange (information sharing) in a separate chapter.

It is also important to realize that one of the outcomes of EHRs will be voluminous healthcare data. As pointed out by Steve Balmer, the CEO of Microsoft, there will be an "explosion of data" as a result of automating and digitizing multiple medical processes.[13] Adding new technologies such as electronic prescribing and health information exchanges will produce data that heretofore has not been available. This explains, in part, why technology giants such as Microsoft, Intel and IBM have entered the healthcare arena. As we begin mining medical data from entire regions or organizations we will be able to make much better evidence based decisions. As we will point out in other chapters, large organizations such as Kaiser Permanente have the necessary information technology tools and large patient population to be able to make evidence based decisions in almost all facets of medicine. Pooling data is essential because most practices in the United States are small and do not provide enough information on their own to show the kind of statistical significance we need to alter the practice of medicine.[14]

The federal government understands the importance of data and information to make evidence based medical decisions. For that reason, they developed http://www.data.gov/communities/health to make datasets from the federal agencies available to a multitude of interested parties, such as healthcare organizations, developers, researchers, etc. Datasets are available through several categories: raw data, special tools and a geodata catalog. As a result of this initiative, a variety of applications, mashups and visualizations have been developed.

Similarly, the Department of Health and Human Services created a Health Indicators Warehouse (HIW) in 2010 that included hundreds of health indicators that will help measure progress towards the Healthy People 2020 program (see info box below). New indicators continue to be added and updated. Importantly, this initiative will be working with technology companies, researchers and others to develop applications and initiatives to improve healthcare.[15-17]

Health Indicators Warehouse

Users can search by:

Topics: chronic disease and conditions, demographics, disabilities, geography, health behaviors, health care, health care resources, health outcomes, health risk factors, hospital referral region, infectious disease initiative, injury and violence, maternal and infant health, mental health and substance abuse, occupational health and safety, oral health, physical environment, population, prevention through healthcare, public health infrastructure, social determinants of health and women's health

Geography: state or county

Initiative: Mobilizing Action towards Community Health, 2008 Community Health Status Indicators, Healthy People 2020, CMS Community Indicators

Data is available to developers via an open application programming interface (API).

www.healthindicators.gov

The most recent and significant event to affect the information sciences in the United States was the multiple programs associated with the HITECH Act of 2009, discussed later in the chapter. The programs include substantial financial support for electronic health records, health information exchange and a skilled HIT workforce. In other chapters we will refer to accountable care organizations and their technology requirements that are part of the Affordable Care Act of 2010.

The introduction of information technology into the practice of medicine has been tumultuous for many reasons. Not only are new technologies expensive, they affect workflow and require advanced training. Unfortunately, this type of training rarely occurs during medical or nursing school or after graduation. More healthcare professionals who are *bilingual* in technology and medicine will be needed to realize the potential of new technologies. Vendors, insurance companies and governmental organizations will also be looking for the same expertise.

Historical Highlights

Information technology has been pervasive in the field of Medicine for only about three decades but its roots began in the 1950s.[18] Since the earlier days we have experienced astronomical advances in technology, to include, personal computers, high resolution imaging, the internet, mobile technology and wireless, to mention only a few. In the beginning there was no strategy or vision as to how to advance healthcare using information technology. Now, we have the involvement of multiple federal and private agencies that are plotting future healthcare reform, supported by health information technology. The following are some of the more noteworthy developments in health information technology:

- Computers. The first general purpose computer (ENIAC) was released in 1946 and required 1,000 sq. ft. of floor space. Primitive computers such as the Commodore and Atari appeared in the early 1980s along with IBM's first personal computer, with a total of 16K of memory.[19] Computers were first theorized to be useful for medical diagnosis and treatment by Ledley and Lusted in the 1950's.[20] They reasoned that computers could archive and process information more rapidly than humans. The programming language known as MUMPS was developed in Octo Barnett's lab at Massachusetts General Hospital in the 1970s. MUMPS exists today in the popular electronic health record know as VistA, used by the Veterans Affairs medical system[21] and Epic Systems Corporation.

- Origin of health informatics. It is thought that the origin of the term medical informatics dates back to the 1960's in France ("Informatique Medicale").[22]

- MEDLINE. In the mid-1960s MEDLINE and MEDLARS were created to organize the world's medical literature. For older clinicians who can recall trying to research a topic using the multi-volume text Index Medicus, this represented a quantum leap forward.

- Artificial Intelligence. Artificial intelligence (AI) medical projects such as MYCIN (Stanford University) and INTERNIST-1 (University of Pittsburg) appeared in the 1970s and 1980s.[23] Since 1966 AI has had many periods where research flourished and where it floundered, known as AI winters.[12] Natural language processing (NLP) is gaining traction in medicine as it has the potential to intelligently interpret free text.

- Internet. The development of the internet began in 1969 with the creation of the government project ARPANET.[24] The World Wide Web (WWW or web) was conceived by Tim Berners-Lee in 1990 and the first web browser Mosaic appeared in 1993.[25-26] The internet is the backbone for digital medical libraries, health information exchanges and web-based medical applications, to include electronic health records. Although the terms *web* and *internet* are often used interchangeably, the internet is the *network-of-networks* consisting of hardware and software that connects computers to each other. The web is a set of protocols (particularly related to HyperText Transfer Protocol or HTTP) that are supported by the internet. Thus, there are many internet applications (e.g. email) that are not part of the web. This is discussed further in the chapter on architectures of information systems.

- Electronic Health Record (EHR). The electronic health record has been discussed since the 1970's and recommended by the Institute of Medicine in 1991.[27] EHRs will be discussed in much more detail in Chapter 3.

- Mobile technology. The PalmPilot PDA appeared in 1996 as the first truly popular handheld computing device.[28] Personal Digital Assistants (PDAs) loaded with medical software became standard equipment for residents in training. They have been quickly supplanted by smartphones like the iPhone. Smartphones and tablets will be discussed in more detail in the chapter on mobile technology.

- Human Genome Project. In 2003 the Human Genome Project (HGP) was completed after thirteen years of international collaborative research. Mapping all human genes was one of the greatest accomplishments in scientific history. Finalizing a draft of the genome is the first step. What remains is making sense of the data. In other words, we need to understand the difference between data (the code), information (what the code means) and knowledge (what we do with the information).[29] Data from mega-databases will likely change the way we practice medicine in the future. The HGP will be discussed in the chapter on bioinformatics.

- Nationwide Health Information Network (NwHIN). The concept was developed in 2004 as the National Health Information Infrastructure and renamed the Nationwide Health Information Network (NwHIN). The goal of the NwHIN is to connect all electronic health records, health information organizations and government agencies in one decade.[30] Achieving interoperability among all healthcare systems and workers in the United States will be a monumental challenge. This will be discussed in more detail in several other chapters.

Key Players in Health Information Technology

Health information technology (HIT) is important to multiple players in the field of medicine. The common goals of these different groups are outlined in Table 1.1.

Table 1.1: Goals of HIT

Goal	Process
Improve	Communication and continuity of care Quality of care Patient outcomes Clinician productivity Return on investment
Reduce	Medical errors and resultant litigation Duplication of tests
Standardize	Medical care by individuals and organizations
Accelerate	Access to care and administrative transactions
Protect	Privacy and ensure security

In the next section we list the key players in HIT and how they utilize health information technology (adapted from *Crossing the Quality Chasm*).[31]

Patients

- Online searches for health information
- Web portals for storing personal medical information, making appointments, checking lab results, e-visits, etc.
- Research choice of physician, hospital or insurance plan
- Online patient surveys
- Online chat, blogs, podcasts, vodcasts and support groups and Web 2.0 social networking

- Personal health records
- Limited access to electronic health records and health information exchanges (HIEs)
- Telemedicine and home telemonitoring

Clinicians and Nurses

- Online searches with PubMed, Google and other search engines
- Online resources and digital libraries
- Patient web portals, secure e-mail and e-visits
- Physician web portals
- Clinical decision support, e.g. reminders and alerts
- Electronic health records (EHRs)
- Smartphones loaded with medical software
- Telemedicine and telehomecare
- Voice recognition software
- Online continuing medical education (CME)
- Electronic prescribing
- Disease management registries
- Picture archiving and communication systems (PACS)
- Pay-for-performance (P4P)
- Health information organizations (HIOs)
- E-research
- Electronic billing and coding

Support Staff

- Patient enrollment
- Electronic appointments
- Electronic coding and billing
- EHRs
- Web-based credentialing
- Web-based claims clearinghouses
- Telehomecare monitoring
- Practice management software
- Secure patient-office e-mail communication
- Electronic medication administration record (e-Mar)
- Online educational resources and CME

- Disease registries

Public Health

- Incident reports
- Syndromic surveillance as part of bio-terrorism program and Meaningful Use program criteria
- Establish link to all public health departments
- Geographic information systems to link disease outbreaks with geography
- Telemedicine
- Disease registries as part of EHRs or health information exchanges
- Remote reporting using mobile technology

Federal and State Governments

- Nationwide Health Information Network
- Financial support for EHR adoption and health information exchange
- Development of standards, services and policies for HIT
- Information technology pilot projects and grants
- Disease management
- Pay-for-performance
- Electronic health records and personal health records
- Electronic prescribing
- Telemedicine
- Broadband adoption
- Health information organizations
- Regional extension centers
- Health IT workforce development

Medical Educators

- Online medical resources for clinicians, patients and staff
- Online CME
- PubMed searches
- Telehealth via video teleconferencing, podcasts, etc.

Insurance Companies (Payers)

- Electronic claims transmission
- Trend analysis

- Physician profiling
- Information systems for quality improvement initiatives
- Monitor adherence to clinical guidelines
- Monitor adherence to preferred formularies
- Promote claims-based personal health records and information exchanges
- Reduce litigation by improved patient safety through fewer medication errors
- Alerts to reduce test duplication
- Member of HIOs

Hospitals

- Electronic health records
- Electronic coding and billing
- Information systems to monitor outcomes, length of stay, disease management, etc.
- Bar coding and radio frequency identification (RFID) to track patients, medications, assets, etc.
- Wireless technology
- E-intensive care units (eICUs)
- Patient and physician portals
- E-prescribing
- Member of health information organizations (HIOs)
- Telemedicine
- Picture archiving and communication systems (PACS)

Medical Researchers

- Database creation to study populations, genetics and disease states
- Online collaborative web sites e.g. CaBIG
- Electronic case report forms (eCRFs)
- Software for statistical analysis of data e.g. SPSS
- Literature searches with multiple search engines
- Randomization using software programs
- Improved subject recruitment using EHRs and e-mail
- Smartphones to monitor research
- Online submission of grants

Technology Vendors

- Applying new technology innovations in the field of medicine: hardware, software, genomics, etc.
- Data mining
- Interoperability
- Certification

Organizations Involved with HIT

Academic Organizations

Institute of Medicine (IOM). One of the leading organizations in the United States to promote health information technology is the Institute of Medicine. It was established in 1970 by the National Academy of Sciences with the task of evaluating policy relevant to healthcare and providing feedback to the Federal Government and the public. In their two pioneering books *To Err is Human* (1999) and *Crossing the Quality Chasm* (2001), they reported approximately 98,000 deaths occur yearly due to medical errors. It is their contention that an information technology infrastructure will help the six aims set forth by the IOM: safe, effective, patient centered, timely, efficient and equitable medical care. The infrastructure would support "efforts to re-engineer care processes, manage the burgeoning clinical knowledge base, coordinate patient care across clinicians and settings over time, support multidisciplinary team functioning, and facilitate performance and outcome measurements for improvement and accountability." They also stress "the importance of building such an infrastructure to support evidence based practice, including the provision of more organized and reliable information sources on the internet for both consumers and clinicians and the development and application of decision support tools."

Two of the IOM's twelve executive recommendations regarding improved healthcare directly relate to information technology:

- "improve access to clinical information and support clinical decision making"
- "Congress, the executive branch, leaders of health care organizations, public and private purchasers and health informatics associations and vendors should make a renewed national commitment to building an information infrastructure to support health care delivery, consumer health, quality measurement and improvement, public accountability, clinical and health services research, and clinical education. This commitment should lead to the elimination of most handwritten clinical data by the end of the decade."

The IOM cited twelve information technology applications that might narrow the quality chasm. Many of these will be discussed in other chapters:

- Web-based personal health records
- Patient's access to hospital information systems to access their lab and x-ray reports
- Access to general health information via the internet
- Electronic medical records with clinical decision support
- Pre-visit online histories
- Inter-hospital data sharing (health information exchange), e.g. lab results
- Information to manage populations using patient registries and reminders
- Patient - physician electronic messaging

- Online data entry by patients for monitoring, e.g. glucose results

- Online scheduling

- Computer assisted telephone triage and assistance (nurse call centers)

- Online access to clinician or hospital performance data.[32-33]

The Association of American Medical Colleges (AAMC). For more than twenty years the AAMC has been an advocate of incorporating informatics into medical school curricula and promoting health informatics in general. In their *Better Health 2010 Report* they made the following recommendations:

- Optimize the health and healthcare of individuals and populations through best practice information management

- Enable continuous and life-long performance-based learning

- Create tools and resources to support discovery, innovation and dissemination of research results

- Build and operate a robust information environment that simultaneously enables healthcare, fosters learning and advances science.[34]

Public-Private Organizations

Bridges to Excellence. This organization consists of employers, physicians, health plans and patients. They currently have multiple care recognition programs incentivized by bonuses: diabetes, cardiac care, congestive heart failure, coronary artery disease, spine care, COPD, asthma, depression, hypertension, physician's office and medical home.[35]

eHealth Initiative. This is a non-profit organization promoting the use of information technology to improve quality and patient safety. Its membership includes virtually all stakeholders involved in the delivery of healthcare. This organization deals with multiple topics related to HIT and has a reports section that provides multiple articles on a variety of HIT topics. They also provide an annual survey of HIOs, starting in 2005. The 2011 survey results are available for a fee, but free for members.[36]

Leapfrog. Leapfrog is a consortium of over one hundred and seventy major employers seeking to purchase the highest quality and safest healthcare. Voluntary reporting by hospitals has made hospital comparisons possible and the results are reported on their website. They also have a hospital rewards program to provide incentives to hospitals that show they deliver quality care. One of their patient safety measures is the use of inpatient computerized physician order entry (CPOE) that will be covered in several other chapters.[37]

Markle Connecting For Health. This organization is a public-private collaboration operated by the Markle Foundation and funded partially by the Robert Wood Johnson Foundation. With over 100 stakeholders, its primary mission is to promote interoperable HIT. They published *Common Framework: Resources for Implementing Private and Secure Health Information Exchange* that helps organizations exchange information in a secure and private manner, with shared policies and technical standards. The Common Framework with nine policy guides and seven technical guides is available free for download on their web site.[38]

National eHealth Collaborative (NeHC). This government-civilian-consumer collaborative took over in early 2009 when the American Health Information Community (AHIC) was dissolved. They are charged with prioritization of HIT standards to promote interoperability. They create *value cases* and refer those for harmonization of standards and once accepted they will be adopted by the certification organizations such as the Certification Commission for Health Information Technology (CCHIT). NeHC is a cooperative agreement partner of the Office of the National Coordinator for Health IT (ONC) and the US Dept. of Health and

Human Services (HHS). NeHC University is a new (2011) online education program to inform stakeholders about multiple HIT issues.[39]

Healthcare Information Technology Standards Panel (HITSP). This panel was a public-private partnership established in 2005 by the Department of Health and Human Services (DHHS). HITSP was charged by the ONC to harmonize standards-based on *use cases* derived from AHIC requirements. Each interoperability specification is a suite of documents that provides a roadmap of how standards and specifications will answer the requirements of the use case. For instance, specifics of the standard for using the Continuity of Care Document (CCD) were released as C32 in March 2008 with a detailed explanation of the technical aspects. The CCD is discussed further in the chapter on data standards. Their contract with the government was terminated in April 2010 and their function was largely replaced by the HIT Standards Sub-Committee discussed in a following section.[40]

The Certification Commission for Healthcare Information Technology (CCHIT) was created by HIMSS and multiple other healthcare professional organizations. Its goals are to: reduce the risk of health information technology (HIT) investment by physicians; ensure interoperability of HIT; enhance the availability of HIT incentives and accelerate the adoption of interoperable HIT. Their initial step was to certify ambulatory electronic health records. By mid-2011 they certified the following categories of HIT: ambulatory EHRs, inpatient EHRs, Health Information Exchanges, Emergency EHRs, Cardiovascular Medicine EHRs, Child Health EHRs, Behavioral Health EHRs, Dermatology, Long Term/Post-Acute Care EHRs, Home Health EHRs and E-prescribing. EHRs that have received certification are listed on the web site. The Commission consists of 20 commissioners from a variety of backgrounds and numerous volunteers in their work groups. CCHIT decided they would offer different levels of EHR certification so more EHRs would qualify for Medicare or Medicaid reimbursement under ARRA: (1) CCHIT certified® 2011, a comprehensive certification that would actually exceed federal standards and includes a usability score, (2) ONC-ATCB Certification 2011-2012 will test EHRs against Meaningful Use regulations, hosted by the National Instititue of Standards and Technology (NIST), (3) EHR vendors can elect to be certified by both CCHIT and ONC-ATCB criteria, and (4) EHR Alternative Certification for Healthcare Providers (EACH) that certifies homegrown technology created by healthcare organizations and not vendors.

As of mid-2011 eighty two ambulatory EHRs had been CCHIT certified with details posted on their site, to include usability ratings. Multiple EHR-related resources are also available. Certification is quite expensive as noted by one reference.[41-42]

National Committee on Vital and Health Statistics (NCVHS) is a public advisory body to the Secretary of Health and Human Services. It is composed of 18 members from the private sector who are subject matter experts in the fields of health statistics, electronic health information exchange, privacy/security, data standards and epidemiology. They have been very involved in advising the Secretary in matters related to the Nationwide Health Information Network (NwHIN).[43]

US Federal Government

The federal government has maintained that information technology is essential to improving the quality of medical care and containing costs; two important aspects of healthcare reform. It is a major financer of health care with the following programs: Medicare/Medicaid, Veterans Health Administration, Military Health System, Indian Health Service and the Federal Employees Health Benefits Program. It is therefore no surprise that they are heavily involved in health information technology and stand to benefit greatly from an interoperable Nationwide Health Information Network. Agencies such as Medicare/Medicaid and the Agency for Healthcare Research and Quality conduct HIT pilot projects that potentially could improve the quality of medical care and/or decrease medical costs. The federal government has recognized the importance of technology in multiple areas and as a result has a new federal chief technology officer and chief technology officer for HHS.

Before specific government agencies are discussed we will outline the new programs included in the American Recovery and Reinvestment Act of 2009 that impact the information sciences.

American Recovery and Reinvestment Act (ARRA). Without a doubt, the most significant recent governmental initiative that affected the field of Informatics was the ARRA. This legislation will impact HIT adoption, particularly EHRs, as well as training and research. ARRA had five broad goals: (a) improve medical quality, patient safety, healthcare efficiency and reduce health disparities; (b) engage patients and families; (c) improve care coordination; (d) ensure adequate privacy and security of personal health information (e) improve population and public health. Title IV and XIII of ARRA, known as the Health Information Technology for Economic and Clinical Health (HITECH) Act was devoted to funding of HIT programs. Table 1.2 summarizes the major pertinent programs that have monies dedicated for these initiatives. The HealthIT website under the DHHS outlines the details of many of the programs listed in the table. In addition to the programs listed in Table 1.2 the following are also important initiatives that were part of the ARRA:

- Privacy and HIPAA changes; to be discussed in chapter on privacy

- The National Telecommunications and Information Administration's Broadband Technology Opportunities Program. This will fund the National Broadband plan discussed in the chapter on telemedicine

- USDA's Distance Learning, Telemedicine and Broadband Program

- Indian Health Services HIT programs

- Social Security Administration HIT programs

- Veterans Affairs (VA) HIT programs [44]

US Department of Health & Human Services (HHS) is the department that serves as an umbrella for most of the important government agencies that impact HIT. The Office of the National Coordinator for Health Information Technology reports directly to the Secretary of HHS and is not an agency. The following are some of the operating divisions under HHS:

- Agency for Healthcare Research & Quality

- Centers for Medicare & Medicaid Services

- Centers for Disease Control & Prevention

- Health Resources & Services Administration

- Indian Health Service

- Food and Drug Administration

- Administration on Aging

- National Institutes of Health [45]

Office of the National Coordinator for Health Information Technology (ONC). The most significant goal of (ONC) is the creation of a universal interoperable electronic health record by the year 2014. To accomplish this goal they are working to harmonize data standards to ensure interoperability and to facilitate health information exchange. ONC reorganized in December 2009, resulting in the following offices: Office of Economic Modeling and Analysis, Office of the Chief Scientist, Office of the Deputy Coordinator for Programs and Policy, Office of the Deputy National Coordinator for Operations and Office of the Chief Privacy Officer. In early 2011 ONC proposed the Goals of the Federal Health IT Strategic Plan and asked for public comment. The following are the broad goals without specific objectives:

Table 1.2: ARRA and HITECH programs that impact the information sciences and HIT

Program	Programmatic Details
ONC	Discretionary money to develop the support for multiple programs. Establish Privacy Officer, HIT Standards and HIT Policy Committees
States	Support for statewide health information exchanges. As of mid-2011, fifty six states, territories and other entities have been funded. Details discussed in the chapter on health information exchange
NIST	Develop HIT standards
HRSA	Upgrade community health centers to include HIT initiatives, such as EHRs
AHRQ, NIH	Develop comparative effective research (CER) programs
Medicare / Medicaid	Medicare and state administered Medicaid will reimburse physicians for Meaningful Use of certified electronic health records (EHRs). Details outlined in the chapter on EHRs
Regional Extension Centers	Create 60 Regional Extension Centers to promote HIT, particularly EHRs for primary care physicians in rural areas. Goal is to support 100,000 clinicians in two years. 100,000 primary care physicians have signed on as of November 2011
HIT Research Center	Collect feedback from the regional extension centers, in order to generate lessons learned
Beacon Community Program	Beacon Program will support 15-20 communities that serve as role models for the early adoption of HIT
Community College Consortia to Educate HIT Professionals	82 participating community colleges throughout all 50 states receive funding to rapidly create or expand H IT training programs that can be completed in six months or less; emphasis is on training the following roles: practice workflow and information management redesign specialists, clinician/practitioner consultants, implementation support specialists, implementation managers, technical/software support, and trainers
Health IT Curriculum Project	ONC Health IT Curriculum Project designated 12 healthcare workforce roles, six of to be educated through 6-month community college programs and six to be educated through one to two year programs at the university level. Five universities were funded as Curriculum Development Centers. The community college curriculum built by the Curriculum Development Centers covers 20 components with 8-12 units within each component and is available to faculty and the public at http://www.onc-ntdc.org/
Competency Exam Program	Support one center to create a competency exam. There will be no charge for the first 10,000 students to take the exam
Program of Assistance for University-based Training	Support for eight institutions to develop programs for HIT professionals requiring university level training. The professional roles targeted by this program are: Clinician/Public Health Leader, Health Information Management and Exchange Specialist, Health Information Privacy and Security Specialist, Research and Development Scientist, Programmers and Software Engineer, and Health IT Sub-specialist
Strategic HIT Advanced Research Projects (SHARP)	Awarded to four centers in 2010. Four focus areas are: HIT security to reduce risk and cultivate technologies of trust, support clinicians to align patient centered care with their practice, improve architectures and applications to exchange information accurately and securely and secondary use of EHR data to improve quality, population health and clinical research

- Goal 1: Achieve Adoption and Information Exchange through Meaningful Use of Health IT
  - Accelerate adoption of EHRs
  - Facilitate information exchange to support Meaningful Use of EHRs
  - Support health IT adoption and information exchange for public health and populations with unique needs
- Goal 2: Improve Care, Improve Population Health and reduce Health Care Costs through the use of Health IT
  - Support more sophisticated uses of EHRs and other health IT to improve health system performance

- Better manage care, efficiency, and population health through EHR-generated reporting measures
- Demonstrate health IT-enabled reform of payment structures, clinical practices, and population health management
- Support new approaches to the use of health IT in research, public and population health and national health security

- Goal 3: Inspire Confidence and Trust in Health IT
  - Protect confidentiality, integrity and availability of health information
  - Inform individuals of their rights and increase transparency regarding the uses of protected health information
  - Improve safety and effectiveness of health IT

- Goal 4: Empower Individuals with Health IT to Improve their Health and the Health Care System
  - Engage individuals with health IT
  - Accelerate individual and caregiver access to their electronic health information in a format they can use and reuse
  - Integrate patient-generated health information and consumer health IT with clinical applications to support patient centered care

- Goal 5: Achieve Rapid Learning and Technological Advancement
  - Lead the creation of a learning health system to support quality, research and public and population health
  - Broaden the capacity of health IT through innovation and research

In summary, ONC is responsible for coordinating all aspects of health information technology in the United States. They are involved with the adoption, standards harmonization, interoperability, privacy/security and certification of electronic health records. In addition they are coordinating the efforts to create the Nationwide Health Information Exchange (NwHIN). They participate with and support multiple private and public health information technology initiatives.

In 2011 Dr. David Blumenthal resigned and Dr. Farzad Mostashari was selected as the National Coordinator by President Obama. The next two federal advisory committees discussed are part of ONC and were created as part of the ARRA. An organizational chart follows below in Figure 1.5.[46]

- Health IT Policy Committee (HITPC). The main goal of this committee is to set priorities regarding what standards are needed for information exchange and establish the policy framework for the development and adoption of national health information exchange. The committee has 20 multi-disciplinary members. In 2011 the working groups were as follows: Meaningful Use, Certification, HIE, NwHIN, Strategic Plan, Privacy and Security, Enrollment, Governance and Quality Measures. The National Coordinator is the chair of the HITPC and their recommendations are posted on their web site. [46]

- Health IT Standards Committee (HITSC). This committee has 27 multi-disciplinary members and chaired by Jonathan Perlin. They are tasked to look at standards, implementation specifications and certification criteria for the exchange of health information. They will likely focus on issues that are prioritized by HITPC. They will use the National Institute of Standards and Technology (NIST) to test standards. Both committees will make recommendations to the

National Coordinator. They have established four working groups: clinical quality, clinical operations, implementation and privacy/security.[46]

Figure 1.5: ONC organization chart (Courtesy ONC)

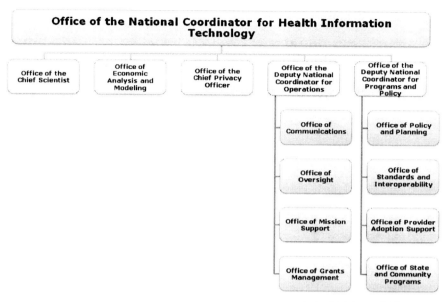

Last updated January 13, 2011

Agency for Healthcare Research and Quality (AHRQ). The AHRQ is "the lead Federal agency charged with improving the quality, safety, efficiency, and effectiveness of health care for all Americans. As one of 12 agencies within the Department of Health and Human Services, AHRQ supports health services research that will improve the quality of health care and promote evidence based decision making."[47] This agency sets aside significant grant money to support healthcare information technology (HIT) each year. Since 2004 AHRQ has invested about $280 million in grants to research HIT. The AHRQ also maintains the National Resource Center for HIT and an extensive patient safety and quality section. They also maintain an extensive HIT Knowledge Library with over 6,000 resources.[47]

Centers for Medicare and Medicaid Services (CMS). CMS is responsible for providing care to 44 million Medicare and 48 million Medicaid patients (2007 data). In an effort to improve quality and decrease costs, CMS has information technology pilot projects in multiple areas, to include pay-for-performance demonstration projects that link payments to improved patient outcomes. They will reimburse for Meaningful Use of certified EHRs. Several projects will be discussed in later chapters.[48-49]

Centers for Disease Control and Prevention (CDC). Although not a primary information technology agency, the CDC has used HIT to promote population health-related issues. Among their programs of interest:

- Public Health Information Network (PHIN), covered in the chapter on public health informatics

- Human Genome Epidemiology Network (HuGENET™) correlates genetic information with public health

- Family History Public Health Initiative is a web site that records family history information and encourages saving it in a digital format so it can be shared. Discussed more in chapter on bioinformatics

- Public Health Image Library contains photos, images and videos on medical topics

- Geographic information systems (GIS) are also covered in chapter on public health informatics
- Podcasts, RSS feeds and web widgets on medical topics
- Online Health Library
- Mobile Pilot Project to text message patients about public health issues [50]

Health Resources and Services Administration (HRSA) is part of HHS with the primary mission of assisting medical care for the underserved and uninsured in the United States, particularly in rural areas. They support federally qualified health centers (FQHCs) and rural health centers (RHCs). As noted in the section on the ARRA, HRSA will support grants for community health centers to include the installation and upgrades of health information technology. They have been a long term grant supporter of telemedicine. On their site they post a variety of health-related data in their HRSA data warehouse. A variety of searchable topics are presented with the ability to present as a table, chart, map or report. [51]

State Governments and HIT

There are a variety of state-based HIT initiatives, evaluating the adoption of technologies such as electronic health records and e-prescribing. State Medicaid offices are anxious to conduct pilot projects aimed at reducing costs and/or improving quality of care. The State Alliance for e-Health was created in 2006 in an attempt to navigate the issues of best practices, policies and adoption obstacles. Support for the Alliance is from ONC as well as a private-public advisory committee. They have three task forces: health information protection, health care practice-health information communication and data exchange taskforces. Their highest priorities are e-prescribing and the privacy and security of health information.[52]

Barriers to Health Information Technology Adoption

According to Anderson, the United States was at least 12 years behind many industrialized nations, in terms of HIT adoption. Total investment in 2005 per capita was 43 cents, compared to $21 for Canada, $4.93 for Australia, $21 for Germany and $192 for the United Kingdom.[53] This situation changed after HITECH implementation. Healthcare organizations tend to spend only 3-4% of their budget on information technology, which is far less than other information dependent industries.[54] Healthcare information technology adoption has multiple barriers listed below and discussed in later chapters:

- Inadequate time. This complaint is a common thread that runs throughout most discussions of technology barriers. Busy clinicians complain that they don't have enough time to read, learn new technologies or research vendors. They are also not reimbursed to become technology experts. They usually have to turn to physician champions, local IT support or others for technology advice.

- Inadequate information. As already pointed out earlier in the chapter, clinicians need information, not data. Current HIT systems are data rich but information poor. This is discussed in detail in the Healthcare Data, Information and Knowledge chapter.

- Inadequate expertise and workforce. In order for the United States to experience widespread HIT adoption and implementation, it will require education of all healthcare workers. According to Dr. Blumenthal (previous National Coordinator for Health Information Technology) the United States will need approximately 51,000 skilled health informaticians over the next five years to create, install and maintain HIT.[46] Dr. William Hersh of the Oregon Health and Science University, echoes the need for a work force capable of leading implementation of the electronic health record and other technologies.[55] Educational offerings will need to be expanded at universities, community colleges and medical, nursing and pharmacy schools. There is a substantial difference between healthcare organizations, in terms of HIT sophistication. The first

Work Force for Health Information Transformation Strategy Summit, hosted by the American Medical Informatics Association (AMIA) and the American Health Information Management Association (AHIMA) made several strategic recommendations regarding how to improve the work force.[56] The American Medical Informatics Association has been the leader in attempting to increase the health information technology workforce with its AMIA 10x10 Program.[57] Their goal is to train 10,000 skilled workers in the next 10 years. The Community College Consortium graduated about 3300 students in mid-July 2011 and it is too early to know how successful job placement will be, given the economic downturn. HIT vendors are looking for applicants with both IT and clinical experience.[58] In addition to skilled informaticians; we will need to educate residents in training and faculty at medical schools, given the rapidly changing nature of HIT. The APA Summit on Medical Student Education Task Force on Informatics and Technology recommended that instead of CME, we need "longitudinal, skills-based tutoring by informaticians."[59] Family Medicine residency programs are generally ahead of other specialty training programs in regards to IT training. They also recommend a longitudinal approach to IT competencies.[60]

- Cost. It is estimated that a Nationwide Health Information Network (NwHIN) will cost $156 billion dollars over five years and $48 billion annually in operating expenses.[61] Technologies such as picture archiving and communications systems (PACS) and electronic health records are also very expensive. The ARRA will help underwrite the initial purchase of some technologies but long term support will be a different challenge.

- Lack of interoperability. Electronic health records and the NwHIN cannot function until data standards are adopted and implemented nationwide. Interoperability and data standards are covered in more detail in other chapters.

- Change in workflow. Significant changes in workflow will be required to integrate technology into the inpatient and outpatient setting. As an example, clinicians may be accustomed to ordering lab or x-rays by giving a handwritten request to a nurse who actually places the order. Now they have to learn to use computerized physician order entry (CPOE). As with most new technologies, older users have more difficulty changing their habits, even if it will eventually save time or money. Poor usability is also an important impediment to good workflow and we will address this in the chapter on electronic health records. According to Dr. Carolyn Clancy, the director of AHRQ:

    "The main challenges are not technical; it's more about integrating HIT with workflow, making it work for patients and clinicians who don't necessarily think like the computer guys do"[62]

- Privacy. The Health Insurance Portability and Accountability Act (HIPAA) of 1996 was created initially for the portability, privacy and security of personal health information (PHI) that was largely paper-based. HIPAA regulations were updated in 2009 to better cover the electronic transmission of PHI or (ePHI). This Act has caused healthcare organizations to re-think healthcare information privacy and security. This will be covered in more detail in the chapters on privacy and security. In the past few years there have been a series of privacy breeches and stolen identities in healthcare organizations, thus adding to the angst.

- Legal. The Stark and Anti-kickback laws prevent hospital systems from providing or sharing technology such as computers and software with referring physicians. Exceptions were made to these laws in 2006, as will be pointed out in other chapters. This is particularly important for hospitals in order to share electronic health records and e-prescribing programs with clinician's offices. Many new legal issues are likely to appear. As an example, there has been discussion of empowering the US Food and Drug Administration to regulate electronic health records and medical devices.

- Behavioral change. Perhaps the most challenging barrier is behavior. In The *Prince* by Machiavelli, it was stated "there is nothing more difficult to be taken in hand, more perilous to conduct, or more uncertain in its success, than to take the lead in the introduction of a new order of things."[63] Dr. Frederick Knoll of Stanford University described the five stages of medical technology acceptance: (1) abject horror, (2) swift denunciation, (3) profound skepticism, (4) clinical evaluation, then, finally (5) acceptance as the standard of care.[64] It is unrealistic to expect all medical personnel to embrace technology. In 1962 Everett Rogers wrote Diffusion of Innovations in which he delineated different categories of acceptance of innovation:

    o the innovators (2.5%) are so motivated they may need to be slowed down

    o early adopters (13.5%) accept the new change and teach others

    o early majority adopters (34%) require some motivation and information from others in order to adopt

    o the late majority (34%) require encouragement to get them to eventually accept the innovation

    o laggards (16%) require removal of all barriers and often require a direct order[65]

    It is important to realize, therefore, that at least 50% of medical personnel will be slow to accept any information technology innovations and they will be perceived as dragging their feet or being *Luddites*.[66] With declining reimbursement and emphasis on increased productivity, clinicians have a natural and sometimes healthy dose of skepticism. They dread widespread implementation of anything new unless they feel certain it will make their lives or the lives of their patients better. In this situation, selecting clinical champions and conducting intensive training are critical to implementation success.

- Health Information Technology Hype versus Fact. The Gartner IT Research Group describes five phases of the hype-cycle that detail the progression of technology from the technology trigger to the peak of inflated expectations to the trough of disillusionment to the slope of enlightenment to the plateau of productivity.[67] Figure 1.6 shows the hype curve for a variety of IT technologies for 2009.

    As already noted, clinicians tend to be leery about new technologies that promise a lot, but deliver little. As a rule, if technology doesn't save time or money physicians are not interested. Importantly, current studies that evaluate HIT are often lacking for multiple reasons, discussed in these articles.[68-69]

    Both the RAND Corporation and the Center for Information Technology Leadership reported in 2005 that HIT would save the US about $80 billion annually.[70] The Congressional Budget Office (CBO), on the other hand, refuted this optimistic viewpoint in May 2008. They published a monograph entitled *Evidence on the Costs and Benefits of Health Information Technology* that reviews the evidence on the adoption and benefits of HIT, the costs of implementing, possible factors to explain the low adoption rate and the role of the federal government in implementing HIT. The bottom line for the CBO is that "By itself, the adoption of more health IT is generally not sufficient to produce significant cost savings."[71]

    There has been several recent articles that called into question the presumption that HIT adoption will generate significant cost saving and one positive review.[72-75] Karsh et al. discusses twelve HIT fallacies that adds a sobering note to the discourse.[76] Finally, Carol Diamond of the Markle Foundation points out that HIT success can't be measured by the number of hospitals that have adopted EHRs or other HIT, but instead whether patient outcomes improve.[77]

Figure 1.6: Gartner Hype Cycle of Emerging Technology 2009 (Courtesy www.Gartner.com)

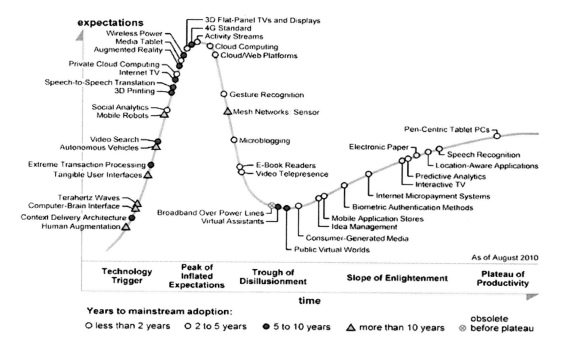

Health Informatics Programs, Organizations and Careers

Health Informatics Programs

One of the best sites to review the various health informatics programs in the United States and overseas can be found on the American Medical Informatics Association's web site. Another excellent site for listing available health informatics programs in the United States and the United Kingdom is the Biohealthmatics web site. Health informatics programs can be degree, certificate, fellowship and short courses. Most programs are part of a university, community college, medical or nursing school and others may be part of a health related organization such as the National Library of Medicine. Courses can be online, taught in a classroom setting or both. Health informatics degree programs are available as follows: associate degree, undergraduate degree, Master's degree, PhD degree or part of another degree program. Master's degrees may be focused on applied training or readying students for a research career. The AMIA program listings will give the reader an idea of how many programs are available in North America and in which category; see table 1.3 (note that 2011 data is not included because the search methodology changed). In addition, it will provide an idea as to the rapid growth of health informatics programs in a relatively short period of time. [57, 78]

As of October 2011, community colleges participating in ONC's Community College Consortia to Educate Health IT Professionals have graduated 5,717 HIT professionals, with 10,065 more students currently in the training pipeline. As of November 2011, universities have produced more than 500 post-graduate and masters-level health information technology professionals, with more than 1,700 expected to graduate by July 2013. The majority of health informatics students in the past have come from healthcare fields. With the current economy and the new monies from the ARRA, IT professionals from other industries are enrolling in health informatics training programs. Often these professionals bring expertise in technology implementation, evaluation and/or user training and programming skills but they often lack clinical experience in healthcare.

Table 1.3: Health informatics programs listed on the AMIA web site

Program Type	2008	2010
Associate degree	1	1
Undergraduate degree	5	9
Master's degree	65	70
PhD degree	26	30
Certificate	36	43
Short courses	13	14
Online courses	28	33
10x10 programs	3	10

Health Informatics Organizations

The following organizations are considered among the most important and influential in health information technology.

American Medical Informatics Association (AMIA)

- Founded in 1990 by the merger of the American Association for Medical Systems and Informatics, the American College of Medical Informatics and the Symposium on Computer Applications in Medical Care
- In 2006 it became a member of the Council of Medical Specialty Societies
- As of 2009 AMIA has greater than 4000 members from clinical, technical and research sectors
- They support five main domains: translational bioinformatics, clinical research informatics, health informatics, consumer health informatics and public health informatics
- Members are from 65 countries
- In 2009 Dr. Ted Shortliffe became the President and CEO of AMIA
- Web site includes job exchange, academic programs, fellowships, grants, and an e-newsletter
- Membership includes subscription to the Journal of the American Medical Informatics Association (JAMIA)
- Opportunity to join a working group (20) to discuss issues and formulate white papers
- Annual national symposium in the fall as well as a spring Congress [57]

International Medical Informatics Association (IMIA)

- Began in 1967 but became officially an independent endorsed organization in 1989
- Membership consists of national, institutional, affiliate members and honorary fellows
- AMIA is the US representative to the IMIA
- IMIA supports the triennial World Congress on Medical and Health Informatics, known as Medinfo
- IMIA supports multiple working groups and special interest groups

- Official journals: International Journal of Medical Informatics, Methods of Information in Science and Applied Clinical Informatics [79]

Healthcare Information and Management Systems Society (HIMSS)

- Founded in 1961
- As of 2011 has over 35,000 individual members and 520 corporate members
- Annual symposium with more than 20,000 attendees
- Professional certification
- Educational publications, books and CD-ROMs
- Web conferences on health informatics topics
- HIMSS Health IT Body of Knowledge resource site
- HIMSS Analytics is a subsidiary that provides data and analytic expertise
- Surveys on multiple topics [80]

American Health Information Management Association (AHIMA)

- Founded in 1928 for medical records librarians
- As of 2011 has more than 61,000 members
- It began as a medical records association but now includes any healthcare worker involved in information management. It offers seven credentials related to four areas: Coding, HIM, privacy and analysis
- "AHIMA supports the common goal of applying modern technology to and advancing best practices in health information management"[56]
- AHIMA web site has an excellent HIT resource section
- AHIMA Journal and Perspectives in Health Information Management are available on their web site at no cost[56]

Alliance for Nursing Informatics (ANI)

- Combines 25 separate nursing informatics organizations
- As of 2007 has more than 3,000 members
- Sponsored by both the AMIA and HIMSS
- Provides a collaborative group for consensus about nursing informatics[81]

American Telemedicine Association

- Established in 1993 to promote telecommunications technology
- Has transitioned to telemedicine, telehealth or eHealth
- Mission is to promote remote access to medical care through telemedicine technology
- Web site has a variety of educational resources and telemedicine forms
- Official journal is Telemedicine and e-Health[82]

Health Informatics Careers

The timing is excellent for a career in health informatics. With the emphasis on increasing adoption of electronic health records and health information exchange, coupled with support from the HITECH Act there has been tremendous interest in health informatics. Healthcare organizations and HIT vendors will be looking for workers who are knowledgeable in both technology and medicine. The Department of Labor estimates that there will be 4% growth in the demand for trained health informatics specialists in multiple areas in the private, federal and military sectors. Informaticians will be needed to design, implement and govern many new technologies arriving on the medical scene, as well as train users. It is anticipated that government reimbursement for EHRs and support for health information exchange will only increase the need for skilled HIT workers. The Biohealthmatics, HIMSS, American Nurse Informatics, Health IT News, AHIMA and the AMIA web sites list multiple interesting health IT jobs. Examples include nurse and physician informaticists, systems analysts, information directors, chief information officers (CIOs) and chief medical information officers (CMIOs).[56-57, 78, 80-81, 83] Recruiting organizations also maintain multiple listings for health IT jobs.

There are a wide variety of jobs available in the informatics realm. The following are just a few of the known positions in a healthcare organization:

- Chief Medical Informatics Officer (CMIO) is usually a physician but could be a nurse who generally reports to the Chief Information Officer (CIO), Chief Executive Officer (CEO) or Chief Medical Officer (CMO). This individual usually works with the CIO to develop a strategic IT plan and to help with the implementation of technologies by clinical staff. They are less IT oriented and more oriented towards overcoming the barriers to adoption and they provide feedback and education to their staff. They evaluate new technologies that may transform healthcare and along with the CIO they help develop policies that affect privacy and security. They commonly have a Master's degree in one of the information sciences.

- Nurse Informaticist (NI) is a nurse who can be the CMIO or can be an individual who works in the nursing department, IT department or is dual hatted. There are three million nurses in the United States, compared to about 800,000 physicians so they are a large pool of knowledge workers. Most nurses are trained to think in terms of systems and process improvement. They are therefore extremely valuable for project management, IT systems managers, data analysts, technology adoption, implementation and training. Nurse Informaticians have had a certification exam since 1995 and have published Scope and Standards of their field in 2008.

- Clinician Informatician (CI) is a clinician who may have formal training with a variety of degrees or simply may have extensive on the job experience and an aptitude for technology. As a result, they are usually early adopters and clinician champions who help the clinical staff in a healthcare organization understand and accept transformational technologies.[84]

The American Medical Informatics Association has been in the process of establishing the medical subspecialty of *clinical informatics*. In September 2011 it was announced that *clinical informatics* was an approved subspecialty, sponsored by the American Board of Preventive Medicine and the American Board of Pathology. The certification will be available to physicians who have a primary specialty designated through the American Board of Medical Specialties (ABMS). In the 2009 March/April issue of the JAMIA, the core content for this new specialty is spelled out.[85-86] The plan is to make board certification available starting in the Fall of 2012. For the first five years practicing informaticians can apply for board eligibility-based on their work experience and criteria to be established by the ABMS which will include board specialty certification in a primary specialty. After five years candidates must complete training in health informatics that is certified by the Accreditation Council on Graduate Medical Education. Similar certification is being discussed for nurses, pharmacists, PhDs and others.

Although physicians can become chief medical information officers in very large organizations, the reality is that nurses have the greatest potential to be involved with IT implementation and training at the average hospital or large clinic. Larger, more urban clinics may have the luxury of in-house IT staff, unlike smaller and more rural practices. Notably, nursing already has an informatics specialty certification.

Health Informatics Resources

Because of the rapidly changing nature of technology it is difficult to find resources that are current. It is also difficult to find resources that are not overly technical that would be appropriate for the health informatics neophyte. There are numerous excellent journals, e-journals and e-newsletters that contain articles that discuss important aspects of health information technology. Because health informatics is gaining popularity in the field of medicine many excellent articles can also be found in major medical journals that do not normally focus on technology. As an example, *Health Affairs*, a bimonthly journal features web exclusives, blogs and e-newsletters of interest to informaticians.[87] Furthermore, several informatics-related web sites link to the major national and international health informatics print and online journals.[88-90]

Books

- Handbook of Biomedical Informatics. Wikipedia Books. 2009[91]

- Guide to Health Informatics. Enrico Coiera. 2003[92]

- Biomedical informatics: Computer Applications in Health Care and Biomedicine. EH Shortliffe and J Cimino 2006[93]

- Sabbatini RME Medical Informatics: Concepts, Methodologies, Tools and Applications. J Tan. Four Volumes. 2009[94]

Journals

- *Journal of the American Medical Informatics Association* is the bimonthly journal of the AMIA. It features peer reviewed articles that run the gamut from theoretical models to practical solutions. The journal is included in the AMIA membership and is most appropriate for medical and IT professionals.[95]

- *International Journal of Medical Informatics* is an international monthly journal that covers information systems, decision support, computerized educational programs and articles aimed at healthcare organizations. In addition to standard articles, they publish short technical articles and reviews.[96]

- *Journal of Biomedical Informatics* was formally known as *Computers and Biomedical Research*. Its editor is Dr. Ted Shortliffe and the emphasis of this bimonthly journal is bioinformatics.[97]

- *Journal of AHIMA* is published 11 months of the year for its members to stay current in health information management-related issues.[98]

- *Computers, Informatics, Nursing (CIN)* is a bimonthly print journal targeting the nursing professional. Also offers PDA downloads, RSS feeds and a newsletter.[99]

E-journals

- *BMC Medical Informatics and Decision Making* is an open-access free online journal publishing peer-reviewed research articles. This journal is part of BioMed Central, an online publisher of 188 online free full text journals. Because it is an open-access model it allows for much more rapid review and publication, a plus for informatics journals. [100]

- *The Open Medical Informatics Journal* is another open-access free online journal that publishes health informatics research articles and reviews. Bentham Science publishes 89 online and print journals as well as 200 online open-access journals. An abstract is available online and the full text pdf copy is downloadable. [101]

- *Journal of Medical Internet Research* (JMIR) is an independent open-access online journal that publishes articles related to medicine and the internet. The articles are free to read in an html format but there is a cost to download articles in a pdf format or to become a member. [102]

- *Electronic Journal of Health Informatics* (eJHI) is an Australian-based international open access electronic journal that offers open access (no fee) to both authors and readers. [103]

- *Applied Medical Informatics* is the fee-based e-journal for the International Medical Informatics Association (IMIA) and the Association of Medical Directors of Information Systems (AMDIS).Its first issue appeared in early 2010. [104]

- *Perspectives in Health Information Management* is the open-access research peer-reviewed e-journal for AHIMA, published four times a year.[105]

Informatics-Related E-newsletters

- *iHealthBeat* is a free daily e-mail newsletter on health information technology published as a courtesy by the California Healthcare Foundation. It is also available through RSS feeds, Twitter and they offer frequent podcasts.[106]

- *HealthCareITNews* is available as a daily online, RSS feed or print journal. It is published in partnership with HIMSS and reviews broad topics in HIT. They also publish the online e-journals *NHINWatch, MobileHealthWatch* and *Health IT Blog*.[83]

- *eHealth SmartBrief* is a free newsletter e-mailed three times weekly. In addition to broad coverage of HIT, they offer RSS feeds, blogs, reader polls and job postings.[107]

- *Health Data Management* offers a free daily e-newsletter, in addition to their comprehensive web site. The web site offers 20 channels or categories of IT information, webinars, whitepapers, podcasts and RSS feeds.[108]

Online Resource Sites

- University of West Florida Health Informatics Program Resource Site augments this book with valuable web links organized in a similar manner as the book chapters. It also includes links to excellent informatics newsletters and journals.[109]

- Agency for Healthcare Research and Quality Knowledge Library is another excellent resource with over 6,000 articles and other resources that discuss health information technology related issues.[110]

- HIMSS Health IT Body of Knowledge is a new site to introduce readers to more than 25 topic categories. Articles, tools and guidelines are offered by HIMSS and other resources. [111]

- Family Medicine Digital Resources Library was created by Dr. Tom Agresta and supported by the Society of Teachers of Family Medicine to promote Informatics education of Family Medicine physicians. In early 2010 they posted 14 presentations that are available to the public.[112]

Informatics Blogs

- *HealthIT Buzz* is a new Blog offered on the HHS HealthIT web site.[113]

- *Life as a Health CIO* by Dr. John Halamka offers insights from his perspective as CIO of Harvard Medical School and Beth Israel Deaconess Medical Center.[114]

- The *Health Care Blog* is hosted by Matthew Holt and considered to be "a free-wheeling discussion of the latest healthcare developments" to include health information technology.[115]

- *E-CareManagement* focuses on chronic disease management, technology, strategy, issues and trends. Content is posted by Vince Kuraitis, a HIT consultant for Better Health Technologies.[116]

- *Health Informatics Forum* is an international forum dedicated to health informatics professionals and students. Has extensive web links.[117]

- *Biological Informatics* was created by Marcus Zillman to compile multiple biomedical informatics sites (100+) into one, as well as a blog.[118]

- *HealthTechtopia* compiles the top 50 health informatics blogs. It is subdivided into General Health Informatics, Anatomy & Physiology, Information Science and Information Technology, Computer Science, Statistics and Radiology and Medical Imaging.[119]

- *Biomedexperts* is a free social network for biomedical researchers. They have created groups based on what articles have been published by the scientists involved. The claim to have profiles on 1.8 million biomedical researchers from 190 countries. Profiles were generated from the last 10 years of PubMed. In this manner research networks can be created.[120]

- *EMR & HIPAA Blog* hosted by John Lynn covers EHRs, HIPAA and HIT issues.[121]

Future Trends

Given the relative newness of health informatics it is not easy to predict the future but some trends seem worth stressing. Many of these points are discussed in more detail in other chapters:

- Regardless of the speed of HIT adoption in medicine, the technology itself will continue to evolve rapidly. Many disruptive technologies such as tablets will present outstanding opportunities. This will require uniquely well trained individuals who understand the technology and have the clinical experience to know how it can be applied successfully in the field of medicine.

- Meaningful Use requirements will continue to evolve (stages 2 and 3) and the bar will be slowly raised. More research is needed to determine what additions are evidence based, worthwhile and will actually impact clinical outcomes.

- New healthcare delivery models such as accountable care organizations will be an experiment well worth watching. If they demonstrate cost savings that are strongly supported by HIT we can expect increased adoption.

- We anticipate more patient centric medical care and associated technologies. For example, more medical apps for smartphones and personalized genetic profiles.

- Mobile technologies will continue to be an important medical platform for patients and clinicians.

- Expect more artificial intelligence in medicine (AIM) to retrospectively and prospectively interpret medical data. WellPoint plans to use IBM's Watson Computer to analyze medical data.[122]

Key Points

- Health informatics focuses on the science of information, as applied to healthcare and biomedicine
- Health information technology (HIT) holds promise for improving healthcare quality, reducing costs and expediting the exchange of information
- The HITECH Act programs have been a major driver of HIT in the United States
- Barriers to widespread adoption of HIT include: time, cost, privacy, change in workflow, legal, behavioral barriers and lack of high quality studies
- Many new degree and certificate programs are available in health informatics
- A variety of health informatics resources are available for a wide audience
- Interoperability and health information exchange is a major priority of the federal government but is challenged by sustainable issues

Conclusion

Health informatics is a new, exciting and evolving field. New specialties and careers are now possible. In spite of its importance and popularity, significant obstacles remain. Health information technology has the potential to improve medical quality, patient safety, educational resources and patient - physician communication, while decreasing cost. Although technology holds great promise, it is not the solution for every problem facing medicine today. As noted by Dr. Safran of the American Medical Informatics Association "technology is not the destination, it is the transportation."[79] We must continue to focus on improved patient care as the single most important goal of this new field.

Research in health informatics is being published at an increasing rate so hopefully new approaches and tools will be evaluated more often and more objectively. Better studies are needed to demonstrate the effects of health information technology on actual patient outcomes and return on investment, rather than studies based solely on surveys and expert opinion.

The effects of the multiple programs supported by the HITECH and Affordable Care Acts will likely be both transformational and challenging for the average practitioner.

Acknowledgements

We would like to thank Dr. Irmgard Willcockson (UT-Houston) for her contributions to the Health Informatics Programs section.

References

1. Ackoff RL. From data to wisdom. J Appl Syst Anal 1989;16:3-9
2. The DIKW Model of Innovation. www.spreadingscience.com (Accessed February 21 2010)
3. Bernstam EV, Smith JW, Johnson TR. What is biomedical informatics? Biomed Inform 2010;43(1):104-10
4. Shortliffe, E .What is medical informatics? Lecture. Stanford University, 1995.
5. MF Collen. Preliminary announcement for the Third World Conference on Medical Informatics, MEDINFO 80, Tokyo
6. UK Health Informatics Society http://www.bmis.org (Accessed September 5 2005)

7.   Center for Toxicogenomics http://www.niehs.nih.gov/nct/glossary.htm (Accessed September 10 2005)

8.   Biohealthmatics http://www.biohealthmatics.com/knowcenter.aspx (Accessed September 5 2008)

9.   Hersh WR. A stimulus to define informatics and health information technology BMC Medical Informatics and Decision Making. 2009;9. www.biomedcentral.com/1472-6947/9/24 (Accessed November 4 2009)

10. The Joint Commission www.jointcommission.org (Accessed September 10 2011)

11. Intel http://www.intel.com/about/companyinfo/healthcare/index.htm (Accessed September 10 2011)

12. Computational Technology for Effective Health Care. Immediate Steps and Strategic Directions. 2009. National Academies Press. Stead WW and Li HS, editors http://books.nap.edu/openbook.php?record_id=12572&page=R1 (Accessed June 16 2010)

13. Balmer S. Keynote Address 2007 HIMSS Conference. February 26 2007

14. Nyweide DJ, Weeks WB, Gottlieb DJ et al. Relationship of Primary Care Physicians' Patient caseload With Measurement of Quality and Cost Performance. JAMA 2009;302(22):2444-2450

15. Data.Gov www.data.gov (Accessed September 10 2011)

16. Community Health Data Initiative http://www.hhs.gov/open/plan/opengovernmentplan/initiatives/initiative.html (Accessed September 10 2011)

17. Healthy People. www.healthypeople.gov/hp2020/Objectives/TopicAreas.aspx (Accessed June 5 2010)

18. Sabbatini RME. Handbook of Biomedical informatics. Wikipedia Books. Pedia-Press. 2009. Germany. http://en.wikipedia.org/wiki/Wikipedia:Books/BiomedicaInformatics

19. A history of computers http://www.maxmon.com/history.htm (Accessed September 30 2005)

20. Hersh WR. Informatics: Development and Evaluation of Information Technology in Medicine JAMA 1992;267:167-70

21. Laboratory of Computer Science. Massachusetts General Hospital www.lcs.mgh.harvard.edu (Accessed December 14 2009)

22. VUMC Dept. of Biomedical informatics http://www.mc.vanderbilt.edu/dbmi/informatics.html (Accessed Oct 1 2005)

23. Health Informatics http://en.wikipedia.org/wiki/Medical_informatics (Accessed September 20 2005)

24. Howe, W. A Brief History of the Internet http://www.walthowe.com/navnet/history.html (Accessed September 24 2005)

25. Zakon, R. Hobbe's Internet Timeline v8.1 http://www.zakon.org/robert/internet/timeline (Accessed September 24 2005)

26. W3C http://www.w3.org/WWW/ (Accessed September 25 2005)

27. Berner ES, Detmer DS, Simborg D. Will the Wave Ever Break: A Brief View of the Adoption of Electronic Health Records in the United States. JAMIA 2005; 12(1):3-7 http://www.ncbi.nlm.nih.gov/pmc/articles/PMC543824/ (Accessed July 6 2005)

28. Koblentz, E. The Evolution of the PDA http://www.snarc.net/pda/pda-treatise.htm (Accessed Oct 3 2005)

29. Human Genome Project. US Dept of Energy http://www.ornl.gov/sci/techresources/Human_Genome/home.shtml (Accessed Oct 5 2005).

30. Nationwide Health Information Network. http://www.hhs.gov/healthit/healthnetwork/background/ (Accessed September 10 2011)

31. Crossing the Quality Chasm: A new health system for the 21st century (2001) The National Academies Press http://www.nap.edu/books/0309072808/html/ (Accessed October 5 2005)

32. Crossing the Chasm with Information Technology. Bridging the gap in healthcare. First Consulting Group July 2002 http://www.chcf.org/documents/ihealth/CrossingChasmIT.pdf (Accessed September 20 2005)

33. To Error is Human: Building a safer Healthcare System (1999) The National Academies Press http://www.nap.edu/catalog/9728.html (Accessed October 5 2005)

34. Association of American Medical Colleges. Better health 2010. http://www.aamc.org (Accessed September 10 2011)

35. Healthcare Initiatives Improvement Institute. Bridges To Excellence http://www.hci3.org/node/1/ (Accessed September 10 2011)

36. E-health Initiative http://www.ehealthinitiative.org/ (Accessed September 10 2011)

37. The Leapfrog Group http://www.leapfroggroup.org/ (Accessed October 5 2011)

38. Markle Connecting for Health www.connectingforhealth.org (Accessed September 10 2011)

39. National eHealth Collaborative. www.nationalehealth.org (Accessed September 30 2011)

40. Health Information Technology Standards Panel www.hitsp.org (Accessed May 10 2009)

41. Certification Commission for Health Information Technology www.cchit.org (Accessed September 10 2011)

42. Cost of New CCHIT EHR Certifications. EMR and HIPAA. www.emrandhipaa.com (Accessed October 1 2009)

43. National Committee on Vital and Health Statistics http://ncvhs.hhs.gov (Accessed September 10 2011)

44. American Recovery and Reinvestment Act of 2009 Public Law 111 – 5. February 17 2009 http://en.wikisource.org/wiki/American_Recovery_and_Reinvestment_Act_of_2009 (Accessed March 2 2009)

45. Department of Health and Human Services. www.hhs.gov (Accessed September 12 2011)

46. Office of the National Coordinator for Health Information Technology. http://healthit.hhs.gov (Accessed September 7 2011)

47. Agency for Healthcare Research and Quality http://www.ahrq.gov/ (Accessed September 11 2011)

48. Centers for Medicare and Medicaid Services. Medicare Demonstrations http://www.cms.hhs.gov/DemoProjectsEvalRpts/MD/ (Accessed September 11 2011)

49. Centers for Medicare & Medicaid www.cms.hhs.gov (Accessed September 11 2011)

50. Centers for Disease Control and Prevention www.cdc.gov (Accessed September 11 2011)

51. Health Resources and Service Administration www.hrsa.gov (Accessed September 11 2011)

52. State Alliance http://www.nga.org/cms/home/nga-center-for-best-practices/center-issues/page-health-issues/col2-content/main-content-list/state-alliance-for-e-health.html (Accessed September11 2011)

53. Anderson GF et al. Health Care Spending and use of Information Technology in OECD countries. Health Affairs 2006;25:819-831

54. EHR and the Return on Investment. HIMSS 2003. www.himss.org/content/files/ehr-roi.pdf. Accessed December 1 2007

55. Hersh W .Health Care Information Technology JAMA 2004; 292 (18):2273-441

56. American Health Information Management Association site http://www.ahima.org (Accessed September 11 2011)

57. AMIA 10 x 10 Program. http://www.amia.org (Accessed September 11 2011)

58. Ackerman K. Jury Still Out on Health IT Workforce Training Programs. iHealthBeat. September 6 2011. www.ihealthbeat.org (Accessed September 6 2011)

59. Hilty DM, Benjamin S, Briscoe G et al. APA Summit on Medical Student Education Task Force on Informatics and Technology: Steps to Enhance the Use of Technology in Education Through Faculty Development, Funding and Change Management. Acad Psych 2006;30:444-450

60. Recommended Curriculum Guidelines for Family Medicine Residents http://www.aafp.org/online/etc/medialib/aafp_org/documents/about/rap/curriculum/medicalinfo rmatics.Par.0001.File.tmp/Reprint288.pdf (Accessed June 12 2009)

61. Basch P et al .Electronic health records and the national health information network: affordable, adaptable and ready for prime time? Ann Intern Med 2005 143(3):165-73

62. Interview with Dr. Carolyn Clancy. Medscape June 2005. www.medscape.com (Accessed November 4 2005)

63. Machiavelli N, The Prince Chapter VI www.constitution.org/mac/prince06.htm (Accessed September 26 2008

64. Knoll, F. Medical Imaging in the Age of Informatics. Stanford University. November 15 2005

65. Rogers EM, Shoemaker FF. Communication of Innovation 1971 New York, The Free Press

66. Luddite www.wikipedia.org/luddite (Accessed November 1 2009)

67. Gartner hype cycle http://gsb.haifa.ac.il/~sheizaf/ecommerce/GartnerHypeCycle.html (Accessed November 21 2007)

68. Shcherbatykh I, Holbrook A, Thabane L et al. Methodologic Issues In Health Informatics Trials: The Complexities of Complex Interventions. JAMIA 2008; 15:575-580

69. Goldzweig CL, Towfligh A, Maglione M et al. Costs and Benefits of Health Information Technology: New Trends From the Literature. Health Affairs 2009 28 (2) w282-w293 www. content.healthaffairs.org/cgi/content/abstract/28/2/w282-w293 (Accessed February 4 2009)

70. Girosi, Federico, Robin Meili, and Richard Scoville. 2005. Extrapolating Evidence of Health Information Technology Savings and Costs. Santa Monica, Calif. RAND Corporation http://rand.org/pubs/research_briefs/RB9136/index1.html (Accessed May 20 2008)

71. Congressional Budget Office Paper: Evidence on the Costs and Benefits of Health Information Technology www.cbo.gov (Accessed May 20 2008)

72. Himmelstein DU, Wright A, Woolhandler S. Hospital Computing and the Costs and Quality of Care: A National Study. Am J Med 2009; 123(1):40-46

73. Black AD, Car J, Pagliari C et al. The Impact of eHealth on the Quality and Safety of Health Care: A Systematic Overview. PLoS Medicine. Jan 2011. www.plosmedicine.org (Accessed January 21 2011)

74. Butin MB, Burke MF, Hoaglin MC, Blumenthal D. The Benefits of Health Information Technology: A Review of the Recent Literature Shows Predominately Positive Results. Health Affairs. 2011;30 (3):464-471

75. Laszewski R. Health IT Adoption and the Other myths of Health Care Reform. January 12 2009 www.ihealthbeat.org (Accessed January 12 2009)

76. Karsh B, Weinger MB, Abbott PA, Wears FL. Health Information technology: fallacies and sober realities. J Am Med Inform Assoc. 2010;17:617-623

77. Diamond CC, Shirky C. Health Information Technology: A Few Years of Magical Thinking? Health Affairs www.healthaffairs.org 2008;27 (5): w383-w390 (Accessed September 3 2008)

78. Biohealthmatics www.Biohealthmatics.com (Accessed September 12 2011)

79. International medical informatics Association www.imia.org (Accessed September 11 2011)

80. Health Information Management Systems Society www.himss.org/jobmine (Accessed September 6 2011)

81. Alliance for Nursing Informatics http://www.allianceni.org/ (Accessed June 12 2009)

82. American Telemedicine Association www.atmeda.org (Accessed November 25 2009)

83. Health IT Job Spot. http://jobspot.healthcareitnews.com/home/5815_rec.cfm?site_id=5815 (Accessed June 16 2009)

84. Introduction to Healthcare Information Enabling Technologies. HIMSS 2010. Chicago IL. Ed. Gensinger RA. pp 57-86

85. Gardner RM, Overhage JM, Steen EB et al. Core Content for the Subspecialty of Clinical Informatics JAMIA 2009;16 (2):153-157

86. Detmer DE, Munger BS, Lehmann CU. Clinical Informatics Board Certification: History, Current Status, and Predicted Impact on the Clinical Informatics Workforce. Appl Clin Inf 2009;1:11-18

87. Health Affairs. http://content.healthaffairs.org (Accessed June 14 2009)

88. Informatics Journals. www.dmoz.org/Health/Medicine/Informatics/Journals/ (Accessed June 18 2009)

89. Health Informatics Journals and Publications www.hiww.org/jou.html (Accessed June 18 2009)

90. Online health informatics journals www.hi-europe.info/library/hi_journals.htm (Accessed June 18 2009)

91. Handbook of Biomedical Informatics.Wikipedia Books. http://pediapress.com (Accessed February 12 2010)

92. Guide to Health Informatics. Enrico Coiera. 2003. Arnold Publications

93. Biomedical Informatics: Computer Applications in Health Care and Biomedicine. EH Shortliffe and J Cimino. 2006 Springer.New York, NY

94. Sabbatini RME. Medical Informatics: Concepts, Methodologies, Tools and Applications. J Tan. Four Volumes. 2008. Information Science Reference. Hershey, PA

95. Journal of the American Medical Informatics Association http://jamia.bmj.com/ (Accessed September 11 2011)

96. International Journal of Medical Informatics http://www.sciencedirect.com/science/journal/13865056 (Accessed September 11 2011)

97. Journal of Biomedical Informatics http://www.elsevier.com/wps/find/journaldescription.cws_home/622857/description#description (Accessed September 11 2011)

98. Journal of AHIMA http://journal.ahima.org (Accessed September 11 2010)

99. CIN: Computers, Informatics, Nursing www.cinjournal.com (Accessed June 18 2011)

100. BMC Medical Informatics and Decision Making. www.biomedcentral.com/bmcmedinformdecismak/ (Accessed September 11 2011)

101. The Open Medical Informatics Journal www.bentham.org/open/tominfoj/ (Accessed September 11 2011)

102. The Journal of Medical Internet Research. http://www.jmir.org/ (Accessed September 11 2011)

103. Electronic Journal of Health Informatics http://ejhi.net (Accessed September 11 2011)

104. Applied Clinical Informatics. http://www.schattauer.de/index.php?id=558&L=1 (Accessed September 11 2011)

105. Perspectives in Health Information Management http://perspectives.ahima.org (Accessed February 15 2010)

106. Ihealthbeat www.ihealthbeat.org (Accessed September 11 2011)

107. eHealth SmartBrief http://www.smartbrief.com/news/EHEALTH/index.jsp?categoryid=7B651A9C-543B-43A9-909D-CC5F80F69335 (Accessed September 11 2011)

108. Health Data Management www.healthdatamanagement.com (Accessed September 11 2011)

109. University of West Florida. Introduction to medical informatics resource page www.uwf.edu/sahls/medicalinformatics/ (Accessed September 11 2011)

110. Agency for Healthcare Research and Quality. Knowledge Library. http://healthit.ahrq.gov/portal/server.pt?open=512&objID=653&parentname=CommunityPage&parentid=10&mode=2. (Accessed September 11 2011)

111. HIMSS Health IT Body of Knowledge http://www.himss.org/asp/topics_HITBOK.asp (Accessed December 29 2011)

112. Family Medicine Digital Resources Library http://www.fmdrl.org/1503 (Accessed September 11 2011)

113. Health IT Buzz www.healthit.hhs.gov/blog/onc (Accessed September 11 2011)

114. Life as a Healthcare CIO http://geekdoctor.blogspot.com (Accessed September 11 2011)

115. The Health Care Blog www.thehealthcareblog.com (Accessed September 11 2011)

116. E-care management http://e-caremanagement.com (Accessed September 11 2011)

117. Health Informatics Forum www.healthinformaticsforum.com (Accessed September 11 2011)

118. Biological Informatics http://biologicalinformatics.blogspot.com (Accessed September 11 2011)

119. HealthTechTopia http://mastersinhealthinformatics.com/2009/top-50-health-informatics-blogs/ (Accessed September 11 2011)

120. Biomedexperts www.biomedexperts.com (Accessed September 11 2011)

121. EMR & HIPAA Blog. www.emrandhipaa.com (Accessed September 11 2011)

122. McNickle M. 5 things to know about Watson's role in healthcare. November 7 2011. www.healthcareitnews.com (Accessed November 8 2011)

2

Healthcare Data, Information, and Knowledge

ELMER V. BERNSTAM

TODD R. JOHNSON

JORGE R. HERSKOVIC

TREVOR COHEN

Learning Objectives

After reading this chapter the reader should be able to:

- Define Data, Information, and Knowledge

- Understand how vocabularies convert data to information

- Describe methods that convert information to knowledge

- Distinguish informatics from other computational disciplines, particularly computer science

- Describe the differences between data-centric and information-centric technology

Introduction

"...current efforts aimed at the nationwide deployment of health care IT will not be sufficient to achieve the vision of 21st century health care, and may even set back the cause if these efforts continue wholly without change from their present course."[1]

In this chapter, we present a framework for understanding informatics. In chapter 1, we introduced the definitions of data, information, and knowledge and now we will build upon these definitions to answer fundamental questions regarding health informatics. What makes informatics different from other computational disciplines? Why is informatics difficult? Why do some health IT projects fail?

In chapter 1, we mentioned the fundamental mismatch between available technology (i.e., traditional computers) and problems faced by informaticians. In this chapter we expand upon these ideas to understand why many health IT (HIT) projects fail. To help organizations appropriately apply HIT, informaticians must understand the limitations of HIT as well as the potential of HIT to improve health.

To illustrate several points, we will begin with a real world example of informational challenges.

Case Study: The Story of E-patient Dave

In January 2007, Dave deBronkart was diagnosed with a kidney cancer that had spread to both lungs, bone and muscles. His prognosis was grim. He was treated at Beth Israel Deaconess Medical Center in Boston with a combination of surgery and enrolled in a clinical trial of High Dosage Interleukin-2 (HDIL-2) therapy. That combination did the trick and by July 2007, it was clear that Dave had beaten the cancer. He is now a blogger and an advocate and activist for patient empowerment.

In March 2009, Dave decided to copy his medical record from the Beth Israel Deaconess EHR to Google Health, a personally-controlled health record or PHR. He was motivated by a desire to contribute to a collection of clinical data that could be used for research. Beth Israel Deaconess had worked with Google to create an interface (or conduit) between their medical record and Google Health. Thus, copying the data was automated. Dave clicked all of the options to copy his complete record and pushed the big red button. The data flowed smoothly between computers and the copy process completed in only few moments.

What happened next vividly illustrated the difference between data and information. Multiple urgent warnings immediately appeared (Figure 2.1). Dave was taking hydrochlorothiazide, a common blood pressure medication, but had not had a low potassium level since he had been hospitalized nearly two years earlier.

Worse, the new record contained a long list of deadly diseases (Figure 2.2). Everything that Dave had ever had was transmitted, but with no dates attached. When the dates were attached, they were wrong. Worse, Dave had never had some of the conditions listed in the new record. He was understandably distressed to learn that he had an aortic aneurysm, a potentially deadly expansion of the aorta, the largest artery in the human body.

Why did this happen? In part, it was because the system transmitted billing codes, rather than doctors' diagnoses. Thus, if a doctor ordered a computed tomography (CT) scan, perhaps to track the size of a tumor, but did not put a reason for the test, a clerk may have added a billing code to ensure proper billing (e.g., rule out aortic aneurysm). This billing code became permanently associated with the record.

After Dave described what happened in his online blog [2] (http://epatientdave.com/), the story was picked up by a number of newspapers including the front page of the Boston Globe.[3] It also brought international attention to the problem of meaning. It became very clear that transmitting data from system to system is not enough to ensure a usable result. To be useful, systems must not mangle the meaning as they input, store, manipulate and transmit information. Unfortunately, as this story illustrates, even when standard codes are stored electronically, their meaning may not be clear.

Figure 2.1: Urgent warning in e-patient Dave's record

Requires immediate attention
Discuss with your doctor soon
Hydrochlorothiazide and Low Amount of Potassium in the Blood
Medications given to people who have certain conditions can lead to an increase in side effects and/or worsening of the condition. Hydrochlorothiazide Oral generally should not be given to people with Hypokalemia. This health profile includes this condition.

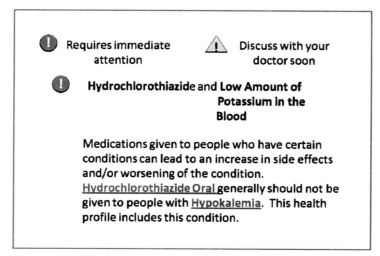

Figure 2.2: e-patient Dave's conditions as reflected in the newly-created personal health record (PHR)

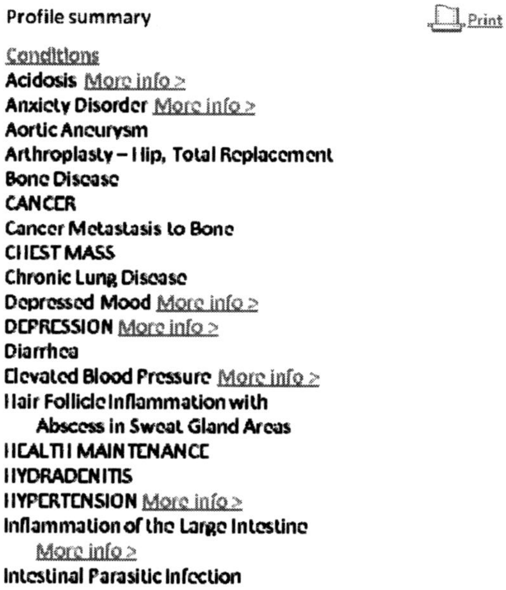

Profile summary
Print
Conditions
Acidosis More info >
Anxiety Disorder More info >
Aortic Aneurysm
Arthroplasty – Hip, Total Replacement
Bone Disease
CANCER
Cancer Metastasis to Bone
CHEST MASS
Chronic Lung Disease
Depressed Mood More info >
DEPRESSION More info >
Diarrhea
Elevated Blood Pressure More info >
Hair Follicle Inflammation with Abscess in Sweat Gland Areas
HEALTH MAINTENANCE
HYDRADENITIS
HYPERTENSION More info >
Inflammation of the Large Intestine More info >
Intestinal Parasitic Infection

Definitions and Concepts

Data, Information and Knowledge

In chapter 1, we defined data, information and knowledge (see Figure 1.1). [4,5] Recall that **data** are observations reflecting differences in the world (e.g., "162.9"). Note that "data" is the plural of "datum." Thus, "data are" is grammatically correct; "data is" is not correct. **Information** is meaningful data or facts from which conclusions can be drawn (e.g., ICD-9-CM code 162.9 = "Lung neoplasm, Not Otherwise Specified"). **Knowledge** is information that is justifiably believed to be true (e.g., "Smokers are more likely to develop lung cancer"). This relationship is shown in Figure 2.3 and we will refer to this diagram later in the chapter.

Figure 2.3: Data, information and knowledge

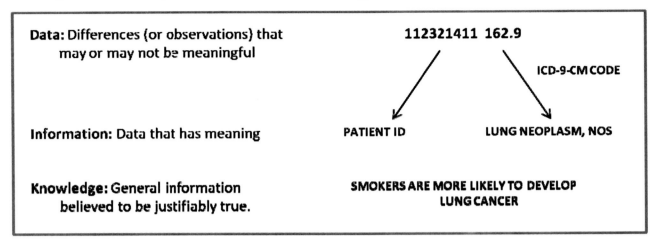

Data

To understand the relationship between data, information and knowledge in health informatics, we must understand the relationship between what happens in a computer and the real world. Computers do not represent meaning. They input, store, process and output zero (off) and one (on). Each zero or one is known as a **bit.** A series of eight bits is called a **byte.** Note that these bits and bytes have no intrinsic meaning. They can represent anything or nothing at all (e.g., random sequences of zeros and ones).

Bits within computers are aggregated into a variety of **data types**. Some of the most common data types are listed below.

- *Integers* such as 32767, 15 and -20

- *Floating point numbers* (or floats) such as 3.14159, -12.014, and 14.01; the floating point refers to the decimal point

- *Characters* "a," and "z"

- *(Character) Strings* such as "hello" or "ball"

Note that these data types do not define meaning. It does not matter whether 3.14159 is a random number or the ratio of the circumference to the diameter of a circle (known as Pi or π).

Data can be aggregated into a variety of file formats. These file formats specify the way that data are organized within the file. For example, the file header may contain the colors used in an image file (known as the palette) and the compression method used to minimize storage requirements. Common or standardized file formats allow sharing of files between computers and between applications. For example, as long as your digital camera stores photos as JPG files, you can use any program that can read JPG files to view your photos.

- Image files such as JPG, GIF and PNG.

- Text files

- Sound files such as WAV and MP3

- Video files such as MPG

Again, it is important to recognize that neither data types nor file formats define the meaning of the data, except for the purpose of storing or display on a computer. For example, photographs of balloons and microscopes can be stored in JPG files. Nothing about the file format helps us recognize the subject of the photograph.

Informatics vs. Information Technology and Computer

Data are largely the domain of information technology (IT) professionals and computer scientists. As computers become increasingly important in biomedicine, biomedical researchers are starting to collaborate with computer scientists. IT professionals and computer scientists concentrate on technology, including computing systems composed of hardware and software as well as the algorithms implemented in such systems. For example, computer scientists develop algorithms to search or sort data more efficiently. Note that *what* is being sorted or searched is largely irrelevant. In other words, the meaning of the data is of secondary importance. It does not matter whether the strings that we are sorting represent names, email addresses, weights, names of cars or heights of buildings.

Though they may be motivated by specific applications, computer scientists typically develop general-purpose approaches to classes of problems that involve computation. For example, a computer scientist may design a memory architecture that efficiently stores and retrieves large data sets. The computer science contribution is the development of the better memory architecture for large data sets; while the memory architecture is not a direct improvement of an EHR per se, it is nonetheless critical to its advancement.

Information and knowledge, on the other hand, are addressed by informatics. To an informatician, computers are tools for manipulating information. Indeed, there are many other useful information tools, such as pen, paper and reminder cards. There are significant advantages to manipulating digitized data, including the ability to display the same data in a variety of ways and to communicate with remote collaborators. From an informatics perspective however, one should choose the optimal tool for the information task – often, but not always, the best tool for the task is computer-based.[4,6]

There are areas that combine computer science and informatics. For example, information retrieval draws on both disciplines. Information retrieval is "finding material (usually documents) of an unstructured nature (usually text) that satisfies an information need by retrieving documents from large collections (usually stored on computers).[7]"

Note that information retrieval is concerned with retrieval of information, not data. For example, finding documents that describe the relationship between aspirin and heart attack (myocardial infarction) is an example of an information retrieval task. The central problem is identifying documents that contain certain meaning. In contrast, retrieval of documents (or records) that contain the string "aspirin" is a database problem (an area of computer science). Importantly, informatics and computer science differ in the problems that they address (see Figure 2.4). We do not mean to imply that computer science is easier or less intellectually challenging compared to informatics.

Figure 2.4: Relationship between informatics and computer science

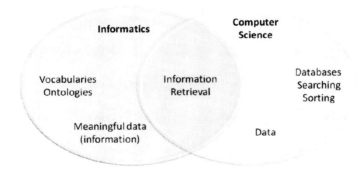

Converting Data to Information to Knowledge

We live in the real world that contains physical objects (e.g., aspirin tablet), people (e.g., John Smith), things that can be done (e.g., John Smith took an aspirin tablet) and other concepts. In order to do useful computation, we have to segregate some part of the physical world and create a **conceptual model**. The conceptual model contains only the parts of the physical world that are relevant to the computation. Importantly, everything that is not in the conceptual model is excluded from the computation and assumed to be irrelevant.

We use the conceptual model to design and implement a **computational model**. In Figure 2.5, the real world contains a person, John Smith. There are many other things in the real world including other people, physical objects, etc. There are many things that we can say about this person, they have a name, height, weight, parents, thoughts, feelings, etc. The conceptual model defines what is relevant; everything that is not in the conceptual model is therefore assumed to be not relevant. In our example (Figure 2.5), we chose name and age. Thus, the height, weight and all other things about John Smith are assumed to be irrelevant. For example, given our conceptual and computational models, we will not be able to answer questions about height. Next we have to define a **representation** (Figure 2.5). We consider a simple example, whole numbers. A representation has three components. The **represented world** is the <u>information</u> that we want to represent (e.g., whole numbers: 0, 1, 2, 3, …). The **representing world** contains the <u>data</u> that represent the information (e.g., symbols "0", "1", "2", "3", …). There must be a **mapping** between the represented world and the representing world. In our example, the mapping is the correspondence between whole numbers and symbols that are used to represent them. Note that the data are, in and of themselves, meaningless.

To do anything useful, we must also have rules regarding the mapping (i.e., relationship between the symbols and the real world) what can be done with the symbols. In our example, these rules are the rules governing the manipulation of whole numbers systems (e.g., addition, multiplication, division, etc.).

The data part of a representational system may also be called its "form", in which case meaning is called its' "content." The word "form" is significant because of its relationship to **formal methods**, which are methods that manipulate data using systematic rules that depend only on form, not content (meaning). These formal methods, including computer programs, depend only on systematic manipulation of data without regard for meaning. Thus, only a human can ensure that the input and output of a formal method (e.g., computer program) correctly capture and preserve meaning.

Figure 2.5: Computational framework

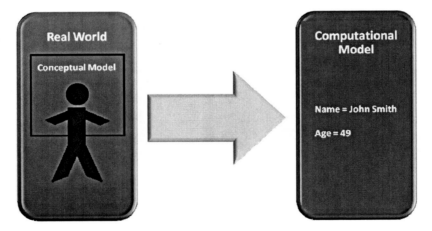

In spite of the fact that formal methods manipulate only form (or data), not meaning, they can be very useful. As long as the formal method does not violate the rules of the physical world, we can apply the method to solve problems in the real world. For example, we can use a whole number representation to determine how many 8-person boats are needed to transport 256 people across the Nile river (i.e., 256 people divided by 8 people/boat = 32 boats).

However, we must be careful because the formal method (division) can easily violate the rules of the real world. For example, suppose that 250 people are in Cairo and six people are in Khartoum (1,000 miles away) and they must cross at the same time. In this case, 32 boats is the wrong answer since 32 boats are needed in Cairo and another boat is needed at Khartoum. In this example, the real world includes location (Cairo vs. Khartoum), but the conceptual model includes only the number of people; location and distance are ignored. Thus, the computational model (based on the conceptual model) gives an inappropriate answer. We can't say that the answer is "wrong." Clearly 256/8=32; the computer did not malfunction. However, in the case where location is important, the numerical answer is not useful.

The distinction between the real (represented) world, the conceptual model (representing world) and the computational model (that which the computer manipulates) is fundamental to informatics. When the real world, the conceptual model and the computational model match, we get useful answers from the computer. When they do not match, such as the case when a critical constraint was left out of the conceptual model, the answers that we get from the computer are not useful.

This is what happened in the case of e-patient Dave. Formal methods (computer programs) were developed that linked fields in the Beth Israel Deaconess EHR to fields in Google Health. Data from one were dutifully transferred to the other. However, the meaning (i.e., that the data being transmitted were billing codes, not actual diagnoses) was lost. Further, there was a flaw in the conceptual model, the computational model or both models that prevented dates from being maintained correctly; perhaps because the dates reflected billing dates, rather than the date when a diagnosis was made.

Data to Information

We are now ready to explain how we convert data into information. Consider the example in Figure 2.1, 162.9 is, in and of itself, meaningless (i.e., it is a data item or datum). However, ICD-9-CM gives us a way to interpret 162.9 as "Lung neoplasm, not otherwise specified." Thus, the vocabulary ICD-9-CM turns the datum into a unit of information.

The computer still stores only data, not information. Thus, only a human can determine whether the meaning is preserved or not. In the case of e-patient Dave, all of the computer systems functioned as they were designed. There were not errors, but upon human review, the meaning was found have been mangled.

However, associating ICD-9-CM 162.9 with a patient record labels the patient record (and thus the patient) as having "Lung neoplasm, not otherwise specified." Of course, we could design systems that turn data into information without using vocabularies. For example, we could design patient records that include a bit for each possible diagnosis. Thus, setting the bit corresponding to lung cancer to 1 would be **semantically equivalent** to associating ICD-9-CM 162.9 with the patient's record. Semantically equivalent is simply another way of stating that the meanings are the same.

Transmission of information, often referred to as **interoperability**, requires consistency of interpretation. The source system (Beth Israel Deaconess EHR for e-patient Dave) and the receiving system (Google Health for e-patient Dave) must share a common way of transforming data into information. However, this is not sufficient. Note that in the case of e-patient Dave, both systems used ICD codes. However, associated information such as dates and most importantly the context: billing code vs. actual diagnosis, was not shared correctly.

Information to Knowledge

Multiple methods have been developed to extract knowledge from information. Note that it would not make sense to directly convert data (which by definition is not meaningful) to knowledge (justified, true belief). Thus, information is required to produce knowledge. Transformation of information (meaningful data) into knowledge (justified, true belief) is a core goal of science.

In the clinical world, most available knowledge is best described as justified (i.e., evidence exists that it is true), rather than proven fact (i.e., it must be true). This is an important distinction from traditional hard sciences such as physics or mathematics.

In this chapter, we focus on informatics techniques that are designed to convert clinical information into knowledge. Thus, we start by describing clinical data warehouses (CDWs) that are often the basis for attempts to turn clinical information into knowledge. We then describe methods for transforming information into knowledge.

Clinical research informatics is becoming increasingly recognized as a distinct sub-field within informatics (see separate chapter on e-research for further information). Clinical research informaticians leverage informatics to enable and transform clinical research. [8,9] By "enable," we mean helping researchers accomplish their goals faster and cheaper than is possible using existing methods. For example, searching electronic clinical data is potentially faster than manually reviewing paper clinical charts. By "transform," we mean developing methods that allow researchers to do things that they cannot do using existing methods. For example, it is not currently possible to use aggregated clinical data to help people make decisions. One cannot ask, in real-time or near real-time, "what happened to patients like me, at your institution, who chose treatment A vs. treatment B?" Although the information required to answer this question is found in the clinical records, a manual chart review cannot be performed in real time. However, before we can realize the benefits of computerized information, we must ensure that meaning is preserved.

Clinical Data Warehouses (CDWs)

The enterprise data warehouse was introduced in chapter 1 (see Figure 1.3). In this chapter, we will focus on clinical, rather than administrative data, thus we will refer to a **clinical data warehouse** or **CDW**.

Increasingly, clinical data are collected via electronic health records (EHRs). Clinical records within EHRs are composed of both **structured data** and **unstructured or (free text)**. Structured data may include billing codes, lab results (e.g., Sodium = 140 mg/dl), problem lists (e.g., Problem #1 = ICD-9-CM 162.9 = "Lung Neoplasm, Not Otherwise Specified"), medication lists, etc. In contrast, free text is similar to this chapter – simply human language such as English, called **natural language**. Clinical notes are often dictated and are represented in records as free text.

From an informatics perspective, structured data is much easier to manage – it is computationally tractable. Ideally, but not always, these data are encoded using a standard such as ICD-9-CM (see chapter on data standards). Thus, retrieving patients with a particular problem is, theoretically, simply a matter of identifying all records that are tagged with a particular code. As we will see later in this chapter, in practice this does not always work. Further, nuances (e.g., similarity to a previous case) or vague concepts (e.g., light-colored lesion, tall man) may be difficult to convey with a "one size fits all" vocabulary.

Similarly, computerized physician order entry (see chapter on electronic health records) can be difficult to implement. If designers allow only structured data, they must anticipate what will be ordered and make choices that constrain the possible inputs. For example, they may choose to use a particular vocabulary for medication orders, allow specific dosing frequencies, etc. Inevitably, however, physicians will want to write unusual orders that will be difficult to accommodate.

Free text, on the other hand, has the advantage of being able to express anything that can be expressed using natural language. On the other hand, it is difficult for computers to process. Indeed, the field of **natural language processing** (NLP) is an active area of research in both computer science and informatics. Within clinical records, the free text notes are critically important. Indeed, as we saw in the case of e-patient Dave, structured data (such as billing codes) may not be clinically accurate. This is not necessarily anyone's fault. Billing codes were assigned for billing, not for clinical care. Thus, it should not be surprising that using billing codes for a different purpose does not yield the desired result. Over 20 years ago, van der Lei warned:

...under the assumption that laws of medical informatics exist, I would like to nominate the first law: Data shall be used only for the purpose for which they were collected. This law has a collateral: If no purpose was defined prior to the collection of the data, then the data should not be used.[10]

Thus, to make sense of clinical records, we must leverage both structured data and free text. This remains an active area of informatics research.

A clinical data warehouse is a shared database that collects, integrates and stores clinical data from a variety of sources including electronic health records, radiology and other information systems. EHRs are designed to support real-time updating and retrieval of individual data (e.g., Joan Smith's age). The general process is shown in Figure 2.6. Data from multiple sources including one or more EHRs are copied into a staging database, cleaned and loaded into a common database where they are associated with **meta-data**. Meta-data are data that describe other data. For example, the notation that a particular data item is an ICD-9-CM term represents meta-data.

Figure 2.6: Overview of clinical data warehousing (ETL = Extract, transform and load)

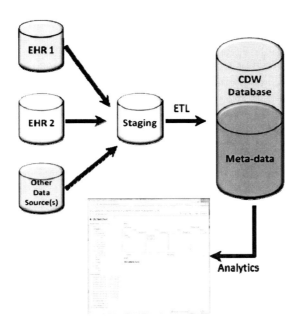

Once loaded into a CDW, a variety of analytics can be applied and the results presented to the user via a user interface. Examples of simple analytics include summary statistics such as counts, means, medians and standard deviations. More sophisticated analytics include associations (e.g., does A co-occur with B) and similarity determinations (e.g., is A similar to B).

In contrast to EHRs, CDWs are designed to support queries about groups (e.g., average age of patients with breast cancer). Although in principle an EHR may contain the same data as a CDW, databases that support EHRs are designed for efficient real-time updating and retrieval of individual data. Thus, a query across patients rather than regarding an individual may take much more time. Further, since EHRs support patient care, queries about groups may be restricted to ensure adequate performance for clinicians. Another important distinction is that CDWs are usually not updated in real-time. Although update schedules differ, daily or weekly updates of the institutional CDW are typical.

CDWs are rapidly becoming critical resources. They enable organizations to monitor quality by allowing users to query for specific quality measures (see chapter on quality improvement strategies) in specific patient populations (e.g., retrieve all women who are 40 years old or older who have not had a mammogram in the past year). Similarly, clinical and translational researchers use CDWs to identify trends (e.g., did screening mammograms detect breast cancer at an early stage?).[11] Comparative effectiveness research (CER) or, more broadly, practice-based research, are increasingly important fields that attempt to link research with clinical practice using CDWs. They complement traditional clinical trials that ask very focused questions. For example, a clinical trial might be designed to compare treatment A vs. treatment B in particular population of patients. In contrast, CER practitioners ask what actually happened in practice. For example, treatment A has been found to be more effective than treatment B in a clinical trial. What actually happened in practice?

Hospital infection control specialists use CDWs to track pathogens within hospitals. Public health agencies traditionally rely on reporting to conduct surveillance for natural or man-made illnesses (see chapter on public health informatics). However, reporting introduces a delay. Accessing aggregated data at the institutional level can be done much faster.

One of the most popular clinical data warehousing platforms is the product of the Informatics for Integrating Biology and the Bedside (i2b2) project based at Harvard Medical School. [12] The open source and very modular i2b2 platform was designed to enable the reuse of clinical data for research, but can also be very useful for non-research tasks such as quality monitoring. As of December 2011, i2b2 has been implemented at 72 academic institutions (60 in the United States alone). [13]

I2b2 relies on a star schema composed of facts and dimensions (Figure 2.7.). *Facts* are pieces of information that are queried by users (e.g., diagnoses, demographics, laboratory results, etc.) and *dimensions* describe the facts. Note that the data model is organized around facts, rather than individual patients, as would be the case for an EHR. Another benefit of organizing the CDW around observations is that data from multiple sources (e.g., different hospitals) can be aggregated into a common data model – new observations are simply added to the table of facts. Meta-data, such as the vocabulary that was used for encoding the fact, is an important component. Thus, the i2b2 data model by itself is not sufficient to ensure interoperability.

Figure 2.7: i2b2 data model [12]

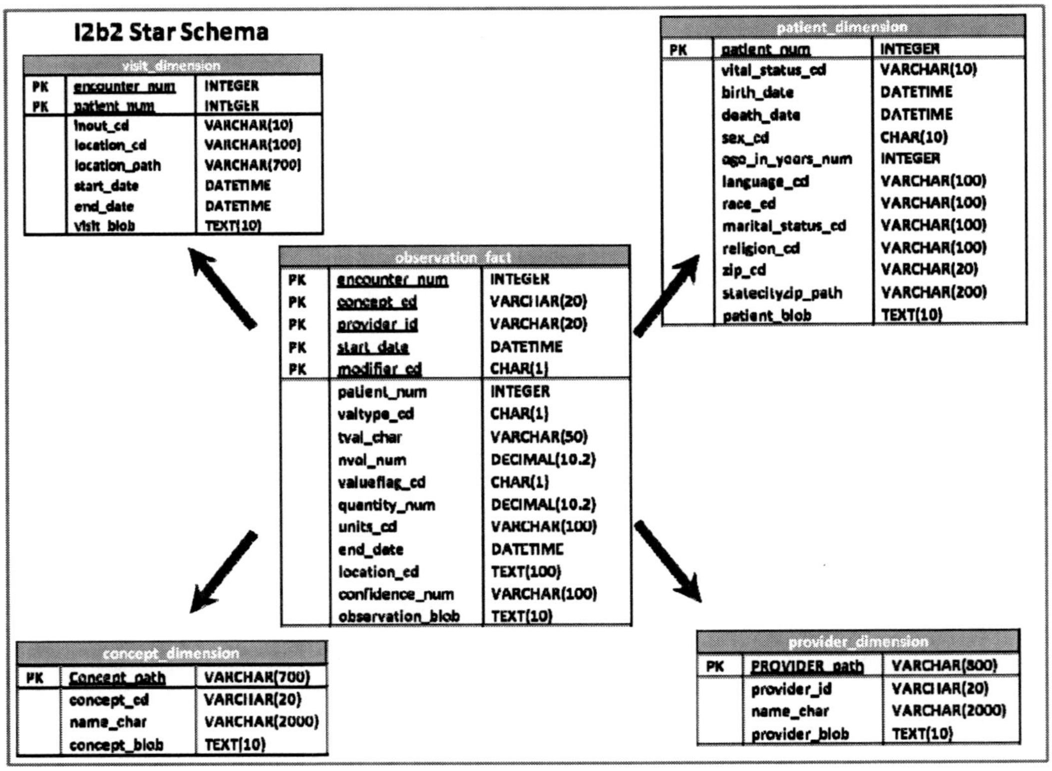

It provides a very usable interface to an institutional CDW that can be used by non-informaticians (see Figure 2.8). Users click and drag concepts from the ontology window (upper left) into the query panes (upper right) and obtain results, such as the number of patients fulfilling certain criteria, in lower right. In addition to the basic i2b2 package, specialized modules have been developed for NLP and other tasks.

In short, clinical data are collected via EHRs and archived in CDWs. As EHRs are becoming increasingly common, CDWs are becoming increasingly important. However, to realize the potential of CDWs to improve health, we must do more than archive data. We must turn these data into information and knowledge. We must be able to "make sense" of clinical data; to make clinical data meaningful (data → information) and then learn from aggregated clinical data (information → knowledge). In practice, many of the benefits of EHRs (see chapter 3) actually require a CDW. The transformation of data into information and knowledge is a core concern of informaticians.

Figure 2.8: i2b2 screenshot

Use of Aggregated Clinical Data

To make use of aggregated clinical information, we must be able to recognize records that belong to patients with specific conditions. For example, we must be able to identify records belonging to patients who have been diagnosed with breast cancer. A simple answer is to rely on billing codes, one of the most common forms of structured data in clinical records. However, as we saw in the case of e-patient Dave, we cannot simply rely on billing codes. Sometimes other structured data are available, problem lists are particularly useful. Unfortunately, problem lists are often out of date or incomplete.[14] Thus, a great deal of interest has focused on extracting information from free text clinical notes.

Concept extraction refers to the problem of identifying concepts within unstructured data, such as discharge summaries or pathology reports. Usually, these concepts are mapped to a controlled vocabulary, such as ICD-9-CM, SNOMED-CT and others. While this may on the surface appear to be a trivial problem, there are many ways in which a single concept might be expressed (for example high blood pressure and hypertension), and it is often the case that a single word or acronym may have multiple medically relevant meanings (for example DM may refer to Diabetes Mellitus or Depressed Mood) that cannot be teased apart without considering contextual cues. Consequently, much effort has been devoted toward the development of systems that aim to map between terms or phrases and controlled vocabularies with accuracy.

Multiple biomedical concept extraction systems exist including MetaMap[15] and cTAKES.[17] Broad-purpose medical language processing systems such as MedLEE [16], have also been adapted to this end. These systems can be tuned to perform well, but require re-tuning when applied to different corpora (e.g., changing institutions) or clinical problems (e.g., breast cancer vs. diabetes mellitus). Table 2.1 summarizes the published performance of these three concept extraction systems; note that the results are not directly comparable to each other due to different tasks and gold standards (a common limitation). [18,19]

Table 2.1: Published performance of three notable biomedical systems

Concept Extractor	Gold Standard	Precision	Recall	F-score (F_1)
cTAKES[17]	Mayo clinic	0.80	0.65	0.72
MetaMap[20]	NLM 500 articles	0.32	0.53	0.40
MEDLEE[21]	Proprietary	0.86	0.77	0.81

Classification refers to the problem of categorizing data into two or more categories. For example, we may want to classify medical records as belonging to patients who have vs. have not been diagnosed with breast cancer. A variety of classification algorithms have been developed, most of which rely on statistical methods. These classification algorithms generally depend on the selection of a set of features, such as the presence or absence of particular terms, concepts or phrases. Once these features have been selected, either manually or through automated methods, medical records can be categorized on the basis of these features. A commonly utilized approach is supervised machine learning, in which an algorithm is used to learn a representation of the features that characterize annotated positive (patients with breast cancer) and negative (patients without breast cancer) cases. New cases can then be categorized automatically based on the extent to which their features are characteristic of previously encountered positive or negative examples.

What Makes Informatics Difficult?

Why are some domains highly computerized, while health care and biomedicine resist computerization? Consider the banking system.[4] It is clearly very complex and involves a vast quantities data and meaning. Why do all banks use computers? In contrast to health care, there are no arguments regarding the suitability of computers to track accounts. We argue that in the case of banking, there is a very narrow "semantic gap" between data and information. In other words, the correspondence between the data (numbers) and information (account balances) is very direct. As we manipulate the computational model, the meaning of these manipulations follows easily.

Consider the differences between banking data and health care data, such as an account at a bank versus a patient (Table 2.2). One difference is that concepts relevant to health are relatively poorly defined compared to banking concepts. The symbols require significant background knowledge to interpret properly. For example, there are multiple ways that a patient can be "sick" including derangements in vital signs (e.g., extremely high or low blood pressure), prognosis associated with a diagnosis (e.g., any patient with an acute aortic dissection is sick), or other factors. Two clinicians when asked to describe a "sick" individual may legitimately focus on different facts. In contrast, a bank account balance (e.g., $1058.93) is relatively objective and is captured by the symbols. Thus, data-manipulating machines (IT) are much better suited to manipulating bank accounts than clinical descriptors.

Table 2.2: Comparison of health and banking data

	Banking data	Health data
Concepts and descriptions	Precise *Example:* Account 123 balance = $15.98	General, subjective *Example:* sick patient
Actions	Usually (not always) reversible *Example:* Move money A → B	Often not easily reversible *Example:* Give a medication Perform procedure
Context	Precise, constant *Example:* US $	Vague, variable *Example:* Normal lab values differ by lab
User autonomy	Well-defined and constrained *Example:* What I can do with my checking account = what you can do	Variable and dependent on circumstance *Example:* Clinical privileges depend on training, change over time, depend on circumstances
Users	Clerical staff	Varied, including highly trained professionals
Time sensitivity	Few true emergencies (seconds)	Many time sensitive tasks, highly variable time sensitivity depending on context
Workflow	Well-defined	Highly variable, implicit

In general, if the problem relates strictly to form (data), or is easily reduced to a form-based problem, then computers can easily be applied to solve the problem. Retrieving all abstracts in PubMed containing the string "breast cancer" is a question related to data and is easily reducible to a form-based data query. On the other hand, retrieving all documents that report a positive correlation between beta blockers (a class of medications) and weight gain is an information retrieval question that depends on the meaning of the query and the meaning of the text in the documents. The latter question is not easily reducible to form and is therefore much harder to automate.

Concepts definable with necessary and sufficient conditions are usually relatively easy to reduce to form, and thereby permit some limited automated processing of meaning. However, concepts without necessary and sufficient conditions (e.g., recognizing a sick patient, or defining pain) cannot be easily reduced to data and are much more difficult to capture computationally. Informatics is interesting (and difficult), in part, because many biomedical concepts defy definition via necessary and sufficient conditions.

Blois argued that, in order to compute upon a system, one must first determine the system's boundaries.[22] In other words, one must define all of the relevant components and assume that everything else is irrelevant. However, this is very difficult to do for biological (or human) systems. If we want to model the circulatory system, can we exclude the renal system? The endocrine system that includes the adrenal glands (releases epinephrine that constricts blood vessels and raises blood pressure)? The nervous system? And so on. With a bank account, it is easy to draw boundaries around the real world concepts that affect an accurate account balance. On the other hand, in biomedicine these boundaries are often impossible to precisely define, so our conceptual and computational models are rarely complete and often lead to inaccurate results, such as we saw above with e-Patient Dave.

Complexity of Knowledge Models

Modeling health care is difficult but this has not stopped informaticians from trying. Notable modeling attempts include the HL7 Reference Information Model or RIM (see chapter on data standards). Work on the RIM started in 1997 and Release 1 was approved by the American National Standards Institute (ANSI) in 2003. The RIM is one of the major differences between the commonly adopted HL7 version 2.x that has been widely used for decades and version 3, which has yet not been as widely adopted. [23] One of the problems is that the RIM is very complex (see Figure 2.9) and does not necessarily match all health care environments. As of December 2011, the HL7 RIM remains somewhat controversial.

Figure 2.9: Overview of the HL7 version 3 RIM (Courtesy HL7 [24])

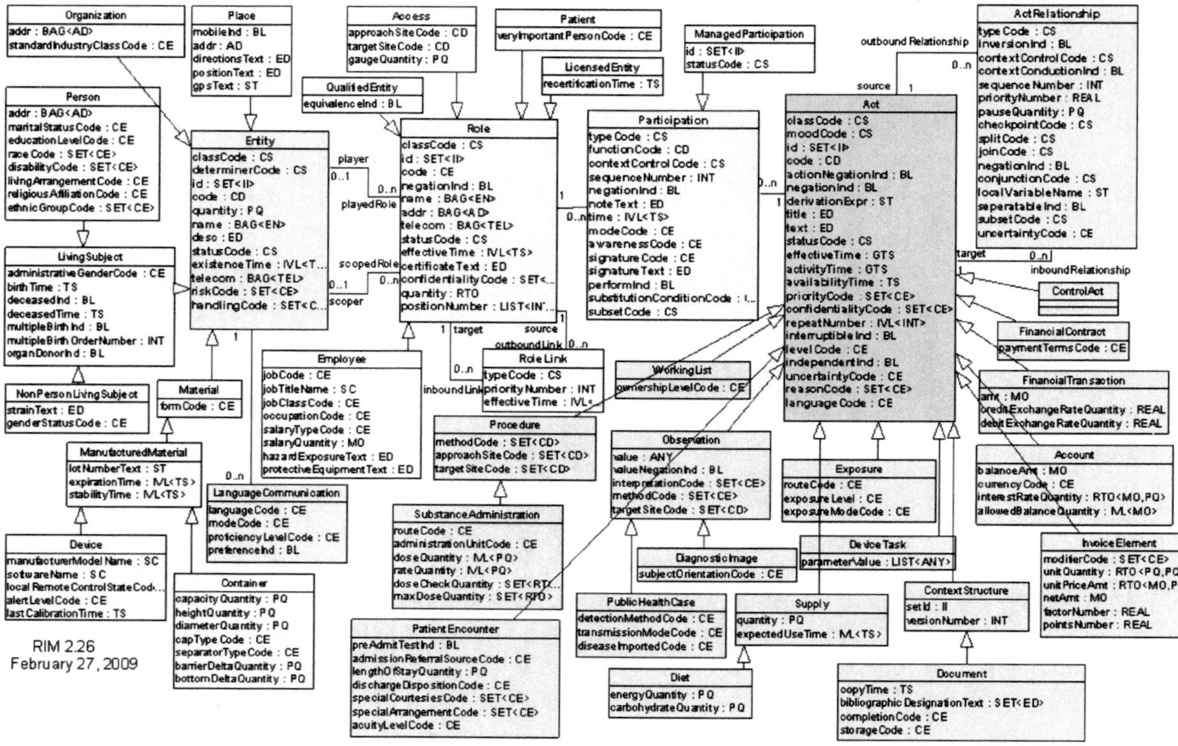

Biomedical informatics is also difficult because biomedical information can be imperfect in a number of different ways:

- Incomplete information: Information for which some data are missing, but potentially obtainable.
  - Example: What is the past medical history of an unconscious patient who arrives at ED?

- Uncertain information: Information for which it is not possible to objectively determine whether it is true or false. This can also be called epistemic uncertainty, because it arises from a lack of knowledge of some underlying fact. This type of imperfection is addressed by probability and statistics.
  - Example: how many female humans are in the US? Although there is a precise answer to this question at any given moment, we can only estimate the answer using statistics.

- Imprecise information: Information that is not as specific as it should be.

- o Example: Patient has pneumonia. This may be precise enough for some purposes, but is not sufficiently precise to determine treatment. For example, antibiotics can treat bacterial pneumonia, but are of little use to a patient with viral pneumonia.

- Vague information: Information that includes elements (e.g., predicates or quantifiers) that permit boundary cases (tall woman, may have happened, large bruise, big wound, elderly man, sharp radiating pain, etc.). Unlike uncertain information, with vague information there is no underlying matter of fact. Even if we knew the age of every female human in the US, we could not precisely answer the question of how many mature women were in the US at that time, because "mature" is a term that has boundary cases; there are women who are clearly mature, those who clearly are not, and a number in between for whom we are unsure that term applies.

- Inconsistent information: Information that contains two or more assertions that cannot simultaneously hold.

  - o Example: Birthdate: 8/29/66 AND 9/17/66

As illustrated in the above examples, all of these imperfections may be found in healthcare information. Humans can deal with these imperfections. For example, we can decide that for clinical purposes, a difference in patient age of a little over two weeks (in itself a vague statement), is insignificant for clinical purposes. Computers, on the other hand, must be explicitly programmed to make such "judgments." However, the number of possible variances and exceptions is effectively infinite. Thus, they cannot all be anticipated and addressed in advance. This is one reason why clinical decision support often gives advice that is, to a clinician, obviously inappropriate to the current patient situation.

In addition, definitions in health care and biomedicine often change over time. Consider the definition of a gene. [25]

Designing systems that adapt to changes in definition that, in turn, can affect other definitions is difficult. Our computers and programming languages process discrete symbols according to precise formal rules. They do not make sense of a highly ambiguous, noisy world or do meaning-based processing. With this background, we can now consider health IT and its various successes and failures in the real world.

Why Health IT Fails Sometimes

"To improve the quality of our health care while lowering its cost, we will make the immediate investments necessary to ensure that within five years all of America's medical records are computerized. This will cut waste, eliminate red tape, and reduce the need to repeat expensive medical tests... it will save lives by reducing the deadly but preventable medical errors that pervade our health care system."

– Barack Obama (Speech on the Economy, George Mason University, January 8, 2009)

Widespread dissatisfaction with health care in America and rapid advancement in information technology has focused attention on Health IT (HIT) as a possible solution. The need for HIT is one of the few topics upon which Democrats and Republicans agree. Both former President Bush and President Obama set 2014 as the goal date for computerizing medical records. To many, HIT seems like an obvious solution to our health care woes. The government's HIT website says that HIT adoption will: improve health care quality, prevent medical errors, reduce health care costs, increase administrative efficiencies, decrease paperwork and expand access to affordable care.[9] However, there is increasing evidence that HIT adoption does not guarantee these benefits. Unmitigated enthusiasm is dangerous for HIT adoption. Similar enthusiasm repeatedly threatens the field of artificial intelligence, resulting in cycles of excitement and disappointment (in artificial intelligence, these cycles are sometimes called "AI winters").

Effects of HIT

HIT is an "easy sell" to an American public increasingly dissatisfied with our health care system. Indeed, there is evidence that HIT can improve health care quality,[26] prevent medical errors,[27] and increase efficiency.[26] Thus, there is reason for optimism. With the American Recovery and Reinvestment Act (ARRA) of 2009, the US government made a multi-billion dollar investment in HIT.[28] Similar investments have been made by the governments of Australia,[29] Belgium,[30] Canada,[31] Denmark,[32] and the United Kingdom.[33]

However, many and perhaps even most HIT projects fail.[34] There is also evidence that HIT can worsen health care quality to the point of increasing mortality,[35] increasing errors[36,37] and decreasing efficiency.[35] In November 2011, the Institute of Medicine issued a report entitled "Health IT and Patient Safety: Building Safer Systems for Better Care" that concluded: "..some products have begun being associated with increased safety risks for patients."[38] There is even a term, "e-iatrogenesis," that refers to the unintended deleterious consequences of HIT.[39] Notably, systems that increase mortality at one institution[35], do not seem to have the same effect at another institution[40]; even though the clinical setting (pediatric intensive care) was similar. Thus, we cannot simply conclude that the system itself is wholly responsible. It is not just the system being implemented, but how it is implemented and in what context that determines the clinical results.

We've Been Here Before: AI Winters

During the 1950s, we were faced with a different problem: the Cold War. Similarly, the government saw IT as a promising (at least partial) solution. If researchers could develop automated translation, we could monitor Russian communications and scientific reports in "real time." There was a great deal of optimism and "...many predictions of fully automatic systems operating within a few years."[41]

Although there were promising applications of poor-quality automated translation, the optimistic predictions of the 1950s were not realized. The fundamental problem of context and meaning remains unsolved. This made disambiguation difficult resulting in amusing failures. Humorous examples include: "the spirit is willing but the flesh is weak" translated English → Russian → English resulted in the phrase "the vodka is good but the meat is rotten."

In 1966, the influential Automatic Language Processing Advisory Committee (ALPAC) concluded that "there is no immediate or predictable prospect of useful machine translation."[42] As a result, research funding was stopped and there was little automated translation research in the United States from 1967 until a revival in 1976-1989.[41]

Similarly, there is currently tremendous interest in HIT. Although there is good evidence that HIT can be useful, some will certainly be disappointed. A recent report by the National Research Council (the same body that published the ALPAC report) concluded that "...current efforts aimed at the nationwide deployment of health care IT will not be sufficient to achieve the vision of 21st century health care, and may even set back the cause if these efforts continue wholly without change from their present course."[43] Thus, there is reason for concern that HIT (and perhaps even informatics, in general) may be headed for a bust. Such an "HIT winter" would be unfortunate, since there are real benefits of pursuing research and implementation of HIT.

The Problem: Health Information Technology is Really Health Data Technology

The fundamental problem is that existing technology stores, manipulates and transmits data (symbols), not information (data + meaning). Thus, the utility of HIT is limited by the extent to which data approximates meaning. Unfortunately, in health care, data do not fully represent the meaning. In other words, there is a large gap between data and information. Since the difference between data and information is meaning (semantics), we call this the "semantic gap."

Social and Administrative Barriers to HIT Adoption. Manipulating data and not information has many consequences for HIT. Note that there is no shortage of computers in hospitals. While most hospitals do not manage their clinical data electronically, all of them manage their financial data electronically. Just like any other organization, many hospitals have functioning e-mail systems and maintain a Web presence. Many clinicians used personal digital assistants, [44] some even communicate with patients using e-mail.

The social and administrative barriers to HIT adoption have been discussed by multiple authors in countless papers. Such barriers include a mismatch between costs and benefits, cultural resistance to change, lack of an appropriately trained workforce to implement HIT and multiple others.[45] To some, clinicians' resistance to computerization appears irrational. However, caution seems increasingly reasonable given the mixed evidence regarding the benefits of poorly-implemented HIT. Thus, the clinical enterprise is not computerized because of rational skepticism regarding the benefit of current HIT, not an irrational resistance to IT or computerization.

Future Trends

Significant research problems must be addressed before HIT becomes more attractive to clinicians. Many of these are outlined in a recent National Research Council report. [43] First, there is a mismatch between what HIT can represent (data) and concepts relevant to health care (data + meaning). This is a very difficult and fundamental challenge that includes multiple long-standing challenges in artificial intelligence (e.g., how computers can be "taught" context or common sense) that have proven very difficult to solve. It seems that until we have true information processing, rather than data processing, technology, the benefits of HIT will be limited.

Second, HIT must augment human cognition and abilities. Friedman recently expressed this elegantly as the "fundamental theorem of informatics": human + computer > human (humans working with computers should perform better than a human alone).[46] The theorem argues that there must be a clear and demonstrable benefit from HIT. In spite of the problems with current HIT, there are clearly situations where HIT can be beneficial. In some ways, human cognition and computer technology are very complementary. For example, monitoring (e.g., waveforms) is much easier for computers than for humans. In contrast, reasoning by analogy across domains is natural for humans but difficult for computers.

How We Will Make Progress

Researchers are exploring multiple promising paradigm-shifting ideas. We can already give examples of approaches that address some of the fundamental problems described in this chapter.

One approach is to recognize the complementary strengths of humans and computers. Humans are good at constructing and processing meaning. In contrast, computers are much better at processing data. We can leverage this understanding to design systems that harness the data-processing power of computers to present (display) data in ways that make it easier for humans to grasp and manipulate meaning. For example, a *word cloud visualization* shows the term frequency in text. [47] The size of the font is proportional to the frequency of the term.

Returning to HIT, we can apply these same principles. For example, Figure 2.10 shows an example of an EHR that integrates clinical decision support. This is not novel, but this example illustrates what could be done by combining multiple types of information on the same screen with an understanding of the user's task.

Defining scenarios when HIT is beneficial with all relevant parameters and demonstrating that using HIT is *reliably* beneficial in these scenarios remains a research challenge. In its present form, HIT will not transform healthcare in the same way that IT has transformed other industries. This is due in part to the large semantic gap between health data and health information (concepts). In addition, many problems with healthcare require non-technological solutions, such as changes in healthcare policy and financing.

Figure 2.10 EHR screen (from John Halamka) showing integration of decision support into the EHR[48]

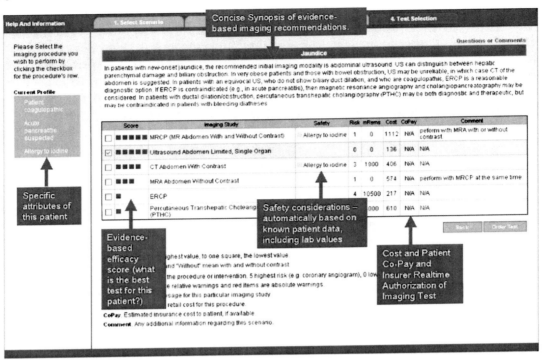

Key Points

- Data are observations reflecting differences in the world (e.g., "162.9") while information is meaningful data or facts from which conclusions can be drawn and knowledge is information that is justifiably believed to be true

- Data are largely the domain of information technology (IT) professionals and computer scientists; information and knowledge is the domain of informatics and informaticians

- Vocabularies help convert data into information

- The transformation of data into information and knowledge is a core concern of informaticians

- When the real world, the conceptual model and the computational model match, we get useful answers from the computer

- Concepts relevant to health are relatively poorly defined compared to e.g. banking concepts

- There is a large "semantic gap" between health data and health information

Conclusion

Problems in healthcare are information and knowledge intensive. Current technology is centered on processing data. This mismatch, or semantic gap, between the problems healthcare IT tries to address and the available technology explains the difficulties that informaticians face every day. It also explains the differences between Informatics and Computer Science. We must advance our information- and knowledge-processing capabilities in order to continue improving healthcare through technology.

References

1. Stead WW, Lin HS, editors. Computational Technology for Effective Health Care: Immediate Steps and Strategic Directions. National Academies Press; Washington, D.C.: 2009. P. 2

2. E-patient Dave http://epatientdave.com/ (Accessed December 13 2011)

3. Electronic Health Records Raise Doubt, Boston Globe, April 13, 2009. (Accessed December 10 2011)

4. Bernstam, E.V., J.W. Smith, and T.R. Johnson, What is biomedical informatics? Journal of biomedical informatics, 2010. 43(1): p. 104-105

5. Floridi, L. Semantic conceptions of information. 2005 October 5, 2005 ; http://plato.stanford.edu/entries/information-semantic/. (Accessed November 13 2011)

6. Bernstam, E.V., et al., Synergies and distinctions between computational disciplines in biomedical research: perspective from the Clinical andTranslational Science Award programs. Academic Medicine : Journal of the Association of American Medical Colleges, 2009. 84(7): 964-70.

7. Manning CD, Raghavan P, Schutze H. Introduction to information retrieval. Cambridge University Press Cambridge 2008.

8. Embi, P.J. and P.R. Payne, Clinical research informatics: challenges, opportunities and definition for an emerging domain. Journal of the American Medical Informatics Association : JAMIA, 2009. 16(3): p. 316-27.

9. Payne, P.R., P.J. Embi, and M.G. Kahn, Clinical research informatics: The maturing of a translational biomedical informatics sub-discipline. Journal of biomedical informatics, 2011.

10. van der Lei, J., Use and abuse of computer-stored medical records. Methods of information in medicine, 1991. 30(2): p. 79-80.

11. Zerhouni, E.A., Translational research: moving discovery to practice. Clin Pharmacol Ther, 2007. 81(1): p. 126-8.

12. Murphy, S.N., et al., Serving the enterprise and beyond with informatics for integrating biology and the bedside (i2b2). Journal of the American Medical Informatics Association : JAMIA, 2010. 17(2): p. 124-30.

13. Informatics for Integrating Biology & the Bedside (I2b2) www.i2b2.org (Accessed December 4 2011)

14. Szeto, H.C., et al., Accuracy of computerized outpatient diagnoses in a Veterans Affairs general medicine clinic. The American journal of managed care, 2002. 8(1): p. 37-43.

15. Aronson, A.R. and F.M. Lang, An overview of MetaMap: historical perspective and recent advances. Journal of the American Medical Informatics Association : JAMIA, 2010. 17(3): p. 229-36.

16. Chen, E.S., et al., Automated acquisition of disease drug knowledge from biomedical and clinical documents: an initial study. Journal of the American Medical Informatics Association : JAMIA, 2008. 15(1): p. 87-98.

17. Savova, G.K., et al., Mayo clinical Text Analysis and Knowledge Extraction System (cTAKES): architecture, component evaluation and applications. Journal of the American Medical Informatics Association : JAMIA, 2010. 17(5): p. 507-13.

18. Chapman, W.W., et al., Overcoming barriers to NLP for clinical text: the role of shared tasks and the need for additional creative solutions. Journal of the American Medical Informatics Association: JAMIA, 2011. 18(5): p. 540-3.

19. Stanfill, M.H., et al., A systematic literature review of automated clinical coding and classification systems. Journal of the American Medical Informatics Association : JAMIA, 2010. 17(6): p. 646-51.

20. Gay, C.W., M. Kayaalp, and A.R. Aronson, Semi-automatic indexing of full text biomedical articles. AMIA ... Annual Symposium proceedings / AMIA Symposium. AMIA Symposium, 2005: p. 271-5.

21. Friedman, C., et al., Automated encoding of clinical documents based on natural language processing. Journal of the American Medical Informatics Association: JAMIA, 2004. **11**(5): p. 392-402.

22. Blois, M.S., Information and medicine: the nature of medical descriptions1984, Berkeley: University of California Press.

23. Smith, B. and W. Ceusters, HL7 RIM: an incoherent standard. Studies in health technology and informatics, 2006. 124: p. 133-8.

24. HL7 Version 3 RIM http://www.hl7.org/documentcenter/public_temp_B69AB426-1C23-BA17-0CA55CBFEF56C9A3/calendarofevents/himss/2011/HL7%20Reference%20Information%20Model.pdf (Accessed December 1 2011)

25. Hopkin, K., The Evolving Definition of a Gene. BioScience, 2009. 59(11): p. 928-31.

26. Chaudhry, B., et al., Systematic review: impact of health information technology on quality, efficiency, and costs of medical care. Ann Intern Med, 2006. 144(10): p. 742-52.

27. Bates, D.W., et al., Reducing the frequency of errors in medicine using information technology. J Am Med Inform Assoc, 2001. 8(4): p. 299-308.

28. American Recovery and Reinvestment Act (ARRA) of 2009.

29. HealthConnect Implementation Strategy v2.1. July 6, 2005; Available from: http://www.health.gov.au/internet/hconnect/publishing.nsf/Content/archive-docs/$File/implementation.pdf

30. France, F.R., eHealth in Belgium, a new "secure" federal network: role of patients, health professions and social security services. InternationalJournal of Medical Informatics, 2011. 80(2): p. e12-6.

31. EHRS Blueprint: An interoperable EHR framework. Version 2. March 2006. [cited 2011 December 11]; Available from: https://knowledge.infoway-inforoute.ca/EHRSRA/doc/EHRS-Blueprint.pdf (Accessed December 3 2011)

32. Protti, D. and I. Johansen, Widespread adoption of information technology in primary care physician offices in Denmark: a case study. Issue brief, 2010. 80: p. 1-14.

33. House of Commons Public Accounts Committee. The National Programme for IT in the NHS: Progress since 2006. Second Report of Session 2008-09. [cited 2011 December 11]; Available from: http://www.publications.parliament.uk/pa/cm200809/cmselect/cmpubacc/153/153.pdf.

34. Littlejohns, P., J.C. Wyatt, and L. Garvican, Evaluating computerised health information systems: hard lessons still to be learnt. BMJ, 2003. 326(7394): p. 860-3.

35. Han, Y.Y., et al., Unexpected increased mortality after implementation of a commercially sold computerized physician order entry system. Pediatrics, 2005. 116(6): p. 1506-12.

36. Levenson, N.G. and C.S. Turner, An Investigation of the Therac-25 Accidents. IEEE Computer, 1993(July): p. 18-41.

37. Koppel, R., et al., Role of computerized physician order entry systems in facilitating medication errors. JAMA, 2005. 293(10): p. 1197-203.

38. Services, C.o.P.S.a.H.I.T.B.o.H.C., Health IT and patient safety: building safer systems for better care, 2011, Institute of Medicine of the National Academies: Washington DC.

39. Weiner, J.P., et al., "e-Iatrogenesis": the most critical unintended consequence of CPOE and other HIT. J Am Med Inform Assoc, 2007. 14(3): p. 387-8; discussion 389.

40. Del Beccaro, M.A., et al., Computerized provider order entry implementation: no association with increased mortality rates in an intensive care unit. Pediatrics, 2006. 118(1): p. 290-5.

41. Hutchins, J., Machine translation: history, in Encyclopedia of language & llinguistics, second edition, K. Brown, Editor 2006, Elsevier: Oxford. p. 375-83.

42. ALPAC, Language and machines: computers in translation and linguistics.Report by the Automatic Language Processing Advisory Committee, Division of Behavioral Sciences, National Academy of Sciences, National Research Council., 1966, National Academy of Sciences, National Research Council.: Washington, DC.

43. Computational technology for effective health care: immediate steps and strategic directions, W.W. Stead and H.S. Lin, Editors. 2009, Committee on Engaging the Computer Science Research Community in Health Care Informatics, Computer Science and Telecommunications Board, Division on Engineering and Physical Sciences, National Research Council of the National Academies: Washington, DC.

44. McLeod, T.G., J.O. Ebbert, and J.F. Lymp, Survey assessment of personal digital assistant use among trainees and attending physicians. J Am Med Inform Assoc, 2003. 10(6): p. 605-7.

45. Hersh, W., Health care information technology: progress and barriers. JAMA, 2004. 292(18): p. 2273-4.

46. Friedman, C.P., A "fundamental theorem" of biomedical informatics. J Am Med Inform Assoc, 2009. 16(2): p. 169-70.

47. Visualizations: JAMIA Content. May13, 2009. http://www-958.ibm.com/software/data/cognos/manyeyes/visualizations/jamia-content (Accessed January 10, 2012)

48. My Life as a CMIO. http://geekdoctor.blogspot.com/2007/11/data-information-knowledge-and-wisdom.html

3

Electronic Health Records

ROBERT E. HOYT

KENNETH G. ADLER

Learning Objectives

After reading this chapter the reader should be able to:

- State the definition and history of electronic health records

- Describe the limitations of paper-based health records

- Identify the benefits of electronic health records

- List the key components of an electronic health record

- Describe the ARRA-HITECH programs to support electronic health records

- Describe the benefits and challenges of computerized order entry and clinical decision support systems

- State the obstacles to purchasing and implementing an electronic health record

- Enumerate the steps to purchase an EHR

Introduction

There is no topic in health informatics as important, yet controversial, as the electronic health record (EHR). In spite of their significance, the history of EHRs in the United States is relatively short. The Problem Oriented Medical Information System (PROMIS) was developed in 1976 by The Medical Center Hospital of Vermont in collaboration with Dr. Lawrence Weed, the originator of the problem oriented record and SOAP formatted notes. Ironically, the inflexibility of the concept led to its demise.[1] In a similar time frame the American Rheumatism Association Medical Information System (ARAMIS) appeared. All findings were displayed as a flow sheet. The goal was to use the data to improve the care of rheumatologic conditions.[2] Other EHR systems began to appear throughout the US: the Regenstrief Medical Record System (RMRS) developed at Wishard Memorial Hospital, Indianapolis; the Summary Time Oriented Record (STOR) developed by the University of California, San Francisco; Health Evaluation Through Logical Processing (HELP) developed at the Latter Day Saints Hospital, Salt Lake City and The Medical Record developed at Duke University[3], the Computer Stored Ambulatory Record (COSTAR) developed by Octo Barnett at Harvard and the De-Centralized Hospital Computer Program (DHCP) developed by the Veterans Administration.[4]

In 1970 Schwartz optimistically predicted "clinical computing would be common in the not too distant future."[5] In 1991 the Institute of Medicine (IOM) recommended electronic health records as a solution for many of the problems facing modern medicine.[6] Since the IOM recommendation, little progress has been made during the last decade for multiple reasons. As Dr. Donald Simborg stated, the slow acceptance of electronic health records is like the "wave that never breaks."[7]

The American Recovery and Reimbursement Act (ARRA) of 2009 was a major game changer for electronic health records, with reimbursement for the Meaningful Use of certified EHRs, as well as other programs that supported EHR education and health information exchange. Reimbursement details will be discussed in more detail later in this chapter.

We will primarily discuss outpatient (ambulatory) electronic health records. Inpatient EHRs share many similarities to ambulatory EHRs but the scope, price and complexity are different. The logical steps to selecting and implementing an EHR are found in Appendix 3.3 at the end of the chapter.

Electronic Health Record Definitions

There is no universally accepted definition of an EHR. As more functionality is added the definition will need to be broadened. Importantly, EHRs are also known as electronic medical records (EMRs), computerized medical records (CMRs), electronic clinical information systems (ECIS) and computerized patient records (CPRs). Throughout this book we will use electronic health record as the more accepted and inclusive term, but either term is acceptable.

Figure 3.1 demonstrates the relationship between EHRs, EMRs and personal health records (PHRs).[8] As indicated in the diagram, PHRs can be part of the EMR/EHR system which may cause confusion.

Figure 3.1: Relationship between EHR, PHR and EMR

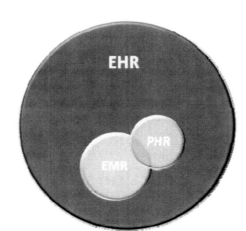

In 2008 the National Alliance for Health Information Technology released the following definitions in an effort to standardize terms used in HIT:

- Electronic Medical Record: "An electronic record of health-related information on an individual that can be created, gathered, managed and consulted by authorized clinicians and staff within one healthcare organization."[9]

- Electronic Health Record: "An electronic record of health-related information on an individual that conforms to nationally recognized interoperability standards and that can be created, managed and consulted by authorized clinicians and staff across more than one healthcare organization."[9]

- Personal Health Record: "An electronic record of health-related information on an individual that conforms to nationally recognized interoperability standards and that can be drawn from multiple sources while being managed, shared and controlled by the individual."[9]

Need for Electronic Health Records

The following are the most significant reasons why our healthcare system would benefit from the widespread transition from paper to electronic health records:

Paper Records Are Severely Limited

Much of what can be said about handwritten prescriptions can also be said about handwritten office notes. Figure 3.2 illustrates the problems with a paper record. In spite of the fact that this clinician used a template, the handwriting is illegible and the document cannot be electronically shared or stored. It is not structured data that is computable and hence sharable with other computers and systems. Other shortcomings of paper: expensive to copy, transport and store; easy to destroy; difficult to analyze and determine who has seen it; and the negative impact on the environment. Electronic patient encounters represent a quantum leap forward in legibility and the ability to rapidly retrieve information. Almost every industry is now computerized and digitized for rapid data retrieval and trend analysis. Look at the stock market or companies like Walmart or Federal Express. Why not the field of medicine?

Figure 3.2: Outpatient paper-based patient encounter form

With the relatively recent advents of pay-for-performance, patient centered medical home model and accountable care organizations there are new reasons to embrace technology in order to aggregate and report results in order to receive reimbursement. It is much easier to retrieve and track patient data using EHRs and patient registries than to use labor intensive paper chart reviews. EHRs are much better organized than paper charts, allowing for faster retrieval of lab or x-ray results. It is also likely that EHRs will have an electronic problem summary list that outlines a patient's major illnesses, surgeries, allergies and medications. How many times does a physician open a large paper chart, only to have loose lab results fall out? How many times does a physician re-order a test because the results or the chart is missing? It is important to note that paper charts are missing as much as 25% of the time, according to one study.[10] Even if the chart is available; specifics are missing in 13.6% of patient encounters, according to another study.[11]

Table 3.1 shows the types of missing information and its frequency. According to the President's Information Technology Advisory Committee, 20% of laboratory tests are re-ordered because previous studies are not accessible.[12] This statistic has great patient safety, productivity and financial implications.

Table 3.1: Types and frequencies of missing information

Information Missing During Patient Visits	% Visits
Lab results	45%
Letters/dictations	39%
Radiology results	28%
History and physical exams	27%
Pathology results	15%

EHRs allow easy navigation through the entire medical history of a patient. Instead of *pulling paper chart volume 1 of 3* to search for a lab result, it is simply a matter of a few mouse clicks. Another important advantage is the fact that the record is available 24 hours a day, seven days a week and doesn't require an employee to pull the chart, nor extra space to store it. Adoption of electronic health records has saved money by decreasing full time equivalents (FTEs) and converting records rooms into more productive space, such as exam rooms. Importantly, electronic health records are accessible to multiple healthcare workers at the same time, at multiple locations. While a billing clerk is looking at the electronic chart, the primary care physician and a specialist can be analyzing clinical information simultaneously. Moreover, patient information should be available to physicians on call so they can review records on patients who are not in their panel. Furthermore, it is believed that electronic health records improve the level of coding. Do clinicians routinely submit a lower level of care for billing purposes because they know that handwritten patient notes are short and incomplete? Templates may help remind clinicians to add more history or details of the physical exam, thus justifying a higher level of coding (templates are disease specific electronic forms that essentially allow you to point and click a history and physical exam). A study of the impact of an EHR on the completeness of clinical histories in a labor and delivery unit demonstrated improved documentation, compared to prior paper-based histories.[13] Lastly, EHRs provide clinical decision support such as alerts and reminders, which we will cover later in this chapter.

Need for Improved Efficiency and Productivity

The goal is to have patient information available to anyone who needs it, when they need it and where they need it. With an EHR, lab results can be retrieved much more rapidly, thus saving time and money. It should be pointed out however, that reducing duplicated tests benefits the payers and patients and not clinicians so there is a misalignment of incentives. Moreover, a study in 1987 using computerized order entry showed that simply displaying past results reduced duplication and the cost of testing by only 13%.[14] If lab or x-ray results are frequently missing, the implication is that they need to be repeated which adds to this country's staggering healthcare bill. The same could be said for duplicate prescriptions. It is estimated that 31% of the United States $2.3 trillion dollar healthcare bill is for administration.[15] EHRs are more efficient because they reduce redundant paperwork and have the capability of interfacing with a billing program that submits claims electronically. Consider what it takes to simply get the results of a lab test back to a patient using the old system. This might involve a front office clerk, a nurse and a physician. The end result is frequently placing the patient on hold or playing *telephone tag*. With an EHR, lab results can be forwarded via secure messaging. Electronic health records can help with productivity if templates are used judiciously. As noted, they allow for point and click histories and physical exams that in some cases may save time. Embedded educational content for clinicians is one of the newest features of a comprehensive EHR. Clinical practice guidelines, linked educational content and patient handouts can be part of the EHR. This may permit finding the answer to a medical question while the patient is still in the exam room. Several EHR companies also offer a centralized area for all physician approvals and signatures of lab work, prescriptions, etc. This should improve work flow by avoiding the need to pull multiple charts or enter multiple EHR modules.

Quality of Care and Patient Safety

As we have previously suggested, an EHR should improve patient safety through many mechanisms: (1) Improved legibility of clinical notes (2) Improved access anytime and anywhere (3) Reduced duplication (4) Reminders that tests or preventive services are overdue (5) Clinical decision support that reminds us of patient allergies, the correct dosage of drugs, etc. (6) Electronic problem summary lists provide diagnoses, allergies and surgeries at a glance. In spite of the before mentioned benefits, a study by Garrido of quality process measures before and after implementation of a widespread EHR in the Kaiser Permanente system, failed to show improvement.[16]

To date there has only been one study published that suggested use of an EHR decreased mortality. This particular EHR had a disease management module designed specifically for renal dialysis patients that could provide more specific medical guidelines and better data mining to potentially improve medical care. The study suggested that mortality was lower compared to a pre-implementation period and compared to a national renal dialysis registry.[17]

It is likely that we are only starting to see the impact of EHRs on quality. Based on internal data Kaiser Permanente determined that the drug Vioxx had an increased risk of cardiovascular events before that information was published based on its own internal data.[18] Similarly, within 90 minutes of learning of the withdrawal of Vioxx from the market, the Cleveland Clinic queried its EHR to see which patients were on the drug. Within seven hours they deactivated prescriptions and notified clinicians via e-mail.[19]

Quality reports are far easier to generate with an EHR compared to a paper chart that requires a chart review. Quality reports can also be generated from a data warehouse or health information organization that receives data from an EHR and other sources.[20] Quality reports are the backbone for pay-for-performance which we will discuss further in another chapter.

Public Expectations

According to a 2006 Harris Interactive Poll for the Wall Street Journal Online, 55% of adults thought an EHR would decrease medical errors; 60% thought an EHR would reduce healthcare costs and 54% thought that the use of an EHR would influence their decision about selecting a personal physician.[21] The Center for Health Information Technology would argue that EHR adoption results in better customer satisfaction through fewer lost charts, faster refills and improved delivery of patient educational material. [22]

Governmental Expectations

EHRs are considered by the federal government to be transformational and integral to healthcare reform. As an example, EHR reimbursement is a major focal point of the HITECH Act. It is the goal of the US Government to have an interoperable electronic health record by 2014. In addition to federal government support, states and payers have initiatives to encourage EHR adoption. Many organizations state that we need to move from the *cow path* to the *information highway*. CMS is acutely aware of the potential benefits of EHRs to help coordinate and improve disease management in older patients.

Financial Savings

The Center for Information Technology Leadership (CITL) has suggested that ambulatory EHRs would save $44 billion yearly and eliminate more than $10 in rejected claims per patient per outpatient visit. This organization concludes that not only would there be savings from eliminated chart rooms and record clerks;

there would be a reduction in the need for transcription. There would also be fewer callbacks from pharmacists with electronic prescribing. It is likely that copying, faxing and mail expenses, chart pulls and labor costs would be reduced with EHRs, thus saving full time equivalents (FTEs). More rapid retrieval of lab and x-ray reports results in time/labor saving as does the use of templates. It appears that part of the savings is from improved coding. More efficient patient encounters mean more patients could be seen each day. Improved savings to payers from medication management is possible with reminders to use the *drug of choice* and generics. [23]

EHRs should reduce the cost of transcription if clinicians switch to speech recognition and/ or template use. Because of structured documentation with templates, they may also improve the coding and billing of claims.

It is not known if EHR adoption will decrease malpractice, hence saving physician and hospital costs. A 2007 Survey by the Medical Records Institute of 115 practices involving 27 specialties showed that 20% of malpractice carriers offered a discount for having an EHR in place. Of those physicians who had a malpractice case in which documentation was based on an EHR, 55% said the EHR was helpful. [24]

Technological Advances

The timing seems to be right for electronic records partly because the technology has evolved. The internet and World Wide Web make the application service provider (ASP) concept for an electronic health record possible. An ASP option means that the EHR software and patient data reside on a remote web server that you access via the internet from the office, hospital or home. Computer speed, memory and bandwidth have advanced such that digital imaging is also a reality, so images can be part of an EHR system. Standard PCs, laptops and tablet PCs continue to add features and improve speed and memory while purchase costs drop. Wireless and mobile technologies permit access to the hospital information system, the electronic health record and the internet using a personal digital assistant, smartphone or laptop computer. The chapter on health information exchange will point out that health information organizations can link EHRs together, in order to share information and services.

Need for Aggregated Data

In order to make evidence based decisions, we need high quality data that should derive from multiple sources: inpatient and outpatient care, acute and chronic care settings, urban and rural care and populations at risk. This can only be accomplished with electronic health records and discrete structured data. Moreover, we need to combine or aggregate data to achieve statistical significance. Although most primary care is delivered by small practices, it is difficult to study because of relatively small patient populations, making aggregation necessary. [25]

EHR is a Transformational Tool

It is widely agreed that US Healthcare needs reform in multiple areas. To modernize its infrastructure we would need to have widespread adoption of EHRs. Large organizations such as the Veterans Health Administration and Kaiser Permanente use robust EHRs (VistA and Epic) that generate enough data to change the practice of medicine. In 2009 Kaiser Permanente reported two studies, one pertaining to the management of bone disease (osteoporosis) and the other chronic kidney disease. They were able to show that with their EHR they could focus on patients at risk and use all of the tools available to improve disease management and population health. [26-27] In a study reported in 2009 Kaiser-Permanente reported that electronic visits that are part of the electronic health record system were likely responsible for a 26.2% decrease in office visits over a four year period. They posited that this was good news for a system that aligns incentives with quality, regardless whether the visit was virtual or face-to-face. [28] Other fee-for-service organizations might find this alarming if office visits decreased and e-visits were not reimbursed.

Need for Coordinated Care

According to a Gallup poll it is very common for older patients to have more than one physician: no physician (3%), one physician (16%), two physicians (26%), three physicians (23%), four physicians (15%), five physicians (6%) and six or more physicians (11%). [29]

Having more than one physician mandates good communication between the primary care physician, the specialist and the patient. This becomes even more of an issue when different healthcare systems are involved. O'Malley et al. surveyed 12 medical practices and found that in-office coordination was improved by EHRs but the technology was not mature enough to improve coordination of care with external physicians.[30] Electronic health records are being integrated with health information organizations so that inpatient and outpatient patient-related information can be accessed and shared, thus improving communication between disparate healthcare entities. Home monitoring (telehomecare) can transmit patient data from home to an office's EHR also assisting in the coordination of care.

Institute of Medicine's Vision for EHRs

The history and significance of the Institute of Medicine (IOM) is detailed in chapter 1. They have published multiple books and monographs on the direction US Medicine should take, including *The Computer-Based Patient Record: An Essential Technology for Health Care*. This visionary work was originally published in 1991 and was revised in 1997 and 2000.[31] In this book and their most recent work *Key Capabilities of an Electronic Health Record System: Letter Report* (2003) they outline eight core functions all EHRs should have:

- Health information and data: In order for the medical profession to make evidence based decisions, you need a lot of accurate data and this is accomplished much better with EHRs than paper charts; *if you can't measure it, you can't manage it*

- Result management: Physicians should not have to search for lab, x-ray and consult results. Quick access saves time and money and prevents redundancy and improves care coordination

- Order management: CPOE should reduce order errors from illegibility for medications, lab tests and ancillary services and standardize care

- Decision support: Should improve overall medical care quality by providing alerts and reminders

- Electronic communication and connectivity: Communication among disparate partners is essential and should include all tools such as secure messaging, text messaging, web portals, health information exchange, etc.

- Patient support: Recognizes the growing role of the internet for patient education as well as home telemonitoring

- Administrative processes and reporting: Electronic scheduling, electronic claims submission, eligibility verification, automated drug recall messages, automated identification of patients for research and artificial intelligence can speed administrative processes

- Reporting and population health: We need to move from paper-based reporting of immunization status and biosurveillance data to an electronic format to improve speed and accuracy [32]

Electronic Health Record Key Components

Many current EHRs have more functionality than the eight core functions recommended by IOM and this will increase as time goes by. The following components are desirable in any EHR system. One of the advantages of certification for Meaningful Use is that it helped standardize what features were important. The following are features found in most current EHRs:

- Clinical decision support systems (CDSS) to include alerts, reminders and clinical practice guidelines. CDSS is associated with computerized physician order entry (CPOE). This will be discussed in more detail in this chapter and the patient safety chapter

- Secure messaging (e-mail) for communication between patients and office staff and among office staff. May include messaging that is part of the Direct Project, explained in the chapter on health information exchange. Telephone triage capability is important

- An interface with practice management software, scheduling software and patient portal (if present). This feature will handle billing and benefits determination. We will discuss this further in the chapter on practice management systems

- Managed care module for physician and site profiling. This includes the ability to track Health plan Employer Data and Information Set (HEDIS) or similar measurements and basic cost analyses

- Referral management feature

- Retrieval of lab and x-ray reports electronically

- Retrieval of prior encounters and medication history

- Computerized Physician Order Entry (CPOE). Primarily used for inpatient order entry but ambulatory CPOE also important. This will be discussed in more detail later in this chapter

- Electronic patient encounter. One of the most attractive features is the ability to create and store a patient encounter electronically. In seconds you can view the last encounter and determine what treatment was rendered

    o Multiple ways to input information into the encounter should be available: free text (typing), dictation, voice recognition and templates

- The ability to input or access information via a smartphone or tablet PC

- Remote access from the office, hospital or home

- Electronic prescribing

- Integration with a picture archiving and communication system (PACS), discussed in a separate chapter

- Knowledge resources for physician and patient, embedded or linked

- Public health reporting and tracking

- Ability to generate quality reports for reimbursement, discussed in the chapter on quality improvement strategies

- Problem summary list that is customizable and includes the major aspects of care: diagnoses, allergies, surgeries and medications. Also, the ability to label the problems as acute or chronic, active or inactive. Information should be coded with ICD-9/10 or SNOMED CT so it is structured data

- Ability to scan in text or use optical character recognition (OCR)

- Ability to perform evaluation and management (E & M) determination for billing

- Ability to create graphs or flow sheets of lab results or vital signs

- Ability to create electronic patient lists and disease registries. Discussed in more detail in the chapter on disease management

- Preventive medicine tracking that links to clinical practice guidelines

- Security and privacy compliance with HIPAA standards

- Robust backup systems

- Ability to generate a Continuity of Care Document (CCD) or Continuity of Care Record (CCR), discussed in the data standards chapter

- Support for client server and/or application service provider (ASP) option [33]

Computerized Physician Order Entry (CPOE)

CPOE is an EHR feature that processes orders for medications, lab tests, x-rays, consults and other diagnostic tests. The majority of articles written about CPOE have discussed medication ordering only, possibly giving readers the impression that CPOE is the same as electronic prescribing. The reality is that CPOE has a great deal more functionality as we will later point out, in this and other chapters. Many organizations such as the Institute of Medicine and Leapfrog see CPOE as a powerful instrument of change. There is limited evidence that CPOE will reduce medication errors, cost and variation of care. This is discussed in the following sections.

Reduce Medication Errors

CPOE has the potential to reduce medication errors through a variety of mechanisms.[34] Because the process is electronic, you can embed rules (clinical decision support) that check for allergies, contraindications and other alerts. Koppel et al. lists the following advantages of CPOE compared to paper-based systems for patient safety: overcomes the issue of illegibility, fewer errors associated with ordering drugs with similar names, more easily integrated with decision support systems than paper, easily linked to drug-drug interaction warning, more likely to identify the prescribing physician, able to link to adverse drug event (ADE) reporting systems, able to avoid medication errors like trailing zeros, creates data that is available for analysis, can point out treatment and drugs of choice, can reduce under and over-prescribing, prescriptions reach the pharmacy quicker. [35]

- Inpatient CPOE: This functionality was recommended by the IOM in 1991. Most studies so far have looked primarily at inpatient CPOE and not ambulatory CPOE. A 1998 study by David Bates in JAMA showed that CPOE can decrease serious inpatient medication errors by 55% (relative risk reduction).[36] Many of the studies showing reductions in medication errors by the use of technology were reported out of the same institution. Other hospital systems are unlikely to experience the same optimistic results. A 2008 systematic review of CPOE with CDSS by Wolfstadt et al. only found 10 studies of high quality and those dealt primarily with inpatients. Only half of the studies were able to show a statistically significant decrease in medication errors, none were randomized and seven were homegrown systems, so results are difficult to generalize.[37]

  With the inception of CPOE we are seeing evidence of new errors that result from technology. A 2005 article reported that the mortality rate increased 2.8%- 6.5% after implementing a well-known EHR. [38] In a 2006 article also from a children's hospital implementing the same EHR they found no increase in mortality. It appears that this was due to better planning and implementation. One of the authors stated that the CPOE system eliminated handwriting errors, improved medication turnaround time and helped standardize care.[39] Nebeker reported on substantial ADEs at a VA hospital following the adoption of CPOE that lacked full decision support, such as medication alerts.[40] On the other hand, another inpatient study showed a reduction in preventable ADEs (46 vs. 26) and potential ADEs (94 vs. 35) compared to pre-EHR statistics.[41] Suffice it to say, clinicians and staff must be properly trained in CPOE; otherwise errors will likely increase, at least in the short term.

- Outpatient CPOE: Americans made 906.5 million outpatient visits in the year 2000. By sheer numbers there is more of a chance for a medication error written for outpatients. According to an optimistic report by the Center for Information Technology Leadership, adoption of an ambulatory CPOE system (ACPOE) will likely eliminate about 2.1 million ADEs per year in the USA. This could potentially prevent 1.3 million ADE-related visits, 190,000 hospitalizations and more than 136,000 life-threatening ADEs.[42] However, a systematic review by Eslami was not as optimistic as he concluded that only one of four studies demonstrated reduced ADEs and only three of five studies showed decreased medical costs. Most showed improved guideline compliance, but it took longer to electronically prescribe and there was a high frequency of ignored alerts.[43] Kuo et al. reported medication errors from primary care settings. He concluded that 70% of medication errors were related to prescribing and that 57% of errors might have been prevented by electronic prescribing.[44]

Reduce Costs

Several studies have shown reduced length of stay and overall costs in addition to decreased medication costs with the use of CPOE.[45] Tierney was able to show in 1993 an average savings of $887 per admission when orders were written using guidelines and reminders, compared to paper-based ordering that was not associated with clinical decision support.[46]

Reduce Variation of Care

One study showed excellent compliance by the medical staff when the drug of choice was changed using decision support reminders.[47] Study conclusions should be interpreted with some note of caution. Many of the studies were conducted at medical centers with well-established health informatics programs where the acceptance level of new technology is unusually high. Several of these institutions such as Brigham and Women's Hospital developed their own EHR and CPOE software. Compare this experience with that of a rural hospital trying CPOE for the first time with potentially inadequate IT, financial and leadership support. It is likely that smaller and more rural hospitals and offices will have a steep learning curve.

On the surface CPOE seems easy, just replace paper orders with an electronic format. The reality is that CPOE represents a significant change in work flow and not just new technology. An often repeated phrase is "it's not about the software, dummy," meaning, regardless which software program is purchased, it requires change in work flow and extensive training.

Adoption of CPOE has been slow, partly because of cost and partly because inputting is slower than scribbling on paper.[48] Although physicians have been upset by new changes that do not shorten their work day, many authorities feel EHRs greatly improve numerous hospital functions. There has been less resistance traditionally in teaching hospitals with a track record of good informatics support. Also, young house staff who work in teaching hospitals and who write the majority of orders are more likely to be tech savvy and amenable to change. It does require great forethought, leadership, planning, training and the use of physician champions in order for CPOE to work. According to some, CPOE should be the last module of an EHR to be turned on and alerts should be phased in to bring about change more gradually. Others have recognized nurses as more accepting of change and willing to teach docs *one-on-one* on the wards.

For more information on CPOE we refer you to a monograph "A Primer on Physician Order Entry" and an article "CPOE: benefits, costs and issues."[49-50]

Clinical Decision Support Systems (CDSS)

Traditionally, CDSS meant computerized drug alerts and reminders to perform preventive tests as part of computerized physician order entry (CPOE) applications. Most of the studies in the literature evaluated those two functions. However, according to Hunt, CDSS is "any software designed to directly aid in clinical decision making in which characteristics of individual patients are matched to a computerized knowledge base for the purpose of generating patient specific assessments or recommendations that are then presented to clinicians for consideration."[51] Therefore, CDSS should have a broader definition than just alerts and reminders.

Two 2005 papers addressed the effects of CDSS on clinical care. Garg and co-authors concluded that overall, CDSS improved performance in 64% of the 97 studies but only 13% of the 52 studies analyzed reported improvement in actual patient outcomes.[52] Kawamoto et al. looked at those factors that contributed to the success of CDSS: automatic CDSS that was part of clinician work flow; recommendations and not just assessments; provision of CDSS at the point of care and computer-based CDSS (not paper-based). When these four features were present, CDSS improved clinical care about 94% of the time.[53]

According to a 2009 article, clinical decision support by nine commercial EHRs was extremely variable and tended not to offer choices.[54] Clearly, the most sophisticated CDSS are developed at medical centers with home grown EHRs and a long record of extensive HIT adoption. With Meaningful Use criteria, certified EHRs will have to conform to CDSS standards which may reduce variability.

Sheridan and Thompson have discussed various levels of CDSS: (level 1) all decisions by humans, (level 2) computer offers many alternatives, (level 3) computer restricts alternatives, (level 4) computer offers only one alternative, (level 5) computer executes the alternative if the human approves, (level 6) human has a time line before computer executes, (level 7) computer executes automatically, then notifies human, (level 8) computer informs human only if requested, (level 9) computer informs human but is up to computer and (level 10) computer makes all decisions.[55] Most EHR systems may offer alternatives and provide reminders but make no decisions on their own. With artificial intelligence and natural language processing becoming more sophisticated, this could change in the future.

Table 3.2 outlines some of the clinical decision support available today. Calculators, knowledge bases and differential diagnoses programs are primarily standalone programs but they are slowly being integrated into EHR systems.

- Knowledge support. Numerous digital medical resources are being integrated with EHRs. As an example, the American College of Physician's PIER resource is integrated into Allscript's Touch Chart.[56] The comprehensive online reference UpToDate has been integrated into six EHRs.[57] iConsult (offered by Elsevier) is a primary care information database available for integration into EHRs. Diagnostic (ICD-9) codes can be hyperlinked to further information or you can use *infobuttons*. Other products such as Dynamed, discussed in the chapter on online medical resources are available as *infobuttons*. Figure 3.3 shows an example of iConsult integrated with the Epic EHR.[58] Another interesting integrated knowledge program is the Theradoc Antibiotic Assistant. The program integrates with an inpatient EHR's lab, pharmacy and radiology sections to make suggestions as to the antibiotic of choice with multiple alerts. Clinicians can be alerted via cell phones, pagers or e-mail. Other modules include Adverse Drug Event (ADE) Assistant, Infection Control Assistant and Clinical Alerts Assistant.[59] A study in the New England Journal of Medicine (NEJM) using this product showed considerable improvement in the prescription of appropriate antibiotics resulting in cost saving, reduced length of stay and fewer adverse drug events.[60]

Table 3.2: Clinical decision support

Type of CDSS	Examples
Knowledge	iConsult®, Theradoc®
Calculators	Medcalc 3000®
Trending/Patient tracking	Flow sheets, graphs
Medications	CPOE and drug alerts
Order sets/protocols	CPGs and order sets
Reminders	Mammogram due
Differential diagnosis	Dxplain®
Radiology CDSS	What imaging studies to order?
Laboratory CDSS	What lab tests to order
Public health alerts	Infection disease alerts

Figure 3.3: iConsult integrated with Epic EHR (Courtesy iConsult)

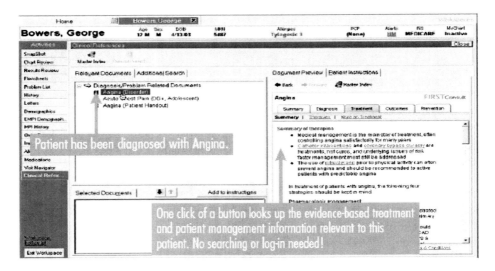

- Calculators. It is likely with time that more calculators will be embedded into the EHR, particularly in the medication and lab ordering sections. Figure 3.4 shows the standalone online and Pocket PC-based program Medcalc3000 with over 100 calculations available. They now offer a *Connect* option that will integrate with EHRs by linking calculators, clinical criteria tools, labs and decision trees.[61] Important calculations, such as kidney function (creatinine clearance) should be calculated on all patients.

- Flow sheets, graphs, patient lists and registries. The ability to track and trend lab results and vital signs, for example, in diabetic patients will greatly assist in their care. Furthermore, the ability to use a patient list to contact every patient taking a recalled drug will improve patient safety. Registries will be covered in more detail in the disease management chapter.

Figure 3.4: Medcalc3000 (Courtesy Medcalc3000)

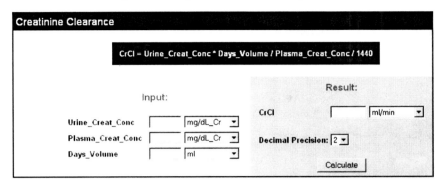

- Medication ordering support. Decision support as part of CPOE possesses several rules engines to detect known allergies, drug-drug interactions, drug-condition and drug-food allergies, as well as excessive dosages. As EHRs and CPOE mature, they will factor in age, gender, weight, kidney (renal) and liver (hepatic) function of the patient, known contraindications based on known diagnoses, as well as the pregnancy and lactation status. Incorporation of these more robust features is complicated and best implemented at medical centers with an established track record of CDSS and CPOE development. As has been pointed out, there are programs that improve antibiotic ordering based on data residing in the EHR.[62] Computerized drug alerts have obvious potential in decreasing medication errors but have not been universally successful to date. According to a systematic review by Kawamoto et al. successful alerts need to be automatic, integrated with CPOE, require a physician response and make a recommendation.[63] Four studies have been published from the Brigham and Women's Hospital showing mediocre compliance, even for black-box type warnings.[64-67] An excellent review by Kuperman et al. describes basic and advanced medication-related CDSS.[68] Further information about alerts is included in the chapters on patient safety and e-prescribing.

- Reminders. Computerized reminders that are part of the EHR assist in tracking the yearly preventive health screening measures, such as mammograms. Shea performed a meta-analysis and concluded that there was clear benefit for vaccinations, breast cancer and colorectal screening, but not cervical cancer screening.[69] A well-designed system should allow for some customization of the reminders as national recommendations change. Reminders are not always heeded by busy clinicians who may choose to ignore them. As a possible solution, preventive reminders could be reviewed by the office nurse and overdue tests ordered prior to the visit with the physician.

- Order sets and protocols. Order sets are groups of pre-established inpatient orders that are related to a symptom or diagnosis. For instance, you can create an order set for pneumonia that might include the antibiotic of choice, oxygen, repeat chest x-ray, etc. that saves keystrokes and time. Order sets can also reflect best practices (clinical practice guidelines), thus offering better and less expensive care. Over one hundred clinical practice guidelines are incorporated into the electronic health record at Vanderbilt Medical Center.[70] For more information on order sets we refer you to this reference.[71]

- Differential Diagnoses. Dxplain is a differential diagnosis program developed at Massachusetts General Hospital. When you input the patient's symptoms it generates a differential diagnosis (the diagnostic possibilities). The program has been in development since 1984 and is currently web-based. A licensing fee is required to use this program. At this time it cannot be integrated into an EHR.[72] In spite of the potential benefit, an extensive 2005 review of CDSSs revealed that only 40% of the 10 diagnostic systems studied showed benefit, in terms of improved clinician performance.[73] Artificial intelligence continues to improve so it is likely that EHRs will have the ability to assist with differential diagnosis in the future.

- Radiology CDSS. Physicians, particularly those in training, may order imaging studies that are either incorrect or unnecessary. For that reason, several institutions have implemented clinical decision support to try to improve ordering. Appropriateness criteria have been established by the American College of Radiologists. Massachusetts General Hospital has had radiology order entry since 2001 and studied the addition of decision support. They noted a decline in low utility exams from 6% down to 2% as a result of decision support.[74]

- Laboratory CDSS. It should be no surprise that clinicians occasionally order inappropriate lab tests, for a variety of reasons. It would be helpful if clinical decision support would alert them to the indications for a test, as well as the price. A Dutch study of primary care demonstrated that 20% fewer lab tests were ordered when clinicians were alerted to lab clinical guidelines.[75]

- Public Health Alerts. The New York Department of Health and Mental Hygiene used Epic EHR's "Best Practice Advisory" to alert New York physicians about several infectious disease issues. The EHR-based alert also hyperlinked to disease specific order sets for educational tips, lab and medication orders.[76]

How well clinicians use CDSS programs such as those discussed, remains to be seen. They will have to be intelligently designed and rigorously tested in order to be accepted. For more information on CDSS, we refer you to the resources cited in references.[77-81]

Electronic Health Record Adoption

Outpatient (Ambulatory) EHR Adoption

In 2006 the adoption rate of ambulatory EHRs was reported to be in the 10% to 20% range, depending on which study you read and what group was studied.[82] Many of the commonly quoted statistics came from surveys, with their obvious shortcomings. It is also important to realize that many outpatient practices may have EHRs but continue to run dual paper and electronic systems or may use only part of the EHR. Furthermore, a significant concern is that small and/or rural practices are more likely to lack the finances and information technology support to purchase and implement EHRs.

In 2008, a seminal article reported on the adoption rate of outpatient EHRs. In this study a sample of 5000 physicians was selected from the AMA master file but Osteopaths, residents and federal physicians were excluded. The most significant finding was that only 4% of respondents reported using a comprehensive EHR (order entry capability and decision support), whereas 13% reported using a basic EHR system. As has been reported before, the adoption rate was higher for large medical groups or medical centers. Responding physicians did report multiple beneficial effects of using EHRs. Given the fact that most experts believe only comprehensive EHRs will impact patient safety and improve the quality of medical care, the 4% adoption rate was disturbing.[83] The 2009 National Ambulatory Medical Care Survey reported that 44% of respondents had a full or partial EHR (that could include paper and electronic) but only 20% had a basic system and 6% had a fully functioning EHR.[84] In summary, very few practices (particularly small rural primary care) have a comprehensive EHR with robust order entry, clinical decision support and reporting capability. With the federal EHR reimbursement program the adoption rate is clearly rising, so follow up studies are needed to determine current adoption rates.

Inpatient EHR Adoption

The American Hospital Association reported on the 2006 use of EHRs with more than 1,500 community hospitals responding. They noted that 68% of the hospitals surveyed had installed inpatient EHRs, but only

11% were fully implemented and these were mainly by large urban and/or teaching hospitals. In only 10% were physicians using computerized physician order entry (CPOE) to order medications, at least 50% of the time.[85]

In 2009 an article by the same authors of the ambulatory NEJM study cited above showed that 7.6% of the respondents reported a basic EHR system and only 1.5% reported a comprehensive EHR. Again, large urban and/or academic centers had the highest adoption rates. User satisfaction rates were not reported.[86]

HIMSS Analytics studies have looked at data from over 5,000 US hospitals to determine the actual level of EHR adoption in 2008, 2009 and 2010. The scale they used rated hospitals from 0, meaning hospitals with an EHR with no functionality installed, to seven indicating a fully functional paperless system that is interoperable and capable of advance reporting.[87] (see Table 3.3) The results indicate that very few hospital systems have achieved an advanced level of EHR sophistication.

One can only speculate why the medical profession has been willing to tolerate the lack of legible and accessible information for so many years. Many physicians believe that purchasing an EHR is not their responsibility and therefore someone else should pick up the tab. Others are concerned that they will purchase the wrong system and waste money and others are simply overwhelmed with the task of implementing and training for a completely different system. Physicians are not noted for embracing innovation but in their defense new technologies should be shown to improve patient care, save time or money in order to be accepted.

There are more than 300 EHR vendors but only about twenty seem to be consistently successful in terms of a large client base. If the selection and purchase of EHRs was easy they would already be universal. As the reader will see later in this chapter, there are issues such as workflow, implementation and training that are just as important as the decision which EHR to purchase.

Table 3.3: EHR adoption statistics by stage in US (Courtesy HIMSS Analytics)

Stage	Cumulative Capabilities	2008	2009	2010
7	Paperless system. Able to generate Continuity of Care Document (CCD). Data warehousing in use	0.3%	0.7%	0.8%
6	Physician documentation (structured templates), full clinical decision support systems (CDSS) and computerized physician order entry. Full picture archiving and communication systems (PACS)	0.5%	1.6%	2.6%
5	Closed loop medication administration	2.5%	3.8%	3.2%
4	CPOE and CDSS	2.5%	7.4%	9.7%
3	Clinical documentation (flow sheets), CDSS, PACS available outside radiology	35.7%	50.9%	50.2%
2	Clinical data repository (CDR), CDSS, document imaging	31.4%	16.9%	15.5%
1	Lab, radiology and pharmacy modules installed	11.5%	7.2%	6.8%
0	Lab, radiology and pharmacy modules not installed	15.6%	11.5%	11.2%

International EHR Adoption

Until recently, the US lagged behind many other developed countries in its adoption of EHRs. In fact, a 2006 study indicated that we were as much as a dozen years behind other industrialized countries in HIT adoption.[88] A 2008 study pointed out that the United Kingdom, Netherlands, Australia and New Zealand had nearly 100% adoption of ambulatory EHRs by primary care physicians (>90% each) while adoption amongst

US primary care physicians was in the 10% to 30% range. No country at that time had significant adoption of comprehensive in-patient EHRs, all being less than 10%.[89] A 2009 study showed that we continued to lag in EHR adoption among primary care physicians.[90]

A major difference between the US and these high EHR adopter countries has been, until recently, the degree of government involvement. Other countries' governments invested heavily in HIT. The United Kingdom, with 20% of the population of the US, committed $17 billion through its National Program for IT (NPfIT). Australia has provided subsidies to adopting physicians and has the National E-Health Transition Authority (NEHTA). Germany has a public-private partnership involved in promoting interoperability standards and certifying EHRs called Gematik. Denmark, long thought to be the international leader in health IT, has a very high EHR adoption rate and the most interoperable system of any country.[91]

All is not wonderful in other countries however. On Sept 23, 2011 UK officials announced that they planned to dismantle their $17 billion health IT project. They stated that some of the nearly $10 billion that they had invested to date was wasted and that their main vendor, Computer Sciences Corporation would not be able to provide the software that was promised.[92]

The HITECH Act of 2009, which created the EHR Incentives Programs in the US, may help us catch up. Since the advent of that program EHR adoption in the US appears to be increasing more rapidly. As of July 2011, 76% of large practices (26+ physicians), 72% of mid-large practices (11-25 physicians), 63% of mid-size practices (six to 10 physicians), 51% of smaller practices (three to five physicians), 42% of two-physician practices and 31% of solo physician practices have adopted EHRs with an overall adoption rate of 40%.[93]

Barriers to Electronic Health Record Adoption

Many of the same barriers to HIT adoption discussed in chapter 1 also pertain to EHR adoption. According to Shortliffe[94] there are four historic constraints to EHR adoption: (1) the need for standardized clinical terminology; (2) privacy, confidentiality and security concerns; (3) challenges to data entry by physicians and (4) difficulties with integrating with other systems. These and other barriers are discussed below.

Financial Barriers

Although there are models that suggest significant savings after the implementation of ambulatory EHRs, the reality is that it is expensive. Multiple surveys report lack of funding as the number one barrier to EHR adoption.[95] In a 2005 study published in *Health Affairs*, initial EHR costs averaged $44,000 (range $14-$63,000) per FTE (full time equivalent) and ongoing annual costs of $8,500 per FTE. These costs included the purchase of new hardware, etc. Financial benefits averaged about $33,000 per FTE provider per year. Importantly, more than half of the benefit derived was from improved coding.[96] This is not a surprise given the fact that studies have shown that physicians often *under-code* for fear of punishment or lack of understanding what it takes to code to a certain level.[97] A 2008 survey reported about one-third of physicians paid between $500-$3,000 per clinician, one-third paid between $3,001-$6,000 and about one-third paid more than $6,000.[98]

It is important to consider that integration with other disparate systems such as practice management systems can be very expensive and hard to factor into a cost-benefit analysis. The web-based application service provider (ASP) option is less expensive in the short term and perhaps in the long term, when you factor in the expenses to maintain and upgrade an office client-server network. According to many studies EHR adoption was far higher in large physician practices that could afford the initial high cost.[99]

Physician Resistance

Prior to EHR reimbursement lack of support by medical staff was consistently the second most commonly perceived obstacle to adoption.[100] Physicians have to be shown a new technology makes money, saves time or is good for their patients. None of these can be proven for certain for every practice. Although you should not expect to go paperless from the beginning, at some point it can no longer be optional. It seems clear that CPOE does take longer than written orders but offers multiple advantages over paper as pointed out previously.[101] Implementation will not fix old work flow issues and will not work if several physicians in a group are opposed to going electronic. We now know that some practices have opted to change or discontinue their use of an EHR. A 2007 survey demonstrated fewer than 20% of respondents had uninstalled their EHR in an effort to step down to a less expensive alternative and 8% had returned to paper.[102] EHRs are not the only important issue for most physicians. They face increases in overhead while reimbursement wanes, along with ICD-10, HIPAA 5010, new healthcare reform and Red Flag rules, just to mention several looming challenges.

Loss of Productivity

It is likely physicians will have to work at reduced capacity for several months with gradual improvement depending on training, aptitude, etc. This is a period when physician champions can help maintain morale and momentum with a positive attitude.

Work Flow Changes

Everyone in the office will have to change the way they route information compared to the old paper system. If planning was well done in advance you should know how your work flow will change. As an example, many offices place the patients chart in the exam room door to indicate that the patient is ready to be seen. How will you do that with an electronic system? Initially, you will have to maintain a dual system of paper and electronic records. Work flow analysis will also determine where you will place computer terminals. Importantly, clinicians will have to maintain eye contact as often as possible and learn to incorporate the EHR into the average patient visit. Use of a movable monitor or tablet PC may help diminish the time the clinician spends not looking at the patient.

Usability Issues

Usability has been defined as the "effectiveness, efficiency and satisfaction with which specific users can achieve a specific set of tasks in a particular environment."[103] Is the software well organized and intuitive such that the user can find what they are looking for with a minimal number of mouse clicks? This is more complicated than what one would expect because there are multiple sub-specialties with unique needs, as well as multiple clinicians who are used to working in a set sequence. Based on several surveys included in this chapter, usability does not necessarily correlate with the amount of money paid for the software. HIMSS now has an EHR usability task force and it is predicted that eventually all certified EHRs will need to pass usability testing.[104]

Integration with Other Systems

Hopefully, integration with other systems like practice management software was already solved prior to implementation. Be prepared to pay significantly for programmers to integrate a new EHR with an old legacy system. An average cost is about $3-$15,000 per interface.[105] Most office and hospitals have multiple old legacy systems that do not talk to each other. Systems are often purchased from different vendors and written in different programming languages. If either the EHR or practice management system's software is upgraded, then interfaces need to be checked and possibly changed. It is now popular to purchase an EHR already integrated with practice management, billing and scheduling software programs.

Lack of Standards

Data standards and medical vocabularies are necessary for interoperability. The initial standards have been proposed by ONC and will be covered in more detail in another chapter. Reimbursement for Meaningful Use will mandate that EHRs demonstrate the ability to exchange information. Although we have numerous standards already accepted (separate chapter) they will likely need to be updated and new standards added based on use cases. Furthermore, computers are based on data and not information, as discussed in the chapter on healthcare data, information and knowledge.

Privacy Concerns

The HITECH Act of 2009 introduced a new certification process for EHRs sponsored by ONC, in addition to CCHIT certification. This new certification ensures that EHRs will be able to support Meaningful Use and that they also will be HIPAA compliant. ONC certification includes requirements on database encryption, encryption of transmitted data, authentication, data integrity, audit logs, automatic log off, emergency access, access control and accounting of HIPAA releases of information. The HITECH Act also strengthened the prior HIPAA requirements as they relate to EHRs, particularly in the areas of enforcement of HIPAA and notification of breaches. Both civil and criminal penalties for Business Associates (as well as covered entities) were introduced. Civil penalties in their harshest form can range up to 1.5 million dollars. If a data breach of PHI (Protected Health Information) occurs, all affected individuals must be notified. If more than 500 individuals are affected, HHS must be notified as well. Sale of PHI is prohibited.[106] Users of EHRs must:

- Use HIPAA compliant technology
- Provide for physical and software security of data systems
- Provide for physical and software security of their network(s) including mobile and remote computing
- Provide access control with defined user roles, passwords and user authentication and auditing
- Monitor and manage user behavior
- Have written security policies and procedures
- Have an effective disaster recovery plan[107]

EHRs pose new potential privacy and security threats for patient data, but with proper technology as well as proper health entity and user behavior, these risks can be mitigated. On the bright side, EHRs offer new safeguards unavailable in the paper record world, like audit trails, user authentication, and back-up copies of records. Further details are available in the chapters on privacy and security.

Legal Aspects

A 2010 Health Affairs article estimated that malpractice costs in the US are around 55 billion dollars annually (in 2008 dollars) or 2.4% of what we spend on health care.[108] Will EHRs increase or decrease that number? Unfortunately the answer isn't in yet. Arguments can be made for either outcome. On one hand, by increasing the quality of care, theoretically EHRs should reduce malpractice risk. Yet that assumes that quality and malpractice are related in a linear fashion, which may well not be the case. On the other hand, EHRs that are poorly designed, or that contain bugs, could promote inadvertent errors. This risk points to a need for monitoring and corrective action related to EHR- generated errors. The Office of the National Coordinator (ONC) for Health IT understands that a system of monitoring and corrective action for EHR-related errors needs to be implemented. ONC outlined its plans for this in a December 2010 statement.[109] As a first step, one can currently report EHR-generated errors to AHRQ-recognized Patient Safety Organizations like PDR Secure.[110]

Two important areas of potential risks and benefits include documentation of clinical findings and clinical decision support. One might expect that the more comprehensive documentation produced by EHRs will improve a physician's defense against malpractice. It certainly may. However the automated way that EHRs carry information forward from one note to the next can also promote errors and potential liability, if a piece of data is recorded incorrectly from the start yet never corrected. E-discovery laws now allow electronically stored data related to patient records to be considered discoverable for the purpose of malpractice, so the metadata and audit trails that supplement EHR documentation can be used both to defend and to impeach a physician in a malpractice case.[111] Will that be a net benefit or liability for physicians? Decision support alerts and guidelines embedded into EHRs could potentially provide a defense against malpractice claims if their advice is followed. But what if alerts or guidelines are overridden? There may be very appropriate reasons to do so, but will physicians be expected to document the reason for each and every alert they override? Will they run the risk of being penalized if they don't?

Improved access to information provided by health information exchanges (HIEs) should improve the coordination of care, the quality of medical information that is available, and thus the quality of medical decision making. But, will clinicians have a tendency to overlook key nuggets of clinical information simply because they are overwhelmed by the volume of information they receive? Will ready access to outside information on a patient make a physician more liable if he or she doesn't always actively search for every piece of potentially relevant information? In additon, user errors can arise as users climb a steep learning curve to become proficient with EHRs. Care especially needs to be taken particularly during the implementation of an EHR to guard against user error.

A study in Massachusetts in 2005 suggested that users of EHRs had fewer paid malpractice claims than non-users (6.1% versus 10.8%).[112] But when the study's results were risk-adjusted by physician demographics including age, specialty and practice size, the statistical difference in paid malpractice claims between the EHR and non-EHR users disappeared. Finally, as EHRs become the standard of care, will practicing without an EHR become a medicolegal liability? At this point in time it is still undetermined whether EHRs will significantly impact the incidence and expense of malpractice in a positive or a negative way.[113]

Inadequate Proof of Benefit

Successful implementation of HIT at a medical center with a long standing history of systemic IT support does not necessarily translate to another healthcare organization with less IT support and infrastructure. A systematic review by Chaudry is often cited as proof of the benefits of HIT, but in his conclusion he states "four benchmark institutions have demonstrated the efficacy of health information technologies in improving quality and efficiency. Whether and how other institutions can achieve similar benefits and at what costs, are unclear."[114]

There have been five recent articles in the medical literature that failed to demonstrate a significant impact of EHRs on medical quality in the US and in Europe.[115-119] A more positive study was published in 2011 of more than 25,000 diabetics in 46 practices that showed achievement of diabetic care was significantly better for practices with EHRs, compared to paper-based practices. They measured intermediate outcomes and not actual patient outcomes, so we don't know the impact on morbidity or mortality.[120]

Several studies have shown increased errors as a result of implementing CPOE[68,121-125] Weiner coined the term *e-iatrogenesis* to mean "patient harm caused at least in part by the application of health information technology."[126] In late 2011 AHRQ released the monograph "Guide to Reducing Unintended Consequences of Electronic Health Records."[127] Eventually, with better training or re-design some of the technology-related errors are likely to be overcome. More research is needed to obtain a balanced opinion of the impact of EHRs on quality of care, patient safety and productivity. Furthermore, we will need to study the impact on all healthcare workers and not just physicians.

The HITECH Act and EHR Reimbursement

Arguably, the most significant EHR-related initiative occurred in 2009 as part of the American Recovery and Reinvestment Act (ARRA). Two major parts of ARRA, Title IV and Title XIII are known as the Health Information Technology for Economic and Clinical Health or HITECH Act. Approximately $20-30 billion was dedicated for Medicare and Medicaid reimbursement for EHRs to clinicians and hospitals. In this chapter we will primarily focus on reimbursement to eligible professionals (EPs) and not hospitals or Medicare Advantage organizations, even though they are also potentially reimbursable. The Centers for Medicare and Medicaid Services (CMS) established a web site www.cms.gov/EHRIncentivePrograms to explain the EHR Incentive Program we will summarize in the following sections.

In order for clinicians to participate in this program they must be: (1) eligible, (2) register for reimbursement, (3) use a certified EHR, (4) demonstrate and prove Meaningful Use, and (5) receive reimbursement.

Eligible Professionals (EPs)

Medicare: Medicare defines EPs as doctors of medicine or osteopathy, doctors of dental surgery or dental medicine, doctors of podiatric medicine, doctors of optometry and chiropractors. Hospital-based physicians such as pathologists and emergency room physicians are not eligible for reimbursement. Hospital-based is defined as providing 90% or more of care in a hospital setting. The exception is if more than 50% of a physician's total patient encounters in a six-month period occur in a federally qualified health center or rural health clinic. Physicians may select reimbursement by Medicare or Medicaid, but not both. They cannot receive Medicare EHR reimbursement <u>and</u> federal reimbursement for e-prescribing. They can receive Medicare reimbursement as well as participate in the Physicians Quality Reporting Initiative (PQRI). If they participate in the Medicaid EHR incentive program they can participate in all three programs.

Medicaid: Medicaid EPs are defined as physicians, nurse practitioners, certified nurse midwives, dentists and physician assistants (physician assistants must provide services in a federally qualified health center or rural health clinic that is led by a physician assistant). Medicaid physicians must have at least 30% Medicaid volume (20% for pediatricians). If a clinician practices in a federally qualified health center (FQHC) or rural health clinic (RHC), 30% of patients must be *needy individuals.* The Medicaid program will be administered by the states and physicians can receive a one-time incentive payment for 85% of the allowable purchase and implementation cost of a certified EHR in the first year, even before Meaningful use is demonstrated.[128]

Registration: Registration began in January 2011. Medicare physicians must have a National Provider Identifier (NPI) and be enrolled in the CMS Provider Enrollment, Chain and Ownership System (PECOS) and National Plan and Provider Enumeration System (NPPES) to participate. Through the end of 2011, approximately 150,000 physicians and 2,800 hospitals have registered for reimbursement and 20,000 physicians and 1,200 hospitals have received reimbursement. Registration details for Medicare and Medicaid are available on the CMS web site.[128]

Certified EHRs: An EHR has to be certified by a recognized certifying organization in order for a physician or hospital to receive reimbursement. As of mid-2011 there are six organizations that can provide certification. The original certification organization CCHIT was discussed in chapter 1 and to date is the only certification organization to test and report usability. Standards and certification criteria are listed on the HHS site as are the currently certified EHRs. Users can view ambulatory and inpatient EHR categories and search by product name. The search should review who certified the EHR, whether it was for a complete or modular EHR and the EHR certification ID number they would need for reimbursement.[129] To date, 1,500 EHRs or EHR modules have received certification.

Meaningful Use (MU): The goals of MU are the same as the national goals for HIT: (a) improve quality, safety, efficiency and reduce health disparities; (b) engage patients and families; (c) improve care coordination; (d) ensure adequate privacy and security of personal health information; (e) improve population and public health. Three processes stressed by ARRA to accomplish this are: e-prescribing, health information exchange and the production of quality reports. As planned, Meaningful Use will occur in three stages:

- Stage 1 (2011): Meaningful Use mandates a *core set* and a *menu set* of objectives. To be a Meaningful Use Stage 1 user, participants must meet all 15 of the core objectives (Appendix A, end of chapter) and select five out of 10 menu objectives (Appendix B, end of chapter). They must choose at least one population and public health measure. The appendices delineate criteria and measures for EPs, not hospitals. For each objective there are reporting measures that must be met to prove Meaningful use

  In 2011 the results of all objectives and measures, to include clinical quality measures will be reported by clinicians and hospitals to CMS and Medicaid clinicians will report to states by attestation. Quality measures are derived from the Physician Quality Reporting Initiative (PQRI) and the National Quality Forum (NQF). Each EP will submit information on three core quality measures in 2011 and 2012 (tobacco use, blood pressure measurement and adult weight screening). They must also choose three other measures that are ready for incorporation into EHRs. Medicare was ready for attestation in early 2011 but as of mid-2011 many states were not ready. Physicians will fill in numerators and denominators for Meaningful use objectives and indicate if the qualify for exclusions and attest that they have met Meaningful use. Details about Meaningful Use and attestation for Medicare and Medicaid are available on the CMS web site. As of October 2011, more than 100,000 clinicians and hospitals signed up for reimbursement. Those who attest in 2011 will not have to meet Stage 2 requirements until 2014.[128]

- Stage 2 and 3: The HIT Policy Committee has proposed MU objectives and measures, viewed at http://healthit.hhs.gov/media/faca/MU_RFC%20_2011-01-12_final.pdf. The proposed changes include increasing the percent compliance with Stage 1 objectives, moving several menu objectives to core and adding new objectives (e.g secure messaging). It is likely that the actual content and time line will change prior to implementation.

Reimbursement

Tables 3.4 and 3.5 list the Medicare and Medicaid reimbursement levels for EHRs. Payment is likely to occur 4-8 weeks after successful attestation. Payments will be held until the Medicare physician meets the $24,000 threshold for allowed charges. Medicare physicians may earn an additional 10% if they practice in a healthcare professional shortage area (HPSA). Payments are based on the calendar year. It is important to note that no monies are paid upfront and contrary to what is published by EHR vendors and others, the amount listed yearly in Table 3.2 is a maximum. Physicians will be reimbursed 75% of allowable Part B charges or up to, for example, $18,000 in the first year. Clinicians are paid in a single annual payment and have to demonstrate Meaningful Use for 90 days of continuous EHR use in the first year and the entire calendar year thereafter.

Medicare physicians who do not use a certified EHR nor demonstrate Meaningful Use will receive penalties of 1% in 2015, 2% in 2016 and 3% in 2017 when they bill Medicare. Penalties could reach 5% in 2018 and beyond if fewer than 75% of physicians are using EHRs at that point. In addition, late adoption might mean that more complex Meaningful Use (Stage 2 or 3) will be required, likely to make purchase and implementation more difficult.

Medicaid is administered by states and will use the same Meaningful Use criteria. In addition to the states being given the reimbursement money by the federal government to give to clinicians and hospitals, they will also receive 90% reimbursement for the cost of administering the program. Medicaid EPs and hospital-based physicians are not subject to possible payment reductions. Unlike Medicare, Medicaid physicians can be paid the first year just to adopt, implement or upgrade an EHR and not yet meaningfully use the EHR. Medicaid EPs must demonstrate Meaningful Use in years two through six. Medicaid physicians are not eligible for the

10% HPSA bonus but can receive the e-prescribing and PQRI bonuses. The last year to begin participation in the Medicaid program is 2016.

Hospitals can also be reimbursed for the purchase of EHRs and can share this technology with the known limits of the *Safe Harbor Act* discussed later in this chapter. Hospitals will start at a base of $2 million annually with decreasing amounts over five years, plus an additional amount dependent on patient volume. Hospitals may receive reimbursement from both Medicare and Medicaid.[128]

Table 3.4: Maximum *Medicare* reimbursement for EHR adoption

Year	2011 (year 1)	2012 (year 1)	2013 (year 1)	2014 (year 1)	2015 (year 1)
2011	$18,000				
2012	$12,000	$18,000			
2013	$8,000	$12,000	$15,000		
2014	$4,000	$8,000	$12,000	$12,000	
2015	$2,000	$4,000	$8,000	$8,000	$0
2016	$0	$2,000	$4,000	$4,000	$0
Total	$44,000	$44,000	$39,000	$24,000	$0

While a substantial number of physicians have stayed on the sidelines, many have taken advantage of the new programs. As of September 2011 about 80,000 physicians were registered for EHR reimbursement according to HHS Secretary Kathleen Sebelius.[130]

Table 3.5: Maximum *Medicaid* reimbursement for EHR adoption

Eligible Clinician	Base Year: Max 85% of EHR cost	Year 1	Year 2	Year 3	Year 4	Year 5	Year 6	Total
Physician	$21,250	$8,500	$8,500	$8,500	$8,500	$8,500	$8,500	$63,750
Dentist	$21,250	$8,500	$8,500	$8,500	$8,500	$8,500	$8,500	$63,750
Nurse mid-wife	$21,250	$8,500	$8,500	$8,500	$8,500	$8,500	$8,500	$63,750
Physician assistant	$21,250	$8,500	$8,500	$8,500	$8,500	$8,500	$8,500	$63,750
Nurse practitioner	$21,250	$8,500	$8,500	$8,500	$8,500	$8,500	$8,500	$63,750
Pediatrician	$14,167	$5,667	$5,667	$5,667	$5,667	$5,667	$5,667	$42,500

Electronic Health Record Examples

There are more than 300 EHRs available in the United States that vary in price from free to about $40,000 with features that range from basic to comprehensive. Importantly, not all have been certified for Meaningful Use. Also, very few EHR vendors have price transparency so only a minority actually post their charges on their web sites. The EHR market has changed rapidly due to Meaningful Use requirements, in addition to advances in technology and user demands.

We will present examples of EHRs in three categories based on size and target audience. Small practice is defined by having one to four physicians and typically do best with subscription service (cloud computing, ASP model, SaaS) where they only need an internet connection. A medium medical practice is defined as having five to 20 physicians that might use a subscription service or have the client-server model with onsite servers which would normally mandate either onsite IT support or contract support services to manage the network. A large practice is defined as having 20 to 99 physicians and most likely will have onsite servers and their own IT staff. A very large practice is defined as having 100+ physicians and will typically utilize the client-server model with their own data center and IT staff as well as programmers and database administrators.

Small Medical Practice

Amazing Charts: This simple and intuitive EHR is ONC ATCB certified and has scored high in usability by multiple reviewers which we will discuss later in the chapter. They offer a three month free trial which is unique among vendors. Currently they claim more than 4000 users.

- Standard package: scheduling, internal secure messaging, charting (template driven), e-prescribing, billing (superbills) and ad hoc reporting are included

- Practice management: a practice management system or web-based model at this time

- Remote access: physicians can access their computers remotely with services such as LogMeIn and they can view but not modify records remotely using an iPhone app

- Pricing: the standard charge is $1,995 per physician (includes training and support for physician and staff) for first year, followed by $995 per physician per year after that for software updates and tech support. For a separate fee they offer offsite backup and a low cost ($500) interface to practice management systems.

Figure 3.5 shows a typical screen shot of a patient encounter from Amazing Charts.[131]

Figure 3.5: Amazing Charts patient encounter screen shot (Courtesy AmazingCharts)

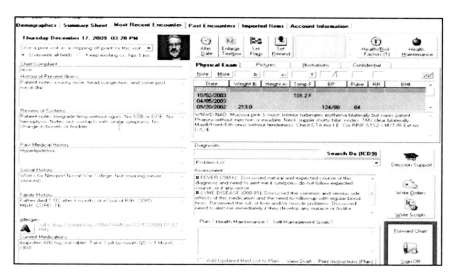

Medium Size Medical Practice

eClinicalWorks: This EHR was selected by the Massachusetts Medical Society because it is multi-featured and well designed with excellent physician acceptance. In 2011 they claimed 55,000 users and they have CCHIT 2011 certification. Their modules are fully integrated and not standalone. The system will operate on Windows or Linux-based servers and is compatible with SQL or MySQL databases. They offer both a web-based and client-server model. It is also one of the few that lists its pricing schedule on their web site. Current modules include:

- EMR module: multiple means of inputting data such as templates, handwriting recognition, voice recognition and free text; tab to access the resource UpToDate; Continuity of Care Record (electronic patient summary) available; patient/disease registries with customizable alerts; referral letters can be automatically generated; e-visit capability

- Practice management: scheduling, billing management, claims scrubbing, business analytics and reporting

- Patient portal: online registration, secure messaging, web consults, prescription refills, online appointing, view of billing statements, lab results, patient education, receive alerts and complete consent forms. (Figure 3.6)

- Clinical messenger: communication with patient via email, text messaging or Voice over IP (VoIP). Patients can confirm appointment with one phone key, receive (normal) lab results and receive individual or group alerts. This is a hosted eClinicalWorks function

- Interoperability: eEHX community health exchange can connect disparate offices, labs and hospitals. Provides master patient index, integration engine, push/pull capability and quality measure reporting and public health alerts

- Mobile: iPhones or BlackBerry smartphones applications to access EHR works are available

Figure 3.6: Patient portal module eClinicalWorks (Courtesy eClinicalWorks)

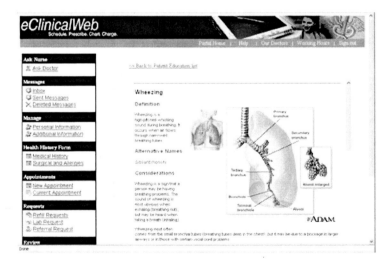

- Pricing:
    - Option 1 (client-server): EHR/PMS combined system is $10,000 for the first physician and $5,000 for the next physician. Maintenance and support (M&S) is additional. EHR only system is $7,500 for the first physician and $4,500 for the next physician. M&S is additional

      o   Option 2 (web-based): EHR/PMS combined system is $400 per physician per month that includes M&S. EHR only system is $250 per physician per month that includes M&S. Hosting by eClinicalWorks is another $100 per physician per month [132]

Large to Very Large Medical Practice

Epic: Epic is the most popular and highest rated EHR for large to very large healthcare organizations like Kaiser Permanente, Geisinger Clinic, Group Health Cooperative and the Cleveland Clinic. They offer an ambulatory EHR for medical practices and an inpatient EHR for hospitals or a system that will work for both. It is interesting that this very intuitive comprehensive EHR is based on early MUMPs programming that is also found in VistA the EHR used by the Veterans Health Administration. The following are their main services:

- EpicCare EHR: approximately 50% of clinicians are specialists so they offer 40 specialty modules that have specialty specific workflow, templates and order sets. Inpatient EHR modules include flow sheets, electronic medication administration record (eMAR), interdisciplinary care plans, hospital outpatient support, clinical pathways, ICU support, ED department, operating room integration, anesthesia and pathology integration, radiology and laboratory information system integration, health information management, nurse triage, home health integration, barcode administration, pharmacy integration and enterprise reporting

- Practice management: registration, scheduling, billing and call management

- Personal Health Record:

  - My Chart is an integrated personal health record (PHR) with the following services: view test results, view upcoming & past appointments, schedule appointments, pay bills securely, get automated health maintenance reminders, view problem-based education materials, request refills, send & receive secure messages with physicians, view a child's records and print growth charts and manage the care of elderly parents

  - Lucy is a standalone PHR not integrated with the EHR

- Information Exchange: EpicCare Link provides a secure web-based portal for read-only access to limited sections of the EHR to community physicians, in addition to secure messaging

- Physician Portal: Epic Web is a physician portal for remote access to the EHR

- Interoperability: Care Everywhere is an interoperability capability for disparate EHRs and can pull in data from Lucy

- Mobile: Epic Haiku is an iPhone app that provides authorized users of Epic's EHR with secure access to clinic schedules, hospital patient lists, health summaries, test results and notes. Haiku also supports dictation and access to inbox. [33]

Future Trends

One doesn't need a crystal ball to determine the direction that EHRs in the US will take over the next several years. The potent force shaping that direction will be the Meaningful Use (MU) criteria of the EHR Incentive Programs. The developer of these criteria is the Health Information Technology Policy Committee (HITPC), a Federal Advisory Committee that advises the Office of the National Coordinator (ONC) and the Department of Health and Human Services (HHS). So far those agencies have closely followed HITPC's recommendations, and it is likely that they will continue to do so in the future. ONC in turn is responsible for creating the EHR certification criteria that ensure that EHRs can perform to specifications that allow for Meaningful Use.

The Meaningful Use program is currently in its first stage (2011-2013), will start its second stage in 2014, and then move to its third stage in 2016. On June 16, 2011, HITPC formally made its recommendations for Stage 2 MU criteria to HHS via ONC. HHS is expected to follow these criteria, announcing its Proposed Notice of Rule Making on Stage 2 requirements in the fall of 2011, and to finalize those requirements in the summer of 2012.

So what direction is HITPC headed? HITPC has designed the MU criteria around five policy areas:

- Improving quality, safety, efficiency and reducing health disparities – goals set out by the Institute of Medicine (IOM)

- Engaging patients and families in their care – another IOM goal

- Improving care coordination

- Improving population and public health

- Ensuring adequate privacy and security protections for personal health information

The proposed Stage 2 criteria, and early suggestions about Stage 3 from HITPC, point to increased care coordination, increased reliance on electronic ordering, more patient portal use, and a greater focus on clinical measurements and quality reporting. Thus we can expect to see EHRs that have more sophisticated analytics, increased standardization, enhanced interoperability, and tight linkages with more sophisticated patient portals than now exist. A desired outcome is that data and information will no longer remain locked in the plethora of EHR silos built by physicians and hospitals, but will electronically flow from one to the other.[134, 135]

Beyond 2016, when the CMS EHR Incentives for the Medicare program end, the direction that EHRs will take is less clear. Experts suggest a number of trends, including an increased reliance on cloud computing,[136] large shared databases used for comparative effectiveness research[137, 138] increasing use of natural language processing[139] more pervasive use of telehealth (virtual visits and consultation),[140] improved clinical decision support, more use of patient registries built into EHR workflow,[141] and greater use and integration of wireless remote outpatient monitoring of patients.[142, 143]

Of course, down the road, one or more unforeseen health IT technologies breakthroughs could alter EHRs in ways that we can currently only barely imagine.

Key Points

- Electronic health records are central to creating health information organizations and a nationwide health information network

- The current paper-based system is fraught with multiple shortcomings

- It is likely that reimbursement for e-prescribing and electronic health records by the federal government will promote adoption

- In spite of the potential benefits of electronic health records, obstacles and controversies persist

- Clinical decision support are still in their infancy and will likely improve in the future with artificial intelligence

- Advance planning and training is mandatory for successful implementation of EHRs

Conclusion

In spite of the slow acceptance of EHRs by clinicians and healthcare organizations, they continue to proliferate and improve over time. Electronic health records have been transformational for large organizations like the VA, Kaiser-Permanente and the Cleveland Clinic, but the reality is that medicine in this country is mostly practiced by small medical groups, with limited finances and IT support. As a new trend, we are seeing outpatient clinicians opt to re-engineer their business model based on an EHR. Their goal is to reduce overhead by having fewer support staff and to concentrate on seeing fewer patients per day but with more time spent per patient. When this is combined with secure messaging, e-visits and e-prescribing the goal of the *e-office* is achievable.[144]

Buyers have a wide choice of features and cost to choose from. At this time cost is a major obstacle as well as the lack of high quality economic studies demonstrating reasonable return on investment. As more studies show cost savings, medical groups that have been sitting on the fence will make the financial commitment.

Without doubt, Medicare and Medicaid reimbursement for EHRs and e-prescribing is the most significant impetus to jump start EHR adoption we have seen. Preliminary studies have shown a significant increase in EHR adoption as a result of reimbursement programs. It is too early to know how well received Stage 1 Meaningful Use objectives and measures will be received, implemented and reported. For those practices that can afford and need complexity, multiple high-end vendors exist. For smaller, rural, primary care practices, simpler alternatives exist. It is also worth noting that purchasing EHRs is only one of multiple difficult challenges facing clinicians and their staff. According to a mid-2009 Medical Group Management Association (MGMA) survey implementing an EHR was ranked third in difficulty preceded by rising operating costs and maintaining clinician salaries in the face of decreasing reimbursement.[145]

Acknowledgement

We would like to thank LeighAnne Cox RNC BSN BA for her input regarding the role of the nurse in EHR implementation.

Appendix 3.1

Meaningful Use Core Objectives and Measures

Stage 1 Objectives	Stage 1 Measures
Use CPOE for medication orders	Measure more than 30% of patients on medications or indicate they are on no meds
Implement drug-drug and drug-allergy interaction checks	Demonstrate functionality is enabled during entire period
Maintain an up-to-date problem list of current and active diagnoses	More than 80% have at least one entry as structured data or indicate they have 0 problems
Generate and transmit electronic prescriptions	More than 40% are transmitted using certified EHRs
Maintain active medication list	At least 80% of patients have at least one entry as structured data
Maintain active medication allergy list	More than 80% of patients have at least one entry as structured data or indicate no allergies
Record demographics: sex, race, ethnicity, date of birth and preferred language	More than 50% of data recorded as structured data
Record and chart vital signs: height, weight, blood pressure, BMI, growth charts for children	More than 50% of patients two years of age or older have vitals recorded as structured data
Record smoking status in patients age 13 older	More than 50% of patients 13 years of age or older have smoking status as structured data
Report quality measures to CMS or states Total of six measures	For 2011, provide aggregate numerator and denominator through attestation; for 2012 submit reports electronically
Implement one clinical decision support rule and track compliance	One clinical decision support rule implemented
Provide patients with a copy of their record upon request: diagnostic tests, problem list, medication lists, allergies	More than 50% of requesting patients receive copy within three business days
Provide clinical summaries for patients for each office visit	Summaries provided for more than 50% of visits within three business days or posted to PHR
Capability to exchange electronically key clinical information among clinicians or patient-approved entities	Perform at least one test of EHR's capability to exchange information
Implement systems to protect privacy and security of patient data in the EHR	Conduct or review a security risk analysis, implement security updates as necessary and correct identified security deficiencies

Appendix 3.2

Meaningful Use Menu Objectives and Measures

Stage Objectives	Stage 1 Measures
Implement drug-formulary checks	EP has enabled this function for entire 90 day reporting period
Incorporate lab-test results into EHR as structured data	At least 40% of lab results entered
Generate patient lists by specific conditions	Generate at least one report with specific condition
Send reminders to patients as per their preference	Send to at least 20% of patients 65 or older or five years or younger
Provide patients with timely electronic access to their lab results, problem list, medication list, allergy list within four business days	At least 10% are provided electronic access within four business days
Use EHR to identify and provide patient specific education resources	At least 10% are able to receive electronic educational resources
Perform medication reconciliation following transitions of care	Perform medication reconciliation at least 50% of the transitions of care
Provide summary of care for any transition of care	Provide summary for more than 50% of transitions of care
Capability to submit electronic data to immunization registry or information system	Perform at least one test of EHR's capability to submit data electronically to immunization registry
Capability to submit electronic syndromic surveillance to public health agencies	Perform at least one test of EHR's capability to submit syndromic data to public health agencies

Appendix 3.3

Logical Steps to Selecting and Implementing an EHR

EHR implementations are complex affairs. They are not simply IT projects. They are practice transformation projects. If approached as simply software to be installed and users to be trained in using the software, an EHR implementation will undoubtedly falter or even fail. Thus, health care organizations involved in implementing an EHR are wise to spend a lot of time planning. A few of the many questions an organization needs to both ask and answer prior to implementing an EHR are: Why are we doing this? Who should be involved? How will this impact end-users and how do we prepare them? What will be the major barriers? What should we start doing now to overcome identified barriers? Are we ready for change? How will the change be managed?

Implementation of an EHR can be divided into three separate yet intertwined phases: Pre-implementation, implementation and post-implementation.[146] While each phase is distinct, the success of subsequent phases depends upon the thorough planning and execution of the prior stages.

Pre-implementation begins with deciding whether to purchase an EHR (it is rare for a health care organization to create one themselves these days) and ends with signing a contract with a vendor for a specific EHR. This requires a thorough understanding not only of the organization's needs and current state but also of the selected software's abilities and limitations. The main activity in pre-implementation is choosing the EHR that will be used, but several steps that might be done during implementation, such as workflow mapping, may be done and some say should be done, during pre-implementation. Workflow mapping involves a detailed step-by-step description, typically utilizing a flowchart of how a particular process is accomplished. For example, how are notes created or how are patient messages handled or how are prescription refills managed? [147]

Implementation of the EHR starts with the signing of the contract and ends with the go-live date. Experts in IT implementations often categorize facets of implementation into People, Process, or Technology issues.[148] Alternatively, they can be termed: Team, Tactics and Technology.

People issues are particularly important in an EHR implementation. Unless the people issues are managed well from the start, later adoption of the varied functionality inherent in an EHR will likely suffer. Key people issues are leadership, change management, goal establishment and expectation setting. An implementation will have three key types of leaders: a project manager, a senior administrative sponsor, and a clinical champion. The clinical champion will invariably be a physician, but hospital settings will typically have a nurse champion as well. The need for a project manager, someone knowledgeable and experienced in managing a complex IT project with overlapping timelines and multiple stakeholders, is obvious. Senior leadership sponsorship and support is also essential, because an EHR implementation will affect nearly all aspects of a hospital or clinic's operations and thus consistent support from the organization's leader or leaders will be required as inevitable bumps in the road are encountered.

Some healthcare organizations have learned the hard way that implementing an EHR without one or more physician champions can be disastrous. When it comes to clinical matters, physicians rely on other physicians. Because an EHR affects clinical practice in so many ways, respected, supportive, influential clinicians are needed to encourage other physicians to accept and utilize the system effectively.[149]

In inpatient settings, a nurse or clinical champion is essential to ensure that decisions made incorporate all disciplines within the facility. When implementing an EHR it is important to view operations from all perspectives (e.g. physicians, nurses, medical assistants, pharmacists, other support personnel and administrators). Without a nurse champion, decisions made might be solely physician-focused. Additionally, nurses commonly drive the change process in hospitals. Commitment to success, engagement of everyone, and a shared interest in improvement is paramount, so attitude is everything.[150]

Because of the degree of change involved in implementing an EHR for the first time, change management skills are needed. This topic is beyond the scope of this book but many good resources can be found on it. One good introductory and classic resource is Kotter's book *Leading Change*.[151] An important part of change management is setting goals and establishing expectations. Be realistic, look at the EHR myths and sins, noted in the info box.

Many specific process (or tactical) decisions are determined during implementation. How will we use the EHR to redesign our workflows? What is our data entry strategy? Which data will we enter discretely, which will we scan and which (if any) will we leave out of the EHR? Who will do this data entry and when? What order sets will we create? What other information systems will the EHR connect to and what kind of interfaces will it require? Will we follow a *big bang* (all personnel/sites and EHR functions at once) or a phased implementation approach (certain user groups and/or certain sites/departments and or certain EHR functions in sequential order)? How will we do our user training? What will we do about note templates? How much customization will we allow? How will we utilize superusers?

What about the technology? EHR software does vary in its complexity. Software designed for larger practices tends to be more customizable but also more complex, requiring more IT support. Small practices may adopt EHRs as a subscription service (SaaS) where they only need to maintain an internet connection and user terminals and everything else is done for them remotely. Large practices may be completely self-contained with their own servers, intranet, backup, terminals and IT staff. Large practice and hospital IT departments will often maintain three software environments for the EHR – production (live), test, and training.

EHR Myths and Deadly Sins

Myths:

- A new EHR will fix everything
- Brand A is the best
- Our software needs to work the way we currently work
- Software will eliminate errors
- Discrete data is always best
- The more templates the better
- Mobile is best
- You must have a detailed plan and stick to it
- You can stop planning

Deadly Sins:

- Not doing your homework
- Assuming the EMR is a magic bullet
- Not including nurses in the planning stages
- Not participating in training
- Thinking you can implement the same processes as paper
- Not asking for extra help
- Being short sighted[152, 153]

Implementation of the EHR is followed by the post-implementation phase which remains in effect for the duration of EHR use. This phase involves maintaining, reassessing and improving the EHR's content and capabilities, facility workflows/processes, and staff training with a focus on continuous improvement and patient safety. In a sense, EHR implementation is never done. As clinical sites learn more about the software from using it, they often learn how to use the software in previously unanticipated ways. And certainly as the

EHR software is periodically upgraded, new functionality is added that increases efficiencies or opens up new possibilities. Post-implementation can also be referred to as maintenance, sustainment or optimization.

Logical Steps

In the next section we present the logical steps towards selection and implementation of EHRs:

- Develop an office strategy. List priorities for the practice. Are you trying to save time and/or money or do you just want to go paperless? Are you looking to be more competitive by offering patient satisfaction-related features like secure messaging, virtual visits, a portal and connectivity with your medical community? Do you need remote computing and remote access for the clinicians? Are you seeking improved workflow to expedite chart pulls and provide easier refills? Do you need more reporting capability than what you currently have? Do you need better integration with your practice management system? Are you trying to integrate disparate programs? Now is the time to study work flow and see how it will change your practice. This is when frequent conferences with your front office staff will be critical to get their input about the processes that need to improve. Make sure physicians are committed to using the EHR. Look for at least one physician champion and be sure your staff is onboard. Do not proceed if there are hold-outs. Factor in your future requirements. Do you plan to add more partners or offices or specialties? Plan for initial decreased productivity.

- Research the EHR topic:
    - Take a short EHR course at a community college or university
    - Utilize expertise from regional extension centers (RECs) (see chapter 1) [154]
    - Read an EHR textbook [155-158]
    - Read important articles, monographs and surveys (EHR articles reside on the University of West Florida HIT Resource Site: http://uwf.edu/sahls/medicalinformatics/EHR.cfm
    - The 2011 EHR User Satisfaction Survey received 2719 responses from family physicians, reporting on 142 EHR systems. Thirty EHRs were used by the majority of respondents (87%) and this served as the corner stone of the survey. The EHRs for the VA, DOD and Indian Health Service were included. A chart correlated the top EHRs by practice size and the number of respondents using the specific EHR. Another chart ranked the 30 EHRs based on 17 dimensions. A third chart ranked EHRs based on whether they were easy and intuitive to use (usability). A fourth chart rated training and support. A fifth chart evaluated whether the EHR enables the user to practice higher quality medicine compared to paper charts and a sixth chart rated the level of overall satisfaction. An average for *all respondents* was included for benchmarking. Overall satisfaction was 50% with 30% dissatisfied. Clearly, cost and EHR size did not correlate with user satisfaction. [159]
    - The 2011 Black Book Rankings of the top 20 EHR vendors for family physicians had similar results to the AAFP survey. [160]

- Utilize HIT Consultants:
    - Consulting firms such as AC Group provide consulting for EHR purchase. In addition they have several fee-based monographs on the subject. [161]
    - KLAS is an independent HIT rating service that vendors pay to join and end-users pay to receive reviews. Their reviews cover EHRs and components based on practice size and include letter grades on implementation, service and product. Their input usually comes from office managers or IT specialists and not necessarily end-users. Physicians can

evaluate survey data on individual vendors free if they are willing to complete an online questionnaire.[162, 163]

- List EHR features need in the practice. Review the key components section of this chapter. Choose the method of inputting: keyboard, mouse, stylus, touch-screen or voice recognition? Don't forget backup systems, e.g. *dual failover.*

- Analyze and re-engineer workflow. Processes such as prescriptions, telephone triage, lab ordering, appointments, scheduling, registration and billing will change with the use of an electronic health record. Healthcare workers must embrace business process engineering (BPR) and business process automation (BPA) to create a digital office. It is wise to map the various processes to see what changes must occur and where you might add computer terminals to execute the process electronically. Some choose to use workflow software to map office workflow. HIMSS offers a toolkit "Workflow redesign in Support of the Use of Information Technology within Healthcare" for its members.[164] Other resources on workflow and process mapping related to EHRs are available.[165, 166]

- Use Project Management Tools. A variety of tools exist that improve organizational skills during the planning process. Consider using standard matrices that are glorified checklists and timelines that help organize your efforts.[167, 168]

- Decide on client-server or the application service provider (ASP) option. One early decision that must be made is whether you want to purchase a standard client-server EHR package which means having the software on your own computers. The other choice is an ASP model which uses a remote server that hosts the EHR software and your patient data. Each has its merits and shortcomings. Almost all EHR vendors now offer both models. Features of an ASP Model:[156]

    o Vendor charges monthly fees to provide access to patient data on a remote server. Fees will usually include maintenance, software upgrades, data backups and help desk support. Monthly fee may be a fixed amount or based on number of users.

    o Lease agreement commitments range from one to five years.

    o ASP may charge a fixed amount or charge for the number of users.

    o ASP can be completely web-based or can require a small software program (thin client) to help share processing tasks.

    o Pros: Lower start-up costs; ASP maintains and updates software; saves money by eliminating or reducing need for local tech support; generaly a better choice for small practices with less IT support; enables remote log-ons, for example, from home or satellite offices.

    o Cons: If your ISP is out of service, then your practice is stalled; security and HIPAA concerns; concerns about who owns the data and cost of monthly cable fees; slower speeds compared to a client-server model; need a fast internet connection, preferablya cable modem, DSL or T1 line.

- Decide on an inputting strategy. Different types of inputting are necessary because clinicians have different specialties, personal preferences and document requirements:

    o Dictation. In spite of the desire by most people who purchase an EHR to avoid dictation, many physicians will not want to give this up because it is part of their routine or they practice in a specialty where the historical narrative is best told with a dictation. Besides cost (10 to 20 cents per line), the disadvantages are the fact it is non-structured data, the physician must proof read and someone must cut and paste the narrative into the EHR, thus causing some delay.

o Speech recognition. Speech recognition is an attractive alternative to standard dictation for many but not all physicians. The cost to purchase, example Dragon Naturally Speaking (DNS) 10®, is approximately $1,600 per physician (on-site training not included) and includes a choice of multiple medical specialty vocabularies. DNS is available for the iPhone and wireless platforms.[169] There is preliminary evidence to suggest speech recognition improves the patient narrative and has a reasonable return on investment.[170] While it is true that speech recognition has improved dramatically in the last few years, it will not be satisfactory for all users. In 2010 Hoyt and Yoshihashi reported a failure rate of 31% in a large scale implementation of voice recognition in a military treatment facility.[171]

o Handwriting recognition. A few EHRs utilizing the tablet PC platform will allow a clinician to write on the tablet and have the information converted to text.

o Digital Pens. Smart (digital) pens are being used as another means of inputting that fits physician workflow.

o Templates. Most EHRs offer a template or point and click option to facilitate inputting history and physical exam data into the EHR. In addition to saving time, templates input data as structured data so it is machine and human readable. Practices can create templates ahead of time before going live and thereby, try to standardize care within a practice. Multiple template designs are available. With MEDCIN every phrase must be located and selected for inputting. Others *document-by-exception* which means there is standard language for most exams; if verbiage does not pertain to a patient, it can be deleted. Most templates can be customized (some on the fly) and shared. Many are disease specific such as *low back pain* or *headache* templates. One concern with templates, besides a potential *robotic note*, is the over use of options such as *auto-negative* where the review of systems can be performed rapidly with the potential for false documentation. Clicking history or physical exam choices that the clinician did not ask or examine is considered fraud. Conversely, submitting an overly detailed history or physical exam that is not justified by the diagnosis could be considered abuse.

o Typing. A minority of physicians will be happy to input their data by typing, particularly if they are tech savvy and excellent typists. Most physicians, however, will complain that typing notes is not why they went to medical school.

o Scribes. Emergency rooms were the first hospital area to hire scribes to shadow physicians and in addition to multiple duties were responsible for inputting information into the EHR by typing, templates, dictation or transcription.[172]

o A blended approach. Medical practices would be wise to offer multiple means to input patient data. As an example, for simple patient encounters for flu, templates may be adequate. For more complex visits dictation or voice recognition may be necessary. Organizations will have to balance the need for productivity by finding better ways to input into an EHR with the needs to have discrete or structured data. As an example, hospitals rated as stage 6 by HIMSS used templates 35%, dictation/transcription 62% and speech recognition 4% for inputting into EHRs. Newer software, using natural language processing, will extract discrete data known as *narradata* from dictations that can be used secondarily for decision support, reporting and billing. This approach is known as discrete reportable transcription (DRT) and may be important for Meaningful Use of EHRs.[173]

- Discuss mobility. Will clinicians need to be wireless? Will they benefit from access of the EHR remotely using a smart phone? Multiple vendors, like Epic, offer their software on, example an iPhone or iPad.

- Decide on EHR / PM Approach. Do you need a combined EHR and Practice Management System? Decide early on if you plan to purchase a combined EHR and practice management system or do you need an interface to be created?

- Survey hardware and network needs. How many more computers will you need to buy? Do you need to hardwire a network and/or are you going wireless? Are you going to need an in-house server with its dedicated closet, air conditioning and backup? Do you need a network switch and commercial grade firewall? Do you need to hire short term or long term IT staff? You will need a data back up and disaster plan. Plan for a commercial grade uninterruptible power supply. Also, plan for a service level agreement if you opt for the ASP model.

  o What interfaces do you need? Will you need interfaces to external laboratory, pharmacy and radiology services or is that part of the package purchased?

  o Do you need third party software? As an example: patient education material, ICD-9 codes, CPT codes, HCPCS database, SNOMED, drug database, voice recognition, etc. Ask if that is part of the purchased package.

- Develop your vendor strategy.

  o Write a simple Request for Proposal (RFP) or Request for Information (RFI). This will cause you to put on paper all of your requirements and will provide the vendor with all of the important details regarding your practice. This formal request will standardize your responses from vendors as they will need to respond in writing how they plan to address your requirements. Exact pricing should be part of the RFP. Sample RFPs are available on the Web. [174]

  o Consider using a web tool to compare EHR vendors. One free web site offers EHR resources, readiness assessments, detailed search engine and vendor comparisons, vendor profiles, EHR top 10 ratings (11 categories).[175]

  o Obtain several references from each vendor and visit each practice if possible. Be sure to select similar practices to yours.

  o The following comprehensive reference by Adler provides an EHR demonstration rating form, questions to ask vendors, RFP advice, EHR references and a vendor rating tool.[176] Create a scoring matrix to compare vendors.

  o The following reference also has a scoring sheet with sections for vendor software, interfaces, third party software, conversion services, implementation services, training services, data recovery services, annual support and maintenance, financing alternatives and terms. It also includes red flags and FAQ's. This reference is intended to compare costs and not EHR functionality between candidate vendors.[177]

  o Obtain in writing commitments for implementation and technical support, including data conversion from paper records; interfacing with practice management (PM) software; exact schedule and time line for training.

- Look for funding:

  o The most obvious choice is Medicare or Medicaid reimbursement under the HITECH Act.

  o As noted before, hospitals can donate EHR systems to physician offices under the *safe harbor* with physicians having to pay 15% of costs.

- o *Physician Quality Reporting Initiative (PQRI)* will reward physicians for quality reports that can be generated by an EHR. We will cover this more in the chapter on pay-for-performance.[178]

- o Check to see if your state has incentive programs

- Select a vendor and develop a contract. Most practices will need to create a contract with legal help. This will ensure the vendor meets their obligations and will define the contract period, duties and obligations, license stipulations, scope of license, payment schedules, termination clauses, upgrades, support, warranties, liabilities, downtime clauses, etc.

- Decide on a strategy to convert paper encounters to electronic format. Most experts advise that key information (medications, allergies, major illnesses, immunizations, lab results, etc.) be keyed in by staff on active patients several months before going live. Decide what documents such as prior encounters, consultations, discharge summaries, etc you need to upload into the EHR. Several resources will help you develop a strategy.[179-182] One vendor posts an approximate charge of 15 cents per page for less than 30,000 pages to scan in paper forms. As an example, for 5000 pages this would amount to a charge of $825.[183]

- Training. It can be said that you cannot train too much. Determine if your vendor has an electronic training database clinicians and staff can use before going live. Assess IT competencies of the clinicians and staff and train for gaps in knowledge.

- Implementation. Consider a phased in approach where clinicians and staff begin with processes such as e-prescribing, internal messages and laboratory retrieval before tackling patient encounters. Develop a go-live plan to determine reduced schedules and frequent debriefs.

References

1.   Weed LL. Medical Records that guide and teach. NEJM 1968;278:593-600

2.   Fries JF, McShane DJ. ARAMIS: a proto-typical national chronic disease data bank. West J Med 1986;145:798-804

3.   Atkinson, JC, Zeller, GG, Shah C. Electronic Patient Records for Dental School Clinics: More than Paperless Systems. J of Dental Ed. 2002;66 (5): 634-642

4.   National Institutes of Health Electronic Health Records Overview. April 2006. MITRE Corp. http://www.ncrr.nih.gov/publications/informatics/EHR.pdf (Accessed October 5 2009)

5.   Schwartz WB. Medicine and the computer. The promise and problems of change NEJM 1970;283:1257-64

6.   The Computer-Based Patient Record: An Essential Technology for Health Care, Revised Edition (1997) Institute of Medicine. The National Academies Press www.nap.edu/books/0309055326 (Accessed October 15 2005)

7.   Berner ES, Detmer DE, Simborg D. Will the wave finally break? A brief view of the adoption of electronic medical records in the United States JAIMA 2004;12

8.   Stead WW, Kelly BJ, Kolodner RM. Achievable steps toward building a National Health Information Infrastructure in the United States JAMIA 2005;12:113-120

9.   Defining Key Health Information Technology Terms April 28 2008 www.nahit.org Organization on longer active. (Accessed May 20 2008)

10.   Tang PC et al. Measuring the Effects of Reminders for Outpatient influenza Immunizations at the point of clinical opportunity. JAMIA 1999;6:115-121

11.   Smith PC et al. Missing Clinical Information During Primary Care Visits JAMA 2005;293:565-571

12.   The President's Information Technology Advisory Committee (PITAC) http://www.nitrd.gov/pubs/pitac/ (Accessed January 28 2006)

13.   Eden KB, Messina R, Li H et al. Examining the value of electronic health records on labor and delivery. Am J Obstet Gynecol 2008;199;307.e1-307.e9

14.   Tierney WM. Computerized Display of Past Test Results. Annals of Int Medicine 1987;107:569-574

15.   Lohr S. Building a Medical Data Network. The New York Times. November 22 2004. (Accessed December 20 2005)

16.   Garrido T, Jamieson L, Zhou Y, Wiesenthal A, Liang L. Effect of electronic health records in ambulatory care: retrospective, serial, cross sectional study. BMJ 2005;330:1313-1316

17.   Pollak, VE, Lorch JA. Effect of Electronic Patient Record Use on Mortality in End Stage Renal Disease, a Model Chronic Disease: A Retrospective Analysis of 9 years of Prospectively Collected Data. Biomed Central http://www.biomedcentral.com/1472-6947/7/38. (Accessed December 20 2007)

18.   US Food and Drug Administration http://www.fda.gov/ola/2004/vioxx1118.html (Accessed Aug 15 2006)

19.   Badgett R, Mulrow C. Using Information Technology to transfer knowledge: A medical institution steps up to the plate [editorial] Ann of Int Med 2005;142;220-221

20.   Housman D. Quality Reporting Through a Data Warehouse Pat Safety & Qual Healthcare Jan/Feb 2009;26-31

21.   Wall Street Journal Online/Harris Interactive Health-Care Poll www.wsj.com/health (Accessed October 24 2006)

22.   Potential benefits of an EHR. AAFP's Center for Health Information Technology. www.centerforhit.org/x1117.xml (Accessed January 23 2006)

23. Center for Information Technology Leadership. CPOE in Ambulatory Care. www.citl.org/research/ACPOE.htm (Accessed November 9 2005)

24. Patient Safety & Quality Healthcare. New Survey Addresses Relationship of EMRs to Malpractice Risk. August 22 2007. www.psqh.com (Accessed August 23 2007)

25. Nweide DJ, Weeks WB, Gottlieb DJ et al. Relationship of Primary Care Physicians' Patient Caseload with Measurement of Quality and Cost Performance. JAMA 2009;302(22):2444-2450

26. Dell RM, Greene D, Anderson D et al. Osteoporosis Disease Management: What Every Orthopedic Surgeon Should Know. J Bone Joint Surg 2009;91:79-86

27. Lee BJ, Forbes K. The role of specialists in managing the health of populations with chronic disease: the example of chronic kidney disease. BMJ. 2009;339:b2395

28. Chen C, Garrido T, Chock D et al.The Kaiser Permanente Electronic Health Record: Transforming and Streamlining Modalities of Care. Health Affairs 2009;28(2):323-333

29. Jacobe D. Worried about...the Financial Impact of Serious Illness. Gallup Serious Chronic Illness Survey 2002 http://poll.gallup.com/content/default.aspx?ci=6325&pg=1 (Accessed January 29 2006)

30. O'Malley AS, Grossman JM, Cohen GR et al. Are Electronic Medical Records Helpful for Care Coordination? Experiences of Physician Practices. J Gen Int Med 2009 DOI: 10.1007/s11606-009-1195-2 22 December 2009

31. The Computer Based Record: An Essential Technology for Health Care. www.iom.edu/Reports (Accessed January 12 2010)

32. Key Capabilities of an Electronic Health Record System: Letter Report.Committee on Data Standards for Patient Safety. http://www.iom.edu/Reports/2003/Key-Capabilities-of-an-Electronic-Health-Record-System.aspx (Accessed January 12 2010)

33. Carter J Selecting an Electronic Medical Records System, second edition, 2008 Practice Management Center. American College of Physicians. www.acponline.org/pmc (Accessed January 10 2009)

34. Bates DM, Teich JM, Lee J et al. The impact of Computerized Physician Order Entry on Medication Error Prevention JAMIA 1999;6:313-321

35. Koppel R et al. Role of Computerized Physician Order Entry Systems in Facilitating Medication Errors. JAMA 2005;293:1197-1203

36. Bates DW et al. Effect of computerized physician order entry and a team intervention on prevention of serious medication errors JAMA 1998;280:1311-1316

37. Wolfstadt JI, Gurwitz JH, Field TS et al. The Effect of Computerized Physician Order Entry with Clinical Decision Support on the Rates of Adverse Drug Events: A Systematic Review. J Gen Int Med 2008;23(4):451-458

38. Han YY et al. Unexpected increased mortality after implementation of a commercially sold computerized physician order entry system Pediatrics 2005;116:1506-1512

39. Del Beccaro MA et al. Computerized Provider Order Entry Implementation: No Association with Increased Mortality Rate in An Intensive Care Unit Pediatrics 2006;118:290-295

40. Nebeker J et al. High Rates of Adverse Drug Events in a highly computerized hospital Arch Int Med 2005;165:1111-16

41. Holdsworth MT et al. Impact of Computerized Prescriber Order Entry on the Incidence of Adverse Drug Events in Pediatric Inpatients. Pediatrics. 2007;120:1058-1066

42. Center for Information Technology Leadership. CPOE in Ambulatory Care http://www.citl.org/research/ACPOE.htm (Accessed April 5 2006)

43. Eslami S, Abu-Hanna, A, de Keizer, NF. Evaluation of Outpatient Computerized Physician Medication Order Entry Systems: A Systematic Review. JAMIA 2007;14:400-406

44. Kuo GM, Phillips RL, Graham D. et al. Medication errors reported by US family physicians and their office staff. Quality and Safety In Health Care 2008;17(4):286-290

45. Mekhjian HS. Immediate Benefits Realized Following Implementation of Physician Order Entry at an Academic Medical Center JAMIA 2002;9:529-539

46. Tierney, WM et al. Physician Inpatient Order Writing on Microcomputer Workstations: Effects on Resource Utilization. JAMA 1993;269(3):379-383

47. Teich JM et al. Toward Cost-Effective, Quality Care: The Brigham Integrated Computing System. Pp19-55. Elaine Steen [ed] The Second Annual Nicholas E. Davis Award: Proceedings of the CPR Recognition symposium. McGraw-Hill 1996

48. Ashish KJ et al. How common are electronic health records in the US? A summary of the evidence. Health Affairs 2006;25:496-507

49. A Primer on Physician Order Entry. California HealthCare Foundation. First Consulting Group September 2000. www.chcf.org (Accessed September 20 2006)

50. Kuperman GJ, Gibson RF. Computer Physician Order Entry: Benefits, Costs and Issues Ann Intern Med 2003;139:31-39

51. Hunt DL et al. Effects of Computer-Based Clinical Decision Support Systems on Physician Performance and Patient Outcomes: A systematic review JAMA 1998;280(15);1339-1346

52. Garg AX et al. Effects of Computerized Clinical Decision Support Systems on Practitioner Performance and Patient Outcomes. JAMA 2005;293(10):1223-1238

53. Kawamoto, K, Houlihan, CA, Balas, EA, Loback DF. Improving clinical practice using clinical decision support systems: a systematic review of trials to identify features critical to success. BMJ. March 2005. http://bmj.com/cgi/content/full/bmj.38398.500764.8F/DC1. (Accessed June 13 2007)

54. Wright A, Sittig DF, Ash JS et al. Clinical Decision support Capabilities of Commerically Available Clinical Information Systems. JAMIA 2009;16:637-644.

55. Sheridan TB, Thompson JM. People versus computers in medicine. In: Bogner MS, (ed). Human Error in Medicine. Hillsdale, NJ: Lawrence Erlbaum Associates, 1994, pp141-59

56. American College of Physicians Physician Information and Educational Resource (ACP Pier) http://pier.acponline.org/index.html?hp (Accessed January 28 2006)

57. UpToDate Partners. http://www.uptodate.com/home/about/emr-partners.html (Accessed September 19 2011)

58. FirstConsult http://info.firstconsult.com/ (Accessed September 19 2011)

59. Theradoc Antibiotic Assistant www.theradoc.com (Accessed September 19 2011)

60. Evans RS et al. A Computer-Assisted management Program for Antibiotics and Other Anti-infective agents NEJM 1998;338 (4):232-238

61. MEDCALC 3000 www.medcalc3000.com (Accessed September 19 2011)

62. Bates DW et al. A Proposal for Electronic Medical Records in US Primary Care JAMIA 2003;10:1-10

63. Kawamoto K et al. Improving clinical practice using clinical decision support systems: a systematic review of trials to identify features critical to success. BMJ 2005 330: 765-772

64. Hsiegh TC et al. Characteristics and Consequences of Drug Allergy Alert Overrides in a Computerized Physician Order Entry System JAIMA 2004;11:482-491

65. Shah NR et al. Improving Acceptance of Computerized Prescribing Alerts in Ambulatory Care JAMIA 2006;13:5-11

66. Bates DW, Gawande AA. Patient Safety: Improving Safety with Information Technology NEJM 2003;348:2526-2534

67. Yu DT, Seger DL, Lasser KE et al. Impact of implementing alerts about medication black-box warnings in electronic health records. Pharm and Drug Safety 2010. www.onlinelibrary.com DOI: 10.1002/pds.2088 (Accessed January 7 2011)

68. Kuperman GJ, Bobb A, Payne TH et al. Medication-related Clinical Decision Support in Computerized Provider Order Entry Systems: A Review. J Am Med Inform Assoc 2007;14:29-40

69. Shea, S, DuMouchel, W, Bahamonde, L. A Meta-Analysis of 15 Randomized Controlled Trials to Evaluate Computer-Based Clinical Reminder Systems for Preventive Care in the Ambulatory Setting. JAMIA 1996;3:399-409

70. Giuse NB et al. Evolution of a Mature Clinical Informationist Model JAMIA 2005;12:249-255

71. Bobb AM, Payne TH, Gross PA. Viewpoint: Controversies surrounding use of order sets for clinical decision support in computerized order entry. JAMIA 2007;14:41-47

72. DxPlain. http://www.lcs.mgh.harvard.edu/projects/dxplain.html . (Accessed September 19 2011)

73. Osheroff JA et al. Improving Outcomes with Clinical Decision Support: An Implementer's Guide. HIMSS Publication 2005. www.HIMSS.org

74. Rosenthal DI, Wilburg JB, Schultz T et al. Radiology Order Entry With Decision Support: Initial Clinical Experience. J Am Coll Radiol 2006;3:799-806

75. Van wijk MAM, Van der lei, J, MOssveld, M, et al. Assessment of Decision Support for Blood Test Ordering in Primary Care. Ann Int Med 2001;134:274-281

76. Lurio J, Morrison FP, Pichardo M et al. Using electronic health record alerts to provide public health situational awareness to clinicians. JAMIA 2010;17:217-219

77. Bates DM et al. Ten Commandments for Effective Clinical Decision Support: Making the Practice of Evidence based Medicine a Reality JAMIA 2003;10:523-530

78. Briggs B. Decision Support Matures. Health Data Management August 15 2005. www.healthdatamanagement.com/html/current/CurrentIssueStory.cfm?Post ID=19990 (Accessed August 20 2005)

79. M.J. Ball. Clinical Decision Support Systems: Theory and Practice. Springer. 1998

80. Clinical Decision Support Systems in Informatics Review. www.informatics-review.com/decision-support. (Accessed January 23 2006)

81. Improving Outcome with Clinical Decision Support: An Implementer's Guide. Osheroff, JA, Pifer EA, Teich JM, Sittig DF, Jenders RA. HIMSS 2005 Chicago, Il

82. Hing, ES et al. Electronic Medical Record Use by Office-Based Physicians and Their Practices: United States, 2006. Advance Data from Vital and Health Statistics. October 26 2007. No.393. www.cdc.gov/nchs. (Accessed December 3 2007)

83. DesRoches, CM et al. Electronic Health Records in Ambulatory Care—A National Survey of Physicians. NEJM 2008;359:50-60

84. Hsiao J, Beatty PC. Electronic Medical Record/Electronic Health Record Use By Office Based Physicians: United States, 2008 and Preliminary 2009. http://www.cdc.gov/nchs/data/hestat/emr_ehr/emr_ehr.pdf (Accessed November 1 2009)

85. Continued Progress: Hospital Use of Information Technology. American Hospital Association 2007. www.aha.org. (Accessed August 18 2007)

86. Jha AK, DesRoches CM, Campbell EG, et al. Use of Electronic Health Records in US Hospitals NEJM 2009;360:1-11

87. HIMSS Analytics www.himssanalytics.org (Accessed April 10 2009)

88. Anderson et al. Health Care Spending and use of Information Technology in OECD Countries. Health Affairs. 2006; 25 (3): 810-831.

89. Jha, AK et al. The use of health information technology in seven nations. International J of Medical Informatics. 2008; 77: 848-854.

90. Schoen C, Osborn R, Doty M et al. A Survey of Primary Care Physicians in Eleven Countries, 2009: Perspectives on Care, Cost and Experiences. Health Affairs 2009 Online 11/5/2009. 10.1377/hlthaff.28.6.w1171 (Accessed October 5 2009)

91. HIMSS Whitepaper. Electronic Health Records – A Global Perspective, 2nd edition. HIMSS Enterprise Systems Steering Committee and the Global Enterprise Task Force. August 2010. Part 1 (1-105).

92. CSC Sued over problems with UKs National EHR Program. http://www.ihealthbeat.org/articles/2011/10/6/csc-sued-over-problems-with-uks-national-ehr-program.aspx (Accessed October 6 2011)

93. SK&A – Physician Office Usage of Electronic Health Records Software. Revised: July 2011.http://www.skainfo.com/health_care_market_reports/EMR_Electronic_Medical_Record.pdf (Accessed October 5 2011)

94. Shortliffe E. The Evolution of electronic medical records Acad Med 1999;74:414-419

95. Wang SJ et al. A Cost-Benefit Analysis of Electronic Medical Records in Primary Care Amer J of Med 2003;114:397-403

96. Miller RH et al. The Value of Electronic Health Records In solo or small group practices Health Affairs 2005;24:1127-1137

97. King MS, Sharp L, Lipsky M. Accuracy of CPT evaluation and management coding by Family Physicians. J Am Board Fam Pract 2001;14(3):184-192

98. Moore P. Tech Survey: Navigating the Tech Maze. September 2008. Physicians Practice www.physicianspractice.com (Accessed October 4 2008)

99. 2003 Commonwealth Fund National Survey of Physicians and Quality of Care. http://www.cmwf.org/surveys/surveys_show.htm?doc_id=278869 (Accessed December 7 2004)

100. Brailer DJ, Terasawa EL. Use and Adoption of Computer Based Patient Records California Healthcare Foundation 2003 www.chcf.org (Accessed February 10 2005)

101. Poissant L, Pereira J, Tamblyn R et al. The impact of electronic health records on time efficiency of physicians and nurses: a systematic review. JAMIA 2005;12:505-516

102. Medical Records Institute's Ninth Annual EHR Survey of Electronic Medical Records, Usage and Trends 2007. http://www.medrecinst.com/MRI/emrsurvey.html. (Accessed December 4 2007)

103. Boone, E. EMR Usability: Bridging the Gap Beween the Nurse and Computer. Nursing Management 2010;41(3): 14-16

104. HIMSS EHR Usability Task Force. June 2009 www.himss.org (Accessed March 15 2010)

105. Sujansky WV, Overhage JM, Chang S et al. The Development of a Highly Constrained Health Level 7 Implementation Guide to Facilitate Electronic Laboratory Reporting to Ambulatory Electronic Health Record Systems. JAMIA 2009;16:285-290

106. HIPAA Survival Guide http://www.hipaasurvivalguide.com/hipaa-survival-guide-21.php (Accessed September 21 2011)

107. Kibbe, DC. 10 Steps to HIPAA Security Compliance. Family Practice Management. 2005; 12(4):43-9

108. Mello,MM et al. National Costs Of The Medical Liability System. Health Affairs. 2010; 29, (9):1569-1577

109. David Blumenthal, M.D., M.P.P. Study of Patient Safety and Health Information Technology Institute of Medicine, National Academy of Sciences. December 14, 2010 http://healthit.hhs.gov/portal/server.pt/gateway/PTARGS_0_0_7739_3171_20795_43/http%3B/wci_pubcontent/publish/onc/public_communities/_content/files/patient_safety_iom_db_statement_121_410.pdf (Accessed October 6 2011)

110. EHR Event http://ehrevent.org (Accessed October 6 2011)

111. Mangalmurti, SS. Murtagh L, Mello MM. Medical Malpractice Liability in the Age of Electronic Health Records. New England Journal of Medicine. 2010; 363: 2060-2067.

112. Miller, Amalia R. and Tucker, Catherine, Electronic Discovery and the Adoption of Information Technology (January 18, 2010). Available at SSRN: http://ssrn.com/abstract=1421244 (Accessed October 6 2011)

113. Virapongse, A et al. Electronic Health Records and Malpractice Claims in Office Practice. Archives of Internal Medicine. 2008; 168(21):2362-2367.

114. Chaudry B, Wang J, Wu S et al. Systematic Review: Impact of Health Information Technology on Quality, Efficiency and Costs of Medical Care. Ann of Int Med 2006;144:E12-22.

115. Yu FB, Menachemi N, Berner ES et al. Full Implementation of Computerized Physician Order Entry and Medication-Related Quality Outcomes: A Study of 3364 Hospitals. Am J of Qual Meas 2009; E1-9doi:10.1177/106286060933626

116. Study: EHR Adoption Results in Marginal Performance Gains. November 16 2009. www.ihealthbeat.org (Accessed November 23 2009)

117. EHR IMPACT. The socio-economic impact of interoperable electronic health record (EHR) and ePrescribing systems in Europe and beyond. Final study report. October 2009. www.ehr-impact.eu (Accessed March 1 2010)

118. Jones SS, Adams JL, Schneider EC et al. Electronic Health Record Adoption and Quality Improvement in US Hospitals. December 2010. www.ajmc.com (Accessed January 3 2011)

119. Romano MJ, Stafford RS. Electronic Health Records and Clinical Decision Support Systems. Impact on National Ambulatory Care Quality. Arch Int Med 2011;171(10):897-903

120. Cebul RD, Love TE, Kain AK, Hebert CJ. Electronic Health Records and Quality of Diabetic Care. NEJM 2011;365(9):825-833

121. Han YY et al. Unexpected Increased Mortality After Implementation of a commercially sold computerized physician order entry system Pediatrics 2005; 116:1506-1512

122. Nebecker JR et al. High Rates of Adverse Drug Events in a Highly Computerized Hospital. Arch Int Med 2005; 165:1111-1116

123. Bates DW. Computerized physician order entry and medication errors: Finding a balance. J of Bioinform 2005;38:259-261

124. Ash JS, Berg M, Coiera E. Some Unintended Consequences of Information Technology in Health Care: The Nature of Patient Care Information System-Related Errors. JAMIA. 2004;11:104-112

125. Berger RG, Kichak, JP. Computerized Physician Order Entry: Helpful or Harmful JAMIA 2004;11:100-103

126. Weiner, JP et al. e-Iatrogenesis: The Most Critical Unintended Consequence of CPOE and other HIT. JAMIA. 2007;14(3):387-388

127. Guide to Reducing Unintended Consequences of Electronic Health Records. AHRQ. September 2011. http://www.ucguide.org/index.html (Accessed September 20 1011)

128. Medicare and Medicaid EHR Incentive Program. www.cms.gov/EHRIncentivePrograms (Accessed June 20 2011)

129. Certified Health IT Product List. CMS. www.cms.gov/EHRIncentivePrograms (Accessed September 16 2011)

130. Daly R. Sebelius reports on EHR incentives. September 12 2011. Modern Healthcare. www.modernhealthcare.com (Accessed September 13 2011)

131. AmazingCharts www.amazingcharts.com (Accessed September 20 2011)

132. EClinicalWorks www.eclinicalworks.com (Accessed September 20 2011)

133. Epic www.epic.com (Accessed September 20 2011)

134. HIT Policy Committee www.healthit.hhs.gov/media/faca/MU_RFC%20_2011-01-12_final.pdf (Accessed September 30 2011)

135. O'Malley. Tapping the unmet potential of health information technology. New England Journal of Medicine. 2011; 364(12):1090-1091.

136. Halamka J. The Rise of Electronic Medicine www.technologyreview.com/business/38473/page1 (Accessed October 1 2011)

137. ONC's new query health initiative—what's in it for e-patients http://e-patients.net/archives/2011/09/query-health.html (Accessed September 30 2011)

138. Narathe,AS and Conway,PH. Optimizing health information technology's role in enabling comparative effectiveness research. The American Journal of Managed Care. 2010;16(12 Spec No.):SP44-SP47.

139. Jha,AK The promise of electronic records – Around the corner or down the road?. JAMA 2011, 306(8):880-881.

140. Kaiser Permanente. http://xnet.kp.org/future (Accessed October 5 2011)

141. Bates,DW and Bitton,A. The future of health information technology in the patient-centered medical home. Health Affairs. 2010; 29(4):614-621.

142. Topol,E. The future of healthcare: Information technology. Modern Healthcare. www.modernhealthcare.com/article/20110725/supplement/110729989/?templat... (Accessed October 5 2011)

143. Kilian,J and Pantuso,B. The Future of Healthcare is Social. www.fastcompany.com/futureofhealthcare (Accessed October 6 2011)

144. Diamond J, Fera B. Implementing an EHR. 2007 HIMSS Conference February 25-March 1.New Orleans

145. MGMA. Medical Practice Today, 2009. www.mgma.com (Accessed August 12 2009)

146. Lorenzi NM, Kouroubali A, Detmer DE et al. How to successfully select and implement electronic health records (EHR) in small ambulatory practice settings. BMC Medical Informatics and Decision Making 2009, 9:15

147. Skolnik,ed. Electronic Medical Records: A practical guide for primary care. New York, NY: Humana Press. 2011

148. Keshavjee, K et al. Best practices in EMR implementation. AMIA: ISHIMR 2006. 1-15.

149. Adler, KG. How to successfully navigate your EHR implementation. Family Practice Management 2007; 14(2): 33-39

150. Adler KG. Successful EHR Implementations: Attitude is Everything. Nov/Dec 2010. www.aafp.org/fpm (Accessed February 20 2011)

151. Kotter,JP. Leading Change. Boston, MA: Harvard Business School Press; 1996.

152. Trachtenbarg DE. EHRs fix everything-and nine other myths. Family Practice Management March 2007 www.aafp.org/fpm/20070300/26ehrs.html (Accessed September 19 2008)

153. McNickle, M. The 7 Deadly Sins of EMR Implementation. Healthcare IT News. September 7 2011. www.healthcareitnews.com (Accessed September 21 2011)

154. Regional HIT Extension Centers. www.healthit.hhs.gov (Accessed September 20 2011)

155. Gartee, R. Electronic Health Records: Understanding and Using Computerized Medical Records. 2011. Pearson Publisher.

156. Carter, J. Electronic Health Records, Second Edition. 2008. ACP Online. www.acponline.org (Accessed September 20 2011)

157. Amatayakul, MK. Electronic Health Records, A Practical Guide for Professionals and Organizations, Fourth Edition. 2009

158. Hamilton, B. Electronic Health Records. 2010. McGraw-Hill

159. Edsall RL, Adler KG. The 2011 EHR User Satisfaction Survey. Fam Pract Man July/August 2011;23-30 www.aafp.org (Accessed September 20 2011)

160. Black Book Rankings. 2011 Client Satisfaction Rankings and Results. www.BlackBookRankings.com (Accessed September 20 2011)

161. AC Group www.acgroup.org (Accessed September 20 2011)

162. Lowes R. Report Names Top Electronic Health Records for Physician Practices. December 2009. www.medscape.com (Accessed March 15 2010)

163. KLAS Research www.klasresearch.com (Accessed March 15 2010)

164. HIMSS Workflow Redesign Toolkit. http://www.himss.org/handouts/HIMSSWorkflowRedesignToolkit.pdf (Accessed September 20 2011)

165. Workflow Analysis. EHR Deployment Techniques. California Healthcare Foundation. January 2011. www.chcf.org (Accessed January 20 2011)

166. Summary Report. Incorporating health information technology into workflow redesign. AHRQ. October 2010. www.ahrq.gov (Accessed January 3 2011)

167. DOC-IT EHR Implementation Roadmap http://www.internetifmc.com/provider/documents/ehr_implementation_roadmap.pdf (Accessed September 20 2011)

168. Build your EHR Timeline http://www.physiciansehr.org/index.asp?PageAction=Custom&ID=48 (Accessed March 16 2010)

169. Nuance www.nuance.com (Accessed September 30 2011)

170. Speech Recognition Improves EMR ROI. Health Management Technology October 2009. www.healthmgttech.com (Accessed February 20 2010)

171. Hoyt R, Yoshihashi. Lessons Learned from Implementation of Voice Recognition for Documentation. Perspectives in Health Information Management. Winter 2010. http://perspectives.ahima.org (Accessed February 1 2010)

172. Conn J. Docs using scribes to ease EHR transition. Modern Healthcare. February 8 2010. www.modernhealthcare.com (Accessed February 20 2010)

173. Cannon J, Lucci S. Transcription and EHRs, benefits of a blended approach. Journal of AHIMA February 10 2010:36-40

174. Sample RFP. http://www.orchardsoft.com/request/rfp/rfp.html (Accessed September 20 2011)

175. AmericanEHR Partners. www.americanehr.com (Accessed September 21 2011)

176. Adler KG. How to Select an Electronic Health Record System. Family Practice Management. February 2005 http://www.aafp.org/fpm/2005/0200/p55.html (Accessed September 20 2011)

177. eHealth Initiative EHR Master Quotation Guide. eHealth Initiative. http://tiny.cc/7RJZg (Accessed September 21 2011)

178. Physician Quality Reporting Initiative (PQRI). www.cms.hhs.gov/pqri (Accessed April 15 2009)

179. California Healthcare Foundation. Chart Abstraction: EHR Deployment Techniques Feb 2010 www.chcf.org (Accessed March 10 2010)

180. Terry K. Technology: EMR Success in 8 Easy Steps. www.physicianpractice.com (Accessed March 10 2010)

181. Nelson R. Getting Data into an EHR. Medpage Today March 3 2008. www.printhis.clickability.com (Accessed October 9 2009)

182. Dinh AK, Kennedy MS, Perkins SG et al.Migrating from Paper to EHRs in Physician Practices. Practice Brief. J AHIMA Nov-Dec 2010:60-64

183. Digital Island http://www.digitisle.com (Accessed March 13 2010)

4

Practice Management Systems

BRANDY G. ZIESEMER

ROBERT E. HOYT

Learning Objectives

After reading this chapter the reader should be able to:

- Document the workflow in a medical office that utilizes a practice management (PM) system integrated with an electronic health record system

- Compare the functionality of a standalone PM system with fully integrated PM software as part of a robust EHR system

- Identify the features of an integrated practice management system

- List the most common integrated PM software packages currently available

- Identify the key advantages and obstacles in converting from paper records and a standalone PM system to an integrated PM with EHR system

- Discuss emerging trends in practice management

Introduction

"In an environment where quality is quantitatively and qualitatively measured, an administrator will be judged to be productive if his or her work leads to an organization performing at a higher level in the metrics of cost, quality, and service."

- Rick Madison and Ken Clarke, *Group Practice Journal*

Most medical offices have had computerized practice management (PM) systems for many years, regardless of whether that office maintains paper medical records, electronic health records (EHRs) or a hybrid of these two. As we will point out, there are many reasons why PM systems have become so prevalent but one of the main reasons is for more rapid claims submission and adjudication. Without an electronic system, time and money would be lost on faxes, phone calls and snail mail. The American Medical Association estimated that inefficient claims submission systems lead to about $210 billion annually in unnecessary costs.[1] A PM system is designed to capture all of the data from a patient encounter necessary to obtain reimbursement for the services provided. This data is then used to:

- Generate claims to seek reimbursement from healthcare payers

- Apply payments and denials

- Generate patient statements for any balance that is the patient's responsibility

- Generate business correspondence

- Build databases for practice and referring physicians, payers, patient demographics and patient encounter transactions (i.e., date, diagnosis codes, procedure codes, amount charged, amount paid, date paid, billing messages, place and type of service codes, etc)

Additionally, a PM system provides routine and ad hoc reports so that an administrator can analyze the trends for a given practice and implement performance improvement strategies based on the findings. For example, a medical office administrator is able to use the PM system to compare and contrast different payers with regards to the amount reimbursed for each given service or the turnaround time between claims submission and payment. The results lead to deciding which managed care plans the practice will participate in versus those plans that the practice may want to consider not accepting in the future. Another example is to analyze all payers for a given service performed in the practice to determine if that service is a good use of the practice's clinical time. This analysis provides one aspect of whether or not the practice should consider continuing to offer a certain service such as case management of a patient who is receiving home health services through an agency. Of course, the administrator has to weigh services that aren't profitable against any negative impact on overall patient satisfaction but the PM system provides a means of analyzing payment performance.

Most PM systems also offer patient scheduling software that further increases the efficiency of the business aspects of a medical practice. Finally, some PM systems offer an encoder to assist the coder in selecting and sequencing the correct diagnosis (International Classification of Diseases, Current revision, clinically modified for use in the United States, or ICD-XX-CM) and procedure (Current Procedural Terminology, fourth edition or CPT-4® and Healthcare Common Procedure Coding System or HCPCS) codes. Even when a physician determines the appropriate codes using a *superbill*, (a list of the common codes used in that practice along with the amount charged for each procedure), there are times when a diagnosis or procedure is not listed on the superbill and an encoder makes it efficient to do a search based on the main terms and select the best code. Furthermore, some encoders are packaged with tools such as a subscription to a newsletter published by the American Medical Association (AMA) known as "CPT® Assistant" that help the practice comply with correct coding initiatives which in turn optimize the reimbursement to which the practice is legally and ethically entitled and avoids fraud or abuse fines for improper coding.

Clinical and Administrative Workflow in a Medical Office

Several steps are common to almost any medical practice with regards to treating patients and getting reimbursed properly for the services provided. The steps are subdivided based on whether or not the patient has been to this practice previously for any type of service. The first step is to get the patient registered. This can be accomplished via a practice website or by the patient calling the office to schedule an appointment. Figure 4.1 demonstrates typical outpatient office workflow.

Patient Registration. This step includes obtaining demographic information, including any healthcare plan or plans the patient has and establishing which member of the patient's household is financially responsible for any balances due either at the time of the visit or after claims adjudication by any healthcare payer(s) the practice agrees to bill for the patient.

Patient Scheduling. The patient is then scheduled for an appointment. If the patient had a previous encounter with the physician, the office receptionist simply has to update any changes to the patient information already on file.

Figure 4.1: Typical Outpatient office workflow (EOB = explanation of benefits)

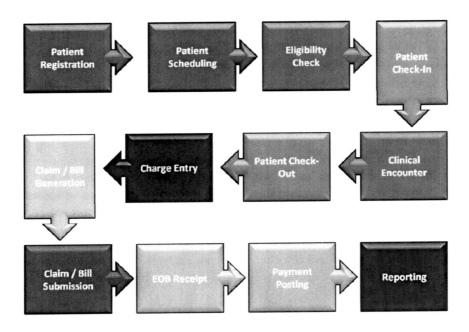

Eligibility Check. For a new patient the insurance information must be verified to ensure that the patient is currently covered by a plan accepted by the practice and the planned services are a covered benefit. If not, the patient must be notified in advance of the visit to determine if they are willing to accept full financial responsibility for the services (i.e. full payment then attempt to get reimbursement from their healthcare plan on their own) or cancel the appointment and find a participating physician. If a practice offers web-based patient registration, there are some choices ranging from designing the website and all applicable online forms internally to contracting with a forms services company. Based on the amount of money the practice is willing to spend, a forms company offers basic forms design for use on the practice's own website. Alternately, they can subcontract to use the company's server and website for forms design, updating, processing and transmitting information to the practice's EHR or PM system. See Medical Web Office services for a sample range of forms and communications services available for medical practices.[2]

Patient Check-In. The patient checks in for the scheduled visit. If already established with the practice the receptionist simply verifies/updates the patient information. If the patient is new, and the data gathered to schedule an appointment was obtained via telephone, the patient is asked to complete a registration form and provide a copy of his or her insurance card(s). Any information not previously obtained is keyed into the computer system for use by the PM system and the source document is added to the paper medical record, if applicable. Scanning the information is an option with an EHR. Most practices that have a PM system that is integrated with an EHR can scan the documents (including bubble sheets completed by the patient at time of registration) into the system once and the information is posted to the appropriate places in both the EHR and the PM system. Sometimes the data that is used by both the EHR and the PM software, such as patient name, is saved to a common database in an integrated system. At other times, however, the shared data is communicated electronically between the EHR and the PM system even though the databases are separate. It is important to know that when the systems have a shared database, this database only contains the part of the clinical record that is used to obtain reimbursement such as the patient demographics, diagnoses and procedures, dates of service, etc. However, the purely financial information is only found in the PM system – such as amount billed and amount paid or information about health plans. This is because it is not advisable to combine the business aspects of health information with clinical aspects. What procedure is done on a given date and the diagnosis that justifies the medical necessity of a procedure is both clinical and financial but how much the procedure costs and how much the patient paid out-of-pocket, etc is purely financial.

Clinical Encounter. The patient is generally first seen by a nurse or medical assistant, to have vitals taken, collect blood and urine samples, if needed, and update the patient's subjective history. The patient is then examined by the physician who takes additional history and completes the objective physical exam and updates the clinical notes in SOAP order – Subjective, Objective, Assessment and Plan. In a paper system, the physician dictates either during the visit or as soon afterward as possible and a transcriptionist creates a paper copy of the notes. Alternately, some physicians use voice recognition technology to dictate directly into a laptop or other device then print out the report generated by the software to file in the paper record. For a sample of a full range of voice recognition software that can be used as a standalone product for creating a paper document or that interfaces with electronic health records, visit Nuance's website.[3]

As discussed previously, in an EHR system clinicians have several options for inputting patient information into the clinical record. They can use voice recognition software, standard dictation or templates. This can be accomplished on a PC or in a wireless mode with a tablet PC. For example, an EHR is formatted with physician workflow in mind and then, customizable by each individual physician to optimize efficiency based on specialty and personal preferences. The customization "initially takes time and patience but is well worth the effort in a practice that sees a lot of patients daily with the same symptoms such as an ear infection for a pediatrician."[4] Therefore, when the physician is face-to-face with the patient, the EHR would have already been started for that encounter by a nurse or other physician extender who would have entered the patient's chief complaint, vital signs and possibly any updates to the patient's subjective history (the subjective portion of the SOAP note).

The physician will continue building the encounter notes by using a series of drop-down menus to indicate body systems examined, tests performed, tests or prescriptions ordered, (the objective portion of the SOAP note), the assessment and the plan. Each selection made by the physician adds to the clinical notes. Clinical notes are a good example of data that is maintained in the EHR but not shared with a PM system. However, EHRs that use computer assisted coding (CAC) technology can convert the standardized notes into codes and the codes are used by both the EHR and the PM system. For example, many EHRs can run the office notes through logic to assign CPT evaluation and management (E&M) codes based on either the 1995 or 1997 guidelines. The EHR system can pass these codes plus many ICD-XX-CM codes over to integrated (same vendor) or interfaced (different vendors) PM system when the systems are compatible. The physician concludes the clinical aspects of the encounter by giving the patient discharge/follow-up instructions and patient education literature. Any lab samples are sent to the lab and, if the patient needs a prescription and the practice uses e-prescribing, a prescription is sent from the EHR to the pharmacy electronically or via Fax. If not, the patient is given a paper copy of the prescription.

Patient Check-Out. The patient is discharged after a receptionist collects any money due and schedules any follow-up visits. If the practice has chosen this feature, the EHR can interface with the PM system scheduler so the physician can schedule a follow-up visit and the patient can take home a printout of the office notes, any education material, the next appointment, plus a paper copy of physician orders or prescriptions for facilities not linked with the EHR.

Charge Entry Claims-Bill Generation. In a standalone PM system, the charges are entered, often from a superbill but sometimes the services are coded from the information in the medical record. In an integrated PM with EHR system, the information needed is sent directly from the EHR to the PM system and a claim is built as described above. However, a person responsible for correct coding and billing must still verify that all applicable codes were brought over to the PM system, add any codes that the system did not assign automatically and scrub codes which means to link the diagnoses to the correct procedures that justify medical necessity and check for obvious errors in order to get them ready to submit as claims to payers.

Claims-Bill Submission. The claims are sent electronically in all but rare cases but they are sent in cycles so once the PM system is updated, the claim is in queue waiting for transmission to a clearinghouse or directly to the payer, such as Medicare.

Remittance Advice (RA) and Explanation of Benefits (EOB) Receipt. Once the claim is sent, the payer electronically (again, there are some exceptions in which the practice will actually get a paper check in the mail) sends a remittance advice (RA) containing the details for each charge paid or denied in that cycle. The RA contains an EOB (payments, denials, denial reason, reduced payments and reasons, patient responsibility, whether or not the claim was sent automatically to a secondary payer, etc.) for each charge by patient.

Payment Posting. The money is electronically deposited into the practice's account. The payer generally mails a paper copy of each individual explanation of benefits to the patient. Billing personnel also have to follow-up when a person has more than one payer, to determine that the claim was transmitted to the appropriate secondary payer. If there is still a balance after the biller has applied payments and written off any charges in excess of the allowed amount for a particular payer, the system moves the balance into a queue to await patient billing. The biller is also responsible for tracking claims and initiating the collections process if a balance due by the patient is not paid in a timely fashion.

Reporting. Daily reports are run and verified to ensure deposits match, all patients who were seen that day have charges in the system, etc. There are both routine reports (daily, weekly, monthly and end-of-year) and ad hoc reports used by the practice.

Telephone calls in a Medical Office. Calls to the practice may be for various reasons from cancelling an appointment to asking if the doctor can see a patient who does not have an appointment. The calls should be prioritized into categories for emergency (in which the practice should advise the patient to hang-up and dial 911 and then confirm the patient is capable of doing that before disconnecting), urgent or non-urgent. Many EHR systems enable the message-taker to route the call directly to the intended recipient instead of having to take a paper message. In this case, protocols exist in which the person answering the phone can take certain actions or make some decisions. For example, the receptionist may be able to determine if outside lab results have been received by the practice or not. If they have, the receptionist can route a message directly to the appropriate clinician requesting the patient be called regarding the results. In more urgent cases, a patient may have a non-emergency, but urgent condition and request to be seen that day. The receptionist may be able to schedule that person in and tell them when to arrive. If the receptionist is uncertain, he or she may route an urgent message to the most appropriate clinician for a decision about whether the patient can be seen that day. The messages go directly to the recipient's attention and may be color-coded to highlight urgent versus non-urgent.[5, 6]

Practice Management Systems and EHRs

When the administrators of a medical practice commit to the conversion of paper-based or hybrid records to an electronic health record system, they should strongly consider converting their existing practice management system to one that is integrated with the EHR they chose. Although this strategy means a higher initial investment of both time and money many practices report that part of their overall success with implementing an EHR with a PM system is due to the increased efficiency and accuracy of the billing process when the systems are integrated. One alternative to discontinuing an existing PM system – especially one that works very well and that everyone in the office who uses it is comfortable with, is to find a reputable EHR vendor that offers interface capabilities with your existing PM system vendor. For example, a statement on the website for Allscripts™ regarding its options for interfacing its EHR with other practice management systems explains, "while Allscripts offers our own integrated, practice management, all of our EHR solutions are built to integrate seamlessly with your existing practice management system, preserving your current investment."[7] One potential setback is anytime a vendor upgrades its software, the interfaces have to be tested and both vendors may need to get involved.

As mentioned in the Electronic Health Records chapter, formal studies of return on investment (ROI) for an EHR in medical practices are very limited at this time and although some of these studies mention how many of the practices surveyed implemented an EHR in conjunction with an integrated PMS, they don't differentiate the results as to the impact on ROI of an integrated EHR/PM system versus a standalone EHR project.

There are some testimonials and customer information pages available on many of the vendor websites that discuss success stories of EHR implementation in which some of the cost savings were realized by the integration of the EHR with a PM system. Some of these websites also list the ROI based on an integrated EHR/PM system.[8, 9]

In 2008, the Medical Society of New York reported on the experience of a solo, internal medicine practice in identifying, purchasing, implementing and maintaining an EHR with a PM system. Some of the highlights from this article include:

- Chose in-office server over application service provider (ASP) because it was more affordable since the practice already had old computers that could be used as workstations and they did not already have business-grade broadband internet service provider necessary for an ASP arrangement.

- Chose to pay a monthly subscription fee to use the EHR/PM software because it was less expensive than purchasing software and hardware required. Lease agreement was approximately $400 per month per physician for software and off-site HIT support.

- HIT vendor provided most of the on-site technical support during the first three months

- Chose the EHR from same vendor as PM because, "to run an efficient office, the two systems should be totally compatible and ideally share the same database."

- The criteria most important to Dr. Volpe included: perform the same functions as a paper record, only better; affordability; navigability for all users between both the EHR and the PM systems; adaptability or the records and templates for information were easily modified and customized as patient information and practice needs changed; track record of vendor; ease of data entry; tight security and, CCHIT certification.

- The transition from paper was easy (by comparison) for the doctor and his staff because they had already used an e-prescribing system and a patient web portal. They transitioned from the old billing system to the new PM system gradually over 30 to 90 days by using the old system for any patient billing that was in-process or occurred before the staff training was complete and then used the new system for all encounters occurring after the staff was trained.

- The overall assessment considering the improvement in efficiency, improved patient care and improved claims coding and submission is: "within one year, Dr. Volpe had paid off the cost of his new hardware, his office had returned to full productivity, and he was earning over $30,000 more than he had the year before, due primarily to reduced overhead costs (space for paper records has been converted to an additional exam room, for example). He, his family, his staff and his patients all love the new system."[10]

According to a recent newsletter offered by the Health Information Management Systems Society (HIMSS), "A tool was recently launched to measure the healthcare industry's move from paper-based processes to electronic business applications. Called the US Healthcare Efficiency Index, this tool is intended to raise awareness about the cost savings that can be achieved by increasing the adoption of electronic business processes. One-third of respondents believed that the largest cost savings would come in the area of claims submission. Nearly half of respondents that work for a healthcare provider organization reported that they submitted claims payments electronically. A similar percent also reported that their organization submitted claims eligibility and claim remittance advice transactions electronically. Another potential area of cost savings would be to have claims payments issued via direct deposit."[11]

There are more than 200 practice management systems on the market with a variety of PM features to include integration with an EHR and the availability as both a client server model and/or an ASP model. In Table 4.1 we have included a list of the better known vendors who offer a combined EHR-PM that is CCHIT certified and available in the ASP model. In the resource section we will direct you to methods to search all available PM systems.

Table 4.1: EHR-PM integrated systems that are currently ONC-ATCB certified and available as an ASP model

EHR-PM System	Web Site	Features
ABELMed EHR-EMR/PM v11	www.abelsoft.com	ASP available
Allscripts MyWay PM **Allscripts Professional PM**	www.allscripts.com	ASP available. Small to medium practices
Athena Clinicals	www.athenahealth.com	Only ASP. Can outsource the billing service
Cerner PowerWorks	www.cerner.com	ASP available. Small to large practices
e-MDs Solution Series	www.e-mds.com	ASP available
eClinicalWorks	www.eclinicalworks.com	ASP available. Small to large practices
GE Centricity Practice Solution	www.gehealth.com	ASP available. Medium to large practices
NextGen Healthcare	www.nextgen.com	ASP available.
Vitera Intergy EHR **Integrated with Medical Manager**	http://www.viterahealthcare.com	ASP available
Waiting Room Solutions	www.waitingroomsolutions.com	ASP available. 26 specialties

Resources

Although there is very little written about the merits and limitations of practice management systems, we can direct you to several helpful resources:

- *Practice Management Systems for Safety-Net Clinics and Small Group Practices: A Primer*, a 2009 monograph by the California HealthCare Foundation, discusses how important PM systems are for safety net clinics but also provides an excellent overview on the subject.[12]

- *Capterra* is a web site that includes a search engine with filters for operating systems (platform), number of users, PM features, inclusion with EHRs, location (example USA) and annual revenue. Over 200 products are included with hyperlinks to the individual web sites, demos and tours.[13]

- *EHRScope* is a search engine for EHRs but includes the ability to filter the search for EHRs that include PM systems, as well as differentiate between systems that are web-based or client-server based. Without filtering, 342 EHRs are included.[14]

- *Selecting a Practice Management System* is a monograph by the American College of Physicians. Members can access this resource that focuses how to go about selecting a PM system, in terms of the steps that are necessary prior to purchase.[15]

- *Medical Practice Management Buyer's Guide* is a 2008 web-based resource that includes pricing and tips before purchase.[16]

- The *Online Consultant* is a fee-based generator of "requests for proposal" (RFP). In the case of PM systems they generate detailed questions about price and functional requirements. Once the RFP is complete they offer the ability to graph and create comparison reports between vendors. Charge is $695 for PM RFP.[17]

- *Wikipedia* offers standard information on PM as well as vendor comparison chart and operating system compatibility chart.[18]

- The *United States Healthcare Efficiency Index* report of "National Progress on Healthcare Efficiency 2010" provides a status report on the healthcare IT initiatives that are part of healthcare reform from the perspective of the billions of dollars that may be saved annually if all paper claims for healthcare reimbursement were eliminated through the adoption of health information technology including practice management systems.[19]

Future Trends

In an effort to make healthcare delivery more cost effective and less prone to fraud and abuse, the Centers for Medicare and Medicaid Services (CMS) continuously assesses and initiates measures to improve the reimbursement process for health care providers. What this means to medical practices, from a revenue cycle management perspective, is both changes that have already been approved and have implementation deadlines and some new demonstration projects.

The most comprehensive change that has been approved with a set deadline is two-fold. First, changing the transmission standards from the current HIPAA transaction set to ASC X12 Version 5010 (Version 5010) which will enable electronic billing systems to accommodate the transition from the International Classification of Diseases-Ninth Revision-Clinical Modification (ICD-9-CM) to the October 1, 2013 implementation of ICD-10-CM. Second, is the actual implementation of ICD-10-CM.

Initially, CMS was set to deny any claims transactions submitted to them on or after January 1, 2012 that were not in the 5010 format, but the enforcement of that rule was postponed to March 31, 2012 due to the number of entities not able to comply by that date.[20] The implementation date did not change from January 1 – only the enforcement of denying claims was postponed.

While the changes to the software in practice management systems, claims clearinghouses and payers does not have much of a direct impact on the users of the practice management system, the October 1, 2013 implementation of ICD-10-CM codes will result in very obvious changes – electronic and paper forms will look different, date entry filed lengths will change and any list of ICD-9-CM codes such as a list of which ICD-9-CM codes justify the medical necessity of which lab tests, just as an example, will all have to be re-designed.

New projects slated for 2012 demonstration projects that impact practice management include the following CMS efforts to help curb improper Medicare and Medicaid payments:

- **Recovery Audit Prepayment Review.** The Recovery Audit Prepayment Review demonstration will allow Medicare Recovery Auditors (RACs) to review claims before they are paid to ensure that the provider complied with all Medicare payment rules. The RACs will conduct prepayment reviews on certain types of claims that historically result in high rates of improper payments. These reviews will focus on seven states with high populations of fraud- and error-prone providers (FL, CA, MI, TX, NY, LA, IL) and four states with high claims volumes of short inpatient hospital stays (PA, OH, NC, MO) for a total of 11 states. This demonstration will also help lower the error rate by preventing improper payments rather than the traditional *pay-and-chase* methods of looking for improper payments after they have been made.

- **Prior Authorization for Certain Medical Equipment.** The second demonstration announced today will require Prior Authorization for certain medical equipment for all people with Medicare who reside in seven states with high populations of fraud- and error-prone providers (CA, FL, IL, MI, NY, NC and TX). This is an important step toward paying appropriately for certain medical equipment that has a high error rate. This demonstration will help ensure that a beneficiary's medical condition warrants their medical equipment under existing coverage guidelines. Moreover, the program will assist in preserving a Medicare beneficiary's right to receive quality products from accredited suppliers.

- **Part A to Part B Rebilling.** The third initiative will allow hospitals to rebill for 90% of the Part B payment when a Medicare contractor denies a Part A inpatient short stay claim as not reasonable and necessary due to the hospital billing for the wrong setting. Currently, when outpatient services are billed as inpatient services, the entire claim is denied in full.[21]

Practice management systems will continue to evolve based on payer initiatives, increased interoperability with electronic health record systems, refinement of pay-for-performance initiatives that require capturing certain ICD-10-CM codes and CPT codes on claims screens for electronic transmission and increased reporting requirements.

Key Points

- Many medical practices are struggling with implementation of EHRs and how to determine the return on investment (ROI) for a new integrated PM system

- As reimbursement methodologies become increasingly more complex and tied to quality measures, it is of utmost importance to ensure a medical office has the tools to obtain full payment to which the practice is legally and ethically entitled by collecting all of the appropriate data that justifies medical necessity and compliance with quality guidelines

- Practitioners not only have to provide high quality and safe patient care but they must do so as efficiently and effectively as to remain competitive

Conclusion

Ambulatory care practices have many options to consider when converting to a robust electronic health record system that is fully integrated with or interfaced with a comprehensive practice management system. The chapter on electronic health records discusses the steps a practice can take to identify the best overall electronic health record and practice management system based on factors such as size and type of practice, degree to which each physician supports the transition, information technology preferences (servers, technical support, purchase versus lease etc.), priority of various features, projected return on investment (ROI) and other considerations. The practice management piece fits in nicely with the overall EHR selection, implementation, training and maintenance process but should be included from the start rather than starting the selection and implementation piece after the EHR system is already in place. The combination of functionality between the clinical and business aspects of medical practice is considered the ideal future direction by many physicians and administrators who have had their combined systems long enough to enjoy an excellent return on their investment, improved efficiency, improved quality of care documentation and reimbursement.

References

1. Pulley J. The Claims Scrubbers. Government Health IT. November 2008 pp10-14

2. Medical Web Office http://www.medicalweboffice.com (Accessed November 19 2011)

3. Nuance Communications, Inc http://www.nuance.com (Accessed November 19 2011)

4. Gartee, R. (2007). Electronic Health Records: Understanding and Using Computerized Medical Records. Upper Saddle River. Pearson/Prentice Hall.

5. Gartee, R. (2007). Electronic Health Records: Understanding and Using Computerized Medical Records. Upper Saddle River. Pearson/Prentice Hall.

6. Torpey, D. Physician Office Workflows. Lecture. UWF 2009.

7. Allscripts™ http://www.allscripts.com/en/solutions/ambulatory-solutions/ehr/Know/BigQuestions.html (Accessed November 19 2011)

8. E-MDs, inc. http://www.e-mds.com/education/articles/roi.html (Accessed November 19 2011)

9. eClinicalWorks http://www.eclinicalworks.com/ (Accessed November 19 2011)

10. HIT Taken to the Next Level: NYS Practices that have Successfully Adopted EMR & PMS. March 2008. Medical Society of the State of New York: News of New York, pp. 5-6.

11. Electronic Claims. Health Information Management Systems Society: Vantage Point. February 2009. http://www.himss.org/content/files/vantagepoint/pdf/VantagePoint_200902.pdf (Accessed November 19 2011)

12. Sujansky W, Sterling R, Swafford R. Practice Management Systems for Safety-Net Clinics and Small Group Practices: A Primer www.chcf.org (Accessed November 19 2011)

13. Medical Practice Management Software. Capterra. http://www.capterra.com/medical-practice-management-software?srchid=135420&pos=1 (Accessed November 19 2011)

14. EHRScope http://www.ehrscope.com/emr-comparison/ (Accessed November 19 2011)

15. Selecting a Practice Management System. (2007) ACP Online. http://www.acponline.org/running_practice/technology/pms/ (Accessed October 6 2009)

16. Medical Practice Management Buyers Guide. May 2008. BuyerZone. http://www.buyerzone.com/software/mpm/buyers_guide1.html (Accessed November 19 2011)

17. Online Consultant. Selecting a new Physician Practice System? http://www.olcsoft.com/physician_practice_management_software_requirements.htm (Accessed November 19 2011)

18. Practice Management Software. Wikipedia http://en.wikipedia.org/wiki/Practice_management (Accessed November 19 2011)

19. U.S. Healthcare Efficiency Index, National Progress Report on Healthcare Efficiency http://www.ushealthcareindex.com/resources/USHEINationalProgressReport.pdf (Accessed November 19 2011)

20. Centers for Medicare and Medicaid Services http://www.cms.gov/ICD10/Downloads/CMSStatement5010EnforcementDiscretion111711.pdf (Accessed November 23, 2011)

21. Centers for Medicare and Medicaid Services Press Release
https://www.cms.gov/apps/media/press/release.asp?Counter=4183&intNumPerPage=10&checkDate=&checkKey=&srchType=1&numDays=3500&srchOpt=0&srchData=&keywordType=All&chkNewsType=1%2C+2%2C+3%2C+4%2C+5&intPage=&showAll=&pYear=&year=&desc=&cboOrder=date
(Accessed November 23, 2011)

5

Health Information Exchange

ROBERT E. HOYT

ROBERT W. CRUZ

NORA J. BAILEY

Learning Objectives

After reading this chapter the reader should be able to:

- Identify the need for and benefits of health information exchange and interoperability

- Describe the concept of health information organizations (HIOs) and how they integrate with the Nationwide Health Information Network (NHIN)

- Compare and contrast the differences between NHIN Direct and NHIN Exchange

- Enumerate the basic and advanced features offered by HIOs

- Detail the obstacles facing HIOs

- Understand the future direction of HIOs and the impact of Meaningful Use

Introduction

Health information exchange (HIE) is a critical element of Meaningful Use (MU) and integral to the future success of healthcare reform at the local, regional and national level. Exchange of health-related data is important to all healthcare organizations, particularly federal programs such as Medicare or Medicaid for several reasons. The federal government determined that HIE is essential to improve: the disability process, continuity of medical care issues, bio-surveillance, research and natural disaster responses.[1] As a result, the federal government has been a major promoter of HIE and the development of data standards to achieve interoperability. Electronic transmission of data results in faster and less expensive transactions, when compared to standard mail and faxes. If the goal of the federal government was only to promote electronic health records, then the end result would be electronic, instead of paper silos of information. Instead, they have created a comprehensive game plan to share health information among disparate partners. Chapter 1 discusses multiple HITECH programs that support HIE and interoperability.

HIE most commonly involves the exchange of clinical results, images and documents. It should be pointed out that it is also important to share financial and administrative data among disparate entities as well. Table 5.1 lists some of the common types of health related data that are important to exchange among the many healthcare partners.

In this chapter we will begin with important HIE-related definitions and then chronicle the evolution of local, state and national organizations created for HIE.

Table 5.1: Common types of health-related data exchanged

Data	Examples
Clinical results	Lab, pathology, medication data, microbiology reports
Images	Radiology reports; scanned images of paper documentation
Documents	Office notes, discharge notes, emergency room notes
Clinical Summaries	Continuity of Care Documents (CCDs)
Financial information	Claims data, eligibility checks
Medication data	Electronic prescriptions, drug allergies
Performance data	Quality measures such as cholesterol levels, blood pressure, etc.
Public health data	H1N1 outbreak data, immunization records
Case management	Management of the underserved/emergency room utilization
Referral management	Management of referrals to specialists

Definitions

The following are commonly cited definitions related to health information exchange.

- Health Information Exchange (HIE) is the "electronic movement of health-related information among organizations according to nationally recognized standards."[2]

- Health Information Organization (HIO) is "an organization that oversees and governs the exchange of health-related information among organizations according to nationally recognized standards."[2]

- Health Information Service Provider (HISP) is an organization that offers "NHIN participants with operational and technical health exchange services necessary to fully quality to connect to the NHIN."[3]

- Nationwide Health Information Network (NHIN or NwHIN) is a *network-of-networks* that establishes standards, services and policies that define how HIOs will engage in the secure exchange of health information over the internet. NwHIN and NHIN are used interchangeably.

- Opt-In and Opt-Out refers to patient privacy and consent policies; the ability for content creators to determine whether or not the personal health record data they create can be shared as well as with whom. Under an opt-in scenario, no health information can be exchanged unless the patient signs a specific informed consent permitting the sharing of data. Opt-out assumes that consumers grant permission for the exchange of personal health information as part of the broader informed consent that they sign when they receive care from a clinician and the halting of data sharing must be triggered by an action from the patient.

- Push and Pull technology relates to the process by which health information is exchanged through the internet. Push technology refers to clinicians sending (pushing) information to another provider mostly by email or other secure messaging process. On the other hand, pull technology is used whenever a clinician sends an electronic request for health information to a server (for example, a server maintained by a HIE), the server searches for the data, and then responds with a match.

- Regional Extension Centers (RECs) were created under the HITECH Act for the purpose of providing technical assistance, best practice information, and education to support providers' implementation and Meaningful Use of electronic health records. Secondarily, RECs are tasked with supporting and enabling nationwide health information exchange.

- Regional Health Information Organization (RHIO) is "a health information organization that brings together health care stakeholders within a defined geographic area and governs health information exchange among them for the purpose of improving health and care in that community."[1]

Note that the term RHIO is inexact because HIOs do not have to be regional; they can include only one city or an entire state. Furthermore, HIOs are being created to exchange health information solely for Medicaid patients or to focus on uninsured populations. In keeping with these new definitions we will use the acronym HIO when addressing health information organizations and RHIO when addressing specific defined regional HIOs. We will use HIE to describe the act of moving or exchanging health information.

History of the Nationwide Health Information Network

In the early 1990s Community Health Information Networks (CHINs) began appearing across the US. Approximately 70 pilot projects were created but all eventually failed and were terminated.[4] Most were thought to fail due to lack of perceived value and sustainable business plan and immature technology. In spite of this early failure, it became apparent that not only would electronic health records need to be adopted, but there would be a need for new local and regional health information organizations (HIOs) to exchange data and eventually connect to a national health information exchange.

In April 2004 President Bush signed Executive Order 13335 creating the Office of the National Coordinator for Health Information Technology (ONC) and at the same time calling for interoperable electronic health records within the next decade.[5] How that would be accomplished was not stated nor was it known at the time of the executive order. In November 2004 ONC sent out a Request for Information (RFI) asking for input on how the Nationwide Health Information Network (NHIN) should be established. In particular, they wanted to know how the NHIN should be governed, financed, operated and maintained.

Based on input obtained through the RFI, the ONC's 2005 report concluded that the NHIN should "be a decentralized architecture built using the internet linked by uniform communications and a software framework of open standards and policies" and a *network-of-networks*.[6] That meant that there would not be a single centralized data repository of patient health information. Creation of the NHIN would require hundreds of HIOs to be interoperable with thousands of individual healthcare entities. (Figure 5.1) It is important to note that the NHIN is not a separate network; instead, it is a set of standards, services and policies that direct how the secure exchange of health information over the internet will occur.

NHIN Prototype Architecture

In 2005 ONC provided $18.6 million in funding towards the NHIN Prototype Architecture initiative. The purpose of this initiative was to demonstrate that a *network-of-networks* approach without reliance on a centralized network could successfully exchange information between regional HIOs. Contracts were awarded to four contractors (Accenture, Computer Sciences Corporation, IBM and Northrop Grumman) to develop the prototype architectures.

Figure 5.1: NHIN Model (Courtesy ONC)

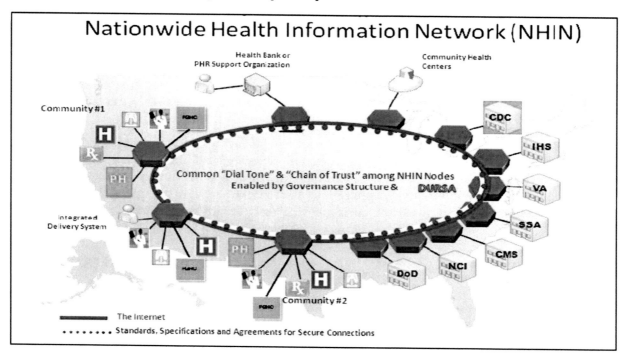

The contractors had to support three *use cases*: (1) EHR-lab use, (2) consumer empowerment and (3) biosurveillance. Additionally, ONC required each contractor to demonstrate the ability to interface with heterogeneous technologies including electronic health records, personal health records, health information organizations (HIOs), and specialized organizations that deal with secondary use of data like public health and research. Interfacing with these diverse users and technologies was intended to demonstrate viability of uniform standards, services and requirements.[7]

The prototype architecture initiatives, which were demonstrated in early 2007, highlighted the issues of security, data standards and technology. Specifics of the four different NHIN architectures can be found in an extensive monograph published by Gartner in May 2007.[8]   According to the report, the contractors validated the following basic NHIN principles:

- A *network-of-networks* approach without a centralized database or services was possible.

- Common standards governing the way exchanges interact with each other are critical.

- The same infrastructure should support both consumers and healthcare providers.

- Consumer controls over the management of information sharing can be implemented.

- An evolutionary approach, rather than a massive replacement or modification of existing health information systems, is desired.

NHIN Trial Implementation

In June 2007 as a follow-up to the successful NHIN Prototype Architecture initiative, the Department of Health and Human Services released a request for proposal (RFP) to participate in phase 2 known as Trial Implementation. Contracts totaling $22.5 million were awarded to nine operational HIOs in October 2007 as part of the NHIN Cooperative: CareSpark, Delaware Health Information Network, Indiana University (Regenstrief Institute), Long Beach Network for Health, Lovelace Clinic Foundation, MedVirginia, New York eHealth Collaborative, North Carolina HealthCare Information and Communication Alliance, Inc. and West

Virginia Health Information Network. In addition, the CDC awarded contracts to study the use of HIOs to support public health information exchange and biosurveillance.[9] In February 2008 ONC announced that 20 federal agencies would connect to the NHIN, as the tenth partner. The Department of Defense and Veterans Administration jointly represent the largest NHIN participants, in terms of patient populations. The other government agencies involved are the Social Security Administration, National Cancer Institute, and the Indian Health Service. This was followed in April 2008 with six additional ONC grants awarded to HealthLINC (Bloomington Hospital), Cleveland Clinic, Community Health Information Collaborative, HealthBridge, Kaiser Permanente, and Wright State University.

Organizations participating in the trial implementation were referred to as Nationwide Health Information Exchanges or NHIEs. This overall effort was known as the NHIN-Connect Gateway (previously referred to as NHIN-C). The purpose of the Trial Implementation was to utilize these NHIEs to test a set of core health information exchange capabilities. The Core Capabilities that were tested by the NHIEs during the Trial Implementation were:

- Look-up, retrieval, and secure exchange of health information

- Application of patient preferences and permissions for sharing of data

- The use of NHIN for other business purposes as authorized by consumers

Eight use cases were developed that would be tested by the NHIEs. For each use case, an interoperability specification which included software services and data structures was developed by the Health Information Technology Standards Panel (HITSP), a public-private sector cooperative partnership. The eight use cases were:

- Authorized release of information to a third-party trusted entity such as the Social Security Administration or Veterans Affairs.

- Bio-surveillance involving the transmission of data to public health entities.

- Consumer control over personally controlled health record information related to registration and medication history.

- Incorporation of laboratory results into an EHR.

- Release of patient health information in response to medical emergencies.

- Transmission of clinical for quality analysis and reporting.

- Specifications for pseudonymization and re-identification.

- Medication management and reconciliation.

In late 2008, HHS hosted a national demonstration of phase 2 of the NHIN, wherein the aforementioned participants exchanged live health information (using test patient data). Specifically, participants tested the ability for a health entity to query a record, compile a patient summary record and send that information back to the person or entity that requested it. The standard used for interoperability by the NHIN was the HITSP C32 specification for Continuity of Care Documents (see chapter on data standards), that included patient demographic and medication information.[10, 11] In summary, the NHIN strategy through the end of 2008 was to establish cross-agency collaboration, identify and develop underlying standards, services and policies, develop gateway tools and participate in trial implementations.

NHIN Exchange

Using the specifications and services developed during the NHIN Trial Implementation, several federal agencies and private sector organizations began exchanging health information in 2009. These current NHIN efforts are known as the NHIN Exchange. The Social Security Administration (SSA), which requests 15 to 20 million medical records each year as part of disability determinations, was selected as the first federal agency to use the NHIN standards and policies to connect to a non-federal entity. In 2009, SSA requested patient information for disability determinations from MedVirginia HIO. The successful exchange with MedVirginia HIO has reduced SSA's time to retrieve disability verification information from an average of 84 days to 46 days. It was announced in February 2010 that the SSA had released $17.4 million to expand their ability to exchange disability-related patient information electronically with 15 additional HIOs.[12] Recognizing that the majority of veterans and active duty service members receive medical care outside their respective systems, the VA and DoD are also involved in the NHIN Exchange. VA's and DoD's health information exchange initiatives with civilian entities will be discussed in another section.

Participants in the NHIN Exchange must submit an application, sign a data use and reciprocal support agreement (DURSA), complete validation testing and be accepted by a coordinating committee. Non-federal entities can participate only through a federally sponsored contract, grant or cooperative agreement. It is anticipated that hospitals, integrated delivery networks, HIOs, state HIOs, Beacon Communities and others will become NHIN Exchange participants in the future.

NHIN CONNECT

The Federal Health Architecture (FHA), which is part of the ONC as well as a collaborative eGovernment Initiative under the Office of Management and Budget (OMB), released the code for an open source NHIN gateway into the public domain in March 2009. Known as CONNECT, the intent of this release was to incentivize and promote adoption of the NHIN by releasing a basic reference implementation of NHIN standard services. With this tool, federal agencies and private sector organizations can use the same gateway to access the NHIN as opposed to each entity developing their own. CONNECT utilizes service oriented architecture (SOA) on a Java-based platform (Figure 5.2). (SOA is discussed further in the chapter on architectures of information systems).

Figure 5.2: Federal Gateway Overview (Adapted from Federal Health Architecture)[14]

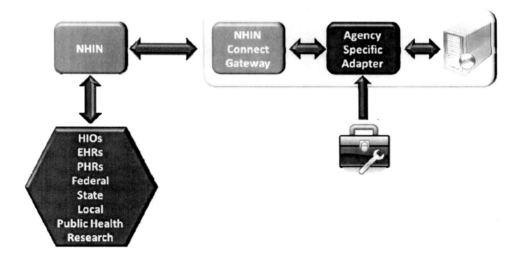

CONNECT is free to download and can be used to: set up a health information exchange within an organization; tie a health information exchange into a regional network of health information exchanges or tie a health information exchange into the NHIN. CONNECT ensures that health information exchanges utilizing CONNECT software are compatible with other exchanges across the country.

Version 2.4 was released in April 2010 and is smaller, requiring less memory and faster. Later in 2010 they offered web services as part of CONNECT to support core services such as secure messaging and patient look-ups. This will allow developers to create new healthcare applications to augment HIE (analogous to iPhone apps). CONNECT releases new versions and updates periodically, with the latest version 3.2.1, released in September 2011.

FHA CONNECT consists of three elements:

- NHIN Gateway implements the core services such as locating patients at other health organizations within the NHIN and requesting and receiving documents associated with the patient. It also includes authenticating network participants, formulating and evaluating authorizations for the release of medical information and honoring consumer preferences for sharing their information

- Enterprise Service Component (ESC) provides enterprise components including a Master Patient Index (MPI), Document Registry and Repository, Authorization Policy Engine, Consumer Preferences Manager, HIPAA-compliant Audit Log and others. This element also includes a software development kit (SDK) for developing adapters to plug in existing systems such as electronic health records to support exchange of health information across the NHIN

- The Universal Client Framework enables agencies to develop end-user applications using the enterprise service components in the ESC[13]

Direct Project (NHIN Direct)

The original concept for the NHIN responded to the mobile nature of our society by recognizing the need of healthcare clinicians to have timely access to patient information across multiple organizations and locations. As initially envisioned, this interoperable exchange of patient data between distant and unaffiliated providers would occur through a *network-of-networks* consisting of HIOs and government agencies. By leveraging existing HIO's and the standards with which they were built, it was believed that these tested and reliable core services would speed the development of the NHIN. The real world implementation of the NHIN, however, has been delayed by issues ranging from technical (deciding on how much of the standard to support), to procedural (agreeing upon vocabularies for proper semantic interoperability, to political (reconciling patient privacy and consent laws between locales).

In response to the complexities of building the *network-of-networks* that have come to light, the NHIN concept has been adjusted by the HIT policy committee's NHIN Working Group to provide more simplistic HIE capabilities via a secure email analogue. This modified version has been renamed NHIN Direct (also referred to by some as NHIN Lite or simply as the Health Internet). The newer model is a set of standards, policies and services that support the exchange of patient data. However, the goal now is a simpler, scalable, more direct exchange to support achieving Stage 1 Meaningful Use. Direct and Connect expose different use cases towards supporting nationwide adoption of secure HIE, similar to the relationship between email and the remainder of services backboned by the internet.

Launched in March 2010, NHIN Direct focused on the deployment of functionality using the lowest cost of entry from a technical and operational perspective. The purpose of NHIN Direct is to supplement traditional fax and mail methods of exchanging health information between known and trusted recipients with a faster, more secure, internet-based method. In other words, Direct helps provider A transmit to provider B patient summaries, reconciliation of medications and lab and x-ray results. Use cases include connecting clinician-clinician, clinician-patient, clinician-health organization, and health organization-health organization exchange. An example of Direct is a primary care physician sending a specialist a clinical summary on a patient that is being referred for care.

The system is based on secure messaging that is managed by a health information service provider (HISP). HISPs can be a healthcare entity, an HIO or an IT organization. The role of the HISP is to provide user authentication, message encryption and maintenance of system security for sending and receiving organizations or clinicians. (see Figure 5.3) By contracting with an HISP, health entities avoid the need for multiple DURSAs or contracts with every provider with whom they exchange data. A list of HISPs will be available on the Direct Project web site.

Figure 5.3: HISP Schema (Courtesy Direct Project)

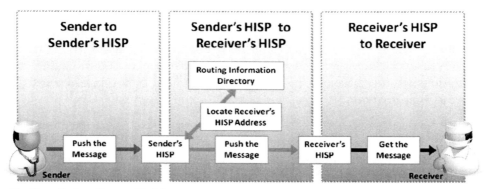

The Direct Project Abstract Model

The Direct Project relies on *push* technology, which refers to sending (pushing) data to a provider. Pushed messages can include attachments, such as lab reports. This push process is much simpler than *pull* technology where a health information exchange database is searched (pulled) for matches to the patient. The HISP maintains a provider directory containing relevant provider demographics including the direct email address that is used to authenticate both the sender and receiver. This process is less complicated than creating and maintaining master patient indices and record locator services that underlie pull technology.

Open source software is being developed so a Direct Project compliant EHR can receive these secure messages and initiate new messages to Direct Project participants. Direct Project providers must obtain a Direct Address and a security certificate. An example of a secure Direct Address would be b.wells@direct.aclinic.org. Direct messages can be received and sent by clinicians regardless of whether they have an EHR. However, most EHR vendors are now working towards Direct Project compliant EHRs so messages can appear in the system's email inbox and output such as Continuity of Care documents (CCDs) can be generated and transmitted seamlessly and securely from one EHR to another. Microsoft HealthVault, a participant in the Direct Project initiative, promotes that it is able to receive a continuity of care document (CCD) via direct secure messaging and parse it into its separate components in the personal health record. By the same token, patients can initiate a secure direct message from HealthVault back to their physician or hospital.

Much work remains to be done before the Direct Project is fully at an operational readiness state. Currently, a number of pilots are underway in Minneapolis, New York, Rhode Island, Tennessee, California, Connecticut, Texas, Missouri, and Oregon. In addition, State health information networks, discussed later in this chapter, are also incorporating Direct Project standards into their service offerings. The current timeline indicates wide-scale deployment will begin in early 2012.[15-17]

Virtual Lifetime Electronic Health Record (VLER)

The VLER initiative followed President Obama's April 9, 2009 direction to the Department of Defense (DoD) and the Department of Veterans Affairs (VA) to create a unified lifetime electronic health record for members of our Armed Services. This initiative is important for several reasons. Injured active duty individuals are likely to receive care in DOD, VA and civilian healthcare systems. Furthermore, about 50% of enrolled VA and DOD patients also receive care routinely in the civilian sector. A unified record integrating military, veterans, and civilian services will streamline record keeping, simplify benefit administration, and ensure continuity of care.

This evolving program has the goal of exchanging health information between DOD, VA and civilian health organizations, using NHIN standards and CONNECT software. As of late 2011 there were eleven pilot sites that will enroll 50,000 to 100,000 veterans consenting to authorize exchange of their records. It is projected that the program will be in place in mid-2012. As an example, the Naval Medical Center San Diego, the VA and Kaiser Permanente are sharing the following types of health related information: allergies, health problems, outpatient medications, lab results, vital signs and immunizations. The goal will be to exchange a Continuity of Care Document (C32 subset) among partners. It is anticipated that VLER will eventually offer additional functionality such as disability tracking, case management, Direct Project capability and benefits status.[18]

The timing of the development of the NHIN and its various components is depicted in Figure 5.4.

Figure 5.4: NHIN and HIE Timeline

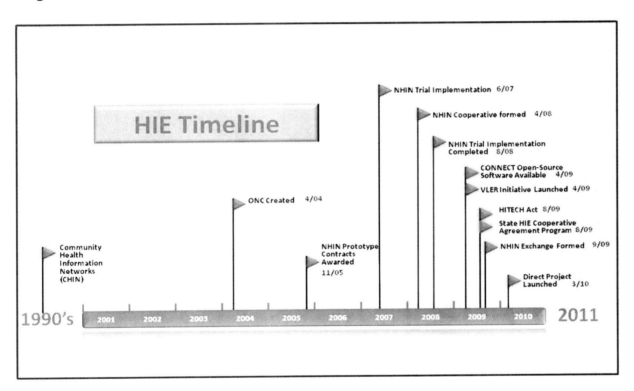

HITECH Act Impact

The 2009 HITECH Act signaled a major federal commitment to expansion of health information technology. Although the HITECH Act focused on incentivizing the expansion of EHRs, it also encouraged the growth of health information exchange through the authorization and funding of the State HIE Cooperative Agreement Program, discussed in a later section. This program closed the state and regional HIE gap by awarding $548 million to 56 state agencies, health information exchange is further supported by incorporating HIE into Meaningful Use stage I objectives necessary for EHR reimbursement.[19]

Health Information Organizations

The late 1990s saw the rise of health information organizations (HIOs) in the United States, largely created with federal startup funds. There was, however, no national game plan as to how to create or maintain HIOs. The National Coordinator for Health Information Technology in 2006 made the following suggestions as to how HIOs might proceed:

- Leverage the internet as the foundation and think web-based

- Build upon existing successes; take advantage of any existing infrastructure

- Have a realistic implementation plan; build incrementally or by phases or modules

- Develop strong physician involvement; involve medical schools and medical societies

- Obtain hospital leadership commitment; much of the information to be shared comes from hospital IT systems

- Do not exclude any stakeholders; HIOs should consist of multiple types of healthcare organizations

- Seek inclusion of local public health officials; the goal is to also develop a public health information network or PHIN

- Obtain support from the business community; vendors who have networking experience will be valuable partners

- Establish a neutral managing partner; a commission or network authority [20]

According to a 2011 national survey there were 85 operational HIOs (actually exchanging clinical information) out of 255 reported HIE entities.[21] It is not known, however, how many HIOs have started and failed. For example, the Santa Barbara County Care Data Exchange was a highly visible HIO that folded in 2007 due to legal, technological and financial issues.[22] An excellent monograph describes the lessons learned from this project.[23] The Pennsylvania RHIO also closed in 2007 due to lack of short and long term financial support.[24]

Most HIOs begin with a collaborative planning process that involves multiple stake holders in the healthcare community. Participation from a broad spectrum of health care entities is necessary for long term sustainability. Potential participants include: insurers (payers), physicians, hospitals, medical societies, medical schools, health informatics programs, state and local government, employers, consumers, pharmacies and pharmacy networks, ambulatory care providers, business leaders, selected vendors and public health departments.

Social capital or an atmosphere of trust is a prerequisite for HIO success. This is particularly true in highly competitive health care regions, where health systems, physician groups, other providers, and payers distrust the motives of the other parties. HIOs are usually complex organizations in which the governing members must reach consensus on governance structure, privacy and security issues, as well as business, technical and legal aspects of HIE. The building of social capital and trust is necessary for sustainability of the HIO.

Multiple functions need to be addressed by an HIO:

- Financing: what will be the sources for short term startup money and on-going revenue? What is the long term business plan? What is the pricing structure?

- Regulations: what data, privacy and security standards will be used?

- Information technology: who will create and maintain the actual network? Who will do the training? Will the HIO use a centralized or de-centralized data repository?

- Clinical process improvements: what processes will be selected to improve? Will the analysis use claims data or provider patient data? Who will monitor and report the progress?

- Incentives: what incentives exist for disparate entities to join?

- Public relations (PR): how will information on the benefits of the HIO be spread to healthcare organizations, physicians and the public?

- Consumer participation: how will the HIO reach out to stakeholders and patients for input?

The planning phase generally takes several years and generally relies on federal and/or state grant support. Upon completion of the planning phase, the HIO is ready to focus on building the technical infrastructure. The web-based infrastructure can be built by local IT expertise or an HIE-specific vendor. HIOs start with simple processes such as clinical messaging (test results retrieval) before tackling more complicated functionality.

Several types of data exchange models exist and determine how data is shared and stored. The following are general categories:

- Federated: decentralized approach where data is stored locally on a server at each network node (hospital, pharmacy or lab). Data therefore has to be shared among the users of the HIO with an import/export scheme

- Centralized: the HIO operates a central data repository that all entities must access

- Hybrid: a combination of some aspects of federated and centralized model

- For further details concerning clinical data exchange models as well as HIOs using these models, we refer you to the article by Just and Durkin. [25]

Table 5.2 outlines some of the pros and cons of the federated and centralized models.

Table 5.2: Pros and Cons of RHIO models (Adapted from Scalese[26])

	Centralized	Federated
Pros	• Simplicity • Data appearance is uniform • Faster access to data • Easier to create	• Greater privacy • Good examples exist • Buy-in may be easier if data is local
Cons	• Higher hardware costs • Higher operating costs • More difficult with very large HIOs	• Data display might not be uniform • Data retrieval delays from others • Potential for node downtime

Although HIOs utilize a variety of web-based infrastructures they tend to utilize the following similar shared services:

- Master patient index (MPI) is a database containing all of the registered patients within the HIO. The MPI assigns a unique patient identifier and uses algorithms to locate the correct patient and any existing records by sorting through a myriad of demographic identifiers. Duplicate records are still a problem in most functioning HIOs

- Record locator service (RLS) directs the inquirer to the physical location of the patient's records once the MPI has identified multiple records. It does not display the patient's results, however.

- Provider directory lists all of the potential data suppliers and users pertinent for the HIO. It is likely to include credentials, address, phone numbers, email addresses and hospital affiliation.

- Data warehouses such as document registry and repository pairs provide both the storage and indexing of patient data accessible via HIE.

The expectation is that HIOs will save money once they are operational. It is presumed that the network will decrease office labor costs (e.g. costs associated with faxing, etc.), improve medical care and reduce duplication of tests, treatments, and medications. Many people feel that insurers are likely to benefit more from HIE than clinicians. It is clear that one of the potential benefits of health information exchange is more cost-effective electronic claims submission. As reported by the Utah Health Information Network, a paper claim costs $8, compared with an electronic claim cost of $1 plus the $0.20 charge by the HIO; therefore a savings of $6.80.[27]

Health Information Organizations may be operated by governmental agencies, private entities or a private-public hybrid organization. They can be for-profit or not-for-profit, however the vast majority are not-for-profit. Operating capital for HIOs in most cases comes from fees charged to participating hospitals, physician offices, labs and imaging centers. Some HIOs charge clinicians a subscription fee (e.g. a flat fee per physician per month), others charge a transaction fee, while others charge nothing. HIOs can address the entire medical arena or simply a sector such as Medicaid patients. HIOs can cover a city, region, an entire state, multiple states or an entire country. Because HIE can be a marketing strategy and important in meeting Meaningful Use as well as new healthcare delivery models such as accountable care organizations (ACOs), integrated delivery networks (IDNs) may be adopting HIE faster than traditional HIOs are being created. Importantly, IDNs can rapidly offer HIE to their networks without the long and difficult process of creating governance and trust between disparate and competitive healthcare organizations.

There are at least four current HIE business models:

- Not for profit HIOs are usually 501(c) 3 tax-exempt organizations that focus on the patient and community and are funded by federal or state funds and rely on tax advantages. An example would be HealthBridge.

- Public utility HIOs are usually created and maintained by state or federal funding. An example is the Delaware Health Information Network.

- Physician and payer collaborative HIOs are created within a defined geographic area and can be either for-profit or not-for-profit. An example is the Inland Northwest Health Services HIE

- For-profit HIOs focus on the financial benefits of exchanging data. An example is the Strategic Health Intelligence HIE.

Furthermore, HIOs can be categorized based on ownership (see Figure 5.5)

HIOs are relatively new so many regions have little experience with the concept and further education is necessary for clinicians and healthcare administrators to convince them to participate in the regional HIO. Studies so far have shown that clinicians and patients are not very knowledgeable about HIOs but support the concept of sharing medical information securely. [28, 29]

There are open source tools available for evolving HIOs. The California HealthCare Foundation donated server software for the master patient index and records locator services. These tools are available through Open Health Tools (OHTs), an international consortium dedicated to open projects across the healthcare information technology domain. Alongside these open source offerings are a wide range of services and toolkits covering the gap from the core services to the edge system nodes in a HIO. This technology assists the EHR vendors attending the yearly Connectathon (interoperability testing event) held at the annual HIMSS conference.[30, 31]

Furthermore, Misys Open Source Solution uses an open source platform for HIE, in spite of the fact that they are a commercial entity. [32] An interesting open source HIE tool (Mirth) is discussed in the info box.[33]

Figure 5.5: Types of HIOs (Courtesy eHealth Initiative 2011 Survey)[21]

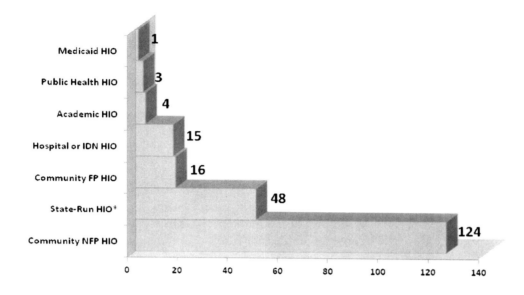

Mirth

Mirth is known as the *Swiss Army Knife* of interoperability. As an open source application it supports all major health data standards and incorporates NHIN Connect. Mirth Meaningful Use Extender (Mux) can operate as a simple HIE to connect area hospitals and medical offices with access via the internet and connectivity to the NHIN. This platform is less expensive than most HIE platforms because it is open source and there is no charge for the software license, however installation, maintenance and support charges pertain. Other exchange products and services are posted on their web site. [55]

According to the 2011 eHealth Initiative survey, which is the authoritative source of information on HIO activity, of 255 HIOs that completed the survey, 24 were termed sustainable: that is, operational, not dependent on federal funding in the past year and at least broke even through operational revenue alone. This compares with 18 sustainable HIOs the year before. Approximately half of operational HIOs charge providers a subscription fee, but multiple revenue models exist. The most common sources of HIO revenue, in order of significance, are: membership fees, federal funds, state appropriations or grants, fees for HIE services and assessment fees.

Most HIOs are not ready for Meaningful Use but many satisfy at least one MU objective such as the exchange of lab results, care summaries, emergency department (ED) episodes or pharmacy summaries. A majority plan to incorporate the Direct Project into their offerings with the most common use case being transitions of care. Eighteen HIOs had behavioral health clinicians contribute data, which is a new trend.

The 2011 eHealth Initiative survey found that HIOs are more likely to adhere to an opt-out policy than to a policy where consumers must actively give permission to the exchange of their health records. Depending on the consent model adopted by the HIO, patient choice can be made by provider, by data type (lab, radiology, etc.), encounter type, by sending organization, by data field or by sensitive data (mental health, etc.). [21] George Washington University reporting to ONC in March 2010 identified the following consent choices:

- No consent: no provision for patient to opt out

- Opt out: patient's data is automatically included but they can revoke permission

- Opt out with exceptions: only select patient data is included (for example, the patient can exclude certain demographic information or sensitive information such as HIV status); patient can withdraw permission to share this limited data set

- Opt in: no patient data is included without permission; patient permits sharing of all or none of information

- Opt in with restriction: patient gives permission to share information but limits which information is included [34]

eHealth Initiative found multiple challenges facing HIOs. Among the challenges identified in the 2011 survey are: developing a sustainable business plan, defining value for providers and consumers, addressing government mandates (e.g. Meaningful Use), addressing technological issues such as integration, governance issues, addressing privacy and security, engaging potential users and accurately linking patient data.

The three most common sources of shared information: hospitals, primary care physicians and community/public health clinics. [21]

Some of the more common HIE functions are listed in Table 5.3

Table 5.3: Health information exchange functionality (Courtesy eHealthinitiative)

Functionality	Functionality
Results delivery	Quality reporting
Connectivity with EHRs	Results distribution
Clinical documentation	Electronic health record (EHR) hosting
Alerts to clinicians	Assist data loads into EHRs
Electronic prescribing	EHR interfaces
Health summaries	Drug-drug alerts
Electronic referral processing	Drug-allergy alerts
Consultation/referrals	Drug-food allergy alerts
Credentialing	Billing

Health Information Organization Examples

The following are local, regional or statewide HIOs that are innovative and successful and can serve as examples to follow.

Utah Health Information Network

- Created in 1993, it has been one of the most financially successful non-profit statewide HIOs in existence.

- They provide administrative (billing and eligibility), clinical (lab reports, medication histories, allergy histories, immunization records and discharge summaries) and credentialing information for physicians and dentists.

- Users can connect to multiple clearinghouses to access payers outside of Utah with one connection.

- Their web site is highly educational and includes their standards and specifications [35]

Nebraska Health Information Initiative (NeHII)

- Statewide roll out began July 2009 and they are now part of the Statewide HIE Cooperative Agreement Program

- Offers a dynamic virtual health record (VHR) for users when they log on that resembles a CCD document

- Also offers a certified EMR-Lite for clinicians who desire an EMR as part of the HIE. Does not include practice management software

- Has a hybrid-federated data storage architecture

- They have experienced a low opt-out rate of about 2%

- 92% of requests are completed in two seconds or less

- E-prescribing available as part of HIE

- Approximately 1,400 physicians are members

- Weekly usage stats are posted on the web site

- $52 per month per clinician for all services [36]

Maine Statewide Health Information Exchange

- One of the largest statewide HIOs

- The network known as HealthInfoNet was launched August 2009 and is now also a Regional Extension Center

- A two year demonstration project with seven healthcare organizations was completed in 2010

- Goal is to link all healthcare entities in the state by 2015 [37]

Indiana Health Information Exchange (IHIE)

Multiple partners helped create this RHIO in 1999, including the Regenstrief Institute that is part of the Indiana University School of Medicine. In 2011, 80 hospitals, long term care and other facilities and 18,000 physicians from within Indiana and adjacent states participated. They opted to use a centralized approach to storing data in one location. They also wanted to be an example for the rest of the country, employ more workers and create more data for better research. The network includes state and local public health departments and homeless shelters. They link to two other HIOs (HealthBridge and Michiana). IHIE is now part of the Central Indiana Beacon Community, the VLER initiative and a Medicare Health Care Quality demonstration project. IHIE is working with a statewide HIE Cooperative Agreement Program to link the state's five HIOs to accomplish statewide HIE. IHIE's disease management program known as QualityHealth First™ supplies monthly reports, alerts and reminders to clinicians, at no charge and is the centerpiece of the Beacon Community program. Their HIO offers the following functions:

- Clinical abstracts

- Physician profiling data

- Results review: radiology results, discharge summaries, operative notes, pathology reports, medication records and EKG reports

- Clinical quality reports

- Research

- Electronic laboratory reports for public health: childhood immunization information and tumor registry

- Syndromic surveillance (looking for syndromes like flu like illnesses to track epidemics or bioterrorism)

- Adverse Drug Event (ADE) detection

- They plan to launch medication reconciliation, diabetes and cholesterol management and breast cancer and colorectal cancer screening [38-41]

HealthBridge

On the following page is a case study of HealthBridge a successful not for profit HIO that is able to provide a multitude of services, compared to many nascent HIOs. Similarly, they have expanded their services to three states or *franchised HIE*. Given their size and maturity, they are also able to point out financial advantages of HIE which is under-reported. Their analysis points out that the manual delivery of lab results costs about $0.75/message compared to $0.12/message for the exchange

They also point out that lab results can be pushed from the HIE directly into their EHR, thus preventing the need for an expensive interface to be built for each lab and hospital. Data is also codified with LOINC, making the data more valuable for quality reporting and analytics. [42]

Claims-based HIOs

SharedHealth. This is a large public-private HIO. They cover 3.5 million privately insured and Medicaid customers and their system is based on claims data. The service is free for participating clinicians and in 2011 they claimed 6,500 users. This is not a true EHR in the sense that it does not include either clinical notes or x-ray results. Some have termed this a payer-based health record (PBHR). They do offer integration with known HIOs as one of their features. In addition:

- Their Clinical Health Record® (CHR) provides medical histories from claims data: patient demographics, problem summary lists, lab results, medication histories, allergies, immunization histories, preventive medicine reminders and electronic prescribing

- They offer a disease registry and population health tools

- Patients can also have access to their CHR

- CHR is being used by East Tennessee State University Family Medicine Department

- They have reported improvements in emergency room utilization, colon cancer screening, pneumonia and flu vaccinations as a result of their HIE [43]

Case Study

HealthBridge is a not-for-profit HIO serving the greater Cincinnati, Ohio as well as parts of Kentucky and Indiana that was founded in 1997. It has been quite successful financially with income not based on federal grants, but rather on monthly subscription fees. HealthBridge provides information exchange for 50 hospitals and 7500+ physicians. They provide access to imaging, fetal heart monitoring and hospital-based EHRs. They were an early NwHIN trial participant and in 2010 they were selected to be a regional HIT extension center and a Beacon Community. Their early technology partner was Axolotl who offered EMR Lite to integrate with their HIE. They have selected Mirth Meaningful Use Exchange (Mirth MUx) as their next interoperability platform. HealthBridge exchanges 3 million messages per month and they have been able to demonstrate an annual return of 5-8% over the past 8 years. Forty nine percent of connections to the HIE are with the EMR Lite option, 38% with other EHRs, 2% print content and 1% faxes. Physicians are not charged with this model for core services. Figure 5.5 demonstrates the architecture used to create the community infrastructure by HealthBridge. They are a HISP and participate in the Direct Project. They also offer workflow redesign and disease registries, data analytics, HIE consulting, quality reporting, public health reporting, syndromic surveillance, claims checks and eligibility verification.[42]

HealthBridge Connected Community Architecture

Availity Health Information Network. This is the first multi-payer based health information exchange. This network uses claims data for patients insured by Blue Cross/Blue Shield of Florida and Minnesota, Health Care Service Corporation, Wellpoint, and Humana with customers in all 50 states. They claim to integrate with EHRs, practice management systems and hospital information systems and most services are available for free. Users can access this site for eligibility/benefit questions, claims clearinghouse, treatment authorizations, referral status, payment collections and to review medications, diagnoses, treatments and lab orders. They claim 600 million transactions per year and offer the following features: eligibility & benefits inquiry, referrals and authorizations, claims submission, claims reconciliation.

- Availity Care Profile® includes:

  - Availity A Continuity of Care Record (CCR) that shows services rendered, lab and x-rays ordered, diagnoses, procedures performed, hospitalizations and immunizations

  - CarePrescribe®, an e-prescribing service

  - An optional patient portal (RelayHealth)

- Availity CareCost®, a cost estimator for patients

- Availity RealMed® revenue cycle management

- Availity CareCollect®, a payment processing service for upfront payments

- Availity CareRead® is a magnetic swipe card with all of the member's ID information [44]

NaviNet. Originally, a large insurance clearing house (reported 800,000 clinicians in their network) they announced in 2009 that they would make their claims-based HIO service free to all state designated entities. A transaction fee would be charged to users who exchange clinical or financial information. They offer access to multiple insurers on one web portal as well as patient access. A patient portal is an option offered by RelayHealth. In 2011 they offered a practice management system and EHR that integrate to their Insurer Connect option and patient portal. This is the first web-based integrated EHR-PMS system that connects to insurers and patients. Cost is unknown. [45]

It is uncertain whether claims-based HIO will catch on and whether new services will be added. Payers stand to gain a lot from electronic data collection and analysis. It should also be noted that the following limitations are associated with this model:

- Model only covers insured patients in the network

- If a patient does not file a claim for a service (pays out of pocket), there will be no record

- A patient can opt-out of sharing data on the HIO

- Patient's employer can opt-out from sharing claims

- Data older than 24 months cannot be retrieved

- Because it is claims-based, there is a lag time between when the test was taken and when the results are posted

Statewide Health Information Exchange Cooperative Agreement Program

In March 2010, fifty-six states, eligible territories, and qualified State Designated Entities (SDE) were funded to build capacity for exchanging health information within and across state lines. This program was created under the HITECH Act to expand HIE/HIO efforts at the state-level while also supporting nationwide interoperability and Meaningful Use. In some states, existing RHIOs expanded to become statewide entities/SDEs. For schema of how this is intended to contribute to HIE overall we refer you to Figure 5.6.

Figure 5.6: Statewide HIE schema (Courtesy ONC State HIE Program[48])

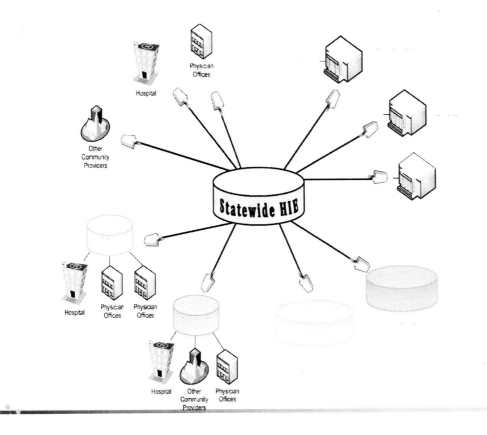

ONC has mandated that State HIE programs ensure that providers will have access to at least one option to satisfy Meaningful Use requirements by the end of 2011. Towards that end, State HIEs and SDEs must address the following priority HIE capabilities:

- E-Prescribing

- Receipt of structured lab results

- Sharing of patient care summaries across unaffiliated organizations

- To receive continued funding, States and SDEs must submit a Strategic and Operational Plans for approval. As of June 2011, all 56 funded entities had received approval for their plans. Each plan must address six key areas:

  o Initiate a transparent process for input from multiple stakeholders

  o Monitor and track Meaningful Use HIE capabilities (e.g. the percent of pharmacies accepting e-prescribing requests)

  o Ensure that the State or SDE framework for privacy and security is consistent with national standards as set by Health and Human Services

  o Address gaps in HIE capabilities to achieve Meaningful Use (example of potential gaps include Medicaid services, rural providers and, small pharmacies)

  o Ensure consistency with national policies and standards including NHIN o   Align HIE strategies with Medicaid and Public Health [46, 47]

The efforts of Florida to meet these ONC mandates are described in the following info box.

Florida Health Information Exchange

The Florida HIE is being managed by the contractor Harris Corp. Its goal is to coordinate the exchange of health information between patients, clinicians, Regional Extension Centers, hospitals, medical offices, HIOs, integrated delivery networks, independent practice associations, long term care facilities, department of health, state immunization registry, federally qualified health centers, labs and electronic prescribing. As of mid-2011 they offered Direct Project connectivity and patient CCD look up services. The backbone for the exchange is based on Mirth® technology, discussed elsewhere in this chapter. Three existing HIOs in Florida are the first participants. Outreach funding is available for rural and financially disadvantaged organizations. http://www.fhin.net/

Health Information Exchange Concerns

There are multiple concerns surrounding the creation and sustainment of a health information organization. The following are just few of the reported concerns:

- Each HIO has a different business model. Do we have enough data to know which model is preferred?

- It is unclear how HIOs will be funded long term. Will funding come from insurers? Clinicians? Employers? Consumers? Federal or state government?

- Approximately $550 million from the HITECH ACT went towards statewide HIE. Have we learned enough at this point to decrease the failure rate?

- Will universal standards be adopted or will different standards for different HIOs prevail?

- Poor cities, states and regions tend to be at a disadvantage. What should be done with geographical gaps in HIOs and what regions should they cover? Should they be based on geography, insurance coverage or prior history?

- Will a NHIN be possible with a low number of sustainable HIOs fail and incomplete adoption of EHRs?

- What are the incentives for competing hospitals and competing physicians in the average city or region to collaborate and share information?

- Will HIOs have to comply with FISMA regulations?

- Will the newest HIPAA regulations (or state personal health information-related laws) become impediments to HIO implementation and operation?

- Opt-in and opt-out patient consent models vary by locality, region, and state. Will one model become standard?

- Is there a strong reason to accredit HIOs?

- How will patient privacy and security rules under Meaningful Use come into play in the HIO domain?

- Very little research has been done to identify which physician specialties are the most frequent requestors of patient data from HIOs. Similarly, little is known about which clinical situations benefit the most from data exchange. This suggests that providers may not value HIE. In the future, will clinicians be comfortable making care decisions based on discrete data elements imported from an external record source?

- Will timely access to patient documentation be realized in the face of technical and procedural hurdles?

- Will physician adoption of the Direct Project standards, in order to meet Meaningful Use Stage 1paradoxically decrease adoption of more formal pull HIE technology?

- How can payers be more consistently involved in support of HIOs? Will providers trust an HIO that is sponsored by or involves payers?

- When will we see more quantitative and qualitative studies to document value and return on investment?

- Will Accountable Care Organizations (ACOs) increase or decrease HIO use?

- Is the current HIO model too complex for success, compared to other models of HIE?

Health Information Organization Resources

It can be argued that creating the technology architecture is the easy part in the life of a HIO. Far more time must be spent planning the governance and financing. It is therefore critical that localities do their homework to research the lessons learned from others who have successfully built a HIO. The following are valuable resources:

- Privacy and Security Solutions for Interoperable Health Information Exchange. Report for the AHRQ, December 2006 [49]

- Guide to Establishing a Regional Health Information Organization. Publication by the Healthcare Information and Management Systems Society. 144 page step-by-step resource. Cost $78 for non-members. [50]

- eHealth Initiative Connecting Communities Toolkit. This non-profit site offers a wealth of web-based information organized into these sections: value and financing module, practice transformation module, information sharing module, technology module and public policy and advocacy module. A resource library and glossary are also part of the toolkit. [51]

- Common Framework: Resources for Implementing Private and Secure Health Information Exchange is published by Connecting for Health that is part of the Markle Foundation. The Framework consists of multiple documents that help organizations exchange information in a secure private manner, with shared policies and technical standards. Using their protocols a tri-state prototype HIO was created. The Common Framework with nine policy guides and seven technical guides is available free for download on their site. [52]

- Characteristics associated with Regional Health Information Organization viability. Authors analyzed data from a large 2008 survey of HIOs. Two factors for success stood out: simplicity in terms of not trying to do too much and early financial commitment from a wide variety of participants. [53]

- Electronic Personal Health Information Exchange. February 2010. Report to Congressional Committees. GAO report on healthcare entities' reported disclosure practices and effects on quality of care. [54]

- Statewide Health Information Exchange. Best Practice Insights From The Field. Bates M, Kheterpal V. March 2010. White Paper. Provides 10 best practices and case studies for those who plan to build a statewide HIE/HIO. [55]

- Secrets of HIE Success Revealed, Lessons from the Leaders. National eHealth Collaborative. July 2011. [56]

Future Trends

While the success of HIOs continues to be uncertain even with extensive HITECH ACT funding, several trends are appearing from the more mature and successful HIOs. First, many are attempting to achieve Meaningful Use by providing HIE to include quality reporting and other advanced functionalities. Second, clinical messaging is being combined with administrative and financial data to give users more of an Expedia.com experience, where multiple disparate functions are available on one web page. It seems likely that we will eventually experience seamless integration of EHRs, practice management systems and claims management as core HIO services. This would offer a single platform to conduct all clinical and financial business and the ability to generate a wide range of reports. Third, we can expect more efforts to use data secondarily for research and as a means of financially supporting HIOs. Fourth, data analytics will likely evolve if the need is perceived and the value proven. Fifth, HIE has no natural or national boundaries. Global Dolphin is a project to exchange medical information between countries [57] while epSOS is European-based interoperability project. Lastly, we can anticipate more mergers of HIE vendors and new vendors appearing if accountable care organizations and Meaningful Use mandate sharing of health information. [58-60]

European eHealth Project

Founded in 2008, epSOS is the European electronic health interoperability project. Its primary purpose is to improve health care of European citizens while travelling abroad through the cross-border exchange of health data. At present 23 European countries participate.

A one-year pilot study is underway. During this pilot, 10 participating epSOS countries are testing the cross-border transmission of patient summary data sets and e-prescriptions. A second phase of testing will address integration of 112 emergency services (similar to the US 911 phone system), integration of the European Health Insurance Card, and patient access to data.

As a pilot project, epSOS is concentrating its efforts on the technical aspects of cross-border interoperability. It is simultaneously addressing the legal, organizational and semantic issues involved with exchanging data between participating nations. http://www.epsos.eu/

Key Points

- Health information exchange is critical for achieving Meaningful Use of electronic health records
- In order to create a Nationwide Health Information Network (NHIN) multiple data standards will need to be reconciled and adopted
- Creating the architecture for a Health Information Organization (HIO) is not difficult; developing the long term business plan is
- Important interoperable demonstrations of the NHIN have taken place with multiple participating civilian and federal partners
- NHIN Direct is a very new fast-track approach to accomplishing Meaningful Use with an uncertain future

Conclusion

Sharing of health-related data is a critical element of healthcare reform and Meaningful Use. Health information exchange among disparate partners is becoming more common in the United States due to evolving HIOs and the NHIN. Federal programs support the creation of exchanges as well as the services, standards and policies that make HIE possible. HIOs are proliferating, largely due to government support but they are often impeded by a lack of a sustainable business model, as well as privacy and security issues. The federal government is moving forward with the Nationwide Health Information Network in an effort to accelerate standards creation and adoption, in spite of the many problems with creating HIOs across the nation and the low adoption rate of EHRs. With the new monies from HITECH ACT for EHRs and HIOs and the new direction of NHIN Direct, the immediate future should be very interesting. At the same time, insurance companies and claims clearinghouses are creating new models based on claims data. Similarly, integrated delivery networks are offering health information exchange as a marketing strategy and so they can participate in new healthcare reform delivery models. It is too early to know what a HIO of the future will look like but it seems clear that we can expect more features and better integration.

References

1. Commonwealth Fund. Perspectives on Health Reform. http://www.commonwealthfund.org/Content/Publications/Perspectives-on-Health-Reform-Briefs/2009/Jan/The-Federal-Role-in-Promoting-Health-Information-Technology.aspx (Accessed July 7 2009)

2. Defining Key Health Information Terms. National Alliance for Health Information Technology www.nahit.org No longer active. (Accessed May 21 2008)

3. Gartner, Summary of the NHIN Prototype Architecture Contracts, A Report for the Office of the National Coordinator for Health IT, May 31, 2007, page 9

4. Soper P. Realizing the potential of community health information networks for improving quality and efficiency through the continuum of care: a case study of the HRSA community access program and the Nebraska Panhandle partnership for HHS. WHP 023A. December 2001. http://www.stchome.com/media/white_papers/WHP023A.pdf (Accessed September 15 2011)

5. Executive Order: Incentives for the Use of Health Information Technology and Establishing the Position of the National Health Information Technology Coordinator

http://www.whitehouse.gov/news/releases/2004/04/20040427-4.html (Accessed February 18 2006)

6. Summary of Nationwide Health Information Network (NHIN). Request for Information (RFI) Responses June 2005. www.hhs.gov/healthit/rfisummaryreport.pdf (Accessed January 30 2006)

7. Summary of the NHIN Prototype Architecture Contracts. www.hhs.gov/healthit/healthnetwork/resources/ (Accessed November 15 2007)

8. Summary of the NHIN prototype architecture contracts. Gartner. http://www.hhs.gov/healthit/healthnetwork/resources/summary report on NHIN Prototype arc hitectures.pdf (Accessed January 3 2008)

9. Monegain B. NHIN contracts awarded to nine exchanges. 10/5/07 Healthcare ITNews. www.healthcareitnews.com Accessed October 6 2007

10. Ferris, N. Federal Agencies begin to build a mini-NHIN. www.govhealthit.com February 2008 (Accessed March 1 2008)

11. Ferris, N. Six more organizations join NHIN demonstration project www.govhealthit.com May 1 2008 (Accessed May 8 2008)

12. Mosquera M. SSA Awards 15 contracts to expand HIE nationwide. Government HIT. www.govhealthit.com (Accessed February 2 2010)

13. Connect Open Source www.connectopensource.org (Accessed March 28 2010)

14. Sankaran, V. The role of SOA on improving health quality http://www.omg.org/news/meetings/workshops/HC-2008/15-02 Sankaran.pdf (Accessed March 26 2010)

15. Halamka J. Introducing NHIN Direct March 10 2010 www.geekdoctor.blogspot.com (Accessed March 23 2010)

16. NHIN Direct. www.NHINdirect.org (Accessed September 12 2011)

17. NHIN Direct. Health IT HHS. http://healthit.hhs.gov/portal/server.pt?open=512&objID=1142&parentname=CommunityPage&par entid=2&mode=2&in hi userid=10741&cached=true (Accessed March 26 2010)

18. VLER Fact sheet VA118-11-R1-0539-002.docx. https://www.vendorportal.ecms.va.gov (Accessed August 31 2011)

19. Kuperman GJ. Health-information exchange: why are we doing it, and what are we doing? J Am Med Inform Assoc 2011;18:678-682.

20. Office of the National Coordinator for Health Information Technology http://www.dhhs.gov/healthit/ (Accessed September 12 2011)

21. Ehealth Initiative. HIE Survey. 2011. http://www.ehealthinitiative.org (Accessed September 12 2011)

22. What Killed the Santa Barbara County Care Data Exchange? March 12 2007. www.ihealthbeat.org (Accessed March 13 2007)

23. The Santa Barbara County Care Data Exchange: Lessons Learned. August 2007. www.cfcf.org. (Accessed September 1 2007)

24. Robinson B. Pennsylvania RHIO to close. June 12 2007 Government Health IT. http://govhealthit.com. (Accessed June 19 2007)

25. Just B, Durkin S. Clinical Data Exchange Models: Matching HIE Goals with IT Foundations. Journal of AHIMA. 2008;79 (2):48-52

26. Scalese D. Which way RHIO? Hospitals and Health Network http://www.hhnmag.com/hhnmag_app/jsp/articledisplay.jsp?dcrpath=HHNMAG/PubsNewsArticle/data/2006June/0606HHN_InBox_Technology2&domain=HHNMAG (Accessed March 3 2007)

27. Sundwall D. RHIO in Utah, UHIN HIMSS Conference Presentation June 6 2005

28. Shapiro JS, Kannry J, KushniruK W, Kuperman G. Emergency Physicians Perceptions of Health Information Exchange. JAIMA 2007;14:700-705

29. Wright A, Soran C, Jenter CA et al. Physician attitudes toward HIIE: Results of a statewide survey. JAMIA 2010;17:66-70 doi 10.1197/JAMIA.M5241

30. Open Health Tools www.openhealthtools.org (Accessed September 12 2011)

31. Vendors Test Open Source HIE Apps. February 26 2009 www.healthdatamanagement.com (Accessed March 2 2009)

32. Misys http://www.misys.com/corp/OpenSource (Accessed September 12 2011)

33. Mirth. www.mirthcorp.com (Accessed September 1 2011)

34. Goldstein MM, Rein AL. Consumer Consent Options For Electronic HIE: Policy Considerations and Analysis. March 23 2010. www.healthit.hhs.gov (Accessed March 24 2010)

35. Utah Health Information Network http://www.uhin.com/ (Accessed September 12 2011)

36. Nebraska Health Information Initiative www.nehii.org (Accessed September 12 2011)

37. Maine Health Information Network. http://www.hinfonet.org/ (Accessed September 12 2011)

38. Indiana Health Information Exchange http://www.ihie.com (Accessed September 12 2011)

39. McDonald CJ et al. The Indiana Network for Patient Care: A Working Local Health Information Infrastructure. Health Affairs 2005; 24:1214-1220

40. Hayes HB. A RHIO that works and pays. Government Health IT. April 2007; pp 38-9

41. QualityHealthFirst http://www.ihie.org/Solutions/quality-health-first.php (Accessed September 12 2011)

42. HealthBridge www.healthbridge.org (Accessed September 12 2011)

43. SharedHealth. www.sharedhealth.com (Accessed September 12 2011)

44. Availity. www.availity.com (Accessed September 12 2011)

45. NaviNet www.navinet.net (Accessed September 12 2011)

46. Statewide HIE Cooperative Agreement Programs. http://www.healthit.hhs.gov/portal/server.pt/community/healthit_hhs_gov__state_health_information_exchange_program/1488 (Accessed September 16 2011)

47. State Health Information Exchange Program. http://statehieresources.org/ (Accessed September 16 2011)

48. 2010 State HIE Leadership Forum Webinair Series. April 15 2010. State Health Information Exchange Leadership Forum. Addressing Key Issues. Building Sustainable HIE Capacity. (Accessed July 15 2011)

49. Privacy and Security Solutions for Interoperable Health Information Exchange. Report for the AHRQ. http://www.healthit.hhs.gov/portal/server.pt/gateway/PTARGS_0...0.../AVAS.pdf(Accessed September 19 2011)

50. Guide to Establishing a Regional Health Information Organization. HIMSS. http://marketplace.himss.org (Accessed September 18 2011)

51. eHealth Initiative Connecting Communities Toolkit www.ehealthinitiative.org (Accessed September 19 2011)

52. Connecting for Health http://www.connectingforhealth.org/ (Accessed September 19 2011)

53. Adler-Milstein J, Landefield J, Jha AK. Characteristics associated with Regional Health Information Viability. J Am Med Inform Assoc 2010;17:61-65

54. Electronic Personal Health Information Exchange. Health Care Entities' Reported Disclosure Practices and Effects on Quality of Care. February 2010. www.gao.gov/new.items/d10361.pdf (Accessed March 10 2010)

55. Statewide Health Information Exchange: Best Practice Insights From The Field. Bates M, Kheterpal V. March 2010. White Paper. http://interest.healthcare.thomsonreuters.com/content/HIEWhitepaper (Accessed March 29 2011)

56. Secrets of HIE Success Revealed, Lessons from the Leaders. National eHealth Collaborative. July 2011. http://www.nationalehealth.org (Accessed August 9 2011)

57. Li J, Zhou T, Chu J et al. Design and development of an international clinical data exchange system: the international layer function of the Dolphin Project. J Am Med Inform Assoc 2011;18: 683-689

58. SureScripts Network. www.surescripts.com (Accessed September 16 2011)

59. Verizon Health Information Exchange. www.verizonbusiness.com/solutions/healthcare (Accessed September 19 2011)

60. Mirth Meaningful Use Exchange www.mirthcorp.com/MirthMUx (Accessed September 16 2011)

6

Data Standards

ROBERT E. HOYT

ROBERT W. CRUZ

Learning Objectives

After reading this chapter the reader should be able to:

- Enumerate the reasons data standards are necessary for interoperability

- Understand the importance of clinical summaries such as Continuity of Care Documents (CCDs) or Continuity of Care Records (CCRs)

- Compare and contrast standards used for electronic health records and Meaningful Use

Introduction

According to the Institute of Medicine's 2003 report *Patient Safety: Achieving a New Standard for Care* one of the key components of a national health information infrastructure will be "data standards to make that information understandable to all users."[1]

In order for electronic health records (EHRs), health information organizations (HIOs) and the Nationwide Health Information Network (NHIN) to succeed there needs to be a standard language; otherwise you have a *Tower of Babel*. We use standards every day but often take them for granted. All languages are based on a semantic language standard known as grammar. The plumbing and electrical industries depend on standards that are the same in every state. The railroad industry had to decide many years ago what gauge railroad track they would use to connect railroads throughout the United States.

Interoperability relies on syntax and semantics. Syntax is a concept that is related to the structure of the communication, e.g. HL7 discussed later in the chapter. Semantics is a concept that denotes meaning of the communication e.g. SNOMED also discussed later in the chapter. Data standards can come in many flavors. Standards that focus on communication between multiple systems are referred to as transport standards. The rules that dictate the format of information as it packaged for transport are known as content standards. Individual segments within a content package are governed by a vocabulary. All of these standards are developed after careful study of real world use cases. [2]

Although we have come a long way towards universal standards, we are not there yet. The progress has been slow in part due to the fact that participation in standards determining organizations is voluntary. Data standards have taken on new significance as a result of Meaningful Use objectives and the need for data sharing.

The next sections will discuss the major data standards and how the standards facilitate the transmission and sharing of data. Not all data standards have been included in the following sections and many standards are still a "work in progress."

Common Data Standards

Extensible Markup Language (XML)

- Although XML is not really a data standard it is a data packaging standard that has served as a structural component for domain specific languages for health information exchange. In order for disparate health entities to share messages and retrieve results, a common data packaging standard is necessary

- XML is a set of predefined rules to structure data so it can be universally interpreted and understood

- XML consists of elements and attributes

- Elements are tags that can envelop data and can be organized into a hierarchy. There are no predefined tags

- Attributes help describe the element

- XML messages have headings and message bodies packaging information by wrapping it in layers of "tags." You must write software to send, receive or display these structures

- Below is a simple example where car-lot is the root element and car is a child element. Each car sibling uses attributes to further define the physical model being represented.[3]

```
<car-lot>
  <car make="Ford" model="Mustang">
    <year>1956</year>
    <vin>9216604</vin >
  </car>
  <car make="Honda" model="Civic">
    <year>1988</year>
    <vin>9335676</vin>
  </car>
</car-lot>
```

Health Level Seven (HL7)

- A not-for-profit standards development organization (SDO) with chapters in 30 countries

- Health Level Seven's domain is clinical and administrative data transmission and perhaps is the most prolific set of healthcare standards

- "Level Seven" refers to the seventh level of the International Organization for Standardization (ISO) model for Open System Interconnection. This serves to communicate that HL7 messaging lives in the application layer of the stack, with subordinate layers serving as items in the overall toolkit

- HL7 provides a set of standards for interactions between healthcare data services

- HL7 is a data standard for communication or messaging between:

    o Patient administrative systems (PAS)

    o Electronic practice management systems

    o Lab information systems (interfaces)

    o Dietary

    o Pharmacy (clinical decision support)

    o Billing

    o Electronic health records (EHRs)

Figure 6.1: HL7 example (The vertical bars, called pipes, separate the fields in this example, while the carets separate components)

```
MSH|^~\&EPIC|EPICADT|SMS|SMSADT|199912271408|CHARRIS|ADT^A04|1817457|
EVN|A04|1999122271408|||CHARRIS
PID||0493575^^^2^ID 1|454721||DOE^JOHN^^^^|DOE^JOHN^^^^|19480203|M
NK1||CONROY^MARI^^^^|SPO||(216)731-4359||EC|||||||||||||||||||||||||||
PV1||O|168~219~C~PMA^^^^^^^^^||||277^ALLEN FADZL^BONNIE^^^^|||||||||||
```

- The most current version of the HL7 standard is 3.0 but version 2.x is still widely in use by all HIT vendors

- HL7 version 2.x separates messages into processable chunks known as segments which contain fields which contain components

- HL7 version 2.x segments are sewn together into messages of a given type (e.g. Admit Discharge and Transfer [ADT] or Pharmacy Administration [RAS])

- HL7 version 2.x messaging is typically performed over minimal lower layer protocol (MLLP)

- HL7 version 3.0 uses XML for packaging its content

- The Clinical Document Architecture (CDA)

    o A HL7 v3.0 content standard that makes documents human readable and machine processable through the use of XML. CDA is used in EHRs, personal health records, discharge summaries and progress notes. CDA delineates the structure and semantics of clinical documents, consisting of a header and body[6,7]

    o In 2007 HL7 recommended (and HITSP endorsed) the use of the Continuity of Care Document (CCD) standard. The CCD is the marriage of the Continuity of Care Record (CCR) (developed by ASTM International) and the CDA (developed by the HL7 organization)

    o A simpler version of CDA is on its way and is referred to as "greenCDA"

- The info box describes the Health Story Project and templated CDA for including narrative notes into EHRs.

Templated CDA

In spite of increasing adoption of EHRs, most patient notes are text and are therefore not discrete data. CDA is a start in the right direction to comply with Meaningful Use.

This HL7 program known as the Health Story Project will match CCD coding patterns and conventions, called "templated CDA." This strategy will help support the transfer of care summaries into an EHR from dictated notes, using CDA templates. [4, 5]

- The CCD and CCR:
  - The electronic document exchange standard for the sharing of patient summary information between physicians and within personal health records
  - The CCD has the advantage over CCR of being able to accept free text and being capable of vocabulary specific semantic interoperability. It contains the most common information about patients in a summary XML format that can be shared by most computer applications and web browsers. It can printed (pdf) or shared as html
  - In 2008 CCHIT required EHRs to generate and format CCD documents using the C32 specification for patient demographic information, medication history and allergies.[8-10] The CCD and CCR are both currently listed as interchangeable content standards for achieving Stage 1Meaningful Use
  - The CCD has 17 data content/component modules as part of the C32 standards as noted in Table 6.1. Each module will have additional data elements
  - Dr. John Halamka has posted a sample CCD as an example for others to view[11]

Table 6.1: Data modules of the C32 standard for the CCD

Data Fields					
Patient Demographics	Advance directives	Functional status	Payers	Vital signs	Social history
Purpose	Problems	Alerts	Results	Medical equipment	Procedures
Medications	Encounters	Family history	Immunizations	Plan of care	

Digital Imaging and Communications in Medicine (DICOM)

- DICOM was formed by the National Electrical Manufacturers Association (NEMA) and the American College of Radiology. They first met in 1983 which suggests that early on they recognized the potential benefits of the storage, sharing, and transmission of digital x-rays.

- As more radiological tests became available digitally, by different vendors, there was a need for a common data standard. Similarly, as more EHRs had PACS functionality, DICOM became the standard for images in EHRs.

- While DICOM is a standard, vendors have modified it to suit their proprietary application resulting in lack of true interoperability

- DICOM supports a networked environment using TCP/IP protocol (basic internet protocol).

- DICOM is also applicable to an offline environment.[12]

- "I do Imaging" is a web site that promotes open source DICOM viewers, DICOM converters and PACS clients.[13]

Logical Observations: Identifiers, Names and Codes (LOINC)

- This is a standard for the electronic exchange of lab results back to hospitals, clinics and payers. HL7 is a *content* standard, whereas LOINC is a *vocabulary* standard.

- The LOINC database has more than 30,000 codes used for lab results. This is necessary as multiple labs have multiple unique codes that would otherwise not be interoperable.

- LOINC is divided into lab, clinical and HIPAA portions.

- The lab results portion of LOINC includes chemistry, hematology, serology, microbiology and toxicology.

- The clinical portion of LOINC includes vital signs, EKGs, echocardiograms, gastrointestinal endoscopy, hemodynamic data and others.

- The HIPAA portion is used for insurance claims.

- As an example:

  o The LOINC code for serum sodium is 2951-2; there would be another code for urine sodium.

  o The formal LOINC name for this test is: SODIUM:SCNC:PT:SER/PLAS:QN (component:property:timing:specimen:scale)

- LOINC is accepted widely in the US, to include federal agencies. Large commercial labs such as Quest and LabCorp have already mapped their internal codes to LOINC

- Other standards such as DICOM, SNOMED and MEDCIN have cross references to LOINC

- RELMA is a mapping assistant to assist mapping of local test codes to LOINC codes

- LOINC is maintained by the Regenstrief Institute at the Indiana School of Medicine.[14] LOINC and RELMA are available free of charge to download from www.regenstrief.org/loinc

- For more detail on LOINC we refer you to an article by McDonald.[15]

EHR-Lab Interoperability and Connectivity Standards (ELINCS)

- ELINCS was created in 2005 as a lab interface for ambulatory EHRs and a further "constraint" or refinement of HL7 standards.

- Traditionally, lab results are mailed or faxed to a clinician's office and manually inputted into an EHR. ELINCS would permit standardized messaging between a laboratory and a clinician's ambulatory EHR.

- Standard includes:

  o Standardized format and content for messages

- o Standardized model for transport of messages
- o Standardized vocabulary (LOINC)
- The Certification Commission for Healthcare Information Technology (CCHIT) has proposed that ELINCS be part of EHR certification.
- HL7 plans to adopt and maintain the ELINCS standard.
- California Healthcare Foundation sponsored this data standard.[16]

IEEE 11073

- Data standards are needed for information to be sent from a medical device to an EHR or hospital information system
- This is a fundamental standard for medical device connectivity and data exchange but is not widely used
- HL7 version 2.x is used for data transfer but only supplies the syntax and not the semantics
- Other initiatives are being developed to solve this interoperability problem
  - o Integrating the Healthcare Enterprise-Patient Care Device (IHE-PCD) Workgroup has developed use case profiles to support integration, alerts and implantable devices
  - o Medical Device Plug and Play Interoperability Program's Integrated Clinical Environment will develop a solution like IHE-PCD that will be based on IEEE 11073
  - o IEC 80001 is standard under development to address devices in a networked environment
  - o Continua Health Alliance focuses on home healthcare devices[17]

RxNorm

- RxNorm is the recommended standard for medication vocabulary for clinical drugs and drug delivery devices, developed by the National Library of Medicine (NLM).
- Each commercial drug vocabulary company e.g. First Data Bank provides medication concept identifiers to the NLM which are then mapped to the concepts in the RxNorm vocabulary.
- Supports interoperability among organizations that deal with clinical drugs.
- The standard includes three drug elements: the active ingredient, the strength and the dose.
- RxNorm is the standard for e-prescribing and will support Meaningful Use.
- RxNorm encapsulates other drug coding systems, such as National Drug Code (NDC).
- The standard only covers US drugs at this point.
- An example of RxNorm: 311642 Methylcellulose 10 MG/ML Ophthalmic Solution.[18]

National Council for Prescription Drug Programs (NCPDP)

- NCPDP is a pharmacy related SDO for exchange of prescription related information
- Script (v10.10) is for communication between physician and pharmacist

- Other standards: batch standard, billing standard, formulary and benefit standard, prescription file transfer standard and universal claim form standard [19]

Accredited Standards Committee (ASC) X12

- A standard for electronic data interchange (EDI) or the computer-computer exchange of business data

- Standard is used in healthcare, transportation, insurance and finance industries [20]

Systematized Nomenclature of Medicine: Clinical Terminology (SNOMED-CT)

- SNOMED is the clinical terminology or medical vocabulary commonly used in software applications, including EHRs

- SNOMED covers diseases, findings, procedures, drugs, etc.; a more convenient way to index and retrieve medical information

- The vocabulary provides more clinical detail than ICD-9 and felt to be more appropriate for EHRs

- SNOMED is also known as the International Health Terminology

- This standard was developed by the American College of Pathologists. In 2007 ownership was transferred to the International Health Terminology Standards Development Organization www.ihtsdo.org

- SNOMED will be used by the FDA and the Department of Health and Human Services

- This standard currently includes about 1,000,000 clinical descriptions

- Terms are divided into 19 hierarchical categories

- The standard provides more detail by being able to state condition A is due to condition B

- SNOMED concepts have descriptions and concept IDs (number codes). Example: open fracture of radius (concept ID 20354001 and description ID 34227016)

- SNOMED CT also defines two types of relationships:

    o "Is a" connects concepts within the same hierarchy. Example: asthma "is a" lung disease

    o "Attribute" connects concepts in different hierarchies. Example: asthma is associated with inflammation

- SNOMED links (maps) to LOINC and ICD-9/10

- SNOMED is currently used in over 40 countries

- EHR vendors like Cerner and Epic are incorporating this standard into their products

- There is some confusion concerning the standards SNOMED and ICD-9; the latter used primarily for billing and the former for communication of clinical conditions [21-23]

- A study at the Mayo Clinic showed that SNOMED-CT was able to accurately describe 92% of the most common patient problems [24]

- SNOMED-CT Example: Tuberculosis

  DE–14800

  . . . .

  . . . .

  . . . Tuberculosis

  . . Bacterial infections

  . E = Infectious or parasitic diseases

  D = disease or diagnosis

MEDCIN®

MEDCIN ® was developed by Medicomp in the 1980s as a proprietary medical vocabulary. In 1997 it was released as a national standard. MEDCIN® cross-references to many of the other standards already discussed. The nomenclature consists of about 270,000 clinical concepts organized into categories: symptoms, history, physical exam, tests, diagnosis and therapy. Each finding is associated with a numerical code, up to seven digits, so the results are structured or codified. Unlike SNOMED, MEDCIN® findings can link to symptoms, exam, therapy and testing. The knowledge base also includes 600,000 synonyms, allowing look-ups under different terms. MEDCIN® is used by several EHR systems, to include the DOD's AHLTA.

The disadvantages of this system are the fact that there is a substantial learning curve to be able to search for all of the necessary MEDCIN® terms in order to create a completely structured note. Second, the note that is created tends to be poorly fluent and not like dictation (Figure 6.2). For that reason, Medicomp developed CliniTalk™ which is a voice to text option that means that a clinician can dictate and the end result is structured data. [25]

Figure 6.2: Simple note created with MEDCIN® (Courtesy MEDCIN®)

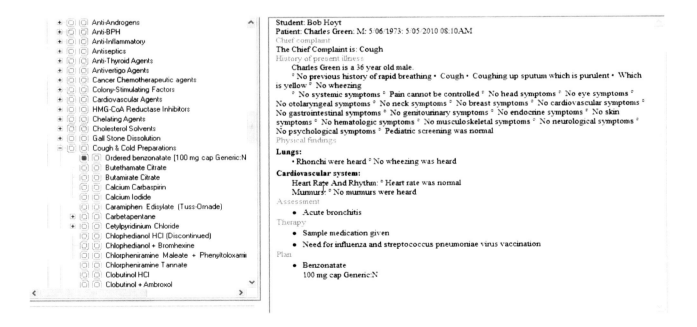

International Classification of Diseases 9th revision and ICD-10

ICD-9 is published by the World Health Organization to allow mortality and morbidity data from different countries to be compared. The basic ICD-9 code, for example diabetes, is 250.00 (three digits). If it covers more detail such as diabetes with kidney complications, it becomes 250.40 (four or five digits). An online web site lists common ICD-9 CM codes. [26]Although it is the standard used to provide a diagnosis for billing over the past 30 years, it is not ideal for distinct clinical diseases.

ICD-10 was endorsed in 1994 but not used in the US. The Federal government set October 2013 as the launch date for ICD-10.[27] ICD-10 will provide a more detailed description with seven rather than five digit codes. ICD-10 CM (clinical modifications) was developed by the CDC for use in all healthcare settings. ICD-10 PCS (procedure coding system) was developed by CMS for inpatient settings only and will replace ICD-9 CM, volume 3. [28] ICD-10 would result in about 141,000 codes instead of the 17,000 codes currently used. A study by Blue Cross/Blue Shield estimated that adoption of ICD-10 would cost the US healthcare industry about $14 billion over the next two to three years. The more digits included generally results in higher reimbursement. [29, 30] Figure 6.3 demonstrates the differences between ICD-9 and ICD-10 formats.

Figure 6.3: ICD-9 and ICD-10 Code Formats

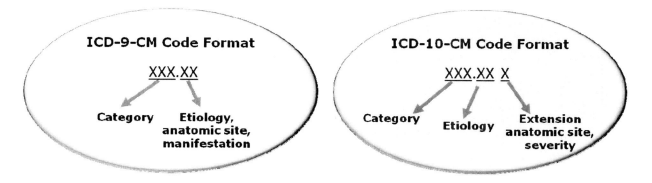

Current Procedural Terminology (CPT)

- CPT is used for billing the level and complexity of service rendered

- Standard was developed, owned and operated by the American Medical Association (AMA) for a fee

- A CPT code is a five digit numeric code that is used to describe medical, surgical, radiology, laboratory, anesthesiology, and evaluation/management services of physicians and hospitals. There are approximately 7,800 CPT codes ranging from 00100 through 99499. Two digit modifiers may be appended to clarify or modify the description of the procedure. The most recent CPT book is version 4

- CPT Codes are published in two versions: the first is for physicians and the second is for hospitals.

- Example CPT code: office visit, established patient, moderate complexity = 99214 [31]

Unified Medical Language System (UMLS)

- Developed by the National Library of Medicine (NLM) to facilitate semantic interoperability among computer systems. Consists of files and software

- UMLS could be used to develop applications, such as electronic health records, dictionaries, language translators, etc.

- UMLS is unique, in that is can link health information, medical terms, drug names, and billing codes across different computer systems

- Access to the UMLS is free but requires registration

- UMLS concept mappings are *as-is* and must be editorialized in order to safely embed in a functional system

- The UMLS Knowledge Sources consist of three components

  o Metathesaurus is a very large multi-lingual vocabulary database with information and relationships (concepts with meaning) related to healthcare: CPT®, ICD-10-CM, LOINC®, MeSH®, RxNorm, and SNOMED CT®

  o Semantic Network categorizes all concepts within the Metathesaurus e.g. anatomical structures, biological function, physical objects, etc.

  o SPECIALIST Lexicon is an English lexicon that includes biomedical terms that support natural language progressing (NLP) [32]

Healthcare Common Procedure Coding System (HCPCS)

- HCPCS are codes used by Medicare and based on CPT codes

- They document medical, surgical and diagnostic services

- HCPCS Level I codes are identical to CPT codes

- HCPCS Level II codes are used by medical suppliers and not clinical services

- More information is available from CMS [33]

Evaluation & Management (E&M) Codes

- In order to bill for a patient visit, ICD-9 CM and CPT codes are selected to best represent the visit. It is up to the clinician to provide documentation to prove the level of the visit

- As an example, if a clinician chooses to select CPT code 99204 for a new outpatient patient visit, they must document that the problems are of moderate 99 to high severity, the physicians spends about 45 minutes face-to-face and the E&M requires these key components: comprehensive history and physical exam and medical decision making of moderate complexity. This implies that an excellent history and physical exam are documented and the problems discussed were moderately complex

- Many EHRs have E&M calculators to help assist the clinician in determining the level of service. This is made easier if templates are used because clicking on history and physical exam elements can calculate an E&M code in the background [34]

- Figure 6.4 shows a typical E&M calculator that is part of an EHR. Note: this is an established patient, the E&M level is in the upper left, the diagnosis and ICD-9 code (462) are in the upper right. Multiple fields are available to input the complexity of the visit so the E&M code can be manually or automatically calculated

Figure 6.4: E&M calculator as part of Healthmatics EHR (Courtesy www.network-systems.com**)**

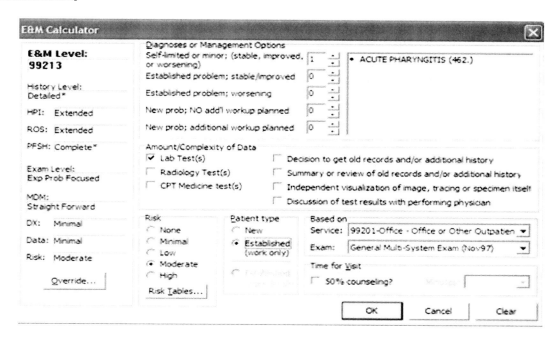

Future Trends

We can expect more data standards as time goes by and further refinement of all existing standards. Subcommittees of ONC are working hard to harmonize data standards to facilitate health information exchange. Decisions will have to be made about what standards will be mandatory for electronic health records. For example, we anticipate that SNOMED will be the primary medical vocabulary of choice for electronic health records in the near future. Meaningful Use stages 2 and 3 may direct all new EHR data standards and vocabularies.

Key Points

- Data standards play a major role in accomplishing interoperability
- We are slowly moving towards industry wide standards, such as the Continuity of Care Document
- Meaningful Use is a strong driver of data standards development

Conclusion

Data standards are critical for interoperability between disparate technologies and organizations. Without agreed upon standards for content and terminology, true semantic interoperability is next to impossible. Multiple standards developing organizations have proposed standards that are being tested, harmonized and updated for application in the field of medicine. Standards are important to exchange clinical data, as well as, administrative and financial data. Standards are essential for exchange of information between electronic health records, health information organizations and the Nationwide Health Information Network. Data standards are on the radar screen as a result of need to meet Meaningful Use and work by groups such as the Health Information Technology Standards Committee.

References

1. IOM. Patient Safety: Achieving a new standard of care. 2004. http://www.nap.edu/books/0309090776.html (Accessed February 22 2005)

2. Kim K. Clinical Data Standards in Health care: Five case studies. California Healthcare Foundation July 2005. http://www.chcf.org (Accessed January 1 2006)

3. Chitnis M, Tiwari P, Ananthamurthy L. Introduction to web services Part 3: Understanding XML http://www.developer.com/services/article.php/1557871 (Accessed June 20 2008)

4. Health Story Project http://www.healthstory.com/ (Accessed October 10 2011)

5. Alschuler L. A Direct Route to Meaningful Use? J of AHIMA Nov-Dec 2010: 52-54

6. Health Level Seven HL7 http://www.hl7.org (Accessed November 11 2011)

7. Dolin RH et al. HL7 Clinical Document Architecture, Release 2. JAIMA. 2006;13:30-39

8. HL7 Implementation Guide: CDA Release 2-Continuity of Care Document. November 1 2007 http://www.himssehrva.org/ASP/CCD_QSG_20071112.asp(Accessed May 1 2008)

9. Bazzoli F. Continuity of Care Document is approved by HL7 and endorsed by HITSP. February 14 2007 http://www.healthcareitnews.com (Accessed May 1 2008)

10. The Continuity of Care Document. Corepoint Health www.corepointhealth.com (Accessed February 29 2010)

11. PPHC Continuity of Care Document http://services.bidmc.org/geekdoctor/summary.xml (Accessed September 24 2011)

12. Digital Imaging and Communication in Medicine. NEMA. 2009 ftp://medical.nema.org/medical/dicom/2009/09_01pu.pdf (Accessed September 15 2011

13. I Do Imaging www.idoimaging.com (Accessed March 24 2010)

14. LOINC www.loinc.org (Accessed January 3 2010)

15. McDonald CJ, Huff SM, Suico JG et al. LOINC, a Universal Standard for Identifying Laboratory Observations: A 5 –Year Update Clin Chem 2003;49 (4):624-633

16. ELINCS. http://www.elincs.chcf.org (Accessed June 8 2009)

17. Day B. Standards for Medical Device Interoperability and Integration. Pat Safe & Qual Health Jan/Feb 2011:20-23

18. RxNorm http://www.nlm.nih.gov/research/umls/rxnorm/index.html (Accessed July 8 2009)

19. National Council for Prescription Drug Programs (NCPDP) http://www.ncpdp.org(Accessed July 8 2009)

20. ASC X12 www.x12.org (Accessed March 14 2010)

21. SNOWMED-CT http://en.wikipedia.org/wiki/SNOMED_CT(Accessed July 8 2009)

22. Joch A. A blanket of SNOMED Federal Computer Week Nov 14 2005;s46-47

23. Lundberg C. SNOMED CT: An Introduction. CAP STS. Presentation. February 22 2010

24. Elkin PL et al. Evaluation of the content coverage of SNOMED-CT: ability of SNOMED Clinical terms to represent clinical problem lists Mayo Clin Proc 2006;81:741-748

25. Medicomp. http://www.medicomp.com(Accessed March 4 2010)

26. ICD-9 Search Engine http://www.eicd.com/EICDMain.htm (Accessed March 4 2010)

27. ICD-9 CMS http://www.cms.hhs.gov/ICD10/01k_2010_ICD10PCS.asp#TopOfPage (Accessed March 20 2010)

28. ICD-10 http://www.who.int/whosis/icd10/othercla.htm(Accessed July 8 2009)

29. Featherly K. ICD-9-CM: An uphill struggle Healthcare Informatics Oct 2004: 14-16

30. Weier S. Letter Encourages Congress to promote ICD-10. http://www.ihealthbeat.orgMay 19 2006 (Accessed February 22 2005)

31. CPT http://www.ama-assn.org/ama/pub/category/3113.html(Accessed November 12 2011)

32. National Library of Medicine. Unified Medical Language System. http://www.nlm.nih.gov/research/umls/ (Accessed October 10 2011)

33. Healthcare Common Procedure Coding System. www.cms.hhs.gov/MedHCPCSGenInfo (Accessed March 24 2010)

34. Evaluation & Management Services Guide July 2009 CMS. http://www.cms.hhs.gov/MLNGenInfo (Accessed March 20 2011)

7

Architectures of Information Systems

ROBERT E. HOYT

ROBERT W. CRUZ

Learning Objectives

After reading this chapter the reader should be able to:

- Understand the internet and World Wide Web
- Discuss why web services are used by HIOs
- List of the components of service oriented architecture
- Understand the importance of networks in the field of medicine
- Compare and contrast wired and wireless local area networks (LANs)
- Describe the newest wireless broadband networks and their significance

Introduction

We believe that the average reader of this book, be they a budding student or seasoned clinician should understand basic architectures and technologies that are commonly part of health information technology. This chapter will focus on three areas: the internet, web services and networks.

The Internet and World Wide Web

Because computers must network in order to exchange data, an internet is a collaboration between many disparate computer networks, and the internet is the largest and arguably most important internet, we will begin with this topic. The internet is a global network-of-networks which use the Telecommunications Protocol / Internet Protocol stack (TCP/IP) as their communications standard. It began as an early (late 1960s) government project with a network known as Advanced Research Projects Agency Network (ARPANET) that could tie together universities securely. The World Wide Web (WWW), on the other hand, operates on top of the internet and was created by Tim Berners-Lee in 1989 with the introduction of the web browser, a software program that allows for connection to web servers over the internet using Hypertext Transfer Protocol (HTTP). The browser is able to request, retrieve, translate and render the content from a remote server on the computer screen for users to view. Web pages are written using Hypertext Markup Language (HTML), making it the language of the web. Here is a simple example of html:

```
<html>
  <body>
    <h1>My First Heading</h1>
    <p>My first paragraph.</p>
  </body>
</html>
```

Interoperability on the internet depends on global standards such as HTTP and the transport standard TCP/IP for transport of HTTP across the internet. Each device (host) must use these standards and must have an Internet Protocol (IP) address. IP addresses can be distributed amongst different tiers of lower layer networks, or "subnetworks." In order for addressing to function properly in the presence of a subnetwork, the machine must both have an IP address and a routing prefix or "subnet mask" (eg. IP address of 192.168.10.1 and subnet mask of 255.254.254.0). Two versions of IP addressing exists today, IP version 4 (IPv4) which has been around for more than 40 years and IP version 6 (IPv6). IPv4 addresses have almost been used up and are being replaced by IPv6. In addition to IP addresses internet servers have domain names based on a domain naming system (DNS). The DNS server maps the IP address to a domain name such as www.uwf.edu. Figure 7.1 demonstrates how this works. Devices can connect to the internet using a modem (dial-up), broadband, wi-fi, satellite and 3/4G cellular data connections.

Figure 7.1: How the internet works to locate web content

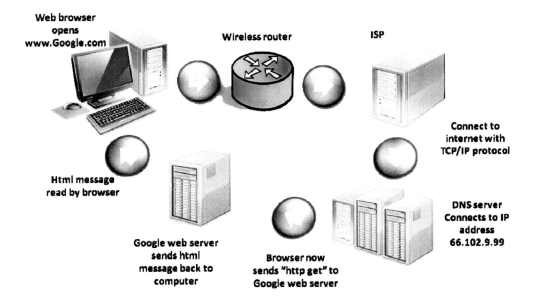

It is useful to think of the internet comprised of two main components, protocols and hardware. The common types of hardware needed are computers, servers, cables and routers. The computer is an end point or client and the connecting points are known as nodes. Computers connect to the internet through an Internet Service Provider (ISP) such as Bell South or AT&T. For example, a viewer uses a web browser (eg. Chrome, Safari, Opera, Internet Explorer, Firefox) to connect to a web site. This sends an electronic request for an IP address to your ISP which routes the request to a DNS server which matches domain name and responds with an IP address. The browser is now capable of sending an HTTP request (again routed through your ISP) to the IP address returned from the DNS request. The result of this set of transactions is an HTTP response with HTML payload from the server which the browser can now render and display on the screen.

In order for this to occur the message must be sent using small packets of information. Packets can arrive via different routes, useful when there is web congestion, and are reassembled back at your computer. Phone calls over the internet (VoIP) and email information are also sent using packets. A router is a node that directs your packets on the internet. The role of ordering these packets and making sure that they make it to their intended recipient in the proper structure is one of the jobs of TCP/IP.

The Uniform Resource Locator (URL) is a specified address to a specific resource. A URL (sometimes also referred to as the Uniform Resource Identifier or URI) can, for example, specify a document provided on the WWW by a web server (e.g. http://www.google.com). The first part of the address is the protocol identifier which indicates what protocol will be necessary to retrieve the resource and the second part, known as resource name, specifies the address of the system to retrieve from as well as the full path to the content to retrieve. The protocol identifier and the resource name are separated by a colon and two forward slashes. As an example, http://uwf.edu/uwfmain/about describes HTTP as the protocol, "uwf.edu" as the server to which the HTTP request will be made, and "/uwfmain/about" as the path to the resource being requested.[1, 2]

Web Services

Prior to the advent of the internet, disparate businesses and health care entities were not able to easily exchange data; instead data resided on a local PC or server and controlled communication links (such as via modem) were required to transport that data to another system. *Web services* are task specific applications which are deployed in a platform independent manner via a series of transactions to and from other web-aware applications/services over a network (such as the internet). Web services can reduce the cost of converting data with external partners, by allowing for a modular component of a larger system to be invoked with little up front effort.

Web services can be broken down into two categories. Representational State Transfer (or RESTful) services are lightweight services which use existing internet infrastructure and World Wide Web (WWW) concepts as their backbone. Simple Object Access Protocol (SOAP) web services utilize a potentially complex series of XML-based ontologies to describe and invoke services over a network. There are obvious pros and cons to each concept, but most often the tradeoff between ease of implementation versus technical depth of field is the main point of comparison struck between the two.

RESTful Services

REST, as a concept, is an aggregate description of the functional model of how HTTP allows for the deployment of the WWW over the internet. It can be utilized to provide non-WWW content delivery over any application protocol, not solely trapped in the realm of the HyperText Transfer Protocol (HTTP). It is important to realize that REST is an architecture, not a standard. As such, there are endless possibilities as to how REST can be applied to act as a service bus. Even though REST itself is not a standard, many standards are utilized when it is used for service interaction. Communication with a RESTful service is a relatively quick to develop process and can utilize any existing content standard for packaging its messaging. Most commonly, a RESTful service will use XML or JavaScript Object Notation (JSON) for this content delivery. RESTful web services require three basic aspects:

- URI (Uniform Resource Identifier). URI is a set of characters defining a specific object, resource, or location. One of the more common uses for a URI is in providing a Uniform Resource Locator (URL) for an object on the WWW. In a RESTful service, a URI can describe the service being invoked or a component within said service.

- Operation Type (GET, DELETE, POST, PUT). These HTTP methods can be extended past their WWW function to provide four different points of access to a RESTful service. If a URI identifies an object, the HTTP operation type defines an accessor method to that object (e.g. GET a list, POST an update, PUT a new record, DELETE a purged record).

- MIME Type (Multipurpose Internet Mail Extensions). MIME is a means of communicating the content type used within a message transferred over the internet. Typically, in a RESTful service, this would be XML or JSON, but it could be any other type.

Web Services using SOAP

SOAP is a protocol standard for interacting with web services. These services require a set of standards for content and a service oriented architecture (SOA) stack, a collection of services. The most common standards used in web services transactions are HTTP, as the internet protocol, with XML as the delivery language (covered in the data standards chapter). SOAP web services require three basic platform elements:

- SOAP (Simple Object Access Protocol): a communication protocol between applications. It is a XML-based platform neutral format for the invocation and response of web services functions over a network. It re-uses the HTTP for transporting data as messages.

- WSDL (Web Services Description Language): a XML document used to describe and locate web services. A WSDL can inform a calling application as to the functionality available from a given service, as well as the structure and types of function arguments and responses.

- UDDI (Universal Description, Discovery and Integration): a directory for storing information about web services, described by WSDL. UDDI utilizes the SOAP protocol for providing access to WSDL documents necessary for interacting with services indexed by its directory.

So how does this work? SOAP acts as the means of communicating, UDDI provides the service registry (like the yellow pages) and WSDL describes the services and the requirements for their interaction. We can begin the process acting as a service requester seeking a web service to provide a specific function. Your application would search a service directory for a function that meets your needs using a structured language. There is a service requester seeking a web service. You search using a search engine that uses a structured language. Once the service provider is located, a SOAP message can be sent back and forth between the service requester and service provider. In reality, a service provider can also be a service consumer so it is helpful to view web services like the *bus* in a PC, where you plug in a variety of circuit boards.

HIOs often require a Master Patient Index (MPI) service to locate and confirm patients and a Record Locator Service (RLS) to identify documentation on those patients. For connecting multiple HIOs you may also require gateways (a network point that acts as an entrance to another network) and adapters (software that connects to applications). [3-5] A valuable recent article, "Improving Performance of Healthcare Systems with Service Oriented Architecture," describes how SOA is the logical backbone for HIOs and electronic health records.[6] Another resource for understanding SOA and healthcare was published in March 2009 by the California HealthCare Foundation, *Lessons from Amazon.com for Health Care and Social Service Agencies.*[7]

Networks

A network is a group of computers that are linked together in order to share information. Although a majority of medical data resides in silos, there is a distinct need to share data between offices, hospitals, insurers, health information organizations, etc. A network can share patient information as well as provide internet access for multiple users. Networks can be small, connecting just several computers in a clinician's office or very large, connecting computers in an entire organization in multiple locations.

There are several ways to access the internet: dial-up modem, wireless fidelity (WiFi), a Digital Subscription Line (DSL), 3G/4G telecommunication, cable modem or T1 lines. The most common type of DSL is Asymmetric DSL (ADSL) which means that the upload speed is slower than the download speeds, because residential users utilize the download function more than the upload function this allows a segmenting of available bandwidth to give the illusion of greater availability. Symmetric DSL is also available and features similar upload and download speeds. Cable modem networks can either be fully coaxial up to a fiber channel node further upstream or can begin with fiber optic transmission to the building, with coaxial cable run internally. Table 7.1 displays data transfer speeds based on the different technologies. Multiple factors influence these speeds, so that theoretical maximum as well as more typical speed ranges are listed.

Table 7.1: Data transfer rates

Transmission method	Theoretical max speed	Typical speed range
Dial-up modem	56 Kbps	56 Kbps
DSL	6 Mbps	1.5 Mbps download/128 Kbps upload
Cable modem	30 Mbps	1-6 Mbps download/128-768 Kbps upload
Wired Ethernet (Cat 5)	1000 Mbps	100 Mbps
Fiber optic cable	100 Gbps	2.5-40 Gbps
T-1 line	1.5 Mbps	1-1.5 Mbps
Wireless 802.11g Wireless 802.11n	54 Mbps 300 Mbps	1-20 Mbps 40-115 Mbps
WiMax	70 Mbps	54-70 Mbps
LTE	60 Mbps	8-12 Mbps
Bluetooth	24 Mbps	1-24 Mbps
3G	2.4 Mbps	144-384 kbps
4G	100 Mbps	10-70 Mbps

Information Transmission via the Internet

Given the omnipresent nature of the internet and faster broadband speeds, the internet is the network of choice for transmission of voice, data and images. It is important to understand the basics of transmission using packets of information. The Internet Protocol (IP) is a standard that segments data, voice and video into packets with unique destination addresses. Routers read the address of the packet and forward it towards its destination. Transmission performance is affected by the following:

- Bandwidth is the size of the pipe to transmit packets (a formatted data unit carried by a packet mode computer network). Networks should have bandwidth excess to operate optimally

- Packet loss is an issue because packets may rarely fail to reach their destination. The IP Transmission Control Protocol (TCP) makes sure a packet reaches its destination or re-sends it. The User Datagram Protocol (UDP) does not guarantee delivery and is used with, for example, live streaming video. In this case the user would not want the transmission held up for one packet

- End-to-end delay is the latency or delay in receiving a packet. With fiber optics the latency is minimal because the transmission occurs at the speed of light

- Jitter is the random variation in packet delay and reflects internet spikes in activity

Packets travel through the very public internet. An encryption technique such as the Federal Information Processing Standard (FIPS) encodes the content of each packet so that it can't be read while being transmitted on the internet. Encryption, however, adds some delay and increase in bandwidth requirements.[8]

Network Types

Networks are named based on connection method, as well as configuration or size. As an example, a network can be connected by fiber optic cable, Ethernet or wireless. Networks can also be described by different configurations or topologies. They can be connected to a common backbone or bus, in a star configuration using a central hub or a ring configuration. In this chapter we will describe networks by size or scale.

Personal Area Networks (PANs). A PAN is a close proximity network designed to link phones computers, PDAs, etc. The most common technology to create a PAN is Bluetooth. Bluetooth technology has been around since 1995 and is designed to wirelessly connect an assortment of devices at a maximum distance of about 30 feet. It does have the advantages of not requiring much power and connecting automatically, but has the disadvantage of being slower, with speeds of 1 Mbps. It operates in the 2.4 MHz frequency range so it can interfere with 802.11g wireless networks. Clearly, the most common application of Bluetooth today is as a wireless headset to connect to a mobile phone, however human interface devices (such as keyboards and mice) are tipping the scales on Bluetooth usage. Many new computers are Bluetooth enabled and if not, a Bluetooth USB adapter known as a dongle can be used or a Bluetooth wireless card. This technology can connect multiple devices simultaneously and does not require "line of sight" to connect. In an office Bluetooth can be used to wirelessly connect computers to keyboards, mice, printers, PDAs and smartphones. This will avoid the tangle of multiple wires. Bluetooth can connect in one direction (half duplex) or in two directions (full duplex). Security must be enabled due to the fact that even though the transmission range is short, hackers have taken advantage of this common frequency. In the near future it is anticipated that low energy Bluetooth devices such as heart monitors will be available with very long battery lives. In addition, faster Bluetooth 3.0 devices are available with speeds in the 24 Mbps range that piggyback on the 802.11 standard.[8]

Local Area Networks (LANs). Generally refers to linked computers in an office, hospital, home or close proximity situation. A typical network consists of nodes (computers, printers, etc.), a connecting technology (wired or wireless) and specialized equipment such as hubs, routers and switches. LANs can be wired or wireless:

- Wired networks. To connect to the internet through your Internet Service Provider (ISP) you have several options.

    o Phone lines can connect a computer to the internet by using a dial-up modem. The downside is that the connection is relatively slow. Digital subscription lines (DSL) also use standard phone lines that have additional capacity (bandwidth) and are much faster network connection than dial up. DSL also has the advantage over modems of being able to access the internet and use the telephone at the same time. Home or office networks can use phone lines to connect computers, etc. Newer technologies include frequency-division multiplexing (FDM) to separate data from voice signals. This type of network is inexpensive and easy to install. Speeds of 128 Mbps can be expected even when the phone is in use. Up to 50 computers can be connected in this manner and hubs and routers are not necessary. Each computer must have a home phone line network alliance (PNA) card and noise filters are occasionally necessary. The downside is largely the fact that not all home rooms or exam rooms have phone jacks

    o Power lines are another option using standard power outlets to create a network. A newer product (PowerPacket®) is inexpensive to install and claims data transfer speeds of 14 Mbps. All that is needed is a power outlet in each room

    o Ethernet is a network protocol and most networks are connected by fiber or twisted-pair/copper wire connections. Ethernet networks are faster, less expensive and more secure than wireless networks. The most common Ethernet cable is category 5 (Cat 5) unshielded twisted pair (UTP). A typical wired LAN is demonstrated in Figure 7.2

    To connect several computers in a home or office scenario, a hub or a network switch is needed. Routers direct messages between networks and the internet; whereas, switches connect computers to one another and prevent delay. Unlike Hubs that share bandwidth, switches operate at full bandwidth. Switches are like traffic cops that direct simultaneous messages in the right direction. They are generally not necessary unless you are running multiple computers on the same network. To handle larger enterprise demands Gigabit Ethernet LANs are available that are based on copper or fiber optics. Cat5e or Cat6 cables are necessary. Greater bandwidth is necessary for many hospital systems that now have multiple IT systems, an electronic medical record and picture archiving and communication systems (PACS).[8]

Figure 7.2: Typical wired local area network (Courtesy Department of Transportation)

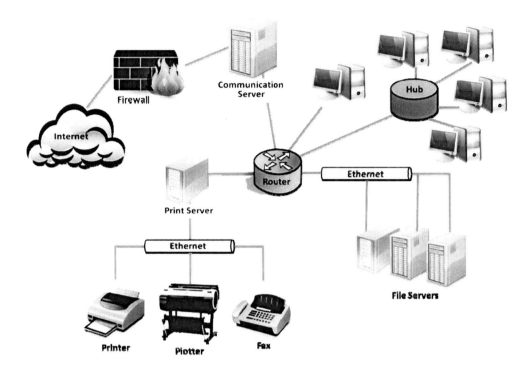

- Wireless (WiFi) networks (WLANs). Wireless networks are based on the Institute of Electrical and Electronics Engineers (IEEE) 802.11 standard and operate in the 900 MHz, 2.4 GHz and 5 GHz frequencies. These frequencies are "unlicensed" by the FCC and are therefore available to the public. Figure 7.3 shows the radio frequency portion of the electromagnetic spectrum where wireless networks function.

Wireless networks have become much cheaper and easier to install so many offices and hospitals have opted to go wireless. This allows laptop/tablet PCs and smartphones in exam and patient rooms to be connected to the local network or internet without the limitations of hardwiring but it does require a wireless router and access points. If an office already has a wired Ethernet network then a wireless access point needs to be added to the network router. A wireless router or access point being used as the hub of communications between systems makes the wireless network be in a state known as infrastructure mode. An ad hoc or peer-to-peer mode means that a computer connects wirelessly directly to another computer and through a routing device.

Figure 7.3: Radio frequency spectrums (Courtesy Commission for Communications Regulation)

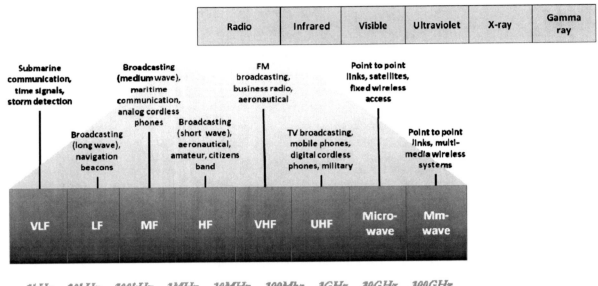

In general, wireless is slower than cable and can be more expensive, but does not require hubs or switches. The standards for wireless continue to evolve. Most people have used 802.11g networks that operate on the 2.4 GHz frequency at peak speeds of 54 Mbps with a range of about 100 meters. Keep in mind that this frequency is vulnerable to interference from microwaves, some cordless phones and Bluetooth. 802.11n is the newest standard that can operate at speeds up to 300 Mbps with a range of about 300 meters. This is accomplished with multiple input/multiple output (MIMO) or multiple antennas that send and receive data much faster and at greater distances. Actual data transfer speeds may be slower than the theoretical max speeds for several reasons. Most modern laptop computers have wireless technology factory installed so a wireless card is no longer necessary. In Figure 7.4 a simple WLAN is demonstrated, with access to the internet over a cable modem and the possibility of both Ethernet and wireless connectivity to different client computers demonstrated. A wireless router will connect the computers, server and printers and has a range of about 90 to 120 feet. For a larger office or hospital multiple access points will be necessary. The network router is usually connected to the internet by an Ethernet cable to DSL or a cable modem. Security must be established using an encryption scheme such as wired equivalent privacy (WEP) 128-bit encryption. Other best-practices for securing a wireless network are the use of a firewall and a unique media access control (MAC) address filtering. Each device on a network has a unique address (MAC) and routers can have security lists which only allow known devices or MACSs into the network.

o An emerging trend for hospitals is to use Voice over IP on a wireless network, referred to as VoWLAN. Hospitals can use existing wireless networks to contact nurses, physicians and employees with any wireless enabled device. Devices such as the Nortel VoWLAN phone or Vocera are frequently used. The chief advantage of this approach is saving local and long distance phone call charges. Using this technology, a patient could directly contact a nurse making rounds so a nurse is not forced to be located near a central nurse-call system. While in the hospital this system could replace landlines, pagers, cell phones and 2-way radios. The downside is that a strong signal is necessary for this system and is more important than that needed with just data.

o Another wireless option is wireless mesh networks that rely on a single transmitter to connect to the internet. Additional transmitters transmit signals to each other over a wide area and only require a power source. Municipalities, airports, etc. are using this type of technology to cover larger defined areas.[9-11]

Figure 7.4: Wireless Local Area Network (WLAN) (Courtesy Home-Network-Help.com)

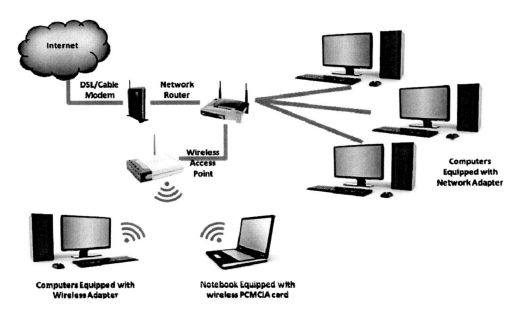

Wide Area Networks (WANs). Cross city, state or national borders. The internet could be considered a WAN and is often used to connect LANs together.

Global Area Networks (GANs). GANs are networks that connect other networks and have an unlimited geographic area. The problem with broadband technology is that it is expensive and the problem with WiFi is that it may result in spotty coverage. These shortcomings have created a new initiative known as Worldwide Interoperability for Microwave Access (WiMax), using the IEEE 802.16 standard. This is a new 4G network will theoretically be about 10 times faster than 3G and it will have much greater capacity which is equally important. The network will be known as a global area network (GAN) with operating speeds in the 54-70 Mbps range. The goal is to be faster than standard WiFi and reach greater distances, such that it might replace broadband services and permit widespread wireless access to the internet. A user would be able to access the internet while traveling or from a fixed location. It would require WiMax towers, similar to cell towers, and a WiMax card in computers, similar to wireless cards. A tower could conceivably cover 3,000 square miles. This would permit a user to connect "non-line of sight" with a weaker lower frequency antenna or line of sight with a higher frequency and stronger dish antennae. WiMax technology also permits voice over IP (VoIP) or phone calls over the internet. WiMax may replace several 3G networks and could be considered a 4G strategy or "broadband wireless." A WiMax network is illustrated in Figure 7.5. In 2008 several companies including Sprint, Intel, Comcast, Cisco and Google collaborated to develop a new company known as Clearwire to promote and develop WiMax technology.[12] In 2011, Clearwire service was available in 80 cities in the US and service was offered via modem for home computers and USB device for laptops. Sprint released a 4G phone (HTC Evo4G) in 2010 that is necessary to take advantage of the 4G network.

The second 4G wireless network being rolled out in several US cities is known as Long Term Evolution or LTE and offered by Verizon, AT&T and US Cellular. Verizon plans to start widespread implementation in 2010 with completion in about five years. Operating in the 700 MHz range it is touted to have maximum download rates of 100 Mbps and upload rates of 50 Mpbs. It is unknown how expensive WiMax or LTE will be over the long haul.[13]

Both 4G wireless approaches transport voice, video and data digitally via Internet Protocol (IP) rather than through switches which will reduce delay and latency. 3G phones will not work on 4G networks.

Figure 7.5: WiMax networks (Courtesy How Stuff Works)

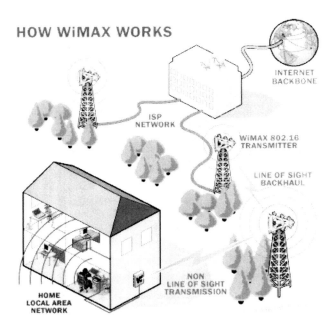

The Commerce Department will establish a lab to test 4G networks so that it can be used for a national public safety network. The lab will specifically test LTE networks because they are supported by a larger number of cellular vendors (80% of cellular market). They plan for the public safety network to be established in the 700 MHz band.[14]

Virtual Private Networks (VPNs). If a clinician desires access from home to his/her electronic health record, one option is a VPN. In this case your home computer is the client and is attached to the network at work by communicating with a VPN server associated with that network. Communicating with nodes over a VPN is akin to working from that network's physical location. The internet can serve as the means of connection with VPN working over both wired and wireless LANs. Authentication and overall security are key elements of setting up remote access to someone else's computer network. (Figure 7.6) "Tunneling protocols" encrypt data by the sender and decrypt it at the receiver's end via a secure tunnel. In addition, the sender's and receiver's network addresses can be encrypted.

Figure 7.6: Virtual private network diagram (Courtesy Cisco)

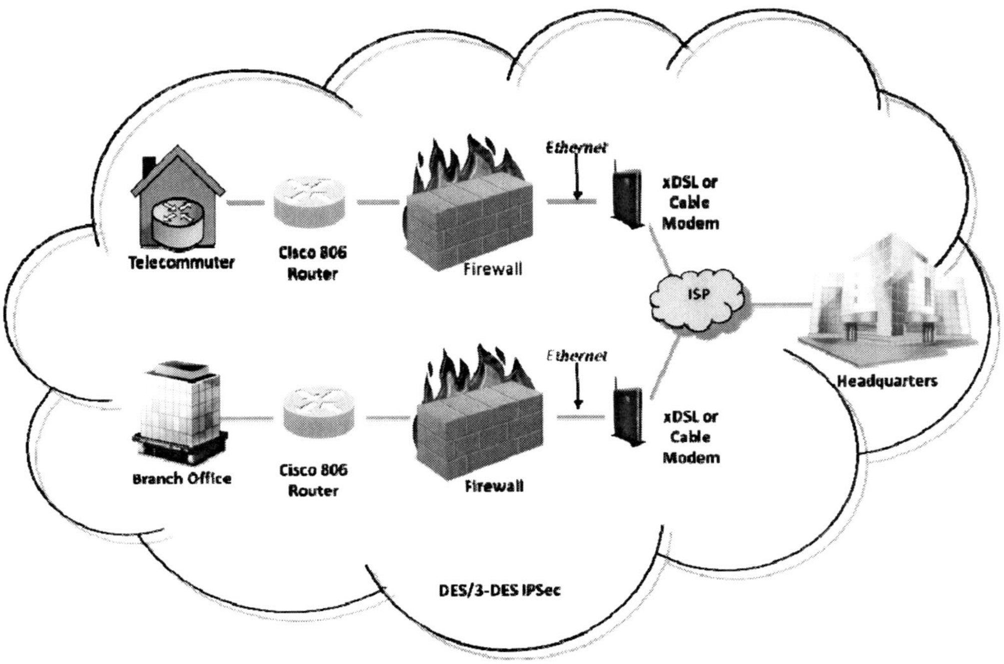

Future Trends

There is a tremendous amount of government and civilian data on the internet but it often is stored in formats such as pdf that are largely non-computable. The Semantic Web will find and interpret the data or create a common framework for data sharing. Data will need to be tagged with metadata tags (data that describes data) and known as "linked data." The World Wide Web Consortium (W3C) has promoted the notion of Resource Description Framework (RDF) as the means to describe documents and images. Another specification will be Web Ontology Language. Better definitions will produce better search results. It will also allow for applications run on the internet to receive and understand data from another application. Sir Timothy Berners-Lee, considered to be the father of the WWW, now promotes the concept of linked data as part of the RDF. He points out that currently you must have application programming interfaces (APIs) and programs like Excel and PDF to interpret data. If the data was linked and encoded by RDF standards, the extra steps would not be necessary. Slowly, organizations such as BestBuy, eBay, BBC and Data.gov have begun participation in web 3.0.[15, 16]

Internet2 is a not-for-profit networking consortium of more than 200 universities, government agencies, researchers and business groups developing applications and a network for the future. The current network is known as *Abilene* and it operates at 10 gigabits per second (100 to 1,000 times faster than Internet1). They have deployed 13,500 miles of dedicated fiber optics as the backbone of the system. They are in the process of upgrading to 8.8 Terabits of capacity and transmission speeds of 100 Gb/s (7,000 times faster than a T1 line). National LambdaRail (NLR) also connects universities (150+) across the nation through fiber optic networks. This unique network connects 28 American cities. Members benefit from using the faster internet to communicate and from the development of interesting middleware. Research is underway to develop programs to support digital video, authentication and security.[17, 18] The Iowa Health System has created a high speed network, known as HealthNet connect for medical sharing described in the info box.[19]

The Institute of Electrical and Electronics Engineers (IEEE) released new standards (802.3ba) for 40 and 100 Gigabit Ethernet network in June 2010. [20] It is anticipated that this network will be used by researchers and others like the Department of Energy (ESnet) who need advanced speed. Verizon recently deployed a

100 Gigabit network in Europe as the first commercial network of its type. There is already talk about bundling 100 Gigabit pipes to create a Terabit Ethernet.

Not all new networks are large and extremely fast. A wireless sensor network protocol known as ANT ™ is available for ultra-low power applications. For example, a sensor can operate for up to three years on a coin cell battery and yet send signals every two seconds every day. The network operates on the 2.4 GHz ISM band. This protocol has wide applicability with wellness, fitness and home monitoring wireless sensors. A variety of chip sets, developer's tools and ANT USB dongles are discussed on the web site. [21]

Ultra-Fast HIE

Created by Iowa Health System it has become a fiber optic network of 60+ Mid-West urban and rural hospital systems based on LambdaRail. The network will provide health information exchange to include large images, education, network services, cloud computing, clinical research and telemedicine.[19]

Key Points

- Most HIOs use web services and service oriented architecture to exchange health related information
- Clinicians who use client-server based electronic health records need to understand wired and wireless office networks
- Wireless networks have become more attractive due to faster speeds and lower prices
- Wireless broadband is here and will make internet access faster and more widely available

Conclusion

Computer systems can use TCP/IP to allow for the transmission of data over multiple different protocols to provide content sharing across a network such as the internet.

Disparate services can be integrated by using web services as part of SOA. This platform provides the greatest degree of flexibility for many businesses, to include HIOs.

Hospitals' and clinicians' offices rely on a variety of networks to connect hardware, share data/images and access the internet. In spite of initial cost, most elements of the various networks discussed continue to improve in terms of speed and cost. Many clinicians' offices will require a network expert to ensure proper installation and maintenance. Wireless technology (WiFi) has become commonplace in many medical offices and hospitals. When wireless broadband (WiMax-LTE) becomes cost effective and widely available it may become the network mode of choice. Network security will continue to be an important issue regardless of mode.

References

1. Internet and World Wide Web. Component 4. Curriculum Development Program. http://www.onc-ntdc.org/home (Accessed September 12 2010)

2. Gargasz Z. How the Internet Works. http://gargasz.info/how_internet_works_i_think.pdf (Accessed September 5 2011)

3. Sankaran V. The role of SOA on improving health quality http://www.omg.org/news/meetings/workshops/HC-2008/15-02_Sankaran.pdf (Accessed March 26 2010)

4. Web services. http://www.service-architecture.com/web-services/articles/index.htm (Accessed June 1 2008)

5. Ananthamurthy L. Introduction to web services. www.developer.com/services/article.php/1485821 (Accessed June 1 2008)

6. Juneja G, Dournaee B, Natoli J, Birkel S. InfoQ. March 7 2008 www.infoq.com/articles/soa-healthcare (Accessed April 21 2009)

7. Lessons from Amazon.com for Health Care and Social Service Agencies. March 2009. California Healthcare Foundation www.chcf.org (Accessed March 20 2009)

8. Gemmill J. Network basics for telemedicine J Telem and Telecare 2005;11:71-76

9. Smith C, Gerelis C. Wireless Network Performance Handbook. McGraw-Hill, Columbus, Ohio, 2003

10. Smith JE. A primer on wireless networking essentials. EDI. www.ediltd.com (Accessed May 24 2009)

11. Lewis M. A Primer on Wireless Networks. Fam Pract Management. Feb 2004. www.aafp.org (Accessed May 25 2009)

12. WiMax www.wimax.com (Accessed May 22 2009)

13. LTE. Gizmodo Feb 18 2009 and Mar 11 2009 www.gizmodo.com (Accessed May 25 2009)

14. Jackson W. New Lab to Pub 4G to the Test. Government Computer News. January 18 2010. p 7

15. Moore J. Waiting for Web 3.0. Federal Computer News. July 20 2009. www.fcw.com (Accessed August 3 2009)

16. Jackson J. The Web's next act. Government Computer News November 9 2009. p.20-29

17. Internet2 www.internet2.edu (Accessed November 7 2011)

18. National LambdaRail www.nlr.net (Accessed November 7 2011)

19. HealthNet connect www.healthnetconnect.org (Accessed November 7 2011)

20. IEEE www.ieee.org (Accessed November 4 2011)

21. ANT Wireless Sensor Network Protocol. www.thisisant.com (Accessed November 7 2011)

8

Health Information Privacy

BRENT HUTFLESS

Learning Objectives

After reading this chapter the reader should be able to:

- Describe privacy and security measures that are part of HIPAA, HITECH Act, and Meaningful Use and how they fit into the Nationwide Health Information Network

- Recognize the importance of data security and privacy as related to public perception, particularly in regards to data breach and loss

- Identify the benefits and pitfalls of local vs. Software-as-a-Service (SaaS) technical security solutions

Introduction

The Health Insurance Portability & Accountability Act (HIPAA) passed in 1996 laid much of the groundwork for the privacy and security measures being adopted within health information technology (HIT) today. The original intent was to direct how patient data was used and made available when patients switched physicians or insurers, and included two major rules covering privacy and security of that data. The American Recovery and Reinvestment Act of 2009 (ARRA), and the HITECH Act which accompanied it, both brought about changes designed to improve privacy and security measures required by modern technologies and close loopholes within the original law. The final rule for Meaningful Use in 2010 also impacted a number of health IT implementations, ultimately leading to a rush to meet deadlines for financial incentives and avoid payment penalties. Before discussing the current state of privacy and security regulation and intent, a primer on HIPAA is needed to show what the original law provides.

HIPAA for the Consumer

Certain organizations, known as *covered entities*, are required to follow the HIPAA Privacy Rule: [1]

- Health Plans

    o Health insurers

    o HMOs

    o Company health plans

    o Government programs such as Medicare and Medicaid

- Health Care Providers who conduct business electronically

    o Most doctors

- o Clinics
- o Hospitals
- o Psychologists
- o Chiropractors
- o Nursing homes
- o Pharmacies
- o Dentists
- Health care clearinghouses

A number of organizations *do not* have to follow HIPAA law despite using personal health data: [1]

- Life insurers
- Employers
- Workers compensation carriers
- Many schools and school districts
- Many state agencies like child protective service agencies
- Many law enforcement agencies
- Many municipal offices

For those organizations that are required to abide by HIPAA, patient data and personal information must be protected according to the Security Rule. Protections apply to all personal health information (PHI), whether in hard copy records, electronic personal health information (ePHI) stored on computing systems, or even verbal discussions between medical professionals. Covered entities must put safeguards in place to ensure data is not compromised, and that it is only used for the intended purpose. The HIPAA rules are not designed to and should not impede the treatment of patients.[2] Covered entities must comply with certain consumer rights; specifically a patient may: [1]

- Ask to see and get a copy of their health records
- Have corrections added to their health information
- Receive a notice that discusses how health information may be used and shared
- Provide permission on whether health information can be used or shared for certain purposes, such as for marketing
- Get reports on when and why health information was shared for certain purposes
- File a complaint with a provider, health insurer, and/or the U.S. Government if patient rights are being denied or health information is not being protected

HIPAA Privacy for Covered Entities

Covered entities have a significant responsibility to protect the privacy and security of patient data and personal information. The U.S. Department of Health & Human Services (HHS) has an excellent website, www.hhs.gov/ocr/privacy/hipaa/understanding/coveredentities/index.html designed to serve this population and inform entities about subjects ranging from patient consent, incidental disclosures, and contracts with business associates to the proper disposal of protected information. For detailed information regarding the HIPAA Privacy and Security rules, HHS, the Office of Civil Rights, and others provide formalized guidance. The following is a summary of highlights.

The Privacy Rule strictly limits how a covered entity and their business associates can use patient data, but there is a method that can be employed to use and release the data without restrictions. The Privacy Rule mandates that organizations de-identify the data by removing 18 identifiers, which reasonably precludes the resulting information from being attributed to a patient. The 18 identifiers follow: [3]

- Names.
- All geographic subdivisions smaller than a state, including street address, city, county, precinct, ZIP Code, and their equivalent geographical codes, except for the initial three digits of a ZIP Code if, according to the current publicly available data from the Bureau of the Census:
  - The geographic unit formed by combining all ZIP Codes with the same three initial digits contains more than 20,000 people.
  - The initial three digits of a ZIP Code for all such geographic units containing 20,000 or fewer people are changed to 000.
- All elements of dates (except year) for dates directly related to an individual, including birth date, admission date, discharge date, date of death; and all ages over 89 and all elements of dates (including year) indicative of such age, except that such ages and elements may be aggregated into a single category of age 90 or older.
- Telephone numbers.
- Facsimile numbers.
- Electronic mail addresses.
- Social security numbers.
- Medical record numbers.
- Health plan beneficiary numbers.
- Account numbers.
- Certificate/license numbers.
- Vehicle identifiers and serial numbers, including license plate numbers.
- Device identifiers and serial numbers.
- Web universal resource locators (URLs).
- Internet protocol (IP) address numbers.
- Biometric identifiers, including fingerprints and voiceprints.
- Full-face photographic images and any comparable images.
- Any other unique identifying number, characteristic, or code, unless otherwise permitted by the Privacy Rule for re-identification.

Covered Entity Permitted Uses and Disclosures of patient data according to the Privacy Rule: [4]

- To the individual
- For treatment, payment or health care operations

- Uses and disclosures with opportunity to agree or object
  - Facility directories
  - For notification and other purposes
- Incidental use and disclosure
- Public interest and benefit activities
  - Required by law
  - Public health activities
  - Victims of abuse, neglect or domestic violence
  - Health oversight activities
  - Judicial and administrative proceedings
  - Law enforcement purposes
  - Decedents
  - Cadaveric organ, eye, or tissue donation
  - Research
  - Serious threat to health or safety
  - Essential government functions
  - Workers' compensation
- Limited data set

Administrative requirements were established by the Privacy Rule to ensure that all covered entities, regardless of size or organization, met a minimum standard for protecting patient privacy and permitting patients to exercise their rights. The scope of the solutions for each of the requirements is left up to the individual organization, but the required categories are as follows: [4]

- Develop and implement written privacy policies and procedures
- Designate a privacy official
- Workforce training and management
- Mitigation strategy for privacy breaches
- Data safeguards - administrative, technical, and physical
- Designate a complaint official and procedure to file complaints
- Establish retaliation and waiver policies and restrictions
- Documentation and record retention - six years
- Fully-insured group health plan exception

HIPAA Security for Covered Entities

Security techniques and safeguards are discussed in the next chapter due to recent changes in HIPAA through the ARRA and HITECH Act, in addition to Meaningful Use recommendations. There are three safeguards that are required by the HIPAA Security Rule that serve as the foundation for these changes, and they are as follows: 5

Administrative Safeguards

- Security management processes to reduce risks and vulnerabilities
- Security personnel responsible for developing and implementing security policies
- Information access management - minimum access necessary to perform duty
- Workforce training and management
- Evaluation of security policies and procedures

Physical Safeguards

- Facility access and control limiting physical access to facilities
- Workstation and device security policies and procedures covering transfer, removal, disposal, and re-use of electronic media

Technical Safeguards

- Access control that restrict access to authorized personnel
- Audit controls for hardware, software, and transactions
- Integrity controls to ensure data is not altered or destroyed
- Transmission security to protect against unauthorized access to data transmitted on networks and via email

Data Storage and Defining Covered Entities

When the Health Insurance Portability & Accountability Act was passed in 1996, most healthcare organizations were still entrenched as paper-based systems. As technology evolved over the past decade, so too did the methods that healthcare entities used to share and store medical information. Electronic billing, patient records and personal data storage was becoming a more common practice, but high profile cases of data loss were increasingly in the news, as occurred when the VA lost a laptop containing more than 26 million veteran records, and later lost another with an additional 38,000 records.[16]

As a result of these and similar breaches, the HIPAA security standard enacted in 2003 needed amending to clarify the requirements for storing and sharing ePHI. In late 2006, HHS released its HIPAA Security Guidance, which identified various forms of remote use, data storage and the requirements for handling and reporting ePHI by covered entities.[17] While it was hoped that this guidance would lead to increased protections, loopholes would remain that would need to be addressed by the ARRA.

By 2009, HIPAA faced new challenges regarding the definition of covered entities. Major software companies had begun aligning new products to the burgeoning field of HIT with a variety of fee-based, open-source, and free solutions and services. Industry leaders Microsoft and Google both offered consumers no-cost, web-based electronic health records that allowed users to share information with physicians, hospitals and pharmacies, and stored vast amounts of medical and personal data. Although this initially appeared to be a wonderful development, both companies asserted that HIPAA did not cover them.[18] Neither giant has come to dominate the no-cost market, and Google Health has left the market as of January 2012.

Recent changes to HIT on a national scale will further the impact of privacy and security on the medical field. Medical organizations and physicians have been bound by HIPAA regulations for more than a decade, but compliance with HIPAA has the potential to impact the financial bottom line beyond fines and penalties now that the new standards have been adopted. The Health IT Policy Committee's recommendations "that CMS withhold meaningful use payment for any entity until any confirmed HIPAA privacy or security violation has been resolved" have tied adherence to a financial obligation.[12] Payment withholdings would be in addition to any potential fines and penalties attributed to the HIPAA privacy or security violations. Those earlier recommendations were later reinforced throughout the Meaningful Use rule published in the Federal Register.[13]

HIPAA, HITECH elements of the ARRA of 2009, and Meaning Use standards all serve to protect privacy and implement security consistency, but these tools alone are not enough to protect the systems, networks and data shares necessary for a national healthcare system. To be able to protect patient data and share medical records securely, other measures must be put in place. Unfortunately, there are far too many examples of what can happen to patient data if not treated appropriately.

Medical Privacy and Security Stories in the News

When assembling material for a privacy and security chapter in 2010, a number of high-profile security breaches in the Healthcare community were just coming to light. The FTC had just notified nearly 100 organizations of data compromise they incurred due to P2P file sharing; losses involving "health-related information, financial records, and drivers' licenses and social security numbers--the kind of information that could lead to identity theft," according to FTC Chairman Jon Leibowitz.[6] Meanwhile, Blue Cross and Blue Shield of Tennessee was still notifying nearly half a million members of identifiable data theft that occurred when 57 hard drives were stolen from one of its facilities. [7] This notice came just one day after Connecticut's Attorney General filed a lawsuit against Health Net of Connecticut, charging HIPAA privacy and security violations stemming from Health Net's loss of 1.5 million member records. [8]

Case Study

Nemours, a multi-state peadiatric health system, announced in 2011 that three encrypted computer backup tapes containing patient and employee financial information for 1.6 million people were missing. The tapes were stored in a locked cabinet and appeared to have been lost during a renovation. Nemours notified appropriate sources in a timely manner and is offering cyber-victims a year of credit monitoring and identity protection services. Nemours plans to move all backup tapes to a remote secure facility.

www.examiner.com/identity-theft-in-nashville/nemours-loses-data-of-about-1-6-million

---

Barely a year later, Health Net revealed once again that they had lost health records when servers were stolen from a California facility, this time numbering 1.9 million – but the organization also waited 3 months to reveal the loss.[9] Privacy Rights Clearinghouse, a consumer privacy organization that tracks breaches, also noted large data breaches and record losses for Sutter Health and Tricare contractor SAIC, with 4.2 million and more than 5 million records respectively. [9] The Ponemon Institute has issued its second annual study showing that health care industry data breaches have increased 32 percent over 2010, and place the average total cost to the industry at $6.5 billion each year.[10] These and other high profile breaches illustrate the importance of securing these records systems and the data they contain. Data security is vital for winning the public trust, the key component to long term success of an electronic health data management system.

Beyond data loss, health-related technology and software services continue to make the news and show that market dominance is no guarantee for success. Two years ago, Microsoft and Google were squaring off in the new consumer-driven, no-cost, personal health record market. Google Health and Microsoft HealthVault were presenting challenges to newly-passed security and privacy laws, with both corporations asserting that neither was bound by HIPAA regulations. Going into 2012, Google Health is no longer offered and personal records are being phased out over the coming year, while competing Microsoft's HealthVault rests on uncertain ground with Redmond's recently announced partnership with GE Healthcare.[11]

HIPAA, Meaningful Use, and the HITECH Act

Many of the core concepts surrounding HIPAA were introduced at the beginning of the chapter, but HHS maintains an excellent 25 page summary of the HIPAA Privacy Rule in PDF format (www.hhs.gov/ocr/privacy/hipaa/understanding/summary/privacysummary.pdf) for quick reference. It is important to remember that the Privacy and Security Rules established through HIPAA were designed for the healthcare system and processes that were in place in the mid-1990s. This system is transforming to technology-driven solutions through the use of electronic medical records (EMRs), clinical decision support systems (CDSS) and other solutions broadly known as Health Information Technology (HIT). Many of the changes are due to a convergence of driving forces; current and imminent government regulations, the need to cut rising insurance costs, calls for healthcare reform, improved technical capability, advanced software solutions, and a higher expectation from consumers to control and manage their own healthcare information.

Today, technology has the opportunity to revolutionize health care, changing hospitals and clinical environments from isolated and unconnected islands of patient treatment records and knowledge to an interconnected system of healthcare. Hospitals and practitioners are taking note of studies advocating the use of electronic health records (EHR) and related systems that have the capability to quickly retrieve patient data and records, saving time, preventing duplication of treatment efforts, reducing drug interactions and contraindication situations; generally improving patient care and reducing administrative costs associated with paper records.[14] However, with new opportunities there often come new risks, and in the case of medicine, an escalating chance to violate HIPAA privacy and security regulations. The number of companies and organizations that are offering data solutions for patients and providers is growing exponentially, increasing the challenge of finding a solution that meets the new reporting, use and billing requirements.

Challenges for the Nationwide Health Information Network

The ARRA provides financial incentives for EHR adoption and use, aiding hospitals, clinics and physicians in the push toward meeting the evolving requirements for electronic capture and tracking of patient data outlined in the Meaningful Use definitions.[12, 20] As pointed out earlier in the stories of data breaches and subsequent lawsuits, a growing number of healthcare organizations are experiencing adoption challenges that impact patients as well as their reputation and bottom line. Perhaps because of this, some of the Stage 2 Meaningful Use criteria are being changed. Despite being little more than a year old, the Stage 1 criteria surrounding security and privacy concerns is already being supplanted by new recommendations by the HIT Standards Committee Privacy and Security Workgroup. This workgroup's most recent recommendations are included in Table 8.1.[13]

Several of the recommendations put forth by the committee are related to data encryption, secure messaging and email, as well as authentication and audit requirements. A number of the proposed standards, for instance the National Institute of Standards and Technology (NIST) special publications 800-63 and 800-53, align to national standards adopted by industry, departments in the federal government, and other industry compliance initiatives, including the FIPS 140-2 encryption standard. [13] While not specifying or mandating particular solutions or product offerings, these newest criteria offer clearer guidance than the recommendations published for Stage 1.

Table 8.1: Stage 2 Privacy and Security Recommendations (Courtesy of HIT Standards Committee Privacy and Security Workgroup)

IWG Ref	HITPC Proposed MU Stage 2 Objective / Measure & Direction to HITSC	HITSC Privacy and Security Workgroup Recommendations	
		Recommended New or Revised Certification Criterion	Recommended Standard(s) and/or Implementation Specifications
46	**Privacy and Security** (NEW) Single Factor Authentication Consumer Web-Based Application	EHR must be able to authenticate the identity of an authorized patient or their personal representative using single-factor authentication (or stronger) based on the standard specified	Standard: NIST SP 800-63 (single-factor authentication)
26	**Secure Messaging** EPs: (NEW) Patients are offered secure messaging online and at least 25 patients have sent secure messages online	EHR must provide the capability to send messages to, and receive messages from, patients using a mechanism that assures that (1) the identity of the patient is authenticated, (2) the identity of the EHR is authenticated, and (3) message content is encrypted and integrity protected.	Example Standards: FIPS PUB 140-2. ANNEX A; IETF RFC 2246 (TLS 1.0); SMTP/SMIME Implementation Spec: NIST SP 800-52 (TLS); NwHIN Transport Specifications
47	(NEW) **Audit Trails for Access to Consumer Web-Based Application**	Covered by general audit criteria	
48	**Privacy and Security** (NEW) Establish Data Provenance for Consumer Web-Based Application	EHR must be able to create and include data-provenance information with any health data downloaded by the patient (e.g. lab that reported test results) or sent to a patient's PHR	

Certifying Compliance

For many HIT vendors, much of the last year was probably spent ensuring that current product lines met the new compliance standards needed for Meaningful Use requirements, particularly with large financial incentives at stake for their respective customer base. Originally, only one body was permitted to certify partial and complete EHR products; the Certification Commission for Health Information Technology (CCHIT). The CCHIT was established to ensure that a product met US Department of Health and Human Services minimum specifications for compliance criterion. Due to the increased workload and number of applications, five additional organizations have been granted approval as certification bodies. Beyond CCHIT, the others include Surescripts LLC, ICSA Labs, SLI Global Solutions, InfoGard Laboratories, Inc., and Drummond Group, Inc. [15] Table 8.2 includes a partial list of fully certified, complete EHR solutions taken from the HealthIT.HHS.Gov website as of December 30, 2011. The most important issue regarding the certification list is the incredible growth of certified complete EHR products in just the past 18 months. In June 2010, only 12 vendors were capable of offering certified products; while the list today stand at nearly 800.[15,19]

The incredible selection of certified EHR offerings is a boon to ambulatory and inpatient organizations looking for the perfect fit, and also offers some assurances that products meet a standard, but this first round of certifications does not yet incorporate the tougher Stage 2 criteria still being decided. With natural market pressures, consolidation and additional certification requirements still forthcoming, the list of available product lines will likely shrink below its current level.

Table 8.2: EHR product listing (Courtesy of http://onc-chpl.force.com/ ehrcert/EHRProductSearch**)**

Certifying ATCB	Vendor	Product	Product Version
InfoGard	Doctor Office Management, Inc.	2011 PhysicianXpress	1
CCHIT	Pulse Systems	2011 Pulse Complete EHR	2011
Drummond Group Inc.	Systemedx Inc.	2011 Systemedx Clinical Navigator	2011.03
CCHIT	Waiting Room Solutions	2011 Waiting Room Solutions Web Based EHR and Practice Management System	4
CCHIT	Health Communication Systems, LLC	2011-14 DirectorMD EHR	10
Drummond Group Inc.	UnisonCare Corporation	2011-2012 Certified UniCharts EMR	version 3
Drummond Group Inc.	VipaHealth Solutions, LLC	24 7 smartEMR	5.1.2
InfoGard	Medaxis Corporation	360EHR	2.12
CCHIT	4Medica	4medica Ambulatory Cloud iEHR (Integrated Electronic Health Record) with 4medica's Integration Engine to connect Ambulatory EMRs	10.4

HIPAA and Data Security in the Cloud and Traditional Client/Server Solutions

Recent changes in technology and product models have thrown an additional element into the mix for organizations to contend with; which type of solution to choose. The traditional practice management or electronic health record solution is based on software that runs on local network infrastructure and is delivered via a client terminal using terminal services or loaded on a workstation. Hospitals and practices maintain the system and equipment locally and worked with vendors for troubleshooting, software change requests, and upgrades. The latest contender is software as a service (SaaS), used to deliver the solution via a Web browser. Oftentimes, SaaS solutions rely on another new technology, cloud computing, to store data and provide the back office computing power traditionally handled by servers and network storage devices. In this type of solution, the hospital contracts with a vendor to provide all of the services which are delivered to the end user. Each solution type has security risks and vulnerability for HIT workers to contend with; whether it is a stolen laptop, missing backup media, or a SaaS service compromise.

Deciding which solution is appropriate for an organization is decided by a number of factors that require careful deliberation and planning. Products like the Application Security Questionnaire (ASQ) from HIMMS can assist organizations performing their own research and planning for HIT solutions. The ASQ is a vendor-neutral, seven page capabilities checklist that hospitals, practices or medical organizations can request that software or services vendors complete for later comparative analysis of the various options being reviewed for selection.[21] Table 8.3 which follows this paragraph was created to compare the SaaS/Cloud and

Client/Server models, indicating some of the advantages and disadvantages of each.[22, 23, 24, 25] While the list is hardly all-inclusive, it provides a primer on what criteria and potential risks need to be considered.

Table 8.3: Cloud versus client-server model

Feature or Attribute	SaaS/Cloud	Client/Server
Integration with current systems	Web-based solution used with browser reduces client integration issues, but may have interoperability issues with other solutions in use.	Client software may have integration issues based on client configuration and may have interoperability issues with other solutions as well.
Software updates/upgrades	Software upgrades and updates are typically seamless, as they occur within the cloud before being delivered to the browser on the client end.	Software upgrades and updates require testing, may require downtime, and can be problematic if some systems are not available during the update window.
Costs	Infrastructure costs tend to be less than client/server, and SaaS solutions are less hardware dependent, but costs for bandwidth availability and service contracts can offset some of the savings.	Infrastructure costs associated with servers, storage, the solution product and support, in addition to life cycle costs of hardware and software that the solutions depend on for new features.
Reliability	Reliability is dependent of the product vendor and the quality and availability of the Internet connection to the provider.	Reliability is dependent of the product vendor and the capability of IT staff.
Availability	24/7 availability dependent upon Internet service	24/7 availability
Scalability	Easily scalable, but highly reliant on amount of bandwidth and signal latency, which serves as the performance bottleneck.	Scalability depends on capability of servers, storage and network infrastructure. Has less network latency to affect performance.
Security	Security in a cloud is still major sticking point, as data is on shared infrastructure and relies on virtual security methods and techniques.	Organization owns equipment and controls network security. Security dependent upon staff and defense measures.
Customization	Customization may be costly or limited due to support requirements in day-to-day operational environment.	Customization may be costly, but organization controls the implementation once complete.
Ownership	No ownership of solution, data is not located on site. Data may be difficult to obtain after contract ends, vendor is absorbed or goes bankrupt.	Organization owns data. Software is still usable in the event the vendor goes bankrupt.
Infrastructure	Requires no changes to infrastructure to support, unless additional bandwidth requirements dictate.	Requires more hardware; application servers and network storage. COOP solutions for redundancy require more equipment still.
Support	Support is almost entirely dependent upon vendor and service level agreement.	Support is dependent upon local IT staff and vendor when needed.

Future Trends

The challenges that Meaningful Use posed to EHR Certification and financial incentives looked daunting when 2011 opened, but the year closed out with a large number of vendors and products that met the new rules and permitted HIT users to cash in on federal incentive programs designed to promote broader electronic adoption in a traditionally hard copy oriented community. Better still is the variety of free or low-cost offerings that can be found at http://onc-chpl.force.com/ehrcert, proving that customers do not have to spend a fortune to offer their patients the latest technologies and the benefits that come with easier information sharing and records distribution.

Unfortunately, there is still a great deal of work to be done securing the medical records and systems access for these new solutions. Of the six largest data breaches that occurred this year, each impacting a million or more people, fully half are directly attributed to the healthcare industry.[26] This is not a trend that can continue if the industry wants to instill patient confidence that practitioners can adequately protect the most sensitive of personal data. Based on Stage 2 Meaningful Use working group discussions, efforts are under way to ensure that appropriate measures are put into place to meet security and privacy requirements going forward.[11]

Key Points

- ARRA and the HITECH Act are designed to supplement the administrative, physical and technical safeguards implemented by HIPAA

- Hundreds of HIT products now meet the 2011 certification requirements

- Data and records security will play a vital role in HIT success or failure

- Emerging technologies offer organizations a choice between traditional client-server or SaaS-Cloud product models

Conclusion

In conclusion, this chapter has sought to evaluate HIPAA and highlight its evolving impact on the future of HIT, the latest healthcare systems such as EHRs, and continuing its intent in securing privacy data. The government continues to form and enforce healthcare privacy and security standards through the passage of the HITECH Act from the ARRA, and the adoption of Stage 1 and Stage 2 Meaningful Use standards. Although there will certainly be more questions and technical hurdles to face as EHRs gain prominence and medical data is collected on a national level, this section identified key topics of importance that should be considered for healthcare data privacy and security.

References

1. For Consumers. www.hhs.gov/ocr/privacy/hipaa/understanding/consumers/index.html (Accessed June 24 2010).

2. Touchet. http://ps.psychiatryonline.org/cgi/content/full/55/5/575 (Accessed November 22, 2011).

3. HIPAA Privacy Rule and Its Impacts on Research. http://privacyruleandresearch.nih.gov/pr_08.asp (Accessed November 22, 2011).

4. Summary of the HIPAA Privacy Rule. http://www.hhs.gov/ocr/privacy/hipaa/understanding/summary/index.html (Accessed November 22, 2011).

5. Summary of the HIPAA Security Rule. http://www.hhs.gov/ocr/privacy/hipaa/understanding/srsummary.html (Accessed November 22, 2011).

6. Widespread Data Breaches Uncovered by FTC Probe. http://www.ftc.gov/opa /2010/02/p2palert.shtm (Accessed November 22, 2011).

7. Goedert J. http://www.healthdatamanagement.com/issues/18_3/security-tenn.-blues-breach-affects-500000-39839-1.html Tenn. Blues Breach Affects 500,000. March 1, 2010 (Accessed November 22, 2011).

8. Goedert J. http://www.healthdatamanagement.com/issues/18_3/security-health-net-sued-for-hipaa-violations-39837-1.html Health Net Sued for HIPAA Violations. March 1, 2010 (Accessed November 22, 2011).

9. The Top Half Dozen Most Significant Data Breaches in 2011. https://www.privacyrights.org/top-data-breach-list-2011 Data Breaches: A Year in Review. December 16, 2011. (Accessed December 29, 2011).

10. Hulme, G. Medical data breaches soar, according to study. CSO Online - Security and Risk. http://www.csoonline.com/article/695521/medical-data-breaches-soar-according-to-study December 1, 2011 (Accessed December 29, 2011).

11. Eastwood, B. Five health IT trends not to watch in 2012. Health IT and Electronic Health information, news and tips - SearchHealthIT.com. http://searchhealthit.techtarget.com/news/2240113145/Five-health-IT-trends-not-to-watch-in-2012 December 29, 2011 (Accessed December 29, 2011).

12. Meaningful Use Documents. http://healthit.hhs.gov/portal/server.pt?open=18&objID=888532&parentname=CommunityPage&parentid=5&mode=2&in_hi_userid=11113&cached=true (Accessed November 22, 2011).

13. DHHS HIT Standards Committee Implementation Workgroup. HIT Standards Committee Privacy and Security Workgroup Stage 2 Meaningful Use Privacy and Security Recommendations. October 21, 2011 (Accessed December 29, 2011).

14. Health Information Technology: Can HIT Lower Costs and Improve Quality. http://www.rand.org/pubs/research_briefs/RB9136/index1.html August 14, 2009 (Accessed November 22, 2011).

15. HealthIT.hhs.gov: ONC-Authorized Testing and Certification Bodies. http://healthit.hhs.gov/portal/server.pt/community/healthit_hhs_gov_onc-authorized_testing_and_certification_bodies/3120Xg&cad=rja (Accessed December 30, 2011).

16. Sullivan, B. VA loses another computer with personal info. http://www.msnbc.msn.com/id/14232678/ August 8, 2006 (Accessed November 22, 2011).

17. Centers for Medicare & Medicaid Services (Redirect to HHS). http://www.hhs.gov/ocr/privacy/hipaa/administrative/securityrule/remoteuse.pdf (Accessed November 22, 2011).

18. Google, Microsoft Say HIPAA Stimulus Rule Doesn't Apply to Them. http://www.ihealthbeat.org/Articles/2009/4/8/Google-Microsoft-Say-HIPAA-Stimulus-Provision-Doesnt-Apply-to-Them.aspx (Accessed November 22, 2011).

19. Products with CCHIT Certified® Comprehensive 2011 certification http://www.cchit.org/products (Accessed June 24, 2010).

20. AMA - American Recovery and Reinvestment Act of 2009 (ARRA). http://www.ama-assn.org/ama1/pub/upload/mm/399/arra-hit-provisions.pdf (Accessed November 22, 2011).

21. HIMSS - Information Systems Security. http://www.himss.org/content/files/ApplicationSecurityv2.3.pdf (Accessed November 22, 2011).

22. Foley, J. Private Clouds Take Shape. http://www.informationweek.com/news/services/business/209904474 August 9, 2008 (Accessed November 22, 2011).

23. Software as a Service. http://www.himss.org/content/files/software_service_presentation091707.pdf (Accessed November 22, 2011).

24. Federal CISOs worry they can't effectively secure cloud computing. http://searchsecurity.techtarget.com/news/article/0,289142,sid14_gci1511668_mem1,00.html (Accessed November 22, 2011).

25. EMR and EHR – ASP Versus Client Server. http://www.emr-match.com/emr-complete-resource/emr-and-ehr-asp-versus-client-server/ (Accessed June 24 2010)

26. Terry, K. Report: Healthcare accounts for 3 of the 6 worst data breaches in 2011 - FierceHealthIT. http://www.fiercehealthit.com/story/report-healthcare-accounts-3-6-worst-data-breaches-2011/2011-12-22 (Accessed December 30, 2011)

9

Health Information Security

BRENT HUTFLESS

Learning Objectives

After reading this chapter the reader should be able to:

- Enumerate the definitions of confidentiality, availability and integrity

- Discuss multiple ways to ensure authentication

- Compare and contrast digital signature and certificate based encryption

- Enumerate different types of security breaches and their causes

- Discuss security standards and the laws intended to protect health data

Introduction

Health information privacy is discussed in the prior chapter, with an emphasis on the Health Insurance Portability and Accountability Act (HIPAA). With a verifiable need to protect health information well established, there is a need to cover the information security aspects. How is health data protected against exposure? How does an increasingly targeted industry turn the tide against the news stories, the hackers, criminals, and identity thieves? More importantly, what mechanisms are medical professionals likely to witness firsthand in the battle to keep the attacks at bay?

This chapter introduces general concepts of information security, and explains the technologies and techniques used by security professionals, without the technical jargon. The major topics include basic security concepts, authentication and identity fundamentals, descriptions of the risk scenarios which may lead to breach, and lastly, the compliance and legal standards being applied to medicine.

Basic Security Principles

The shift towards electronic health records, personal health records, health information exchanges, and web-based health applications creates a whole new problem of incredible proportions. How do you secure the most private of personal information, health data, for just over 300 million people? More difficult still, how does an industry spotlighted in the media for failing to protect this data instill confidence with the public whose data is being collected and used? Without better assurances and solutions by vendors, insurers and health care organizations, it may be difficult to win and keep the public trust. The resolution may rely on a set of security principles that are the foundation for current solutions in other industries, such as banking, retail industries or the airlines.

Definitions

According to the International Information Systems Security Certification Consortium (ISC²), among others, there are three pillars of information security (confidentiality, availability, and integrity) that are fundamental to protecting information technology solutions such as health information technology (HIT).[1] Security measures are instituted collectively to meet one or more of these primary goals, with the end result being one where confidentiality, availability and integrity are all covered.

- *Confidentiality* as it is termed here refers to the prevention of data loss, and is the category most easily identified with HIPAA privacy and security within healthcare environments. Usernames, passwords, and encryption are common measures implemented to ensure confidentiality.

- *Availability* refers to system and network accessibility, and often focuses on power loss or network connectivity outages. Loss of availability may be attributed to natural or accidental disasters such as tornados, earthquakes, hurricanes or fire, but also refer to man-made scenarios, such as a Denial of Service (DoS) attack or a malicious infection which compromises a network and prevents system use. To counteract such issues, backup generators, continuity of operations planning and peripheral network security equipment are used to maintain availability.

- *Integrity* describes the trustworthiness and permanence of data, an assurance that the lab results or personal medical history of a patient is not modifiable by unauthorized entities or corrupted by a poorly designed process. Database best practices, data loss solutions, and data backup and archival tools are implemented to prevent data manipulation, corruption, or loss; thereby maintaining the integrity of patient data.

Security Tools and Solutions

Information security is in many ways analogous to physical security techniques employed at a residence or place of employment. Some solutions are used to deter and prevent access, such as locks on doors and windows, use of shrubs, bright or motion-sensitive lighting, video cameras, guard shacks, fencing and gates. Similarly, business networks and information resources are protected by access control lists (ACL), firewalls, intrusion detection and intrusion prevention systems, authentication systems, and monitoring and auditing services designed to mimic their physical counterparts around the building or home. Instead of a key, one uses a username and password or token to gain access. The firewall acts as the barrier designed to keep out those who do not belong. Intrusion systems take the place of video surveillance; and similar to footage used for evidence in a crime, these systems can help forensics investigators track an intrusion back to its source. Monitored services imitate physical alarm systems, and forensics specialist track intruders who may unwittingly leave a trail of evidence, ultimately leading to real-world arrests and convictions. One real-world example of this is the case of Army PFC Bradley Manning, who leaked untold quantities of classified data to WikiLeaks founder Julian Assange, and then failed to cover the digital tracks which led investigators to the evidence used to try him.[2] It is the correlation between the physical and the electronic that much of this chapter builds on.

Organizational Roles

Information security roles and responsibilities can vary widely from organization to organization depending on size, industry, compliance mandates and laws, technology initiatives, maturity, private or public, and even profit model. Policy regarding information security practices is often set by chief information officers (CIO), chief technology officers (CTO), information technology (IT) directors or similar; often with input from chief medical informatics officers (CMIO), HIPAA compliance officers, or the like. Depending on resources, the information technology teams may consist of network, system administration, security and data personnel, or could be the very same technical staff relied upon for all office or clinic IT needs. No matter the titles, this supporting staff is often tasked to defend key systems, networks, and patient data from risk, and assist with any investigations resulting from a data breach.

Authentication and Identity Management

Who are you? More importantly, can you prove who you say you are? These are the chief tenants of authentication, and are supported by photo identification, biometrics, smart card technologies, tokens, and the old standard; user name and password. Authenticating users, patients and staff is essential for providing system access, ensuring only those with need to know have access, protecting important data, and lending legal credence to actions and records.

Basic Authentication

The devices and methods people use to gain access to systems, data, and web solutions vary depending upon the sensitivity of the data, the capabilities of the systems, resource constraints - both technical and monetary, and the frequency of access. All of the methods discussed here rely on what is known as two or multi-factor authentication. The factors fall into three categories – something one knows, something one has, or something that one is.[10] The most basic of these is the tried and true username and password combination still employed by a majority of users today, combining two things that a user knows. Another option is utilizing a grid card, smart card, USB token, one time password (OTP) token, or OTP service in combination with something a user knows, such as a passphrase or PIN. All of these rely on something one has; either a card, token, or in the event of the OTP service, some mechanism to view a message that contains the one time use character string or passphrase to be used. By combining something a user has with something he or she knows, two-factor authentication occurs. Figure 9.4 contains a selection of these authentication tools, showing a grid card, smart card, OTP card and OTP smartphone service application.

Figure 9.4: Various Authentication Tools (Image Sources – Entrust.com)[11]

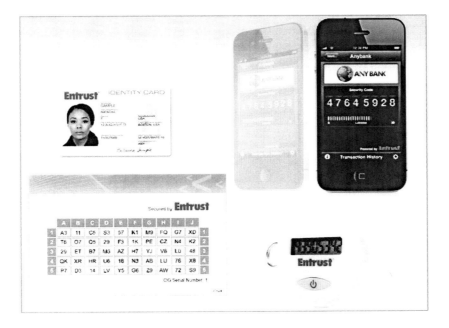

Single Sign On

Anyone who has used a computing device more than a few times quickly learns that most systems, whether physical workstations or web based solutions, require some method of authentication, typically in the form of a username and password. Before long, users find themselves with a growing list of usernames and passwords for any number of devices, email accounts, banking access, social networks, retail websites, and even a few dedicated to work resources.

What if there was a way to use one set of credentials, or one mechanism, to easily access many of the resources one uses every day, but with security that identical usernames and passwords cannot provide? This is the practice known as single sign on (SSO), and when implemented correctly, it allows users to access a variety of disparate systems using one set of stored credentials. SSO can be utilized for more than system and network access, enabling users to authenticate to the web and software as a service (SaaS) solutions as well. One common example of SSO is a service offered by Google to partner organizations which use Google Apps, such as Gmail, Google Docs, and Calendar. The partner organizations, perhaps a small business or school, "control usernames, passwords and other information used to identify, authenticate and authorize users for web applications that Google hosts" through the SSO solution, offering seamless transitions between local resources and hosted applications.[3]

Although the above example employs a username and password, other mechanisms such as smart cards, tokens and even biometrics are capable of offering SSO capability for a wide range of client, software, and web-based solutions. This permits the use of a single token or card used for workstation or network access to also connect with web-based application and other solution software without additional user logon.[4]

Digital Signature

Part of the problem that arises when shifting from hard copy medical record documentation to electronic format is signing new records. Beyond the obvious improvement in discerning the signatory compared to a handwritten signature, there needs to be some additional contrivance that provides some assurance that the digital signature is valid and that it was placed by the person it is attributed to. In the case of patient records, this digital signature also acts as the legal signature of the practitioner. As such it can serve as non-repudiation for electronic messaging and records access for audit purposes, and in some cases meets compliance controls or measures for identity management. The strength of this technology is such that email correspondence containing an electronic signature is sufficient to prove that the originator and the signatory are one in the same. This is possible because the originator is the only one with the unique key required to produce the electronic signature.[5] The example shown in Figure 9.1 provides some insight into what occurs when a user send a digitally signed email, a common form of official electronic correspondence.

Figure 9.1: Electronic Signature (Source - Microsoft X.509 Technical Supplement)[6]

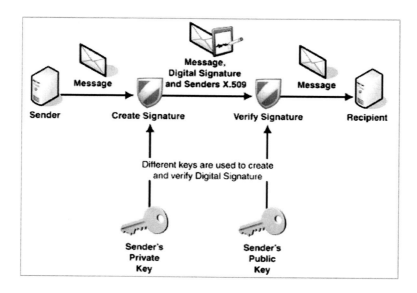

Certificate Based Encryption

An advanced form of digital certificate technology is certificate based encryption. Whereby a digital signature is used to provide assurances and non-repudiation from a given party, encryption is intended to completely obscure the contents of a message, preventing compromise of sensitive information in the event that a message is intercepted en route. Although the algorithms used for encrypting data are somewhat complex, the practical applications are easily understood. By having pre-shared public keys, individuals can send correspondence to each other taking comfort in the knowledge that the contents are protected from prying eyes. Figure 9.2 displays two employed keys, similar to the digitally signed message. In the case of encryption though, it is the recipient's public key that is used by the send to encrypt the message, not the sender's. Since the recipient has the lone private key, only he or she will be able to decipher the message and view the contents.

Figure 9.2: Public Key Encryption (Source - Microsoft X.509 Technical Supplement)[6]

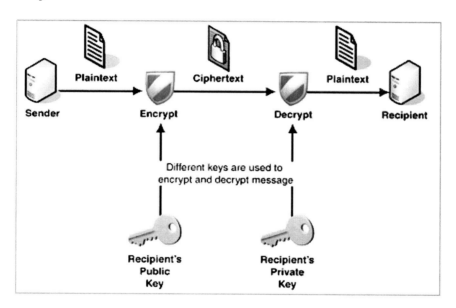

Given the constraints placed upon organizations trying to meet HIPAA, Sarbanes-Oxley Security (SOX), or Payment Card Industry (PCI) compliance mandates, encryption provides a much-needed layer of security designed to protect the most sensitive of data. A practical example where this could be employed is correspondence involving patient records. As an alternative to de-identification of a patient record, a record could be sent with all identifying data to a qualified recipient using the data encryption mechanism described above. This type of data protection mechanism is permitted for sensitive data in motion, as described in the Health Information Technology for Economic and Clinical Health (HITECH) Act[7] and is a component of secure messaging through NHIN Direct as described in the chapter on health information exchange.

Digital & Information Rights Management

Beyond identifying the user and authenticating against systems and web solutions, users can also be controlled for roles, permissions, and access in fine detail. Digital Rights Management (DRM) and Information Rights Management (IRM) are related data access concepts that are gaining in adoption as compliance initiatives and risk management practices take hold in organizations across many industries. One common application of DRM and IRM functionality is with content management systems, such as Microsoft SharePoint or EMC Documentum product lines.[8] While it is essential to secure sensitive health data from unauthorized access, it is increasingly important to limit any unnecessary access to patient records. DRM and

IRM allow organizations to limit user or system access to data only when it is needed, place time constraints upon said access, limit how and where data can be viewed, modified and moved, as well as create records for auditing and forensics purposes. Figure 9.3 illustrates the permissions function within Microsoft Word.

Figure 9.3: Microsoft Word 2010 IRM (Source: Microsoft Office)

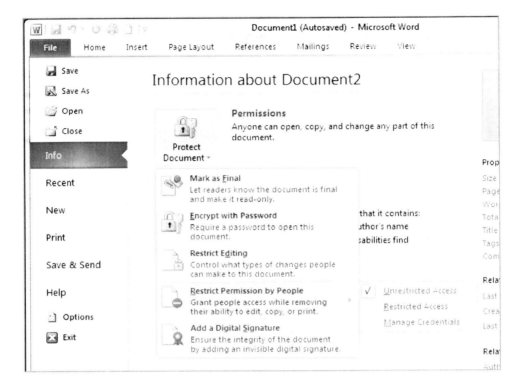

The DRM and IRM mechanisms employed on an organizational content management system (CMS), electronic health record (EHR) product data repositories and the like allow for setting granular rights permissions to the user level. Records of the activity can be used for compliance audits, but also serve as evidence in cases where unauthorized access is suspected, an example of which is a case in Florida where three non-medical hospital employees accessed medical records of accident victims and forwarded this information to a law referral service.[9] Though the three individuals were eventually caught, this activity could have been prevented if DRM and IRM controls had been in place.

Biometric Authentication

In addition to authentication mechanisms which rely on something the user has (e.g. grid card or USB token), there are now biometric authenticators based on physical user identifiers. Biometric authentication typically uses a fingerprint, retinal scan, or voice imprint, although iris, vein and even heart rhythm based ECG scans have been proposed as solutions in recent years.[12] When combined with passphrases or the tokens, cards, and OTP solutions discussed previously, a two or multi-factor authentication solution can be employed.

The key take-away from the examples of two-factor authentication is the difficulty these present to would-be attackers or data thieves, as it greatly increases the complexity required for user access over a simple username and password combination. Although usernames and passwords are not likely to fade away anytime soon, increased adoption of other more secure methods is almost certain, particularly in the face of increased data breaches, attacks, and even industry regulations.

Security Breaches and Attacks

The privacy chapter provides a number of current cases where healthcare related organizations had been the victims of attack or negligence which led to a significant breach of patient or personal data. Fully half of the major data breaches in 2011 were attributable to the healthcare industry,[13] a trend which cannot continue without impacting further adoption and acceptance of electronic health records by the public and providers alike. The information box summarizes some of the security trends noted in 2011. The following sections describe some of the issues which lead to data compromise.

2011 Healthcare Security in Review

Ponemon Institute released a 2011 research report on patient privacy and security. Data was obtained from interviews with executives from 72 healthcare organizations. The following are key findings:

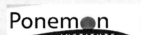

- Healthcare data breaches are on the rise; 32% rise over the previous year, in spite of new regulations

- Widespread use of mobile technology is putting data at risk

- In spite of breaches, many organizations have not set data privacy and security as a priority

- Financial consequences of data breaches are very significant, as is patient trust and productivity; the average number of lost or stolen records was 2,575 and the cost of a breach was $2.2 million

- Medical identify theft is a major problem; 35% of breaches were detected due to patient complaints

- The average time to notify data breach victims was 7 weeks

http://thielst.typepad.com/files/2011-ponemon-id-experts-study.pdf

---

Physical Theft

Several years ago a laptop was stolen from a VA employee containing the records of millions of veterans. Although that laptop was later recovered, theft of devices and storage continue to result in data loss, and more specifically, patient data loss despite the advent of protection mechanisms which would render such information irretrievable to the thieves.

Computing devices such as laptops, desktops, and even servers are stolen each year out of cars, homes, and places of business. Although servers are usually considered relatively safe due to their back office location out of public view, facilities still fall victim to break-ins where thieves take valuables such as servers without directly targeting those resources for their data.[14] Unfortunately, in such cases the burglars make off with the entire databases, exposing patients and facilities alike to grave risks. Beyond computing devices, storage also presents risks to organizations when not encrypted and treated with an appropriate level of care. A multi-billion dollar lawsuit stemming from lost archive tapes stolen from an employee vehicle has the potential to financially ruin the company, SAIC, who was managing the storage media for Tricare Management Activity.[15] Portable and removable storage media is, by its very nature, more susceptible to theft but its limited commercial application may have thus far prevented large breaches.

To put the impact of medical data theft into perspective, we refer readers to the HHS website that lists all of the reported data breaches affecting over 500 users from 2009-2011. The site lists the covered entity, the number of breach victims, the type of breach and the location of data (laptop, server, paper, etc.). The link is located at http://www.hhs.gov/ocr/privacy/hipaa/administrative/breachnotificationrule/breachtool.html and is available in a variety of formats. Of all the thefts listed, 86 involved a laptop computer, showing that portable machines are still used to store sensitive data despite industry recommendations to avoid such behavior.

Theft Countermeasures

Outside of recovering lost and stolen devices and performing forensic analysis to determine whether data has been accessed by unauthorized individuals, there are a number of measures that can be used to render data unusable to the thieves. Encryption standards such as FIPS 140-2 are being applied to storage mandates in order for organizations to adhere to HIPAA data protection criteria in the event of loss or theft.[16] Simply put, encryption renders the stored data irretrievable or otherwise indecipherable to those who attempt to access it without the proper decryption key. Some of these solutions exist at a hardware level, and the FIPS 140-2 validation that often accompanies it is applied to physical storage devices such as drives found in servers workstations and portable computing devices as well as hardened portable storage devices, such as thumb drives. Figure 9.5 shows a common encrypted personal storage device.

Figure 9.5: IronKey Encrypted Drive[17]

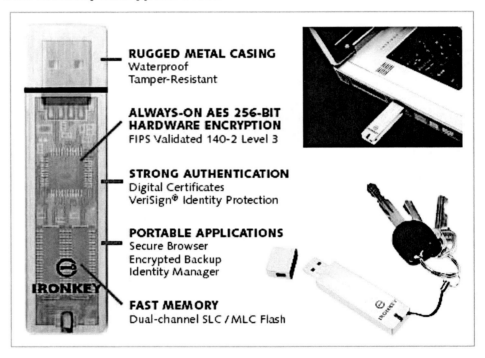

Other software-based encryption techniques, such as whole disk encryption, can be applied to computing platforms after the fact to similarly protect these devices. When encryption is configured appropriately, an organization could theoretically leave their most sensitive data in a box outside of their doors with no worry of data compromise, although few would recommend this practice.

Physical or Logical Access

In the information security profession, there is a common expression that states that if one can touch a system, one can own it. In the health care environment, this is more difficult to accomplish for the outsider who has to bypass the greater physical security mechanisms often in place. For the insider however, this is far easier to achieve. Insider threat for the purposes of this discussion can be defined as those employees or staff with physical or logical access to systems in an organization who seek to steal, damage, or compromise data. Whether it is a disgruntled employee seeking retribution, someone performing data collection for personal gain by identity theft, or an insider seeking data for competitive or corporate espionage purposes, organizations need to be aware of the potential risks from within. For hospitals this threat could extend to data corruption, similar to what happened at a Minnesota hospital after a logic bomb planted by an angry former employee disabled a core program.[18]

There are a wide variety of data collection techniques that can be used by insiders, ranging from simple storage devices and key-loggers to technologies that seemingly spring from the latest spy movie, where malicious users create encrypted connections to destinations outside of organizational networks to pass the information. A new technique employs a feature common to most smartphones to capture keyboard keystrokes with 80 percent accuracy – the phone merely needs to sit on the same surface and does not employ the microphone.[19] In spite of this seemingly complex onslaught of technology designed to gather the most sensitive of data, most can be mitigated through identification and remediation of vulnerable systems by IT and security staff within an organization. The key to preventing incidents often comes down to user awareness, training, policies designed to limit access, and layered security practices know as defense in depth.

Accidental or Negligent Disclosure

Large data breaches and disclosures make the news and fill the pages of Health and Human Services (HHS) notification website, but many of the routine disclosures are accidental or negligent in nature. Luckily, HHS added a threshold for harm that permits covered entities and business units to perform a risk assessment and determine whether an incidental exposure poses any risk to the affected patient.[20] This type of leak can occur through inadequate control of paper records, inadvertent release of sensitive information to unauthorized parties, through overheard conversations, or even poor housekeeping practices around copiers, fax machines, and recycling bins. Protected health information that is transmitted in non-secure means across networks, or through email messages without the proper levels of protection or adequate de-identification are also examples of electronic disclosure.

Intrusions and Attacks

Theft has accounted for a number of HIPAA-related data breaches, but a growing risk is data loss or exposure due to intrusions and attacks by groups such as Anonymous and LulzSec which seek to expose wide swathes of private data to public scrutiny.[21] These and other groups rely on a combination of intrusion vectors more commonly known as advanced persistent threat (APT). APT is not an attack type in and of itself, but a methodology that employs as many mechanisms as possible to gain a foothold inside of an organization; attacking organizations on physical and wireless networks, attempting to compromise machines and user accounts through disguised email messages, corrupted PDF files and exploited webpages, social networking sites such as Facebook, and by using social engineering through phone calls and brazen onsite attempts.[22] As could be expected, these attacks are significantly more difficult to defend against, but diligent staff and awareness can reduce the risks to organizations.

Standards, Compliance and Law

Outside of avoiding the bad publicity and financial implications that attacks and data breaches pose, organizations often have additional incentives to maintain or increase their security posture, which often originates with laws and compliance mandates. HIPAA and subsequent laws have placed breach notification reporting requirements on covered entities and business units. Recent guidance places a 60 day requirement on breach reporting, but responsible organizations should be encouraged to visit the instructions pages hosted by HHS.[23] Failure to meet the requirements, acknowledge responsibility, and provide notification to

HHS and news outlets for breaches affecting 500 or more individuals could result in severe fines and penalties.

HIPAA, the HITECH Act and Meaningful Use all play prominent roles in compliance initiatives for many healthcare organizations, but depending on the size, complexity, and public or private standing of a company, other compliance initiatives and laws exist which have direct bearing on how data must be protected, reported, and even audited. Instead of covering these topics in depth, they are listed in table 9.1 with simple descriptors. Each standard, regulation, best practice or governance is too complex to cover in detail here.

Table 9.1 Security Standards and Law

Security Standard/Law	Brief Description
ISO 20000/27000	International IT Governance and IT Security standards
COBIT	IT Governance framework
ITIL	Information Technology Infrastructure Library , IT service management
NIST SP 800-53	National Institute of Standards and Technology, IT security controls
SOX	Sarbanes–Oxley Act; Public company accounting law
HIPAA	Health Insurance Portability and Accountability Act of 1996
Meaningful Use (HITECH Act)	Identifies the technical capabilities required for certified EHR technology, and bonus qualifiers for organizations which meet them
PCI-DSS	Payment Card Industry Data Security Standard
FISMA	Federal Information Security Management Act

Future Trends

Given the slow adoption of information security practices by the medical industry and current prevalence of data loss, data breaches are likely to remain in the forefront of healthcare security for the next few years.

Multifactor authentication technologies will gain acceptance and offer acceptable performance to become more common in larger practices and hospitals. Patient privacy and security may drive this further, particularly if medical care and access is tied to benefit confirmation through more secure mechanism to limit medical identity theft.

Mobile records, access and security will become more common, with ever-capable smart phones and consumer expectation sparking innovation and leading change toward portable data sharing.

Data encryption will likely become a de facto posture, with requirements mandated in updates Meaningful Use regulation or follow-on legislation.

<table>
<tr><td colspan="1">Key Points</td></tr>
</table>

- The security of Healthcare data is extremely important for successful current and future adoption of HIT

- Unfortunately, security breaches of healthcare data are not falling, they are on the rise

- All healthcare organizations must adhere to the latest security mandates in order to stem the tide of security breaches

- Security measures will continue to improve and proliferate, but so will the efforts of criminals who seek illicit access to protected health data and identity theft

Conclusion

In this chapter we offer an overview of health information security with emphasize on authentication and other basic security concepts. Until appropriate security measures are adopted, embraced and enforced healthcare organizations will be faced with a variety of old and new breaches. The end result will be expensive due to law suits and fines and there will be an erosion of patient confidence in any new technology that stores protected patient information. It seems likely therefore that we will have to see new security laws and new penalties for those who fail to adopt and enforce the appropriate security of healthcare data. We can expect more security-related objectives and measures in meaningful use stages 2 and 3.

References

1. Tipton, H. & Henry, K. Official (ISC)2 Guide to the CISSP CBK. (ISC)2 Press Series, 2007. Chicago: Auerbach.

2. Gould, J. Agents say Manning didn't fully cover tracks - Air Force News | News from Afghanistan & Iraq - Air Force Times. http://www.airforcetimes.com December 19, 2011. (Accessed January 9 2012).

3. "Google Apps - Google Code." Google Code. http://code.google.com/googleapps/domain/sso/saml_reference_implementation.html (Accessed January 9, 2012).

4. Clercq, J. & Guido G. Microsoft windows security fundamentals. Burlington, MA: Elsevier Digital Press, 2007.

5. Bradford, Andrew. The Investment Industry for IT Practitioners an Introductory Guide. Hoboken: John Wiley & Sons, 2011.

6. "X.509 Technical Supplement." http://msdn.microsoft.com/en-us/library/ff647097.aspx December 2005. (Accessed January 10 2012).

7. "HIPAA.com - Transmission Security Encryption: What to Do and How to Do It." HIPAA.com. http://www.hipaa.com/2009/07/transmission-security-encryption-what-to-do-and-how-to-do-it (Accessed January 10 2012).

8. Kundurmutt, J. Comparing Core ECM Features of Documentum and SharePoint 2010 – Which Platform is Right for You?. TCS White Papers. www.tcs.com/SiteCollectionDocuments/White%20Papers/EntSol_Whitepapers_Core-ECM-Features-Documentum-SharePoint_11_2011.pdf (Accessed January 10 2012).

9. Santich, K. & Breen, D. Privacy breach at Florida Hospital: The hospital says employees improperly accessed ER patients' personal information - Orlando Sentinel.

http://articles.orlandosentinel.com/2011-09-30/news/os-florida-hospital-privacy-breach-20110930_1_identity-theft-patients-access September 30, 2011. (Accessed January 10 2012).

10. Harris, Shon, and Eric Ouellet. Security+ certification all-in-one exam guide. Berkeley, Calif.: McGraw-Hill/Osborne, 2003.

11. Digital Certificates, PKI, & Two Factor Authentication Solutions - Entrust. http://www.entrust.com (Accessed January 11 2012).

12. Stavroulakis, P. & Stamp, M. Handbook of Information and Communication Security. New York: Springer, 2010.

13. Terry, K. Report: Healthcare accounts for 3 of the 6 worst data breaches in 2011 - FierceHealthIT. http://www.fiercehealthit.com/story/report-healthcare-accounts-3-6-worst-data-breaches-2011/2011-12-22 (Accessed December 30 2011)

14. Stolen server contains medical info on 40,000 eye patients. Office of Inadequate Security. http://www.databreaches.net/?p=11688 (Accessed January 12 2012).

15. Risk to Patients from Data Breach Met with Proactive Response. TRICARE - Media Center. http://www.tricare.mil/mediacenter/news.aspx?fid=738 (Accessed January 12 2012).

16. Guidance to Render Unsecured Protected Health Information Unusable, Unreadable, or Indecipherable to Unauthorized Individuals. United States Department of Health and Human Services. http://www.hhs.gov/ocr/privacy/hipaa/administrative/breachnotificationrule/brguidance.html (Accessed January 12 2012)

17. IronKey Flash Drive|IronKey S200|D200|Basic & Personal http://www.usbmemorysticks.net/ironkey-flash-drives (Accessed January 12 2012).

18. Undercover. CSO, April 2008.

19. Lemos, R. Smart Phones Could Hear Your Password. Technology Review: The Authority on the Future of Technology. http://www.technologyreview.com/computing/38913/page1/ October 18, 2011. (Accessed January 12 2012)

20. Murphy, M. & Waterfill, M. New HIPAA guide for 2010: 2009 ARRA Act for HIPAA security and compliance law & HITECH Act, your resource guide to the new security & privacy requirements. Bloomington, Ind.: Authorhouse, 2010.

21. Pelofsky, Jeremy. Hacking groups say they are back after FBI arrests| Reuters. Business & Financial News, Breaking US & International News | Reuters.com. http://www.reuters.com/article/2011/07/21/us-cybersecurity-response-idUSTRE76K66G20110721 (Accessed January 12 2012)

22. DHS. A-0010-NCCIC-BULLETIN. "Anonymous" And Associated Hacker Groups Continue To Be Successful Using Rudimentary Exploits To Attack Public And Private Organizations. www.fbiic.gov/public/2011/jul/A-0010-NCCIC-BULLETIN.pdf (Accessed January 12 2012).

23. Instructions for Submitting Notice of a Breach to the Secretary. United States Department of Health and Human Services. http://www.hhs.gov/ocr/privacy/hipaa/administrative/breachnotificationrule/brinstruction.html (accessed January 13 2012).

10

Health Informatics Ethics

KEN MASTERS

Learning Objectives

After reading this chapter, the reader should be able to:

- Describe the 20th century medical and computing background to health informatics ethics
- Identify the main sections of the IMIA Code of Ethics for Health Information Professionals
- Describe the complexities in the relationship between ethics, law, culture and society
- Describe different views of ethics in different countries
- Summarize the most pertinent ethical principles in health informatics ethics
- Discuss the application of health informatics ethics to research into pertinent areas of health informatics, including behavior as a medical student

Introduction

"It is immaterial for the experiment whether it is done with or against the will of the person concerned."

- Dr. Karl Brandt, Final Statement, Nuremberg Trials, 19 July 1947.[1]

As is obvious from the subject in this text book, health informatics combines themes from medical fields and from informatics fields. It is to be expected, then, that health informatics ethics will combine information from medical ethics and from informatics ethics. This section details the recent history of these two fields so that the reader can understand the context within which modern health informatics ethics is to be discussed. Because of the unique nature of medicine, this chapter will spend considerable space examining recent medical ethics first.

The Road from Nuremberg

Nuremberg (alternate spelling *Nuernberg*, German spelling *Nürnberg*) is a town in Germany. Before and during World War II, Germany's National Socialist Party (Nazi Party) had controlled Germany, and had occupied much of Europe. During this time, at least 11 million people (mostly Jews, Poles, Romani ("Gypsies"), Eastern Europeans, and others regarded by the Nazis as "sub-humans" or "undesirables"[2]) were systematically murdered in what is now referred to as "The Holocaust." Many of the victims were murdered in large camps called concentration camps.

At the end of World War II, a series of legal trials was held in Nuremberg to examine crimes against humanity that had been committed in Germany and German-occupied countries.[3] To its shame, the German medical profession had cooperated with the Nazi Party on such a scale that medical and other health professionals were tried separately, and the abridged transcripts of these trials (referred to as the "Medical Case") make up more than 1,300 pages of testimony and supporting documentation.[1, 4]

Crimes committed by medical professionals had included widespread euthanasia and sterilization of mentally and physically handicapped people (referred to as "useless eaters" (*unnützen Esser*) and lives "unworthy of life" (*lebensunwerten Lebens* or *Lebensunwerts*)[4-7]), and also a large number of medical and biological experiments conducted on concentration camp inmates. Victims included men, women and children. Most died extremely painfully as a result of the experiments, and many of those who survived were later murdered by the camp authorities. Permission or consent for the medical experiments was almost never obtained from the inmates. Where "permission" was obtained, it was usually only as an alternative to death, or with the promise of release. None of the surviving victims was ever released by their captors, nor were any death sentences commuted.[4] Ironically, many of the experiments would have been illegal if they had been conducted on animals, as the Nazis had introduced strict laws governing the use of animals in medical experiments.[1]

At the Nuremberg medical trial, a code of conduct, which later became known as "The Nuremberg Code," was presented. The Nuremberg Code was in direct response to the medical crimes.[1] (See the NIH site at http://ohsr.od.nih.gov/index.html for the code). The Code emphasized the need for experimental subjects' voluntary consent to the experiment, regard for their safety (including mental suffering), balance of risks, and right to withdraw from the experiment if they wish. In addition, the Code noted that the responsibility for performing the experiment lay with the qualified medical experimenter, and this responsibility could not easily be transferred.

Figure 10.1: Children's Memorial, Mauthausen Concentration Camp

Figure 10.2: Dissection Room, Mauthausen Concentration Camp

As the prosecution at Nuremberg had noted, the medical professionals who had performed these procedures had violated a basic medical principle of "First, do no harm" (*primum non nocere*).[4] Because the Code was in direct response to the Nazi medical experiments, it focuses on the rights of patients and experimental subjects; it is not a broader code dealing with general ethics of medical practitioners in other situations.

World Medical Associations' (WMA) Declaration of Helsinki

After the publication of the Nuremberg Code, several countries reviewed their medical ethics (see the section below on International Considerations). In 1964, the World Medical Association drew up the first version of the "Declaration of Helsinki" (DoH), which broadened the concerns of the Nuremberg Code, and was to be applicable across the globe. Since then, the DoH has been through several reviews, and the current version was adopted in Seoul in 2008.[8]

Although it is not the same as the Nuremberg Code, the DoH is similar to it, as it deals with the human subjects' safety, consent, risks, and right of withdrawal. Amongst the significant additions (significant in the light of health informatics) is the right to "privacy, and confidentiality of personal information of research subjects" (Article 10 and 23) and the privacy regarding the use of identifiable human information (Article 25). We return to these issues later.

Informatics Ethics

Although one may argue that the history of information ethics begins with the ancient Greeks,[9] it is only in the latter half of the 20th century that machine-based information and ethics were viewed together for the first time. At roughly the same time that the Nuremberg Code was being developed, Norbet Wiener first published his book *The Human Use of Human Beings* in which he considered the social and ethical implications in the relationship between machines and humans.[10] From the 1970s onwards, work by people like Kostrewski and Oppenheim[11] and Robert Hauptman[12] dealt with ethical questions in information research. In 1997, Severson[13] introduced four principles of Information Ethics: (1) Respect for information property; (2) Respect for privacy; (3) Fair representation; (4) Non-maleficence (or "doing no harm").

As computers have developed further, codes of ethics for professional organizations have also evolved. Two examples are the Association for Computing Machinery's *Code of Ethics and Professional Conduct*[14] and the Canadian Information Processing Society's *Code of Ethics and Professional Conduct.*[15] These codes also refer directly to honoring the rights of the individual, respecting privacy and confidentiality, and doing no harm.

Figure 10.3: Health Informatics Ethics formed from Medicine, Ethics and Informatics

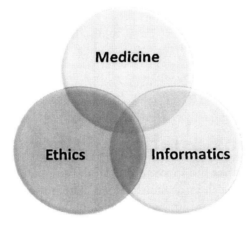

With similarities between the principles of ethics in medicine and ethics in informatics (especially in areas of respect for subjects, privacy, and doing no harm), it is to be expected that these would be issues contained in health informatics codes that were developed in the late 20th century and early 21st century. Indeed, these principles are contained in the International Medical Informatics Association's (IMIA) Code of Ethics for Health Information Professionals.[16] Although the code is aimed at health informatics personnel, it should be remembered that any medical person working with electronic data, will also be a Health Informatics Professional (HIP) or a Clinical Informatics Professional (CIP).

The IMIA code is extensive, going much further than the Nuremberg Code and the Declaration of Helsinki, and has the following components:

- Part I: Principles
    - o Fundamental Ethics Principles: autonomy, equality and justice, beneficence, non-malfeasance, impossibility (recognizing that some things are impossible), and integrity.
    - o General Principles of Information Ethics: information-privacy and disposition, openness, security, access, legitimate infringement, least intrusive alternative, and accountability.
- Part II: Rules of Ethical Conduct
    - o Subject-centered duties: these focus on electronic records, and are aimed at ensuring that subjects of electronic records are protected from abuse of their information.
    - o Duties towards other Health Care Professionals (HCPs): these focus on proper support, keeping HCPs informed of relevant information, maintaining standards of data storage, and intellectual property.
    - o Duties towards institutions/employees: these include integrity, loyalty, ensuring the safety of the institution's data, evaluation of systems' security, alerting and informing the institution of problems in good time and working within their scope of competence.
    - o Duties towards society: these include the proper collection, storage and safe-guarding of appropriate data, informing the public, and not participating in work that violates human rights.
    - o Self-regarding duties: these include recognizing one's own limitations, maintaining competency and avoiding conflict.
    - o Duties towards the profession: these include not bringing the profession into disrepute, impartiality, and assisting and maintaining standards of professionalism amongst colleagues.

The IMIA Code is continually under review, and serves as a useful guide for all people who work in health informatics fields.

International Considerations: Ethics, Laws and Culture

The first part of this chapter describes the medical ethical developments from World War II to the present day, and then the development of health informatics ethics. The impression is one of a great tragedy created by one country's lapse of medical ethics, internationally punished for breaking the widely-accepted ethical practices, resulting in a neat and linear path towards a set of near-perfect and internationally-accepted codes of ethics in the medical and health informatics fields.

While useful, this impression is a deceptive over-simplification, and the student of health informatics needs to be aware of greater complexities, especially with regards to national and international practices, and the relationship between ethics and the law. Part of the reason for the conflict is that ethics in general is strongly influenced by a country's laws and culture, but the relationship between ethics, law, culture and society is unclear, is not fixed internationally, and may be fluid even within a given country over time.

Different Views of Ethics

While there are many theories of ethics, for our purposes, we can look at three broad views regarding the relationship between ethics, law, culture and society:

- Ethics does not exist outside the law, and exists only for the good of a properly ordered and legal society. Therefore, a society's needs and the prevailing laws define ethical behaviour.

- Ethics is usually strongly informed by the law, society, and the prevailing culture, and are extensions of these. There are ethical requirements that are not necessarily required by law, but what is ethical can never conflict with what is legally required.

- Ethics exists entirely outside of the law, and is a matter of personal conscience. Because ethics grows from within social practices, there is usually correspondence between ethics and the law; where there is conflict, the ethical viewpoint must always prevail.

Significance of Different Views

In the codes and activities outlined above, we can see the different views being expressed. When these views are translated into practices, the difficulties of implementing ethics become more apparent. Some examples are:

- Among some Western doctors, there was the feeling that the Nuremberg Code was useful for "barbarians," but unnecessary for civilized physicians.[17]

- Although the Declaration of Helsinki says that local laws must be respected (article 10), it points out that "no national or international ethical, legal or regulatory requirement should reduce or eliminate any of the protections for research subjects set forth in this Declaration."

- A code of ethics is only a code of ethics. It carries no legal weight *at all*. If a person is found to be acting unethically, then their organizations and institutions may take actions such as revoking licenses, and refusing permission to practice, but that is the extent of their powers. A person must be guilty of committing a crime in order to be punished in a court of law. The codes of ethics referred to always place their ethics in the context of law (for example, the confidentiality requirement exists, unless otherwise demanded by law[14, 15]).

- The medical experiments carried out by the Nazi doctors were almost always in compliance with the law and legal instructions from superiors, and were usually meticulously documented in reports.[1, 4] The counter-argument was that, even if a legal order were received, the physician should refuse to obey an order that he believes to be morally unjustified.[1]

- A strong argument for supporting the Nazi medical experiments was for the good of society, especially considering the saving of soldiers' lives during war.[1, 4]

- Medical staff from Japan had also conducted many experiments on the Chinese population, and had used live prisoners in training procedures for their doctors.[18]

- There are many instances where a second person's rights might override the confidentiality rights of a patient.[19]

- Many of the principles in the Nuremberg code were not universally followed as standard procedures, even in prosecuting countries. The Nuremberg Code had not existed before World War II; rather, it emerged as a response to the atrocities witnessed. Part of the defense was that, at the time, there had been many international medical experiments performed on condemned prisoners (including conscientious objectors), who received no pardon or reduction in sentence, and it was also questionable whether all international medical experiment subjects (or their parents, in the case of minors) had given their permission.[1, 20] Some of the most important medical experiments (such as those by Edward Jenner) had been performed without any indication of consent.

Although not widespread, other countries, including the USA, had conducted medical experiments on humans who were not fully informed, and so, could not have given informed consent. Some had been conducted before the war, but many continued well after.

Figure 10.4: Doctor examining a Tuskagee Syphilis Experiment subject (Source: United States National Archives and Records Administration)

At the very time of the Nuremberg trials, the Tuskegee Syphilis Experiment was being conducted in the USA, and ended only in 1972 when it was reported in the press.[21-23] Other experiments included the U.S. syphilis tests in Guatemala,[24] the Sonoma State Hospital experiments on disabled children,[25] and the radiation experiments on American citizens.[26] In all of these cases, the central tenets of the Nuremberg Code had not been followed. (For further reading on this topic, see Anthony Clare's *Medicine Betrayed*.[27])

Codes of Individual Countries

The American Medical Informatics Association. With the international differences in medical ethics, it is to be expected that there are also differences in health informatics ethics. Indeed there are, and several individual countries have their own health informatics codes. A complication is that much of the activity covered by health informatics may also be conducted in other fields, so different codes may exist for workers in those fields. This section highlights a few.

The American Medical Informatics Association (AMIA) has a code of ethics for its members.[28] It is significantly shorter than the IMIA code, but also looks the ethical relationship between doctor and patient (and the patients' family), colleagues, institutions and employers, and society in general. Regarding patients, there is an emphasis on confidentiality and security of information, and using all information for the intended purpose only. In the area of research, the code notes that "Basic human rights, especially as articulated and regulated in conducting research, must remain the highest ethical standard,"[28] although there is no specific mention of issues like informed consent and right of withdrawal from the experiment or trial.

The AMIA document is also cognizant of difficulties, however, and makes it clear that the code "is not intended to be prescriptive; it relies on the commonsense of the membership."[28]

United Kingdom. In the United Kingdom, the UK Council for Health Informatics Professions (UKCHIP) has the *UKCHIP Code of Conduct*,[29] which "sets out the standards of behaviour required of health informatics professionals registered with the" UKCHIP. The code has four short sections, dealing with "Working to professional standards," "Respecting the rights and interests of others," "Protecting and acting in the interests of patients and the public," and "Promoting the standards and standing of the profession."

European Parliament Directive. Similarly, there is the European Parliament Directive (95/46/EC) of 1995[30] which is binding on all member countries of the European Union, deals with the protection of data, and covers a wide range of issues from privacy to security. The most pertinent principles of this directive have been synthesized by de Lusignan et al.[31] into these:

- Personal data shall be processed fairly and lawfully.

- Personal data shall be obtained and processed for one or more specified and lawful purposes and not in any manner incompatible with those purposes.

- Personal data shall be adequate, relevant and not excessive in comparison to the purpose that it was collected for.

- Personal data shall be accurate and up-to-date where necessary.

- Personal data should not be kept longer than is deemed necessary.

- Personal data shall be processed in accordance with the rights of individuals as set out in the act.

- Personal data shall have appropriate security measures in place.

- Personal data shall not be transferred outside of the European Economic Area (EEA) unless adequate protections exist for the rights and freedoms of data subjects.

While this is a useful guide, the actual legally applicative directive is difficult for the lay person to understand, and appears extremely difficult to apply. For example, Article 8 of the directive states that "Member States shall prohibit the processing of personal data revealing racial or ethnic origin, political opinions, religious or philosophical beliefs, trade-union membership, and the processing of data concerning health or sex life." It is then followed by a series of exceptions where this does *not* apply, and is also followed by the statement that "Subject to the provision of suitable safeguards, Member States may, for reasons of substantial public interest, lay down exemptions in addition to those laid down in paragraph 2 either by national law or by decision of the supervisory authority." This means that member states may make laws that override the main paragraph of the article.

There are also several international guides, developed by different international medical associations that deal with use of specific health informatics activities, such as electronic health records and email communication between doctor and patient. These guides cover ethical issues such as data privacy and protection. Some examples are American Medical Association's (AMA) *Guidelines for Physician-Patient Electronic Communications* [32] and the *Guide to Australian electronic communication in health care*.[33]

Finally, there are many ethics guides from other disciplines that impact upon ethics in health informatics.[34]

Pertinent Ethical Principles

In this rather strange mixture of ethics, laws, and cultural influences, there are some principles that appear to be common. Given the complexities outlined above, it is useful for the student of health informatics to have a summary of the most pertinent ethical points, and that summary is supplied here. Principles such as the right to privacy, informed consent in research, and the non-transferability of ethical responsibilities (accountability) will be discussed. With the understanding that the importance of each of these will be viewed differently in different circumstances, these are useful guides from medical ethics and health informatics ethics. In these descriptions, any reference to patients would refer to research subjects and to their families. In health informatics education, these will extend to students.

- Right to privacy. The patient has a right to privacy, which means that information that the HIP has obtained must not be shared with others unless there is reason to believe that it is in the best interests of the patient.

- Guard against excess. There should be safeguards against excessive personal data collection; only data specifically needed should be collected.

- Security of data. The right to privacy, and maintaining patient safety, also means that there is a responsibility on the researcher to keep the data as secure as possible, in order to prevent unauthorised access to it. As an extension of this, and incorporating ethical operations of the institutions in which the HIP works, if the HIP becomes aware of security problems, even those that are not under his/her direct control, the responsible persons must be informed. The emphasis on security, however, must not be so strong that it impedes the patient's right to access that data.

- Integrity of data. This is also related to security, and the HIP has to ensure that data are kept current and accurate. In addition, data cannot be presented in such a way that it presents an untrue picture of reality, or is specifically designed to mislead the reader.

- Informed consent. While the patient should be aware of what is to happen, that awareness can only be complete if the patient is truly informed. Similarly, the researcher may do only what has been consented to, and if the researcher wishes to do anything else (e.g. use data for any other purpose), then new consent must be obtained. Crucial aspects of informed consent are: "(1) competence of the subject to consent...,(2) disclosure of information, (3) the subject's understanding of the information being disclosed, (4) volition or choice in the consent, and (5) authorization of the consent."[35]

- Laws. The HIP needs to be aware of the laws that apply. Where there is a conflict between the law and the professional ethics, the HIP will have to make difficult decisions. In addition to the discussion above, this issue is explored further a little later in this chapter.

- Medical ethics. The HIP needs to understand that health informatics ethics is a sub-set of medical ethics. This means that all issues that apply to medical ethics in general, such as the physical and psychological safety of the patient, also apply in health informatics ethics.

- Sharing the data. If it is necessary to share the data with anyone else (e.g. for further research, temporary or permanent storage, or data transmission), then the HIP must be sure that all the above principles are also being adhered to.

- Wider responsibility. HIPs have ethical responsibilities towards their employers and the wider community regarding the protection of the data and maintaining professional standards.

- Beneficence and non-maleficence. Implicit in all these are the principles of beneficence and non-maleficence. This means that the ethics must be beneficial to the patient, and must be consciously aimed at preventing harm to the patient.

- Non-transferability. The responsibility and accountability for adhering to these rests with the HIP, and cannot simply be transferred.

Difficulties Applying Medical Ethics in the Digital World

The previous section of this chapter traced the recent history of health informatics ethics, and showed the various principles involved. At this stage, it is obvious that the issues are extremely complex. It is now the time to turn to some practical examples to see how some of these principles can be applied.

Ethical Issues with Large Databases: Informed Consent and Confidentiality

A difficult issue when conducting research on large databases, including hospital data bases of Electronic Health Records or Electronic Medical Records (EHRs or EMRs), is how to obtain informed consent for the use of patient data. In addition, one needs to ensure that patient information remains anonymous. These issues are always important, but even more so when working with conditions (such as HIV/AIDS or psychiatric conditions) that have social stigmas.[36]

One way of obtaining informed consent for use in research is to obtain "broad informed consent"[37] at the time that the information is gathered. This is an idea borrowed from the study of large biological samples, and is regarded as the most practical and economically viable approach for researchers. A variation is to grant consent for the database to be used for specific types of research only.

Some databases may be small (such as from a clinic or hospital) while others may be large (a national database), and several countries are grappling with problems of informed consent for researchers while protecting patients from abuse.[38, 39] In one instance, Iceland created a national database with "presumed consent," but allowed individuals to opt out of the program, thereby removing their information from the database. This solution is not always simple, and the legal relationship between presumed and informed consent, especially around issues of identifiable genetic material, continues today.[39, 40]

Any approach will be influenced by the national culture, so will differ from country to country, and may also differ depending on the nature and purpose of the database.[38]

Because obtaining general or presumed consent could be open to abuse, it is extremely important that the researcher ensures that the research does not conflict with other areas of ethics, such as exposing the patients to stress or exposing any identifying information.[36, 37] One should also guard against corporate ownership of such databases, as these organisations may work to different ethical guidelines, and there may be conflicts of interest in the research and research outcomes. Where such ownership cannot be avoided, then researchers should not be unethically influenced in their work.[41]

As addressed in the Chapter on Privacy, Health Care Workers in the US need to be aware of the HIPAA Privacy Rule, [42] as this applies to all doctors who transmit any patient data electronically.

Research on electronic postings: privacy and disclosure. The internet is full of information simply waiting to be analysed. One area that has received attention has been online environments in which users create postings in conversations. These might be in discussion lists (sometimes called "listserves"), forums, bulletin boards, and social networking sites, such as Facebook.

In many of these sites, medical information, sometimes very personal, is exchanged. Even in sites where personal information is not exchanged, the context may be a medically-oriented site. The prime ethical questions for the researcher researching these sites revolve around informed consent and the privacy of the information that is shared on these sites. In short, the question is this: are these electronic postings to be treated with the same level of confidence and anonymity that one would apply to patients in a self-help group?

Resolving the issue depends on whether the 'human-subject' model or the 'textual' model is applied. These are explained further below.

- The human subject model. Briefly, the human subject model is an extension of the medical view of patient information, and it views the electronic postings as reflective of real people, and so all the ethical rules regarding informed consent, privacy, and ensuring that there is no psychological or physical harm to participants must apply.[43, 44] This means that, before quoting from or referring to a site, the researcher must obtain informed consent.

- The textual object model. An opposing view is that a posting in a bulletin board is merely a piece of text, and is subject only to the laws and ethics that govern any piece of text. These might include rules regarding plagiarism and copyright, but do not involve anything to do with a human patient. The text has been placed into a publicly-accessible area (the internet) and any expectation of privacy and confidentiality is unwarranted. As Walther argues, this is much like a conversation in a park, and that "people do not expect to be recorded or observed although they understand that the potential to do so exists."[45]

If the person has not posted personally identifiable information, then there is even less concern regarding privacy; after all, the only problem that might exist is that the person can be identified. For example, in the USA, a "Human subject means a living individual about whom an investigator (whether professional or student) conducting research obtains:

- o data through intervention or interaction with the individual, or
- o identifiable private information."[46]

If people are concerned about identity, however, they can create pseudonyms and usernames that make it difficult to identify them. One may argue that, if they have not taken such precautions, then they are willing to have themselves publicly identified.

Finally, the textual object model is supported by much 20-century literary theory[47, 48] which clearly separates any discussion of text from the discussion of the author or even the author's intention.[49, 50]

Problems with the textual object model. There are several problems with applying the textual model to medical research, and some of these are:

- The arguments are frequently based on traditions from other fields (e.g. sociology, or literary theory). We are working in a medically-related field (specifically, health informatics), and so should give greater credence to ethical rules in that field of study.

- Although a specific posting may not contain information that can identify a person, when many of these postings are combined, it may be easy to form a picture that can be used to identify the person.

- There is a strong tradition in medicine that, even when objects are researched, they are not specifically identified.

- In the many examples above, we have seen that laws should not be taken as a standard of ethics. At best, they may set a *minimum*, from which the ethical HIP works.

The difficulties and disadvantages of applying the human subject model to electronic postings. Having said this, the researcher must be aware that there are difficulties with applying the human subject model when performing research on electronic postings. This subsection identifies some of these, and suggests solutions.

- Establishing informed consent can be difficult, if not impossible. With a group of several thousand, where people join and leave continually, how does one establish informed consent?

- One approach is to attempt to determine whether these are necessary. This depends largely upon the rules of registration and public access to the list.[43,51-53] If the list is very tightly controlled, where members have to be a member of a particular organisation, are have to supply corroborating evidence of their identity, then the researcher is advised to obtain full informed consent. If, however, the list is large, registration requires only an arbitrary user name and password (and, perhaps an email address), and the site is searched and indexed by general search engines, then informed consent is less important.

- How does one preserve privacy and anonymity?

- Again, one can be guided by the amount of privacy that is assumed in the group. In addition, however, unless informed consent from individuals has been obtained, the researcher should avoid referring to specific postings or individuals. The researcher should even avoid anonymous quoting, as this can be used through a search engine to identify the original piece of text. Rather, the research should use aggregated data (i.e. totals, means, etc.) to give an overall impression.

- Finally, if the researcher wishes, she/he may wish to disguise the site. This is discussed in more detail below.

Researcher's Responsibility for Data Security

In July 2009, The University of North Carolina at Chapel Hill discovered that a computer had been hacked as far back as 2007. The data from the Carolina Mammography Registry containing some 180,000 mammography records (including 114,000 social security numbers) had been potentially exposed.[55] One of the prime concerns was of responsibility and culpability.

The chief researcher, Professor Bonnie C. Yankaskas argued that the university IT security staff were responsible for the security of the file server.

The university argued that the chief researcher was to blame.

Initially, Prof. Yankaskas was demoted, and had her salary reduced. After a legal fight costing Prof. Yankaskas some $350,000, her position and salary were restored, but she was forced to retire, effective at the end of 2011.[56]

Transferring Ethical Responsibility

A tempting route to reducing the researcher's ethical responsibilities would be to transfer the responsibility for ethical behaviour to others, allowing the researcher to concentrate on the task at hand: the research. This might be done in three ways:

- As long as the researchers are obeying the law, they are safe from prosecution, as the laws of the State are there as a guide.

- If the researchers work at an institution that has an Ethics Committee or an Institutional Review Board (IRB), then they submit a protocol that describes the research beforehand, and then receive ethics approval for the research. The researchers may feel that they are now 'covered' and so can do whatever they like, as long as they stick to the protocol.

- Keeping data secure is a complex technical process, so one should simply have a database manager who takes full responsibility for the data. If the data are then mistakenly made public, it is the database manager who has to deal with the problem.

These, however, are not effective solutions:

- We have already seen that handing this responsibility over to the law or State, highlighted during the Nuremberg Trails is not an acceptable solution, as laws do not establish ethics.

- Because of the newness of the field of health informatics, IRBs may not have representatives that are fully aware of the ethical issues and technical applications (e.g. that that simply searching on a quotation from a forum allows you to find that forum immediately), or the extent to which informed consent is required.

- Ultimately, you are responsible for your data. Technical staff may be responsible for the storage systems, but the overall responsibility for the material cannot be transferred to anyone else. In cases where data breaches have occurred, all parties (including the institution) may face legal prosecution, as was the case in which a clinic's data regarding HIV patients was compromised because of peer-to-peer file sharing software on their computers.[54]

Electronic Communication with Patents and Caregivers

All medical students know that it is relatively easy to find most people's email address. When you are a practising doctor, your name, work address and telephone number will already be known to your patients. Using that information, finding your email address is a small step. Because of the convenience of email communication (to both you and your patients), your patients will wish to email you on a range of topics. One of the most important benefits is that instructions you give can be clearly laid out, and can be referred to later by the patient; this greatly reduces the risks to the patient.

There are, however, ethical issues that need to be considered when medical personnel interact online with patients. Two guides that have already been mentioned, the AMA's *Guidelines for Physician-Patient Electronic Communications*[32] and the *Guide to Australian electronic communication in health care*[33] have useful information for the practising doctor. In addition, the AMA has another guide that refers to email communication.[57]

The AMA's guide begins by explaining the value of email communication, and then gives useful advice about setting up the communication channels and some medico-legal issues. This includes things like making the patient aware of who is reading the email, the types of email topics that are acceptable, use of language, and tips for the patients to ensure they can quickly reference the relevant emails. In addition, the guide advises that the physician should not use email communication with new or prospective clients with whom no personal contact has yet been established, should maintain the same ethical standards that apply to other areas of medicine, should ensure that permission has been obtained for email communication, and should ensure that the email has a disclaimer dealing with breaches of security and privacy, identity of corresponding parties and possible delays in responses.

Practical Steps

Measures to Ensure Consent Forms and Other Documents are Understood

This chapter has discussed informed consent at length, and research usually has to be accompanied by a consent form that is signed by the research participant. But how sure are we that the participant has actually understood the contents of the consent form and other documents (e.g. survey forms)? In face-to-face research, the researcher can pose questions to ensure that the information has been understood; in online research this is not always possible. (Even in face-to-face research, the use of questions can be embarrassing to the research subject, and time-consuming.) A useful approach is to reduce the complexity level of the language so that the person can understand the form.

For English, there are several tests used to determine the complexity of language in a text, although the most popular are the *Flesch Reading Ease Test*, and the *Flesch-Kincaid Test*. You can find out more about these tests, but, essentially, the tests check various characteristics of a document, such as the average number of words in a sentence and the number of syllables in each word. The *Flesch Reading Ease Test* assigns a value of one to 100 (where one is most difficult, and 100 is easy), and the *Flesch-Kincaid Test* assigns a number that corresponds to the US school grade. This means that a document with a Flesch-Kinkaid Test score of eight could be understood by an 8th grader, while a score of 14 would be at university level.

There are several computer applications that can perform the test for you. If you are using the Microsoft Office suite, you can implement the test in MS-Word, by making the following changes to your settings:

- MS-Word 2003:

- In the pull-down menu **Tools,** select **Options | Grammar**

- Under **Grammar**, select **Show readability statistics**

- MS-Word 2007:

  - Click on the **Office Button**

  - Select **Word Options**

  - Select **Proofing** | When correcting spelling and grammar in Word

  - Select **Show readability statistics**

- MS-Word 2010:

  - Click on **File**

  - Select **Options**

  - Select **Proofing** / When correcting spelling and grammar in Word

  - Select **Show readability statistics**

From now on, when you run a spelling and grammar check, the final dialog box will display the readability scores (see Figure 10.5).

In addition to the percentage of passive sentences, the dialog box will also give the Flesch-Reading Ease and the Flesch Kincaid Grade scores. You can use these statistics to modify your documents, and re-run the test until you are satisfied with the results.

Note that it is easy to 'trick' the tests. The aim is not to trick the tests, but rather to use the results of the test as a guide for your own research. For instance, if the document reported about in Figure 10.5 were a document for university students, it would probably be suitable. If it were a consent form to be given to children, it would probably require extensive revision.

Figure 10.5: Readability statistics from a document in MS-Word 2007

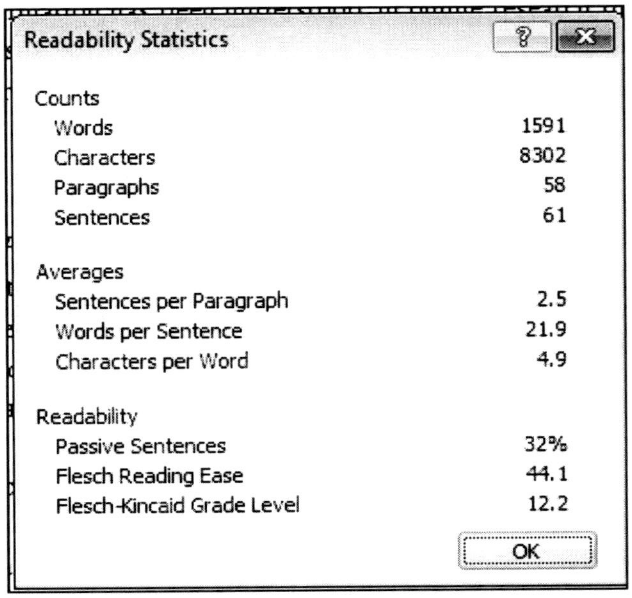

Simple Data Protection

Security breaches at medical facilities occur on an almost daily basis.[58] While your network administrators will probably implement several strategies to assist with security, you can also do your part. This is particularly important if you are using a computer in a shared location (such as at home) or using a laptop, which has a high risk of being stolen.

- Passwords: most operating systems allow you to password-protect your computer. In addition, most word processing, spreadsheet and database programs have built-in password protection. This allows your documents to be password-protected.

- There are several encryption programs which can allow for entire folders to be encrypted. That way, any documents stored in the folder, whether password-protected or not, will be protected. A free encryption tool is *TrueCrypt* (available from http://www.truecrypt.org/).

- You should have an anti-virus program on your computer, and keep it up-to-date. There are several good, free anti-virus programs, such as *AVG Free* (available from http://free.avg.com/ww-en/homepage).

- You should also have anti-spyware software installed on your computer. A good, free anti-spyware program is *SUPERAntiSpyware Free Edition* (available from http://www.superantispyware.com/)

Disguising Researched Web Sites or Bulletin Boards

Unfortunately, no matter how much you try to protect your research subjects, there will be some people who will attempt to discover their true identity.[44, 49] This may because they view it as their right, or do not work to the same ethical model, or who simply see it as a detective game to be played. In medical research, there is a tradition of intentionally changing information in order to prevent people using it to identify the subjects. For example, you may change people's names, cities or even experiences and medical conditions, as long as it does not directly impact on the nature of the research. This is considered "heavy disguise,"[49] and is also employed by some researchers who research web sites.[59-61]

If you are performing research on bulletin boards in a web site, and you wish to disguise your research web site so that it has little chance of being discovered, you may wish to create a dummy web site that is designed to lure investigators away from the real site. Such a web site is called a "Maryut site."[44] (The term "Maryut site" is a reference to the story of the creation of a decoy site at Maryut Lake to prevent Alexandria Habour's being bombed during World War II.) The process of using a Maryut site would be the following:

- The researcher creates a fake (or "Maryut") web site that has a structure similar to the research site.

- The researcher then populates the Maryut site with plausible information. This would include structures (e.g. names of forums) that are found in the research site, plus additional forums. The new information would need to be of such a nature that it does not detract from the validity of the research. An example would be the name of a forum that one might expect to find on such a site but that does not exist in the research site.

- In the research paper, amongst the real information listed, the researcher lists the fake information that is found only in the Maryut site. This information must not materially affect the research, in much the same way that alterations to patients' experiences do not materially the research.

- The Maryut site is taken off-line, and archiving sites may list only top-level pages that will hold the Maryut site's general information (e.g. forum names and number of participants).

- If the information from the published research is then identified as "clues," this information will more closely match the Maryut site than it will match the research site. If this information is used to search for the research site, the information on the Maryut site will divert "detectives" away from the research site, and will point them either to the discontinued Maryut site or to the archived site.

- The researcher might even post information into blogs, discussion lists, online forums and social networking sites, under pseudonyms, directing detectives to the Maryut site.

- Naturally, it is possible to create more than one Maryut site, all similar, but with small differences, to be found under different but similar search strategies. Some may never be found.

- An essential part of the disguise may be for the researcher never to disclose whether or not Maryut sites were used during the research.

The ethics of making Maryut sites, in the interests of safeguarding the non-disclosure, may be a point of discussion by IRBs and ethics' committees. A danger with this method is that the Maryut site may inadvertently point the detectives to a valid secondary site that has nothing at all to do with the research.

Ensure IRBs, Ethics Committees and other administrative structures are aware of health informatics ethics issues.

This chapter has already referred to the fact that representatives on IRBs may not be familiar with ethical issues in health informatics. It is the responsibility of researchers and other HIPs to inform their IRBs of the health informatics ethics issues. This will allow the IRB to better understand the preventative actions taken by HIPs, and also to understand the motivations behind such actions. This information sharing can take the form of workshops and reports, supported by practical implementation when applying for research and grant approval.

Health Informatics Ethics and Medical Students

The final portion of this chapter deals with the medical student. The medical student is already a health professional, simply at the early stages of his/her career. Medical students are usually bound by the rules of their national medical professional organisations; in the same way, they should feel bound by the ethical rules of health informatics.

Online Behaviour in Social Networking and Other Interactive Sites

A major change brought about by Web 2.0 are the tools that provide the facility for every web user to contribute to web pages. The most obvious of these tools are wikis, blogs, and social networking sites like Facebook. As health professionals, however, students need to remember that everything they post online may stay online for a long time – even if it is deleted, it may be stored in electronic archives. With this in mind, students should be extremely careful about online comments and photographs of themselves, colleagues and patients. (This applies to all students, not only medical students).[62]

- A survey amongst US medical schools recently found that as many as 60% reported incidents in which students had posted inappropriate material online. Students were usually given warnings, but, in some cases, were dismissed.[63]

- There is something of a tradition of medical students posing with their cadavers, sometimes merely as illustrations for teaching, but very often for other purposes, such as humour.[64] Times have changed, however, and, in the digital age, the posting of such photographs on to the internet is generally not accepted by medical schools.[65]

Figure 10.6: Dissecting Room, Jefferson Medical College, Philadelphia (Source: US Library of Congress)

Figure 10.7: Students and teachers of the Civil Medical School, Constantinople (Source: US Library of Congress)

Other Student Activities

Research projects. As a student, you will be involved in several research projects. Some of these may be small projects performed by yourself (e.g. surveys of fellow students), or they may be parts of larger projects set by other researchers. In all cases, it is important to remember that you are bound by the same ethical rules that are raised in this chapter. In cases of doubt, speak to the staff, your supervisor, or to members of your Ethics Committee or IRB.

Plagiarism. Similarly, professional conduct extends to plagiarism. Although plagiarism is not specific to health informatics, because of the ease with which information can be copied and pasted, there is a temptation to plagiarise others' material. At most institutions, students found guilty of plagiarism may be expelled from their institutions.

Use of Paper Mills. It may be tempting to make use of "Paper Mills." Paper Mills are web sites that allow you to submit your assignment details, and somebody else will write your assignment for you (for a fee). Again, however, if students are caught, they are usually expelled immediately.

Manipulation of electronic files. Electronic files (whether text, audio, video, or still graphics) are easy to manipulate. There are many acceptable reasons for doing so, such as for editing purposes, or for protecting the identity of research subjects. When performing such manipulation, you need to ensure that you are not breaking any copyright or other laws. In addition, you need to ensure that the finished product does not present false information.

Recording of lectures and other class activities. It may be tempting to video- or audio-record classes or lectures. This may be useful for students, so that they can watch and/or listen to the lectures afterwards. Before doing this, you need to ensure that you have the lecturer's permission. In most cases, you will also need the permission from fellow students. As a guide, refer to the discussion on informed consent above.

Accessing documents illegally. Frequently, while performing research, you will find an abstract to an article, and you wish to read that article. Unfortunately, a great deal of information is available in books or journals that charge subscription fees (i.e. you have to pay to access the journal or individual articles or books). In most cases, your university library will already have paid a fee, and will be able to grant you legal access to these resources. In some instances, however, they have not, and so you do not have legal access to the resource.

Because students (and qualified doctors) want access to these resources, they are tempted to use sites that share such articles.[66,67] Alternately, they use websites that share usernames and passwords to library databases.[68] The justification for doing so is that, ultimately, patients will benefit from the knowledge that the health professional has gained. Unfortunately, this practice is usually both illegal and unethical.

There are other, both legal and ethical, methods that one may wish to try. These include:

- Searching for the resource in a legitimate site. Frequently, publishers allow authors to place copies of their articles on their own web sites, and in publicly accessible repositories. These can be searched and the articles freely downloaded.

- Contacting the authors. Authors are usually permitted to send copies of their articles to a limited number of people who request them. You can contact the author and make such a request. (The author's contact details will usually be visible on the same page that showed the abstract).

Future Trends

Because health informatics ethics relies on practices from diverse and continually evolving fields, it is difficult to make predictions about future trends. That said, however, based on the recent history, there are a few likely scenarios:

- Because medicine and informatics are diverse fields, balancing the ethical practices of health informatics will always be difficult for the healthcare provider (HCP). The various codes of ethics will be continually updated to take technological developments into account, but will always lag some way behind these developments.

- Digitized medical data will play an increasingly important role as a commodity in patients' lives.

- Because of these changes and the continual emergences of new technologies, HCPs (including students) will be faced with new ethical challenges. They will need to use the basic principles as guides, and their consciences where the principles do not take these developments into account.

- The tension between ethics, culture and law will not become easier in the short term.

- Health informatics ethics is likely to emerge as a field of study by itself.

Key Points

- Health informatics ethics stems from medical ethics and informatics ethics, and combines principles from both fields.

- The IMIA Code of Ethics for Health Information Professionals contains guidelines in a range of categories, namely: fundamental ethics principles, general principles, subject-centred duties, and duties towards HCPs, institutions / employees, society, the profession and oneself.

- The relationship between ethics, law, culture and society is extremely complex and fluid, and varies internationally and chronologically. The HCP must become acquainted with the issues that have a direct bearing on his or her practices.

- The most pertinent ethical principles in health informatics ethics relate to: right to privacy, guarding against excess, security and integrity of data, informed consent, data sharing, wider responsibilities, beneficence and non-maleficence and non-transferability of responsibility. All these must be seen within the legal and medical ethics context.

- There are several examples of the application of the principles to research and other situations. These applications can be used as a guide for the HCP, beginning with the HCP as a student

Conclusion

At Nuremberg, a total of 23 defendants (of whom 20 were medical doctors) were tried for medically-related crimes. Seven were acquitted of all charges. Of the 16 found guilty, seven were sentenced to death. These seven, including Dr. Karl Brandt, were executed on 2nd June 1948.

From one of the darkest periods of medical history, codes of ethics evolved. From these codes and codes in informatics ethics, health informatics ethics codes have further evolved. Although these codes have varying degrees of effectiveness, they do provide essential principles for the medical student who will work with electronic data. It is essential that these principles are understood and applied as conscientiously as possible.

References

1. Nuernberg Military Tribunals. Trials of war criminals before the Nuernberg Military Tribunals under Control Council Law No. 10, Volume II: "The Medical Case" and "The Milch Case." Washington: US Government Printing Office, 1949.

2. Nuernberg Military Tribunals. Trials of war criminals before the Nuernberg Military Tribunals under Control Council Law No. 10, Volume IV: "The Einsatzgruppen Case" and "The Rusha Case." Washington: US Government Printing Office, 1949.

3. The Library of Congress. Nuremberg Trials. http://www.loc.gov/rr/frd/Military_Law/Nuremberg_trials.html (Accessed January 30 2011), 2010.

4. Nuernberg Military Tribunals. Trials of war criminals before the Nuernberg Military Tribunals under Control Council Law No. 10, Volume I: "The Medical Case." Washington: US Government Printing Office, 1949.

5. Hohendorf G, Rotzoll M, Richter P, Eckart W, Mundt C. Die Opfer der nationalsozialistichen "Euthanasie-Aktion T4." Nervenarzt 2002;73:1065-74.

6. Mostert MP. Useless Eaters: Disability as genocidal marker in Nazi Germany. Journal of Special Education 2002;36(3):155-68.

7. Proctor R. Racial Hygiene: medicine under the Nazis. Cambridge: Harvard UP, 1988.

8. World Medical Association. World Medical Association Declaration of Helsinki - Ethical Principles for Medical Research Involving Human Subjects, 2008. http://www.wma.net/en/30publications/10policies/b3/index.html (Accessed January 12 2010).

9. Capurro R. Towards an ontological foundation of information ethics. Information Ethics: Agents, Artifacts and New Cultural Perspectives. Oxford, 2005. http://www.capurro.de/oxford.html (Accessed March 02 2011).

10. Wiener N. The Human Use of Human Beings: Cybernetics and Society. Cambridge, Massachusetts: Da Capo Press (Originally published in 1950), 1988.

11. Kostrewski B, Oppenheim. Ethics in information science. Journal of Information Science 1979;1(5):277-83.

12. Hauptman R. Ethical challenges in librarianship. Phoenix, Arizona: Oryx Press, 1988.

13. Severson R. The Principles of Information Ethics. New York: M.E. Sharp, 1997.

14. Association for Computing Machinery. Code of Ethics and Professional Conduct, 2011. http://www.acm.org/about/code-of-ethics (Accessed February 10 2011)

15. Canadian Information Processing Society. Code of Ethics and Professional Conduct, 2007. http://www.cips.ca/?q=system/files/CIPS_COE_final_2007.pdf (Accessed February 10 2011)

16. International Medical Informatics Association. The IMIA Code of Ethics for Health Information Professionals, 2011. http://www.imia-medinfo.org/new2/pubdocs/Ethics_Eng.pdf (Accessed January 31, 2011)

17. Katz J. The consent principle of the Nuremberg Code: Its significance then and now. In: Annas G, Grodin MA, editors. The Nazi doctors and the Nuremberg Code: human rights in human experimentation. New York: Oxford UP, 1992.

18. Lafleur WR, Böhme G, Shimazono S. Dark Medicine: Rationalizing Unethical Medical Research. Bloomington: Indiana UP, 2007.

19. Richards E. Transcript of "Tarasoff vs Regents of University of California, 17 Cal. 3d 425, 551 P.2d 334," 2002. http://biotech.law.lsu.edu/cases/privacy/tarasoff.htm (Accessed February 19 2011)

20. Harkness JM. Nuremberg and the Issue of Wartime Experiments on US Prisoners. JAMA 1996;276(20):1672-5.

21. Reverby S. Tuskagee: Could it happen again? Postgrad. Med. J. 2001;77:553-4.

22. Reverby S. A new lesson from the old "Tuskagee" study. Huffington Post 2009;December 3: http://www.huffingtonpost.com/susan-reverby/a-new-lesson-from-the-old_b_378649.html (Accessed February 12 2011)

23. Reverby S. Examining Tuskagee: the infamous syphilis study and its legacy. Chapel Hill: University of North Carolina Press, 2009.

24. McNeil DG. U.S. Apologizes for Syphilis Tests in Guatemala. New York: New York Times, 2010. http://www.nytimes.com/2010/10/02/health/research/02infect.html?_r=1&th&emc=th (Accessed October 3 2010)

25. Leung R. A Dark Chapter In Medical History: Vicki Mabrey On Experiments Done On Institutionalized Children. *60 Minutes,* 2005. http://www.cbsnews.com/stories/2005/02/09/60II/main672701.shtml (Accessed February 8 2011)

26. Advisory Committee on Human Radiation Experiments. Final Report. U.S. Government Printing Office, 1995. http://www.hss.energy.gov/healthsafety/OHRE/roadmap/achre/report.html (Accessed February 12, 2011)

27. Clare A. Medicine Betrayed: The Participation of Doctors in Human Rights Abuses. London: Zed Books, 1992.

28. American Medical Informatics Association (AMIA). AMIA Code of Ethics, 2010. https://www.amia.org/inside/code (Accessed February 7 2010)

29. UK Council for Health Informatics Professions. UKCHIP Code of Conduct, 2011. http://www.ukchip.org/ (Accessed February 10 2011)

30. European Parliament. Directive 95/46/EC of the European Parliament and of the Council of 24 October 1995, 1995.

31. De Lusignan S, Chan T, Theaom A, Dhoul N. The roles of policy and professionalism in the protection of processed clinical data: A literature review. Int. J. Med. Inf. 2007;76:261-8.

32. AMA. Guidelines for Physician-Patient Electronic Communications, 2008. http://www.ama-assn.org/ama/pub/category/2386.html (Accessed November 20 2008)

33. Standards Australia. Guide to Australian electronic communication in health care. Sydney, Australia: Standards Australia, 2007.

34. European Commission. Research ethics and social sciences, 2010. http://ec.europa.eu/research/science-society/index.cfm?fuseaction=public.topic&id=1433 (Accessed October 8 2010)

35. Waltz CF, Strickland OL, Lenz ER. Measurement in Nursing and Health Research (4th ed.). New York: Springer, 2010.

36. Simon GE, Unützer J, Young BE, Pinjcus HA. Large medical databases, population-based research, and patient confidentiality. Am. J. Psychiatry 2000;157:1731-7.

37. Petrini C. "Broad" consent, exceptions to consent and the question of using biological samples for research purposes different from the initial collection purpose. Soc. Sci. Med. 2010;70:217-20.

38. Chadwick R, Berg K. Solidarity and equity: New ethical frameworks for genetic databases. Nature Reviews Genetics 2001;2(4):318-21.

39. Gulcher JR, Stefánsson K. The Icelandic healthcare database and informed consent. N. Engl. J. Med. 2000;342:1827-30.

40. Gertz R. An analysis of the Icelandic Supreme Court judgement on the Health Sector Database Act. Script-ed 2004;1(2):241-58.

41. Bernard GR, Artigas A, Brigham KL, Carlet J, Falke K, Hudson L et al. The American-European Consensus Conference on ARDS. Am J Respi Care Med 1994;149:818-24.

42. United States Department of Health & Human Services. Summary of the HIPAA Privacy Rule. Office for Civil Rights, 2003.

43. Eysenbach G, Till JE. Ethical issues in qualitative research on internet communities. BMJ 2001;323:1103-5.

44. Masters K. Non-disclosure in internet-based research: The risks explored through a case study. Internet Journal of Medical Informatics 2010;5(2). http://www.ispub.com/journal/the_internet_journal_of_medical_informatics/volume_5_number_2_51/article_printable/non-disclosure-in-internet-based-research-the-risks-explored-through-a-case-study.html (Accessed April 16 2010).

45. Walther JB. Research ethics in internet-enabled research: Human subjects issues and methodological myopia. Ethics and Information Technology 2002;4:205-16.

46. National Institutes of Health Office of Human Subjects Research [USA]. Code of Federal Regulations, Title 45, Public Welfare, Part 46: Protection of Human Subjects. Bethesda, 2005.

47. Wimsatt W, Beardsley M. The Intentional Fallacy. In: The Verbal Icon: Studies in the Meaning of Poetry. Lexington: University of Kentucky Press, 1954. p 3-18.

48. Barthes R. Image Music Text. New York: Hill and Wang, 1977.

49. Bruckman A. Studying the amateur artist: A perspective on disguising data collected in human subjects research on the internet. Ethics and Information Technology 2002;4:217-31.

50. Bassett EH, O'Riordan K. Ethics of internet research: Contesting the human subjects research model. Ethics and Information Technology 2002;4:233-47.

51. Eysenbach G, Wyatt J. Using the Internet for Surveys and Health Research. J Med Internet Res 2002;4(2):e13.

52. Frankel MS. Ethical and legal aspects of human subjects research on the Internet. American Association for the Advancement of Science, 1999.

53. Jankowski NW, van Selm M. Research ethics in a virtual world. Guidelines and illustrations. In: Carpentier N, Pruulmann-Vengerfeldt P, Nordenstreng K, Hartmann M, Vihalemm P, Cammaerts B et al., editors. Media Technologies and Democracy in an Enlarged Europe. Tartu University Press, 2007.

54. Circuit Court For the 16th Judicial Circuit KC Illinois. Complaint: John Doe1 et al. vs. Open Door Clinic of Greater Elgin, 2010.

55. Carolina Mammography Registry (University of North Carolina). The Carolina Mammography Registry Server Compromise: Frequently Asked Questions, ND. http://www.unc.edu/cmr/breach_faq.pdf (Accessed November 7 2011)

56. Ferreri E. Breach costly for researcher, UNC-CH. NewObserver.com, 2011. http://www.newsobserver.com/2011/05/09/1185493/breach-costly-for-researcher-unc.html (Accessed November 7 2011)

57. American Medical Association. Code of Medical Ethics [Physician Version], ND. http://www.ama-assn.org/amam1/pub/upload/mm/369/ethics-in-hand-physician-version.pdf (Accessed February 15 2011)

58. DataBreaches. Databreaches.Net: Office of Inadequate Security., 2011. http://www.databreaches.net/?cat=22 (Accessed February 14 2011)

59. Turkle S. Constructions and Reconstructions of Self in Virtual Reality. Mind, Culture, and Activity 1994;1(3):158-67.

60. Turkle S. Life on the Screen: Identity in the Age of the Internet. New York: Simon & Schuster, 1995.

61. Turkle S. Multiple subjectivity and virtual community at the end of the Freudian century. Sociological Inquiry 1997;67(1):72-84.

62. Krebs B. Court Rules Against Teacher in MySpace 'Drunken Pirate' Case. The Washington Post, 2008. http://voices.washingtonpost.com/securityfix/2008/12/court_rules_against_teacher_in.html (Accessed February 14 2011)

63. Chretien KC, Greyson R, Chretien J-P, Kind T. Online Posting of Unprofessional Content. JAMA 2009;302(12):1309-15.

64. Warner JH, Edmonson JM. Dissection: photographs of a rite of passage in American medicine, 1880-1930. New York: Blast Books, 2009.

65. Einiger J. Cadaver photo comes back to haunt resident. ABC News, 2010. http://abclocal.go.com/wabc/story?section=news/local&id=7253275 (Accessed February 14 2011)

66. Masters K. Opening the non-open access medical journals: Internet-based sharing of journal articles on a medical web site. Internet Journal of Medical Informatics 2009;5(1). http://tinyurl.com/kmoajournals (Accessed October 28 2009).

67. Masters K. Articles shared on a medical web site - an international survey of non-open access journal editors. Internet Journal of Medical Informatics 2010;5(2). http://www.ispub.com/journal/the_internet_journal_of_medical_informatics/volume_5_number_2_51/article_printable/articles-shared-on-a-medical-web-site-an-international-survey-of-non-open-access-journal-editors.html (Accessed April 16 2010).

68. Masters K. Opening the closed-access medical journals: internet-based sharing of institutions' access codes on a medical web site. Internet Journal of Medical Informatics 2010;5(2). http://www.ispub.com/journal/the_internet_journal_of_medical_informatics/volume_5_number_2_51/article_printable/opening-the-closed-access-medical-journals-internet-based-sharing-of-institutions-access-codes-on-a-medical-web-site.html (Accessed April 16 2010).

11

Consumer Health Informatics

M. CHRIS GIBBONS

ROBERT E. HOYT

Learning Objectives

After reading this chapter the reader should be able to:

- Identify the origin of consumer health informatics

- Identify and discuss consumer health informatics (CHI) tools

- Discuss the features and formats of personal health records

- Identify electronic tools for patient to physician communication

- Outline barriers to CHI adoption

- Discuss the future of consumer health informatics

Introduction

Considerable interest is emerging concerning the potential of information and communications technologies that are tailored to consumers and used within the context of managing health or healthcare issues. This emerging focus has been referred to as consumer health informatics (CHI). The federal government is fully supportive of CHI as demonstrated by the Federal Health IT Strategic Plan 2011-2015 goal to "[e]mpower individuals with health IT to improve their health and the health care system."[1] Also, several Meaningful Use objectives address health information technology's (HIT) impact on patients, addressed in the chapter on electronic health records.

This chapter discusses several consumer health informatics topics: patient health information applications, home telemedicine devices, patient portals, personal health records and electronic patient - physician communication. It should become obvious after reading that these topics are interrelated and not separate. In addition, many of these features may be integrated with electronic health records and health information organizations. Figure 11.1 displays multiple interrelationships extant in consumer health informatics.

We will first begin with a discussion of the origins and driving forces behind consumer health informatics.

Figure 11.1: Consumer health informatics interrelationships

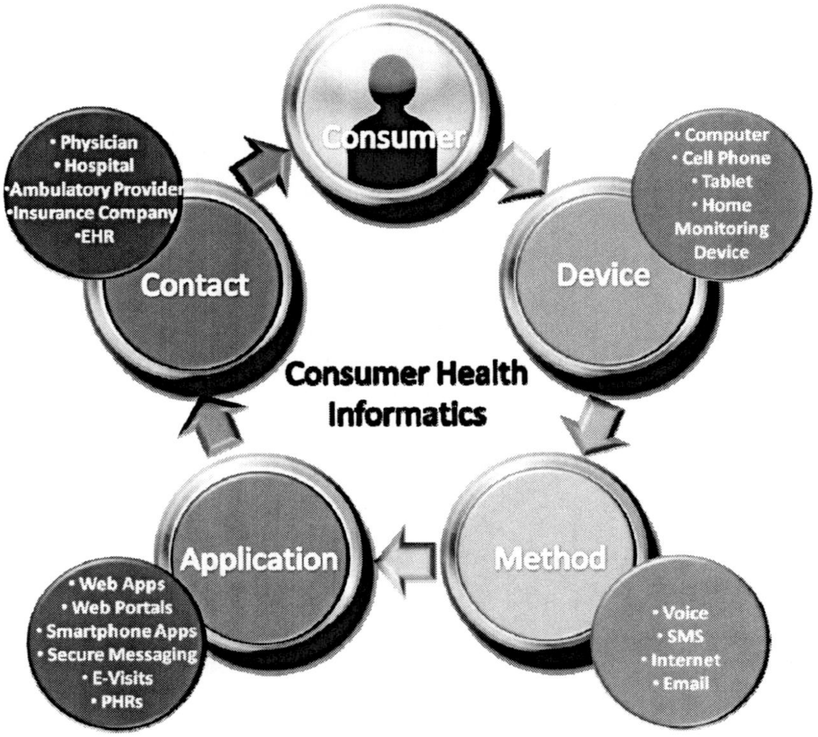

The Origins of Consumer Health Informatics

Several factors including the widespread availability of the internet, the spread in home broadband adoption and the growth of wireless/mobile internet access have contributed to the growth and interest in consumer health informatics. On October 13, 1994, the internet became available, for the first time, to thousands of individuals.[2] Seventeen years later, the internet has reached into just about every facet of life.[2] In the early days just 15% of individuals were using the internet and they were doing so primarily via Bulletin Board Services or proprietary businesses like CompuServe and Prodigy. Today, internet use continues to soar with 78% of Americans over the age of 18 being internet users.[3] This includes approximately 95% of individuals between the age of 18 and 30, three quarters (74%) of individuals over the age of 50 and almost half (42%) of all seniors over the age of 65.[3] Among adult internet users, the top three activities accomplished on line include reading email (92%), searching for information/using a search engine (92%), and looking for health or medical information (83%).[4]

The increase in wireless/mobile internet access is another phenomenon facilitating rapid internet adoption and the growth of consumer health informatics. At least 34% of internet users have logged onto the internet via a wireless device.[5] Like broadband utilization, wireless users especially when checking email and getting news, tend to be more deeply engaged in cyber activities than those without wireless access.[5] In terms of where and how people are wirelessly accessing the internet, overall more than one quarter (27%) of internet users access the net wirelessly at a place other than home or work. In addition almost 40% of internet users have a laptop computer and almost 90% of laptop users connect to the internet via a home wireless network.[5] Currently 83% of US adults have a cell phone of some kind, with 42% of these cell phone owners (or 35% of all adults) owning a smartphone.[6] Smartphone adoption is highest among the affluent and well-educated, the (relatively) young, and minorities with 44% of African-Americans and Latinos reporting use of a smartphone.[6] The majority (87%) of smartphone owners access the internet or email on their handheld, with two-thirds (68%) doing so on a typical day.[6] One-fourth (25%) of smartphone owners mostly go online using their phone, rather than with a desktop or laptop computer.[6] While many of these individuals have other sources of online access at home, roughly one-third of these "cell mostly" internet users do not have a high-speed broadband connection at home.[5] Smartphone owners under the age of 30, smartphone users from racial and ethnic minority groups, and smartphone owners with relatively low income and education levels are particularly likely to say that they mostly go online using their phones.[6]

Finally, another important contributor to the growth of the internet and the emergence of consumer health informatics was the increase in home broadband adoption. Over the last decade, broadband adoption has gone from being the province of the elite to a mainstream behavior by the majority of Americans.[7] After several consecutive years of modest but consistent growth, broadband adoption seems to have plateaued in 2009. Currently, two-thirds of American adults (66%) have a broadband internet connection at home which is statistically unchanged from 2009.[7] Interestingly the growth in broadband adoption has been particularly strong in middle-income households, especially among African Americans and those with little education.[8]

Broadband internet connections appear to draw people deeper into internet use. More than half of online users say they spend more time online since getting a high-speed internet connection at home.[7] In the early days of broadband access as many as 35% (48 million adults) of all internet users reported posting online content to the web with home broadband users accounting for 73% of those who post content to the internet. As such, very quickly a significant association between having a home broadband connection and users' putting content online began to emerge among early adopters.[9]

User-generated content did not, however, stop with these early adopters. By as early as 2006, 44% of home broadband users had their own blog or webpage, worked on group blogs or web pages, remixed digital content and re-posted or shared it online.[9] While user generated content is dominated by young people, 31% of those over age 50 with a broadband connection at home have engaged in at least one of these activities. More significantly, user-generated content seems to have shaped broad expectations about the primary purpose and uses of the internet. It helped to diminish the dominant role of traditional "experts", in favor of the collective judgments of ordinary people. This is evidenced by the fact that at least 45% of online users have indicated that the internet has played an important or crucial role in important life decisions including helping someone with a health or medical decision. In the past, the vision of a broadband world focused largely on specialists expanding their audience through two-way video and email.[10] Increasingly though, users are driving the evolution of what online interactivity means. The value of the internet, broadband and increasingly wireless connectivity then, extends beyond access to information, to active participation and decisionmaking among people with shared interests, problems or concerns.

Within the context of health and healthcare, these national trends were also seen among consumers, patients and caregivers. Since the early days of the internet, the proportion of online health seekers have swelled to approximately 80% of all online users.[11] As a result, online health seeking is the third most popular online activity following email and using a search engine. Since one-quarter of adults do not go online, the percentage of health information seekers is 59% among the total U.S. adult population.[11] The most likely groups to look online for health information include: adults who in the past 12 months have provided unpaid care to a parent, child, friend, or other loved one, women, Caucasians, adults between the ages of 18 to 49, adults with at least some college education, and adults living in higher-income households.[11]

Symptoms and treatments continue to dominate internet users' health searches. Two thirds (66%) of internet users look online for information about a specific disease or medical problem, 56% of internet users look online for information about a certain medical treatment or procedure while 44% of internet users look online for information about doctors or other health professionals. Finally 36%, 33% and 22% of internet users look online for information about hospitals or other medical facilities, information related to health insurance, or for information about environmental health hazards respectively (see Table 11.1).[11] As such, the internet has enabled these individuals to become increasingly informed and empowered.

Table 11.1: Health topics searched[11]

Topics Searched	% Users Who Searched
Specific Disease	66%
Medical Treatment	55%
Physicians & Health Professionals	44%
Hospitals & Medical Facilities	36%
Health Insurance, including Medicare & Medicaid	33%
Food Safety	29%
Drug Safety	24%
Environmental Health	22%

As with the broader national trends outlined above, the rapid growth in online health activities was fueled in part by significant increases in home broadband and wireless access which in turn enabled many health seekers to engage in much more intense health information seeking and communication activities.[8] In addition, patients report being drawn to the cost, convenience and anonymity of online health information. These individuals report that they have generally been able to find what they are looking for and that it is not just about clinical information, lab tests, diagnostics or scheduling doctor visits. Rather the internet appears to increasingly help patients and caregivers connect to emotional support, decision support and practical help for dealing with their health issues or the health concerns of loved ones under their care.[8] The internet also provides many online the opportunities to provide support by helping others who are also online keep up with the latest information and health news. As the internet has matured there has been an increasing interest in wellness activities, information and resources in addition to disease oriented information and resources.[8]

Finally, within the healthcare community there is broad agreement regarding the existence of significant variability in the quality of healthcare provided across the country. As such, there is also consensus regarding the need to (1) increase the quality of care provided to all patients, (2) increase patient engagement with their providers and the healthcare system and (3) make healthcare services more patient centered.[12-14] It is within this general context of improving access to high quality health care for all that several authorities have suggested that health information technologies may play an important role in addressing these issues for providers, patients, caregivers and consumers.[12, 13]

It is on this backdrop of national trends in the widespread availability of the internet, growth in home broadband adoption and increased demand for wireless/mobile internet access and improvements in healthcare quality and access that the term *e-Health* was born. First used by industry leaders and marketing executives in 1999, it was an attempt to convey the promises, principles and excitement around the application of e-commerce (electronic commerce) to the health arena. The earliest definition indicated that "e-health represents the intersection of medical informatics, public health and business and referrers to health services and information delivered or enhanced through the internet and related technologies. In a broader sense, the term characterizes not only a technical development, but also a state-of-mind, a way of thinking, an attitude, and a commitment for networked, global thinking, to improve health care locally, regionally, and worldwide by using information and communication technology."[14] This definition clearly suggested that the emergence of the internet and related electronic technologies presented new opportunities for healthcare to enable (1) consumers to interact directly with the healthcare system online; (2) improved possibilities for institution-to-institution transmissions of data; and (3) new possibilities for peer-to-peer communication among patients, caregivers and consumers.[14] Soon thereafter the term *consumer health informatics* emerged to distinguish the explicit and primary incorporation of the needs and perspectives of the patient in emerging electronic tools from those of healthcare providers in the development of emerging "medical tools." The field was originally defined in 2001 by Eysenbach as a branch of medical informatics that "analyzes consumers' needs for information, studies and implements methods of making information accessible to consumers, and models and integrates consumers' preferences into medical information systems."[15] Since that time however the field and the definition itself has undergone significant evolution. In 2001, Houston et al. wrote that CHI incorporated a broad range of topics, the most common being patient decision support and patient access to their own health information.[16] Currently, the Agency for Healthcare Research and Quality defines consumer health informatics applications as "any electronic tool, technology or electronic application that is designed to interact directly with consumers, with or without the presence of a healthcare professional that provides or uses individualized (personal) information and provides the consumer with individualized assistance to help the patient better manage their health or health care." [17]

Classification of Health Informatics Applications

Many consumer health informatics tools have been developed over the last decade. In fact, as of December 2, 2011, the total number of individual mobile apps in the marketplace is approaching one million.[18] Approximately 2,000 apps become available each day. Across all platforms (Apple, Android, Blackberry, Windows), approximately 3.28% (more than 32,000) are health & fitness apps.[18] This does not include the growing number of health websites and applications developed for consumers but not designed for mobile applications nor offered in an online "app" store. This is covered in detail in the chapter on mobile technology.

While many CHI tools perform more than one function or task, generally this very large number of consumer health informatics tools may be functionally classified, classified based on the types of health processes addressed or classified by the types of outcomes they are designed to achieve. Finally, they may also be classified categorically based on the disease or condition they have been designed to address.

Functionally CHI tools may be further subdivided into those which are designed to enable consumers to assess or monitor some health parameter. Body Mass Index calculators and heart attack risk calculators are two examples of CHI tools in this category. Functionally CHI tools may also be subdivided into those which primarily provide health education or knowledge building. Disease or condition trackers (pregnancy tracker) that provide information about a specific condition fall into this category. Finally, CHI tools may be further functionally subclassified as those which provide health or healthcare management assistance. Personal health records and emergency response applications are two examples of CHI tools in this category. Another example might be web sites to compare physicians or hospitals, such as www.healthgrades.com or www.hospitalcompare.hhs.gov .

CHI tools may also be classified according to the types of health processes they address. Typical health processes addressed by CHI tools include diagnostic, (heart rate monitor), therapeutic (herbal remedies), lifestyle applications (stress reduction), and tools that enable or facilitate coordination between activities (calendar apps, personal health records). CHI tools may also enable consumers to monitor their conditions (weight and exercise trackers) or they may be primarily motivational (Zumba DVD's, Wii Fit).

A third way in which CHI tools may be classified is based on the type of outcomes associated with use of the tool. Within this scheme these tools may address clinical outcomes (blood pressure, glucose), relationship centered outcomes (shared decision making), intermediate outcomes (kept follow-up appointments) or other health care processes (prevention, wellness).

Obviously, one does not have to look very far to find evidence of new patient-oriented (consumer-centric) healthcare IT innovations:

- United Healthcare offers a fifteen minute teleconference with a physician to anyone with a computer, web cam and $45. Anywhere, anytime of day or night.

- Grocery store and food chain flu vaccination kiosks.

- Smart bicycles which measure calories burned, weight of the rider, duration and intensity of exercise achieved while riding.

Health Education & Information Applications

As addressed earlier, two-thirds of adults use health related web sites as their premier resource for health information. One reason for this is that patients are becoming more discriminating in their choices of all aspects of healthcare. No longer do they automatically accept the opinion of their physicians. Or, in some cases they look on line for answers they were too timid to ask their physician or forgot to ask. In a Harris poll it was shown that 57% of patients discussed their internet search with their physician and 52% searched the internet after talking to their physician. Eighty-nine percent (89%) felt their search was successful demonstrating confidence in the internet as the new health library.[19] Excellent medical web sites exist but searches can yield low quality answers, particularly when personal web sites are searched. As an example, in one study of internet searches for the treatment of childhood diarrhea, 20% of results failed to match guidelines published by the American Academy of Pediatrics.[20]

Patient Education/Health Information Web Sites

The following are only a sample of the many valuable patient education web sites available today.

WebMD. With more than 30 million people visiting this site monthly, it should be considered one of the true standard bearers. They have an extensive health library with top topics listed for men, women and children. Treatment and drug information is available, as is medical news. A symptom checker tool provides a patient with a simple differential diagnosis based on their symptoms, age and gender. A daily e-mail newsletter is offered that can be customized to a patient's concerns. The only negative about this site is the commercial influence of advertisements. http://www.webmd.com/

Revolution Health. This free web site offers disease information, forums, CarePages, health calculators, a physician finder and symptom checker. Personal health records were discontinued in 2009. Members can rate physicians and hospitals, in addition to treatments and medications. They also offer an insurance marketplace to discuss and compare insurance options. Although these services are free, they also offer a fee-based membership that will allow a member to call and discuss a health concern 24 hours a day. There is limited commercial influence in the form of ads. http://www.revolutionhealth.com/

MedlinePlus. This is the premier free patient education site developed by the National Library of Medicine and the National Institutes of Health that links to the best and most respected web sites, such as the Mayo Clinic. MedlinePlus was ranked as the top information/news web site on the American Customer Satisfaction Index of federal government web sites.[21]In spite of its high marks, many patients and clinicians do not know about this site and many healthcare organizations pay for patient education content that could be obtained for free. Figure 11.2 shows the results of a search for abdominal pain, showing the high quality references and the convenient folders on the left. Features of the web site include: 800+ health topics , drug information, health encyclopedia, 165 tutorials, videos of surgical procedures, topics available in 40 languages, tool such as quizzes, calculators and self-assessments, health dictionary, physician and hospital locator, link to clinical trials.gov and health news. www.medlineplus.com

Figure 11.2: Search results for abdominal pain   (Courtesy MedlinePlus)

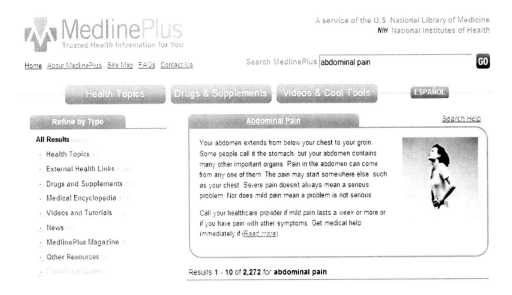

Healthfinder. This government funded web site provides resources on a wide range of health topics selected from over 1,600 government and non-profit organizations. It is coordinated by the Office of Disease Prevention and Health Promotion and its health information referral service, the National Health Information Center. http://healthfinder.gov/

Everyday Health. This web site with more than 100 participating health centers provides information on the diagnosis, management and prevention of diseases and conditions, as well as on healthy lifestyles. It has an *ask-the-expert* question and answer section and multiple patient communities. It is part of Everyday Health, Inc. that has Everyday Health and multiple partnership web sites that also provide patient information and services. http://www.everydayhealth.com/

Healthwise. Multiple companies sell patient education for use on commercial medical web sites. Healthwise is a not-for-profit company that provides more than 6,000 medical topics in their knowledgebase. Other features include decision making tools, *take action tools* for chronic diseases and over 1,000 illustrations. http://www.healthwise.org/

UpToDate. This extremely popular physician education site also has a patient education site, aimed at college educated patients, unlike many sites that are aimed at high school educated patients. There is no charge for limited access to this site that covers more than 20 medical categories. For access to all articles posted, there is a charge. http://www.uptodate.com/index

FamilyDoctor. The American Academy of Family Physicians sponsors this comprehensive free site. They cover all age groups as well as over the counter (OTC) drugs and a large library of health videos.[22] http://familydoctor.org/familydoctor/en.html

Lab Tests Online. This free site allows for searching by test, disease condition or screening. The site is well organized with excellent resources for those seeking more information about clinical tests, why they are drawn, the results and what abnormal results mean. http://labtestsonline.org/

Home Telemedicine Devices

Home telemonitoring is an important aspect of telemedicine or the remote delivery of care, which is discussed in much greater detail in the chapter on telemedicine. A myriad of new devices are being produced that are capable of wireless uploads to the internet, electronic health records and personal health records. Data can then be tracked and trended, analyzed and shared. It remains to be seen who will be reimbursed to manage this growing volume of data. In all likelihood, medical practices will eventually be reimbursed for telemedicine but it will be a nurse who is part of a disease management program who will manage the data.

We are also witnessing convergence in technologies, for example, there are digital scales and blood pressure monitors that not only display on a smartphone but also are backed up via WiFi to a web server as a dashboard for others to view and analyze. Results can also be sent to Microsoft HealthVault. This is an attractive combination for both patient and physician for disease management.[22]

There are multiple comprehensive home telemonitoring units available. For example, Well@home system consists of several physiologic monitors that synchronize to a web site on the internet or to the Meditech electronic health record. Figure 11.3 shows the patient monitor and peripheral devices.[23]

Figure 11.3: Home telemonitoring unit (Courtesy well@home)

The following is a list of features that many home telemonitoring/telemanagement systems have in common:

- Medication and health related task reminders with compliance documentation.
- Educational content about medications or disease entities with interactive instruction.
- Touch screen main monitor.
- Patients can report symptoms and this information is forwarded to caregiver or medical office.
- Data from monitors (blood pressure, weight, oxygen levels and blood sugar) with transmission by WiFi or 3G.
- Alerts can be created and sent to physician if the monitoring results are out of range.

Patient Web Portals

Web portals are web-based programs patients can access for health related services. A web portal can be a standalone program or it can be integrated with an electronic health record. Patient portals began as a web-based entrance to a healthcare system for the purpose of learning about a hospital, healthcare system or physician's practice. They clearly began as a marketing ploy to attract patients who are internet savvy, but have now become a patient expectation. Common features of patient web portals are listed in Table 11.2.

Table 11.2: Web portal features

Feature	Comments
Online registration	Allows patients to complete information before an office visit or hospitalization
Medication refills	Secure messages can be left for physicians to refill or renew medications, instead of telephone calls
Laboratory results	Patients can find results on recent tests as well as an explanation
Electronic visits	Portals exist that facilitate e-visits and the payment process
Patient education	Links to common educational sites
Personal health records (PHRs)	Allows patients and their families to create and update their PHR
Online appointments	Allows patients to see what appointments are available and when
Referrals	Patients can request referrals to specialists, e.g. OB-GYN
Secure messaging	More convenient than playing phone tag
Bill paying	Online payment using credit card is faster than "snail mail"
Document uploading	Several portals allow uploading of medical records to their site
Tracking function	Portal allows patients to upload diet, blood sugars, blood pressures, etc.

Most patient portals offer multiple services, whereas others like TeleVox offer a specific service, like lab results notification. This secure web-based program, known as LabCalls™ enables patients to access a web site and obtain lab results. The nurse or doctor leaves the results along with a canned explanatory message. Patients can also receive a text message on their cell phone that lab results are ready. This program integrates with the practice management system or the EHR.[24]

A minority of web portals actually integrate with an EHR, which means that most patient data has to be manually inputted. In the future when EHRs become more widespread and comply with Meaningful Use, selected patient lab results will automatically upload to the patient portal, thus saving time and money. Patients will also be able to access parts of their electronic records. A 2006 Harris Interactive study showed that 83% of patients wanted lab tests online and 69% wanted online charts to manage chronic conditions.[25]

Although several studies have shown patient interest in having access to lab results it remains to be seen if that would change consumer behavior or clinical outcomes. Are these patients primarily college educated and tech savvy? Do they desire results because physicians' offices are too slow to provide results? In a study at Beth Israel hospital, patients who accessed their portal PatientSite were younger and with fewer medical problems. They tended to access lab and x-ray results and use secure messaging more than a non-enrollee group.[26]

In another survey by Connecting for Health, the following are the responses to what consumers thought their health information online would do: clarify doctor instructions (71%); prevent medical mistakes (65%); change the way patients managed their health (60%); and improve the quality of care (54%). [27]

Little is written about the benefits of patient web portals for the general consumer. McLeod Health System in Florence, South Carolina used online scheduling as part of the portal NexSched and was able to demonstrate fewer *no shows* and claims denials. They predicted a savings of about $1 million dollars yearly as a result of this program.[28]

Group Health Cooperative, a large mixed-model healthcare organization studied the effect of integrating its new comprehensive patient portal MyGroupHealth with its Epic electronic health record. The highest monthly user rates per 1,000 adult members were: test results, med refills, after-visit summaries and patient to provider messaging. A patient satisfaction survey revealed that the satisfaction rates were: 94% were satisfied overall, 96% for med refills, 93% for patient to provider messaging and 86% for test results. Although early use of the web portal was low there was a steady increase over time. Attrition rates were not reported.[29, 30]

Robert Wood Johnson Foundation has awarded $2.45 million dollars to six organizations to study the effect of patient portals on chronic disease.[31]

Patient Portal Examples

MySaintAls is the portal for Saint Alphonsus Medical Center located in Boise, Idaho. This comprehensive portal offers all of the standard features as well as the unique Patient Vault. They charge $10 monthly to upload (scan) and store patient records on their server. Lab results are accompanied by a separate program that explains the significance of the results and likely reduces the number of routine questions. www.MySaintAls.com

Epic MyChart is a patient portal integrated with a well-established electronic health record system. They also offer a standalone PHR known as *Lucy*. MyChart functions: view test results, view and schedule appointments, pay bills, receive health maintenance reminders, view educational material, request refills, secure messaging, view child's record and manage care of elderly parents. An interactive demo is available on the web site. http://www.epic.com/software-phr.php

Intuit Health Patient Portal is a portal that offers multiple features for patients and medical offices. They claim to have a user base of 4,000+ physician offices. They also have an extensive knowledge library of 6,000 medical conditions to expedite an e-visit. A new area of involvement is the patient centered medical home model where they supply the technology to connect patient with physician. A free return on investment (ROI) calculator is available on their site. www.medfusion.net. Additional features include:

- Front office solution to deal with patient registration, forms, appointments, check-in and patient messaging

- Back office solution for online bill payment, billing messaging and a virtual credit card payment system

- Clinical solution includes medication renewal and refills, secure messaging, personal health records, referral management, virtual visits, symptom assessment, laboratory results and reminders

ReachMyDoctor. This site is aimed at improving communication with the doctor's office. www.reachmydoctor.com. The program offers two options:

- Free: schedule appointments, request medication refills, request a referral and address billing and insurance issues

- Subscription: for $8.95 monthly a patient can ask the physician non-urgent questions via secure e-mail. Physicians must be part of the network

My HealtheVet. This portal integrates with the Veterans Health Administration's EHR (VistA) and offers lab wellness reminders, appointments, a personal health record (PHR), medication refills, patient education and online monitoring of: activity, food intake, oximetry, blood pressure, glucose and weight. In the 2011 time frame they added secure messaging, online appointments and lab results (hematology and chemistry). The *blue button* feature on the web site permits veterans to view, print or download results. http://www.myhealth.va.gov/

Personal Health Records (PHRs)

PHR Definitions

According to the American Health Information Management Association (AHIMA) the **personal health record (PHR)** is: "an electronic, universally available, lifelong resource of health information needed by individuals to make health decisions."[26]

The National Alliance for Health Information Technology defines a PHR as follows:

"an individual's electronic record of health-related information that conforms to nationally recognized interoperability standards and that can be drawn from multiple sources, while being managed, shared controlled by the individual"[27]

Introduction

The Institute of Medicine promotes PHRs by stating "patients should have unfettered access to their own medical information."[28] The first principle endorsed by the Personal Health Technology Council is that "individuals should be able to access their health and medical data conveniently and affordably."[29]

Interest in PHRs comes from multiple sources. In 2002 the Markle Foundation established *Connecting for Health*, a public-private collaboration to promote better information sharing between doctors and patients. In their 2004 position paper they suggested: PHR development should be increased, PHRs will educate patients about their health and common data standards are a logical starting point. In one of their surveys 61% of respondents agreed that they should have access to their medical information *anytime, anyplace*.[30] A 2004 Harris Interactive survey demonstrated 42% kept personal or family health records, but only 13% stored their records electronically.[31] The Centers for Medicare and Medicaid Services (CMS) released a Request for Information (RFI) about PHRs in 2005 to determine its future direction. They are aware of the need for better patient information sharing and storage in older patients who are on multiple medications and have multiple physicians. Medicare patients can now download claims data.[32]

Ideal PHR Features

In spite of the fact that PHRs are new and available in many formats, experts believe that PHRs should have the following features in order to be successfully adopted:

- Portable, i.e. information will transfer even when there is a job, insurer or clinician change
- Interoperable, i.e. standardized PHR format can be shared among disparate partners, such as the Continuity of Care Document (CCD) discussed in the chapter on data standards
- Auto-populated with clinical and test results that would be inputted automatically
- Controlled by the patient
- Longitudinal record and not just a snapshot
- Private and secure
- Integrated into the clinician's workflow and not be a separate process

The reality is that no organization has the ideal solution, with all of the above features.[33-35] In the following section we will discuss PHR formats, based on connectivity.

PHR Formats

Tethered. The word tethered implies that the PHR is connected to one platform and not interoperable. The earliest and most common examples of this would be claims-based PHRs from insurers and healthcare organizations. Other examples would be PHRs tethered to an EHR or standalone patient portal. Payer-based PHRs have the advantages of being free to patients and easy to populate with claims data. They have the disadvantages of not being portable or interoperable and not controlled by patients. Moreover, claims data is usually several weeks old and indicates that a test was ordered but may not provide the actual results. Furthermore, the payer-based PHR is not longitudinal because it only covers patient encounters insured by their company.

In 2006 America's Health Insurance Plans (AHIPs) and Blue Cross/Blue Shield Association announced a comprehensive plan to supply PHRs to their members. Importantly, the established core data standards determined that much of the information would come from claims and administrative data. With established data standards PHRs can be shared between different insurance companies, should a patient move or change coverage.[36, 37] Aetna has offered its PHRs to 6 million of its 37 million members. They also offered their PHR to Medicare plan enrollees. Of interest, they will use *CareEngine,* a software rules engine that reviews the claims-based PHRs and gives personalized alerts (called *care considerations*) to patients and physicians about how to improve medical care.[38]

Untethered PHRs. Untethered PHRs imply they are not connected to one platform and they have more interoperability potential. PHR programs are available in multiple mobile and static platforms. There are more than fifty untethered personal health record products on the market, giving consumers many choices, but obvious limitations.[39]

- Web-based. Most are commercial sites that are secure and can be accessed from a distant site. A minority of PHRs reside in patient portals that connect to an electronic health record system but this is likely to increase as a result of Meaningful Use.

- Mobile technology. Patient health information can be downloaded to:

    o Secure digital cards and USB drives. Most USB programs synchronize to a web-based portal where patient information is also stored. Mobile technology offers several unique advantages. It is not dependent on the internet for operation and is truly portable, but not interoperable.

    o Smart Phones. Given the soaring popularity and expanding features of smartphones they may become the mobile storage of choice.[40] Blue Cross of Northeastern Pennsylvania provided members secure electronic access to their medical records via smartphones. The information was derived from claims data and patient input. The program will use *MobiSecure Wallet and Vault* to access the internet and authenticate the user.[41] The US Army has seen merit in smartphone PHRs (see info box). Another example is MyRapidMD that downloads a client's emergency medical information to any Java enabled cell phone. Membership includes a wallet card, phone sticker, windshield sticker and key chain cards.[43]

    o Smart Cards. The United States has been slow to use smart cards in the field of medicine, unlike countries like France and Germany. Smart cards are used for patient authentication and can hold 64 KB to 144KB of information (60 plus pages of single spaced text). These cards have a small processor that can be programmed to perform tasks such as encryption and the cards are re-writable. Cards can be read by contact or be contactless, using radio frequency identification (RFID). They have the potential to speed up electronic claims submission by decreasing clerical errors. An issue with current healthcare cards is that they are not standardized and as such are not readable by all readers. Medical Group Management Association (MGMA) is promoting an industry-wide effort known as ProjectSwipeIT to standardize these cards, even though they are not truly smart cards.[44] For more details about smart cards in healthcare we refer you to a 2009 monograph.[45]

Army mCare

The Army has a five year program to evaluate the use of smartphones for case management of wounded warriors. The device agnostic app provides bidirectional encrypted messaging, appointment reminders, health questionnaires, injury tips and general military news. Participants receive daily messages to measure mood states, with negative mood scores monitored by case managers. This project is actually part of a randomized controlled trial so outcomes will be compared to those wounded warriors without smartphone technology. By mid-2011 they had 700 participants.[42]

Personal Health Record Systems is an arbitrary term to indicate an untethered PHR that is interoperable. The following are the major examples of these systems:

- Microsoft HealthVault. Program includes a PHR and interfaces with other third party health applications. This platform released the source code of the HealthVault.NET Software Development Kit and the XML interfaces under the Open Specification Promise (OSP). This will enable third party developers to develop HealthVault compatible applications. In 2011 they became a health information service provider so they participate in the Direct Project, explained in the chapter on health information exchange. They plan to be able to receive Continuity of Care Document (CCDs) from physicians or hospitals and then parse the information into pertinent data fields in the PHR. Mobile connectivity is planned for the 2011-2012 time frame. Pilot projects with integrated delivery networks are underway but results are unknown.[46-48]

- Dossia. The system was founded by Applied Materials, BP America, Intel, Pitney Bowes and Walmart with data derived from insurers, pharmacies and physicians. The system known as Indivo is hosted by Children's Hospital in Boston and consists of a free open source, open standards platform. Application programming interfaces (APIs) are available to developers for customization. Indivo handles both CCR and CCD documents. They plan to interface with EHRs, patients, physicians, researchers, health information organizations and public health services.[49] More details about this application were reported by Mandl et al. in the medical literature.[50]

- Travelers Electronic Health Record Template (TrEHRT): An international partnership with the AMIA, IMIA and others created a free web-based repository of personal health information that should prove helpful for patients who travel internationally. This information can be inputted and accessed on the web and viewed on an Android smartphone. Emergency medical information can be viewed without a log-on. Categories of information include (1) Basic patient data; 15 data elements (2) Present medications (3) Medical history that includes diagnoses, allergies, smoking and drinking status and vaccinations (4) Test reports; 3 data elements (5) Travel history (6) Family heredity disease history and (7) Physician contact information. Available for English, Chinese, French, Spanish and Russian speaking patients.[51,52]

PHR Barriers and Issues

Kaelber et al. published a very helpful article in titled "A Research Agenda for Personal Health Records." They point out that there is very little published about the impact of PHRs on patient behavior and outcomes. They believe research is needed to answer the following questions:

- What PHR functionality is needed in the areas of data collection, sharing, exchange and self-management?

- What is needed to improve adoption of PHRs by patients and clinicians? Research should focus on specific populations like the elderly, patients with chronic diseases, etc.

- What is needed to ensure privacy and security?

- What PHR architecture or model is likely to be most effective? Tethered? Untethered?

- What is the business case for PHRs and are the incentives aligned for patients and physicians?[53]

Thus far, personal health records have been voluntary, placing the burden of downloading and maintaining health information on the patient. A busy physician's office is not likely to want this additional responsibility without reimbursement. Meaningful Use requirements for reimbursement will likely make interaction with PHRs more palatable for physicians. The vision is to have records stored in repositories like Microsoft HealthVault in a format (XML) that is compatible with EHRs, HIOs, etc., but this will likely take years to accomplish, for those patients who are interested. As PHRs develop more user friendly features, perhaps the appeal to the average consumer will increase. Some PHRs, for example, will provide alerts such as *medications about to expire* or *upcoming medical appointments*. An ideal business model for personal health records does not exist. Some studies suggest patients are willing to purchase their own PHRs if the price is low and others suggest insurance companies are the entity most likely to play a major role. Theft of personal health information (PHI) is a definite concern to the average patient and personal health records is just one more platform of concern.

At this time it appears that patients are lukewarm about personal health records, regardless of format or who picks up the bill. A 2006 survey of more than 11,000 American and Canadian consumers demonstrated that 22% of patients have not used the PHR feature offered by their insurance company and 53% had never accessed their health plan's web site. Thirty-four percent (34%) interviewed did not trust the site's security and 29% did not see benefit from the concept of the PHR.[54] In a similar 2007 survey by Aetna, only 36% of those surveyed were familiar with PHRs and of those only 11% used one to track their medical history.[55] It will take time to determine the impact of PHRs and what incentives actually are important to patients and other members of the healthcare team. There are healthcare consultants who believe that PHR systems have a brighter future than HIOs for future data interoperability, but both suffer from inadequate business models and consumer enthusiasm.[37] A 2010 survey of adults showed that only 7% of those surveyed have used a personal health record, an increase from 2.7% in 2008.[56]

Other countries such as England are dabbling with the issue of PHRs for its citizens. They created HealthSpace in 2003 to store health notes, book appointments, store physiological parameters such as blood pressure and access National Health Service (NHS) contacts and health links. By late 2008 they allowed access to the NHS Summary Care Record that included allergies and medication histories.[57]

PHR Research Project Examples

Medicare. There are plans to determine whether PHRs change patient outcomes or save costs. A one year pilot program known as Medicare PHR Choice began in 2009 for patients located in Arizona and Utah. CMS will transfer up to two years of Medicare claims data upon request. Medicare partnered with HIP USA, Humana, Kaiser Permanente and the University of Pittsburgh Medical Center. Each plan has a unique PHR that allows patients to access their own information. It is anticipated that the information will derive from hospital and physician claims data. Medicare will also pilot a PHR program in South Carolina in this same time frame using claims data. Both pilots were completed in 2011 and the final results have not been published. All Medicare recipients can now download their claims data results using a *blue button* at MyMedicare.gov.[58, 59]

Robert Wood Johnson Foundation created Project HealthDesign to study PHRs that deal with chronic diseases such as breast cancer, diabetes, chronic pain, etc. using nine research teams. The final report for round one is posted on their web site. Current projects include the monitoring of elderly activities with a PHR and a PHR for adults with asthma and depression and or anxiety.[60]

Electronic Patient to Physician Communication

"Digital Rx: Take two aspirin and e-mail me in the morning"

- *New York Times*, March 2, 2005

Secure Patient to Physician E-Mail

The vast majority of Americans today use e-mail, but few use it to communicate routinely with their physicians.[61] Manhattan Research reported in 2010 that a survey of 1900 US physicians showed that 39% of physicians used electronic messaging to communicate with their patients. According to this survey dermatologists and oncologists were the two specialties most likely to use this technology.[62] On the other hand, the 2009 National Health Interview Survey of over 7000 adult patients noted about 5% of adults had communicated with a clinician in the past 12 months.[63] Importantly, in a 2001 survey 41% of respondents found it very frustrating to see a physician in person when they thought a telephone call or e-mail would suffice.[64] Physicians cite the following reasons for not using e-mail: liability (56%), poor compensation (45%), privacy issues (43%), staff not trained (30%) and the feeling that face-to-face visits are better (27%).[65]

Multiple benefits of e-mail communication have been pointed out. The communication is asynchronous so physicians can answer at their convenience and avoid *phone tag*. Overhead is lower for electronic messages compared to phone messages and they are self-documenting with a good audit trail that includes time stamps. Patients tend to lose less time from work for minor issues. It is preferable to communicate with patients using secure messaging through a web-based third party where authentication is validated; not true for routine e-mail. A 2007 article in Pediatrics reported that physician to parent e-mail communication was 57% faster than telephonic communication and resulted in high consumer satisfaction. Forty percent (40%) of e-mails occurred afterhours but only amounted to one to two e-mails per day. It should be noted that the only physician involved in this study was a pediatric rheumatologist and not a generalist. Also, only one-third of families offered e-mail communication took advantage of it. It is therefore difficult to generalize these results to the average physician's office.[66]

Potential disadvantages of e-mail messaging might include: indigent and elderly patients less likely to use the service; inability to examine the patient; potential for communication errors; possible slow responses; security issues and the potential to be overwhelmed.[67, 68] Clinicians are likely to use secure messaging in the future if EHRs become widespread, there is clear evidence on the return on investment or there is a reimbursement strategy. There will need to be new guidelines to cover secure messaging. Cliff Rapp, a medical risk manager, recommends the following: develop clear cut policies, provide a disclosure statement, publish disclaimers about emergencies and privacy, use encryption, comply with HIPAA, designate a privacy officer and obtain confirmation of message delivery.[69]

Electronic Visits

Electronic visits (e-visits or virtual visits) are an example of telehealth or telemedicine where medical care is delivered remotely (telemedicine is covered in much more detail in another chapter). Virtual visits are available as a continuum of care (Figure 11.4).

Figure 11.4: Remote patient communication continuum

The web-based choices for patient - physician communication include:

- Secure messaging and templates to input a health concern and wait for a response by a clinician

- Telephonic communication (audio) to communicate with clinicians along with secure messaging in a web application

- Audio and video using standard web cam and secure messaging on a web application

Secure messaging: E-visits require secure messaging and not routine e-mail. In this section we are addressing a virtual visit and not just a simple e-mail communication. Virtual visits have the advantages of much better security and privacy and the ability to have a third party involved in the billing process. Patients and physicians must utilize a username and password to log onto a secure web site in order to conduct an e-visit. Numerous vendors such as Intuit Health Patient Portal provide the platform for e-visits in addition to their patient portal features. Some authorities feel the e-visits have a bright future. A Price Waterhouse study estimated that 20% of outpatient visits could be eliminated by using e-visits.[70] A new CPT code 0074T was developed specifically for e-visits.[71] Guidelines need to be established to define what constitutes an e-visit in order for insurance companies to reimburse for the electronic visit.

Several reports address how e-mail and e-visits might impact a physician's productivity.[72-74] A 2007 report from the Kaiser Permanente system suggested that e-mail communication decreased office visits by about 10%, compared to a control group. This would be good news for a health maintenance organization but bad news for fee-for-service practices, unless e-visits are reimburseable.[75] The consensus is that minor complaints can be dealt with more efficiently electronically, thereby allowing sicker patients to be seen in person. Furthermore, patients miss less time from work for minor issues. It has also been pointed out that if the patient provides a history during the e-visit and still has to be seen face-to-face, the physician has the advantage of knowing why the patient is there, therefore saving time.

In spite of the enthusiasm for virtual visits, most patients are not willing to pay more than minimal co-pays for an e-visit.[76] In one study it was reported that, compared to controls, patients who had e-visits had lower insurance claims. The profit more than paid for the $25 physician charge and the $0 to $10 patient co-pay. Importantly, 50% were less likely to miss work and 77% said it only took 10 minutes for the e-visit. Patient and physician satisfaction levels were good.[77] Pilot projects and studies are underway to evaluate e-visits. BlueCross/Blue Shield of Tennessee and other regions are reimbursing physicians for electronic visits.[78-81] The University of California at Davis Health System has been performing e-visits since 2001 and states that 80% of insurance companies in their region support the concept. Participating physicians seem to be more cost effective and physicians are receiving bonuses.[82] Cigna's HealthCare for Seniors program now covers electronic visits for older patients using a standalone web portal. Non-seniors must pay a fee for e-visits.[83] In 2008, Cigna and Aetna announced that they would expand their virtual visit pilot programs to the rest of the nation. Cigna paid $25 to $35 for an online consultation and the patient paid a co-pay with a credit card via the web portal. Aetna will have a similar program that will include 30 specialties. These payers are also looking at discounts to physicians as an incentive to offer virtual visits.[84]

A new free secure messaging service is available and known as HouseDoc. It supports a virtual asynchronous visit, a request for medication refills, a request for appointments and test results. Security is provided by SSL encryption. If the clinician charges the patient, the web service charges $2 and services are paid for by credit card.[85]

An excellent review of patient to provider communication can be found in a 2003 monograph by the First Consulting Group.[86]

Telephonic visits: The concept of virtual visits has spawned innovation in the delivery of healthcare. As an example, *TelaDoc* is a telephone-based consult service that is intended to supplement the care delivered by the primary care physician. This web-based application guarantees a clinician will return a phone call in three hours and the average charge is $35. They claim to have one million members and offer services 24/7. The clinician will prescribe and handle refills but not prescribe narcotics or order labs. Interestingly, they save the patient encounter as a Continuity of Care Record (CCR) that can be shared with others and accessed at the next visit. Aetna offered this service to its patients in Texas and Florida in 2011; it is therefore likely we will have data to measure the impact of this innovation.[87]

Audio-Video Televisits: Another innovative virtual visit service worth mentioning is *American Well*. Patients can interact with clinicians using web-based videoconferencing, as well as secure chat, secure messaging, voice over IP and telephonic communication. They are promoting 24/7 access for patients from home and aim to coordinate care with the primary care clinician (PCM) and insurance company. The service locates an appropriate clinician (including specialists), initiates a live audio-video conversation with a clinician and forwards the results to the PCM. For the clinician there is automatic claims submission and a per-consultation malpractice insurance coverage is offered. In addition, clinical practice guidelines are promoted for standardized care, known as *online care insight*. This vendor is promoting this application for the patient-centered medical home model and accountable care organizations. In 2010, they introduced *Team Edition* with the goal of supplying on-demand specialty care as part of the team. Delta Airlines will make American Well services available to all employees and Rite-Aid will use the platform for in-store consultations with its pharmacist network. The approximate cost for an e-visit is $45.[88]

MDLiveCare is another telehealth initiative that provides real time virtual visits by secure messaging, audio or audio-video visits using a web cam. Features include:

- Visits available 24/7 including for mental health

- Lab testing through Lab Corp

- E-prescribing with option to have drugs mailed to home

- Visit information can be shared with Microsoft HealthVault

- Member card provides access to emergency medical information

- Physicians are encouraged to join to supplement practice incomes

- With membership costs of $9.95/month, a visit costs $39.95 and patients have unlimited free e-mail advice; a virtual visit by a non-member is $59.95[89]

3G Doctor is a United Kingdom-based audio-visual consulting service available for the smartphone. The charge for the visit is £ 35.[90]

Little has been reported about the medical value of e-visits. A 2010 article did confirm that e-visits seem to be a successful alternative to standard care for the follow-up treatment of acne by dermatologists. The intervention group used a web portal, aided by digital photographs sent to the physician every six weeks. Patient and physician satisfaction was high. The intervention saved time for patients and was time neutral for Dermatologists.[91]A variety of medical studies have been published recently using a simple web cam and a free software program a variety of medical studies.[92-94]

Barriers to CHI Adoption

Despite the explosion in the development of CHI tools, a recent review of the evidence suggest that many barriers may still exist to the widespread utilization of these tools and strategies.[16] User barriers can pertain to either the clinician or the consumer. Although providers do not generally use CHI tools, clinician endorsement affects consumer choice, and thus negative attitudes of clinicians may be a barrier to consumer use.[17] Other consumer barriers include lack of home internet access, concerns about privacy, limited literacy and knowledge, language hurdles, cultural issues, and lack of technologic skills.[17] Application usability or user-friendliness, patient knowledge, literacy, and lack of needed computer skills have all been identified as barriers to CHI utilization. Privacy concerns, control of information, lack of trust, lack of consumer acceptability, usefulness, credibility, expectations are common barriers to CHI use.[17] Finally, physical or cognitive disability, computers use anxiety, lack of built in social support, lack of personal contact with clinicians and the belief that IT would not be an improvement to current care have all been cited as barriers to the adoption and utilization of CHI tools and applications.[17]

Future Trends

The large number of CHI tools available to consumers might be taken to suggest the value of these applications. Unfortunately, in the overwhelming majority of cases, the efficacy of the CHI tools has not been evaluated.[17] Among those that have been evaluated, most tend to focus on one or more domains of chronic disease management.[17] While this is very important and clearly needed, insufficient attention has been given to the role of CHI applications in the acute exacerbation of symptomatology or other urgent and emergent problems that may occur in home and community-based settings. Thus, the role of CHI applications in primary, secondary, and tertiary prevention needs to be more adequately explored.[17]

Given the prevalence of mental health and psychiatric issues, the value of CHI applications in the context of mental health, coping, and stress should also be thoroughly evaluated. Sociocultural factors are increasingly important determinants of health care outcomes. The potential impact on social factors including social isolation and social support and perhaps even broader social determinants of health need to be evaluated and may prove useful in helping patients address select health concerns in the home- and community-based setting.[17]

Key Points

- Healthcare consumers are becoming more sophisticated and more demanding
- For some healthcare processes patients would like to have the same convenience of an ATM machine
- Patients are using the World Wide Web as the medical library of choice
- Patient web portals are now available that are standalone or integrated with electronic health records that offer a multitude of patient-oriented services
- Everyone is talking about personal health records but it is unknown who will pay
- Secure patient - physician e-mail and e-visits have great potential to expedite acute care visits, once reimbursement becomes standard

Conclusion

Despite a large and growing number of CHI tools and applications, overall the CHI field is new and still evolving, particularly as it relates to evaluation and documentation of the effectiveness of these tools.[17] The evidence from those tools that have been evaluated suggests that while there may be a role for CHI applications to reach consumers at a low cost and obviate the need for some activities currently performed by humans, it is likely that a more important role is to enhance the efficacy of interventions currently delivered by humans. The literature also suggests that at least three critical elements are most often found in those CHI applications that exert a significant impact on health outcomes. These three factors are (1) individual tailoring, (2) personalization, and (3) behavioral feedback.[17] Personalization involves designing the intervention to be delivered in a way that makes it specific for a given individual. Tailoring refers to building an intervention, in part, on specific knowledge of actual characteristics of the individual receiving the intervention. Finally, behavioral feedback refers to providing consumers with messages regarding their status, wellbeing or progression through the intervention.[17]

These messages may come in many different forms. They can be motivational (You did great today!) or purely data driven (You completed 80% of your goal today). Interestingly, it is not clear from this literature that CHI-derived behavioral feedback is any better than feedback originating from human practitioners or others. Rather, it appears that the feedback must happen with an appropriate periodicity, in a format that is appealing and acceptable to the consumer, not just the provider.[17]

Generally speaking, the scientific literature also suggests that CHI applications may positively impact healthcare processes such as medication adherence among asthmatics. CHI applications may also positively impact intermediate outcomes across a variety of clinical conditions and health behaviors, including cancer, diabetes mellitus, mental health disorders, smoking, diet, and physical activity. CHI applications may not have much impact on intermediate outcomes among individuals who are obese or suffer with asthma or COPD. In addition, the evidence appears relatively strong in support of the positive impact of CHI on selected clinical outcomes, particularly mental health outcomes.[17]

To facilitate uniform reporting and improve the quality of the work in this field, consideration should be given to development of a national CHI applications design and development registry and CHI applications trials registry with uniform reporting requirements. However, the developers of these applications come from a wide and diverse array of backgrounds. Some have significant technical expertise while others are clinicians. Research in this multidisciplinary field would be greatly enhanced by an accepted vocabulary, nomenclature or ontology. Currently, there is much confusion among the varied developers of CHI tools, between the platform upon which the application is built, the technical specifications of the CHI application and the educational or behavioral content of the messages included in the application. While a strict rendering of the current definitions of these elements allows for little conceptual overlap, the literature is replete with examples of investigators who describe the technical platform on which the CHI application (cell phone) runs yet provide no further technical specifications regarding the applicaion.[17]

More work will need to be done to explicate the role of human factors, socio cultural factors, human computer interface issues, literacy and gender. Currently most CHI research is being primarily conducted among white/Caucasian adult patients, and it is not clear how the findings apply to non-white populations. The importance of this limitation is heightened by the fact that the internet will be the primary means of the consumer's ability to use and take advantage of CHI tools. While technological platforms may vary, most CHI applications will, in one way or another, rely on the internet to perform its functions. Consumer internet familiarity and utilization trends will have significant impact on the ability of CHI applications to be successful across all consumer populations.[17]

Interestingly most of the evaluative research being done is being conducted among middle aged adult populations; significant opportunities exist for additional research among other age groups of consumers. It may even be that the impact of CHI applications may be greater among non middle aged adult consumers because these consumers may be most likely to adopt CHI applications (children, adolescents and young adults) and they may have the most to gain from using effective CHI applications (elderly).[17]

Also most CHI applications that have been evaluated to date are designed to run on desktop computers. More work will need to be done to understand the role of other technological platforms including cell phones,

PDA's, TV, satellite, on Demand, health gaming platforms (Wii, XBOX, Gamecube etc). Related to technological platforms used for CHI applications is the potential role of social networking applications. Very few currently evaluated CHI applications explored the dynamics and potential utility of using social networking applications (Skype, Twitter, MySpace, Facebook, You Tube, blogs, Second life, Yoville and Farmville, Patients like Me, etc.) to support behavior change or improve health outcomes. While it may be challenging to envision the elderly twittering, use of these applications may open opportunities to address health problems impacted by trust, social isolation, cognitive stimulation and low literacy). This type of research may inevitably lead to a broader array of interactivity among patients and their caregivers with measurable psychological and physiological health benefits for users and patients. In so doing, CHI applications may accrue greater appeal and effectiveness among patients because these applications are assisting patients to address real life issues that in the past may have been unrecognized barriers to achieving optimal health.[17]

References

1. Federal Health IT Strategic Plan 2011-2015. www.healthit.hhs.gov (Accessed November 30 2011)

2. Rainie L, Fox S, Horrigan J, Fellows D, Lenhart A, Madden M et al. Internet: The mainstreaming of online life. 2005. Washington, DC. Pew Internet and American Life Project. http://www.pewinternet.org/~/media/Files/Reports/2005/Internet_Status_2005.pdf.pdf (Accessed December 4 2011)

3. Internet Adoption 1995-2011. Pew Charitable Trusts. 2011. www.pewinternet.org/trend-data/internet-adoption.aspx (Accessed December 4 2011)

4. What Internet users do online. 2011. Pew Internet. www.pewinternet.org/trend-data/online-activities-total.aspx (Accessed December 4 2011)

5. Horrigan J. Wireless Internet access. 2007. Pew Internet and American Life Project. www.pewinternet.org (Accessed December 4 2011)

6. Smith A. Smartphone use and adoption. 2011. www.pewinternet.org (Accessed December 4 2011)

7. Smith A. Broadband adoption. 2010. www.pewinternet.org (Accessed December 4 2011)

8. Gibbons M C. E-Health Solutions for Healthcare Disparities. New York. Springer Pubs, 2008.

9. Horrigan J. Home broadband adoption. 2006. www.pewinternet.org (Accessed December 4 2011)

10. Horrigan J. Broadband: What's all the fuss about? 2007. www.pewinternet.org (Accessed December 4 2011)

11. Fox S. Health Topics. 2011. Pew Internet. http://www.pewinternet.org/~/media//Files/Reports/2011/PIP_HealthTopics.pdf (Accessed December 4 2011)

12. Committee on Quality of Health Care in America. To Err is Human: Building a Safer Health System. 1999. Wash DC. National Academies Press

13. Committee on the Quality of Health Care in America. Crossing the Quality Chasm: A New Health System for the 21st Century. 2011. Wash DC. National Academies Press

14. Eysenbach G. What is ehealth? J Med Internet Res 2001;3:E20

15. Eysenbach G. Consumer health informatics. BMJ 2000;320:1713-1716

16. Houston JD, Fiore DC. Online medical surveys: using the Internet as a research tool. MD Computing. 1998:15:116-120

17. Gibbons MC, Wilson RF, Samal L et al. Impact of consumer health informatics applications. Evid Rep Technol Assess (Full Rep) 2009;1-546

18. Scott J. Mobile apps approaching major milestone-totaling 1 million in days. December 2 2011. http://148apps.biz/mobile-apps-approaching-major-milestone-totaling-1-million-in-days/ (Accessed December 4 2011)

19. Harris Poll http://harrisinteractive.com/harris_poll/index.asp?PID=584 (Accessed January 10 2006)

20. McClung H. The Internet as a source for current patient information. Pediatrics 1998;101: p. e2

21. American customer satisfaction index of fed govt web sites. December 2004. www.theacsi.org/index.php?option=com_content&task=view&id=27&Itemed=62 (Accessed January 5 2005)

22. Withings www.withings.com (Accessed October 14 2011)

23. Well@home www.wellathome.com (Accessed October 14 2011)

24. Televox. www.televox.com (Accessed October 11 2011)

25. First Health, HarrisInteractive. Consumer Benefits Health Survey. Executive Summary http://www.urac.org/savedfiles/URACConsumerIssueBrief.pdf (Accessed January 12 2006)

26. American Health Information Management Association www.ahima.org (Accessed March 22 2010)

27. National Alliance for Health Information Technology. Defining Key Health Information Technology Terms April 28 2008 www.nahit.org (Accessed May 1 2009)

28. Crossing the Quality Chasm: A New Health System for the 21st Century Institute of Medicine 2001 The National Academies Press p. 8

29. Personal Health Technology Council www.markle.org (Accessed October 14 2011)

30. Connecting for Health. Working group on policies for sharing information between doctors and patients. July 2004 http://www.connectingforhealth.org/resources/wg_eis_final_report_0704.pdf (Accessed October 1 2005)

31. HarrisInteractive market research http://www.harrisinteractive.com/news/newsletters/healthnews/HI_HealthCareNews2004Vol4_Iss1_3.pdf (Accessed October 1 2005)

32. Centers for Medicare and Medicaid Services www.cms.hhs.gov and http://www.gcn.com/vol1_no1/health_IT/36422-1.html (Accessed November 1 2005)

33. Terry K. Will PHRs rule the waves or roll out with the tide? Hospitals & Health Networks. www.hhnmag.com (Accessed September 2 2009)

34. Grossman JM, Zayas-Caban T, Kemper N. Information Gap: Can Health Insurer Personal Health Records Meet Patients' and Physicians" Needs? Health Affairs 2009;28(2):377-389

35. Kuraitis V. Birth Announcement: The Personal Health Information Network (PHIN). E-CareManagement Blog March 8 2008. http://e-caremanagement.com (Accessed July 30 2008)

36. Insurers to Provide Portable, Interoperable PHRs. www.ihealthbeat.org December 14 2006. (Accessed December 14 2006)

37. Industry Leaders Announce Personal Health Record Model; Collaborate with Consumers to Speed Adoption. http://bcbshealthissues.com December 13 2006 (Accessed December 14 2006)

38. Aetna Broadens PHR Availability. August 28 2008. www.healthdatamanagement.com (Accessed August 0 2008)

39. Personal Health Records in the Marketplace http://library.ahima.org/xpedio/groups/public/documents/ahima/pub_bok1_027459.html (October 20 2005)

40. Gartner Says Sales of Mobile Devices Grew 5.6 Percent in Third Quarter of 2011; Smartphone Sales Increased 42 Percent. November 15 2011. http://www.gartner.com/it/page.jsp?id=1848514 (Accessed December 5 2011)

41. Diversinet's mobisecure wallet http://www.diversinet.com/ (Accessed October 14 2011)

42. Army.mil Mobile. http://www.army.mil/mobile/ (Accessed November 20 1011)

43. MyRapidMD www.myrapidmd.com (Accessed October 14 2011)

44. Project SwipeIT. www.mgma.com/swipeithome (Accessed October 14 2011)

45. A Healthcare CFO's Guide to Smart Card Technology and Applications Smart Card Alliance February 2009. www.smartcardalliance.org (Accessed October 14 2011)

46. Lohr S. Kaiser Backs Microsoft Patient Data Plan. June 10 2008. The New York Times

47. Microsoft HealthVault. www.healthvault.com (Accessed October 13 2011)

48. Cleveland Clinic Tests Microsoft's HealthVault PHR System November 10 2008 www.ihealthbeat.org (Accessed November 10 2008)

49. Indivo. http://indivohealth.org (Accessed October 14 2011)

50. Mandl KD, Simons WW, Crawford WCR et al. Indivo: a personally controlled health record for health information exchange and communication. www.biomedcentral.com BMC Medical Informatics and Decision Making 2007;7:25

51. Travelers Electronic Health Record Template www.trehrt.com (Accessed January 4 2012)

52. Li Y, Detmer DE, Shabbir S et al. A global traveler's electronic health record template standard for personal health records. JAMIA 2012;19:134-136

53. Kaelber DC, Jha AK, Johnston D et al. A Research Agenda for Personal Health Records JAMIA 2008;15:729-736

54. Survey: Consumers Have Concerns About Insurer-Provided PHRs. January 31 2007. www.ihealthbeat.org. (Accessed January 31 2007)

55. Aetna. www.aetna.com (Accessed October 14 2011)

56. Urdem T. Consumers and Health Information Technology: A National Survey. April 2010. www.chcf.org (Accessed April 10 2010)

57. Pagliari C, Detmer D, Singleton P. Potential of electronic personal health records. BMJ 2007;335: 330-333

58. CMS and the Defense Department Pilot Projects Could Jump Start PHR Use. March 2009 www.hhnmag.com p.9

59. Centers for Medicare/Medicaid http://www.cms.hhs.gov/perhealthrecords/ (Accessed October 14 2011)

60. Project HealthDesign www.projecthealthdesign.org (Accessed October 13 2011)

61. Slack WV. A 67 Year Old Man Who e-mails his Physician JAMA 2004;292:2255-2261

62. Physicians in 2012: The Outlook on Health Information Technology. Manhattan Research. 2010. www.manhattanresearch.com (Accessed March 23 2010)

63. National Health Interview Survey. January-June 2009. Reported February 2010. NCHS. CDC. www.cdc.gov (Accessed March 23 2010)

64. Study Reveals Big Potential for the Internet to Improve Doctor-Patient Relations. Harris Interactive 2001 www.harrisinteractive.com (Accessed September 24 2006)

65. WSJ examines physician's reluctance to e-mail patients. www.ihealthbeat.org June 3 2003 (Accessed October 2004)

66. Rosen, P, Kwoh, CK. Patient-Physician E-mail: An Opportunity to Transform Pediatric Health Delivery. Pediatrics 2007; 120(4): 701-706

67. Car J, Sheikh A. E-mail consultations in health care: scope and effectiveness. BMJ 2004;329:435-438

68. The Changing Face of Ambulatory Medicine—reimbursing physicians for computer-based care. American College of Physicians Medical Service Committee Policy Paper March 2003 www.acponline.org/ppvl/policies/e000920.pdf (Accessed March 15 2003)

69. Rapp C. Liability Issues Associated with Electronic Physician-Patient Communication. Internat Ped March 2007;22(1)

70. Healthcast 2010: Smaller world, bigger expectations. Price Waterhouse Cooper. November 1999 www.pwc.com (Accessed February 3 2006)

71. Broder C. What's in a code? www.ihealthbeat.org January 14 2004 (Accessed January 14 2004)

72. Liederman EM .Web Messaging: A new tool for patient-physician communication JAMIA 2003;10:260-270

73. Chen-Tan Lin .An Internet Based patient-provider communication system: randomized controlled trial JMIR 2005;7 (4): e47

74. Leong SL .Enhancing doctor-patient communication using e-mail: a pilot study J Am Board of Fam Med 2005;18:180-8

75. Zhou YY et al. Patient Access to an Electronic Health Record with Secure Messaging: Impact on Primary Care Utilization. Am J Manag Care 2007;13:418-424.

76. Juniper Research October 2003 www.juniperresearch.com (Accessed December 10 2005)

77. The RelayHealth web visit study: Final Report www.relayhealth.com (Accessed December 2 2007)

78. Tennessee Hospital Pilots two e-mail programs www.ihealthbeat.org March 14 2005 (Accessed March 14 2005)

79. Blue Shield of California web communications pilot to enroll 1,000 physicians www.ihealthbeat.org April 20 2004 (Accessed April 20 2004)

80. Health Plan to pay Doctors for web visits www.ihealthbeat.org May 24 2004 (Accessed June 30 2004)

81. Microsoft Pilot Project to Test Online Physician Visits www.ihealthbeat.org January 10 2006 (Accessed February 2 2006)

82. UC Davis Virtual Care Study. Eric Liederman. Presented at AMDIS/HIMSS 2004

83. Cigna Offers Seniors Free Online Health Services. April 11 2007. www.ihealthbeat.org. (Accessed April 12 2007)

84. Lowes R. Aetna and Cigna want to pay you for online visits. Medical Economics. January 25 2008. www.memag.com (Accessed January 29 2008)

85. HouseDoc. http://housedoc.us (Accessed October 14 2011)

86. Online Patient-Provider Communication Tools: An Overview. First Consulting Group. November 2003. California HealthCare Foundation www.chcf.org (Accessed September 20 2006)

87. TelaDoc www.teladoc.com (Accessed October 14 2011)

88. American Well www.americanwell.com (Accessed October 14 2011)

89. MDLiveCare http://mdlivecare.com (Accessed October 14 2011

90. 3G Doctor. http://3gdoctor.com (Accessed October 14 2011)

91. Watson AJ, Bergman H, Williams CM et al. A Randomized Trial to Evaluate the Efficacy of Online Follow-up Visits in the Management of Acne. Arch Derm 2010;146(4):406-411

92. Kelechi T, Green A, Dumas B. Online Coaching for a lower limb physical activity program for individuals at home with a history of venous ulcers. Home Healthcare Nurse. 2010;28(10):596-605

93. Nield MA, Soo Hoo GW. E-BREATHE, an Internet based program. Am J Respir Crit Care Med. 2010;181:A3451

94. Hori M, Kubota M, Ando K. et al. The effect of videophone communication (with skype and webcam) for elderly patients with dementia and their caregivers. Gan To Kagaku Ryoho. 2009;36 Suppl 1:36-8

12

Mobile Technology

ROBERT E. HOYT

REYNALD FLEURY

ROBERT W. CRUZ

Learning Objectives

After reading this chapter the reader should be able to:

- Describe the evolution from personal digital assistants to smartphones and the emergence of mHealth

- List the various ways mobile technology is currently being used in healthcare today

- Compare and contrast mobile technology for clinicians and patients

- Identify the limitations of mobile technology

Introduction

In the first edition of our textbook we primarily discussed the use of personal digital assistants (PDAs) in the field of medicine. With the advent of smartphones and tablets, such as the iPad, we feel mobile technology is a broader and more appropriate term. Handheld technology or mobile electronic devices are also acceptable. Mobile technology is a logical transitional step from the personal computer. With improving speed, memory, wireless connectivity and shrinking form factor (size and shape), users desire a mobile platform for their information and applications as well as phone capability, e-mail and access to the internet. Without a doubt, mobile technology is the fastest evolving topic in this textbook and could be considered disruptive technology. Because of its pervasive and popular nature mobile technology has become a component of many consumer informatics, disease management and telemedicine strategies. We will discuss standalone mobile software programs (apps) but the larger picture is system wide integration of mobile technology into healthcare.

Some would argue that we have entered the mHealth (mobile health) era in which mobile technology will play a much larger role in healthcare. mHealth, a subcategory of eHealth, can be simply defined as the "delivery of healthcare services via mobile communication devices."[1] Currently, we are including cell phones, smartphones and tablets as mobile electronic devices, but this is arbitrary and subject to change.

We will discuss mHealth in more detail, but first we will start with the history of mobile technology.

History of Mobile Technology

Cellular Mobile Telephony

The history of modern mobile technology is relatively recent. Primitive mobile phones arose in the 1970s but didn't gain popularity until 2G cellular networks appeared in Finland in the early 1990s. Worldwide adoption with 3G cellular networks became a reality in the 2001 time frame.[2] Figure 12.1 demonstrates worldwide mobile cellular telephone subscriptions from 2000 to 2010.

Figure 12.1: Global telecommunications (Courtesy International Telecommunications Union)[3]

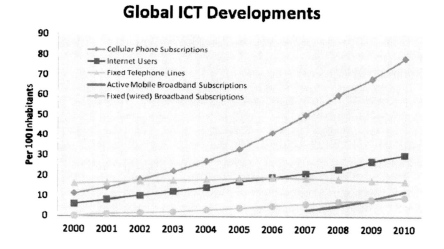

Figure 12.2 compares the usage of various technologies worldwide and confirms the astronomical adoption and growth of cellular telecommunications in the past decade. This growth has also occurred in developing countries; in fact, some countries have "leapfrogged" ahead and skipped the landline phase with widespread wireless cellular adoption. We will address how some countries have leveraged cellular technology in another section.

Figure 12.2: Technology usage worldwide [3-6]

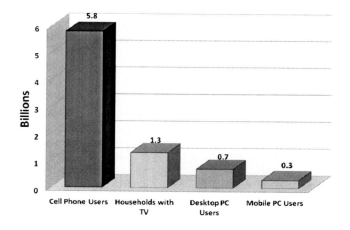

Personal Digital Assistants (PDAs)

In the early 1990s the Apple Newton PDA appeared with a monochrome screen, a weight of .9 lbs., measurements of 7.25 x 4.5 x .75 inches, 150 K of SRAM, a processor speed of 20 MHz, short battery life and a cost of $700.[7] It obviously did not succeed because it was too big, heavy, slow and costly for the average consumer.

The next handheld product to catch the public's attention was the Palm Pilot 1000 released in 1996.[8] It was smaller, less expensive and had 128K of memory but did not become popular with the medical profession until the *killer application* Epocrates was released in 1999.[9] First, there was the excitement of knowing that drug facts could be retrieved much more rapidly with the PDA compared to the Physician Desk Reference (PDR) and secondly, the program was free. The PDA was also a platform to store all medical "pearls" rather than stuffing notes into the pockets of a white coat. Other companies got on the bandwagon rapidly to produce PDAs. This was followed by PDAs with phone capability, internet access, WiFi and multimedia capability. As wireless cellular communication caught on worldwide, consumers desire for one platform resulted in a convergence of technologies. The shift from PDAs to smartphones was rapid, occurring over only six to eight years.

Smartphones

There is no industry-wide definition of a smartphone. Some define it as having an operating system that can support the execution of third party applications and others define it as simply having more functionality than conventional cell phones. For the purpose of this chapter we will use the term smartphone and include only those that have operating systems capable of hosting medical software. With the evolution of cloud computing, more and more medical programs will be hosted in the cloud and not on the device. Therefore, one could eventually state that a smartphone is one that is internet capable. There are likely to be further convergence of handheld technologies, as devices such as the Apple iPad blur the distinction between smartphones and laptop/tablet/slate computers. It is too early to know if these devices will be used frequently by physicians making rounds or evaluating patients in an exam room. There is a paucity of evidence to prove this type of mobile technology is superior to the standard desktop computer, but this is likely to change over time.[10]

Pew Research Center reported in 2011 that 42% of cellphone users owned a smartphone. The highest users were the financially well off and well educated young individuals, however, a significant number of minorities were also smartphone users.[11] Manhattan Research reported in 2011 that 81% of physicians used smartphones, compared to 30% in 2001.[12]

Slate/Tablet PCs

The first generation of tablet PCs were all Windows operating sytem-based and required a stylus and keyboard for input. Additionally, they were heavy, expensive and had short battery life. Clinicians tried to use this technology in exam rooms and on hospital rounds but few continued, due to the stated limitations.

There has been an avalanche of slate or tablet computers, such as the Apple iPad and Android devices, since 2010. Gartner has predicted that 108 million tablets will be sold in 2012.[13] The new tablets are light weight, have prolonged battery life, excellent screen resolution, extensive medical apps, instant-on capability and a convenient form factor. The landscape over the past year has been volatile with offerings from HP running WebOS being completely discontinued, and a bleak outlook for the BlackBerry Playbook from RIM. The two top contenders are currently the iOS powered iPad line and the vast array of devices running the Android operating system. According to Manhattan Research, 30% of surveyed physicians use the Apple iPad to view electronic health records (EHRs) or digital images and communicate with patients.[12] The Veterans Affairs Department will begin a test of 1,000 iPads in 2011 for their physicians, with possible roll out for home telemedicine in the future. Physicians are both self-adopting (purchasing the devices for themselves) as well as having their organizations purchase and roll out the devices.

Mobile Health (mHealth)

Conceptual framework

In order to evaluate the rapidly changing mobile technology landscape an organizational schema is required. We will use, but modify, the conceptual framework of Caroline Free et al., displayed in Figure 12.3 to organize our discussion about mHealth (mobile technology in healthcare).[14] The three themes are "tools for health research," "improving health services" and "improving health outcomes." The tools for health research will be discussed in the chapter on e-research. This chapter will begin with interventions for patients.

Figure 12.3: Conceptual framework for mobile technology (adapted from Caroline Free[14])

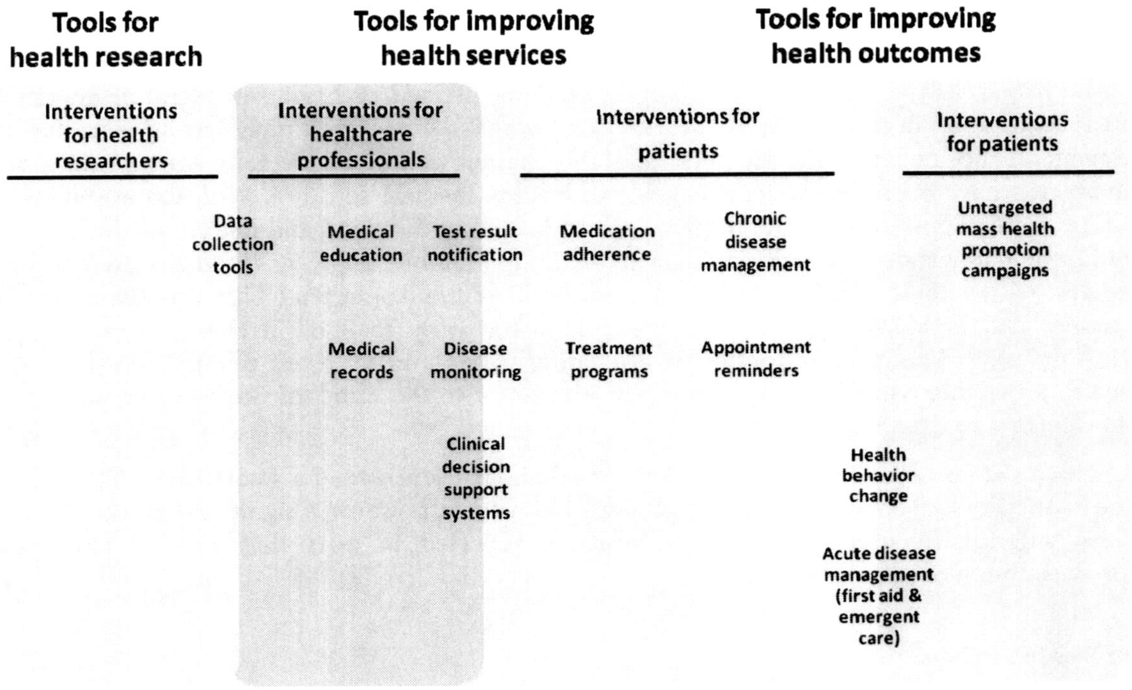

Tools for health research	Tools for improving health services		Tools for improving health outcomes
Interventions for health researchers	Interventions for healthcare professionals	Interventions for patients	Interventions for patients
Data collection tools	Medical education	Test result notification	Medication adherence	Chronic disease management	Untargeted mass health promotion campaigns
	Medical records	Disease monitoring	Treatment programs	Appointment reminders	
		Clinical decision support systems		Health behavior change	
				Acute disease management (first aid & emergent care)	

Mobile technology and patients

Devices. As already stated, most adult patients worldwide own a cell phone and the percent that own a smartphone or tablet PC is steadily increasing. It is important to keep in mind that smartphones and tablets have access to the internet so in addition to software applications (apps) that can be downloaded to the device, multiple web-based programs can be accessed. Access to the internet is extremely important as it has become the default health library, as discussed in the chapter on consumer informatics. Many patient-oriented web programs will be discussed in other chapters e.g. telemedicine. We will begin with a discussion of text messaging and healthcare.

Text messaging. Short Message Service (SMS) or text messaging began in 1992 and consisted of 160 character messages sent between two cell phones. It is a service available for the simple cell phone and smartphone. SMS is a worldwide telecommunication service phenomenon because the technology is inexpensive and ubiquitous. It has been shown to be more cost effective than phone or mail communication. While the United States has been behind the rest of the world, in terms of text messaging, and one of the few countries to charge for receiving text messages, it is catching up. According to a 2011 Pew Research poll 83% of American adults own cell phones and 73% send and receive text messages. Cell phone owners between the age of 18 and 24 exchange more than 50 text messages daily and would prefer SMS over voice calls.[15] Importantly, minority groups are heavy users of this technology as well.[16]

Text messaging has been utilized to help solve multiple healthcare issues worldwide. In general, SMS use in healthcare falls into the following categories:

- Appointment reminders. Several international studies have shown improved outpatient clinic attendance with text messaging, compared to no reminder or other technologies.[17, 18]

- Education. A variety of educational programs have been well received, such as Text4Baby to educate pregnant mothers and SEXINFO that provides sexual health messages to teenagers.[19, 20]

- Disease management. SMS has been shown to improve chronic disease management for diabetes mellitus and asthma. [21, 22]

- Behavior modification. Text messaging has been successfully used for smoking cessation reminders to improve quit rates. [23]

- Medication compliance. According to Manhattan Research, 49% of ePharma consumers would be interested in email or text reminders for medications and refills. [24] One study showed a very significant improvement in compliance with HIV medications after receiving text messages. [25]

- Laboratory results notification. SMS has been used to notify patients and physicians about lab results. One study showed much quicker time to treatment for patients with chlamydia (sexually transmitted infection), when notified by SMS. [26] A second study showed a much faster clinical response time to an elevated serum potassium using SMS. [27]

- Public Health. Text messaging is a new and interesting approach that can be used by local, state and federal public health programs worldwide. For instance, the Seattle and King County Public Health department is exploring the use of SMS to notify patients for emergencies and routine issues such as immunization reminders. [28] In 2009 The Centers for Disease Control and Prevention (CDC) embarked on a SMS project to send alerts to patients in the United States covering topics in general health, as well as emergency preparedness and response. A subscriber enters their age, gender, role and zip code and they receive tailored messages weekly and can alert users in that zip code about public health emergencies. [29] A HHS Text4Health Task Force released recommendations to leverage text messaging to improve US health in September 2011. [30]

Because individuals in most developing countries own a cell phone and not a smartphone they utilize as many features of voice communication and SMS as possible. The 2009 report by the United Nations and Vodafone Foundations *mHealth for Development* stressed the following objectives using this technology: increase access to healthcare, particularly remote populations; improve ability to diagnosis and track illness; more actionable public health information and expand patient education and training of health workers. [31] In the side bar is an example of mobile technology projects in Mexico.[32]

There are simply too many health-related text messaging initiatives internationally to mention, so we will refer readers to these additional resources [33-35] and the chapter on public health informatics.

Medical software categories for patients. In this section and the following section we have elected to mention categories of medical software that are located in the Apple iTunes App Store and the Android Market because the vast majority of apps, popular with patients and clinicians are located there. While we know that thousands are available for download and we know the download statistics, we don't know how many are regularly used and their actual impact on behavior or patient outcomes. Software is also available for the iPad and Android tablets, but the app choices are not as extensive. The tablet device is clearly superior when a larger field of view is required.

Mexico mHealth

CardioNET is a text messaging service to remind patients to diet and exercise as obesity is a national issue. It also provides a cardiac risk assessment tool.

VidaNET(LifeNET) is a network to educate and remind HIV patients about their illness and the importance of appointment, medication and lab testing compliance.

DiabeDiario is their most recent and ambitious mobile technology project to tackle diabetes that affects 10% of Mexicans. It will combine web, email and text messages to monitor and treat diabetes. [32]

Table 12.1: Medical software categories for patients and examples (AS = App Store, AM = Android Market) [36, 37]

Software Category	Examples	Functionality
Connect with healthcare system	Group Health (AS,AM)	Patient portal to check on appointments, lab results, etc.
Personal health record	MobiSecure (AS, AM)	Mobile PHR platform that backs up to the cloud
Telemedicine	Skype Mobile	With forward facing camera video-teleconferencing is possible
Medication reminders	MedCoach (AS)	Medication and refill reminder. Can connect to your pharmacy
Fitness coach	RunKeeper (AS, AM)	Tracks activity and fitness
Mind fitness	Brain Trainer (AS)	Cognitive training
Immunization guides	Shots 2011(AM)	Guide for what immunizations are needed based on age and gender. Other important info.
Disease management	Diabetes Manager (AS),	Monitors meals, blood sugar and insulin doses
	iHealth BPM (AS)	Blood pressure cuff or weight scale sends results to iPhone/iPad
Prevention guide	AHRQ ePSS (AS, AM)	Guide for national recommendations for preventive care based on age, gender, smoker, etc.
Diagnostic	MelApp (AS)	Phone camera takes picture of skin lesion and estimates risk of malignant melanoma

Mobile Technology and Cinicians

Devices. We will not discuss routine cell phones and clinicians, although there are few initiatives that use text messaging to alert physicians. As previously defined, smartphones have an operating system that allows medical software to be installed. Additional capabilities include phone service, e-mail, internet access, calendars, contact lists, task lists, cameras and video capability. Synchronization to a computer can be by Bluetooth, WiFi, USB connection or even via the cloud with apps such as Evernote or via features built into the operating system such as Apple's iCloud on iOS. Touch screens and speech recognition have made data entry easier, compared to a stylus. Internal memory is no longer an issue with smartphones because most have slots for mini SD cards, available in the one to 32 GB range. Physicians who may have carried a pager, cell phone and PDA can have a single multi-purpose device to receive routine phone calls, text messages or voice mails. Moreover, with much faster internet access we can anticipate more interest in using smartphones to e-prescribe, access online resources, access EHRs, access images and many more functions.

More web sites are producing mobile versions of their web sites to accommodate the smaller screen size of most smartphones. Web sites can detect the browser version requesting a page from the server, and can redirect the browser to a mobile-only page or use style sheets made specifically for the mobile platform to display the content in a more compact and easier to browse manner.

Medical Software categories for clinicians. Table 12.2 categorizes popular medical software programs that are free or fee-based that can be obtained from the iTunes App Store or Android Market and provides examples of each category. This list is not intended to be exhaustive as both the iTunes App Store and Android Market have thousands of health, fitness and medical software programs listed. The iTunes App Store lists which apps are available for the iPhone, iPad or both.

Table 12.2: Medical software categories for clinicians and examples (AS = App Store, AM = Android Market) [36, 37]

Software Category	Examples	Functionality
Drug information	Epocrates (AS, AM), Medscape Mobile (AS, AM), Mobile Micromedex (AS, AM)	Extensive drug library, drug interactions, pill ID, disease reference, calculators, etc.
Calculators	MedCalc (AS, AM), Archimedes (AS, AM), Calculate QxMD (AS, AM)	Perform multiple common calculations used by most physicians
	Framingham risk scores (AS, AM)	Calculates 10 year risk of heart disease based on risk factors
	ABG Interpreter (AM),	Blood gas interpretation
	Infusion rate (AM)	Calculates IV infusion rates
Database programs	HanDBase (AS, AM)	Relational database
	GoCanvas (AM)	Mobile forms with geolocation that back up to server
Immunization guides	Shots 2011(AM)	Guide for what immunizations are needed based on age and gender. Other important info.
Medical resources	5 Minute Clin. Consultant (AS, AM)	Covers 715 topics succinctly
	Sanford Guide (AS, AM), Johns Hopkins Guide (AS, AM)	Popular guides to direct care for infectious disease
	UpToDate (AS, AM), DynaMed (AS, AM)	Extensive resources covering most sub-specialties
	iCXR (AS)	Chest x-ray educational resource

Software Category	Examples	Functionality
	Heart EKG Guide (AS, AM)	EKG educational resource
	Derm101: Point of Care	Dermatology resource
	LabDx (AS), Pocket Lab Values (AS)	Laboratory results resources
	Procedures Consult (AS)	Procedures resource by specialty
	mTBI Pocket Guide (AM)	Pocket resource for TBI
	Relief Central (AS, AM)	Extensive resource for relief workers
	WISER (AS)	Hazardous material responder resource
Prevention guide	AHRQ ePSS (AS, AM)	Guide for national recommendations for preventive care based on age, gender, smoker, etc.
Diagnostic	Eye Chart Pro (AS)	Electronic eye chart
	iExaminer (AS)	Uses iPhone plus hardware to take pictures of retina
	Diagnosaurus (AS, AM)	Search 1,000+ differential diagnoses
Image viewer	Resolution MD Mobile (AS)	Mobile access to image server
	Mobile MIM (AS)	Mobile image viewer that can be used by radiologists
Journal	BMJ (AS), Chest (AS), NEJM (AS, AM)	Provides access to major medical journals, podcasts and videos
Medline search	PubMed Mobile (AS, AM)	Mobile means to access Medline
Monitoring	AirStrip OB, AirStrip Cardiology (AS, AM)	Mobile views of multiple physiologic parameters
Coding	MD Coder (AS, AM), Hospital Rounds (AS), E/M Code Check (AS)	Charge capture apps
Medical translator	Medibabble (AS)	Translates history and physical exam elements in five languages
EHR access	Epic Haiku (AS, AM), Quest360 mobile (AS)	Mobile access to EHRs
Telehealth	Online Care Mobile (AS)	Mobile means to access American Well's Online Care Suite for e-visits
	Skype Mobile (AS, AM)	Can be used for virtual visits
Dictation	Dragon Dictation (AS, AM)	Mobile means to dictate

mHealth Developed Countries. By and large, what we have presented in this chapter represents standalone solutions and applications. That is likely to change as more healthcare organizations and clinicians adopt mobile technology and security issues are adequately managed. As mobile electronic devices, such as the iPad improve, it is likely they will become the platform of choice for the exam room and hospital rounding. This will be influenced by comprehensive integration with the healthcare system's EHR, lab and radiology information systems, financial systems and data warehouse. It will also be influenced by better voice recognition to augment the touch screen and integration with technologies that provide real-time vital sign, cardiac and OB monitoring. [38] We are already seeing enterprise solutions appearing that are, in part, based on a mobile solution and offer charge capture, patient schedules, mobile dictation and clinical results (lab, xray, medications, vitals, office/hospital notes). [39] The case study: iPhysician below describes the mobile physician who benefits from mobile technology with enterprise connections.

mHealth Developing Countries. Mobile technology is being used nationally and internationally as part of a variety of healthcare initiatives. mHealth has the potential to empower patients with medical information so they can control their own health and wellness. It also has the potential to connect patients with medical offices and public health which they might not be able to do otherwise. If mHealth programs succeed with increasing access to medical information and care they will reduce costs and morbidity and thus transform medical care. mHealth programs are very diverse and usually deal with epidemiology, chronic disease monitoring and treatment and research usually as part of public health or population health. Most developing countries have an intact telecommunications network for routine cellphone technology. As economic circumstances improve so will the adoption of smartphone technology with its enhanced ability to access the internet and download medical apps. More information about mHealth in developing countries can be found in this reference. [40]

Case Study: The iPhysician

A physician uses an iPad to connect to his EHR while seeing patients in the exam room. This same platform displays digital x-rays and other images that are useful for patient education. It's lightweight with long battery life, qualities missing in other mobile electronic devices he has tried. He can easily access medical apps like Epocrates for simple drug related questions, the 5 Minute Clinical Consultant for straight forward issues and UpToDate or DynaMed for more complex medical questions. He dictates into his iPad using voice recognition software. His mobile device helps display anatomical drawings for patients so they better understand e.g. coronary artery disease. He is able to use a digital camera feature to record a large skin nevus he intends to follow closely.

He relies on this same technology when he makes hospital rounds, except now he uses software such as Hospital Rounds to keep track of the patients he sees and intends to bill. Because he is a hospital attending he is credentialed to have access to the hospital EHR and its information systems that now includes real time monitoring of vital signs, EKGs, O2 sats, etc. While making rounds an outpatient calls him to discuss an acute problem and he is able to access office charts on his iPad because his EHR is web-based.

That evening when he returns home he reviews a digital medical newsletter and the latest online issue of the New England Journal of Medicine on his tablet. He receives an email from a very concerned patient and decides to have a virtual visit with the patient using MedFusion and his tablet. After a round of *Angry Birds*, the iPad is set up to charge and he heads to bed.

Mobile Telemedicine Projects

Electronic Mobile Open-Source Comprehensive Health Application (eMocha) is a free open source initiative developed by Johns Hopkins Center for Clinical Global Health Education. The programs consists of two components: (1) *Android Os phone*: Uses XML-based forms to collect data in multiple formats (text, pictures, bar codes, audio and video) that is geo-stamped and backed up to a server. eMOCHA can deliver multimedia courses and lectures (MP4 format) that can be accompanied by quizzes on the phone. Phone can be used for voice calls and e-consultations using the phone camera. (2) *Remote server:* Data from the phone is sent to a MySQL database which can also send information to the mobile device such as forms, videos, webcasts and lectures. Data is then available for mining. This platform is currently being used for HIV care in Uganda. In late 2011 an Android app for tuberculosis was released that has a symptom algorithm and educational material. [41]

Sana Mobile is a similar project developed by volunteers from many departments at the Massachusetts Institute of Technology (MIT). Remote healthcare workers can input data into the phone, including images and they are sent to the EHR OpenMRS where specialists can view the record and respond back to the healthcare worker. Projects are underway with this platform in Africa, Brazil, Greece, India, Philippines and Columbia. [42]

Mobile Technology Resources

- iMedicalApps www.imedicalapps.com

- Skyscape www.skyscape.com

- Downloadable Apps for Mobile Devices. Uniformed Services University. www.lrc.usuhs.mil/local/MobileDevice/devices.html

- mHealthInitiative www.mobih.org

- mHealth Alliance www.unfoundation.org/global-issues/technology/mhealth-alliance.html

- Management and Security of Health Information on Mobile Devices www.ahima.org

- Mobilehealthnews www.mobilhelalthnews.com

- Wireless Healthcare www.wirelesshealthcare.co.uk

Limitations of Mobile Technology

Smartphones and slate computers were not initially intended to replace PCs or laptops, in spite of their impressive evolution. However, with better performance, more features, longer battery life, better input methods, to include portable keyboards, this is no longer the case. As with all technologies, there are universal limitations that need to be intelligently managed:

- Cost is a factor, but in spite of the initial charge for hardware and the monthly data charges, healthcare workers are purchasing smartphones and tablets in large numbers

- Inputting information is slow but improving constantly with technologies such as voice recognition and pattern recognizing soft keyboards

- Small screen size is an issue for smartphones but not tablets so clinicians are likely to own both, unless they converge in the future

- Security will always be an issue so additional protection is necessary

- Interoperability is an issue but most medical software is now available for multiple platforms

- In spite of the lack of proof that mobile technology improves patient outcomes or significantly impacts clinician productivity this has not stopped many clinicians from embracing new mobile technology. We need better research to objectify the use of mobile technology

Future Trends

Mobile technology is extremely popular in developed and developing nations, based on its pervasive nature, convenient form factor, affordability, expanding applications and scalability. A myriad of smartphone and tablet applications (1 million+) are available for patients and clinicians alike. Integration with electronic health records and other hospital information systems is already occurring. New uses for mobile technology in healthcare are arising at an unprecedented rate. As new peripheral devices are adapted and integrated with smartphones we can expect innovations heretofore never conceived. Similarly, tablet PCs may become the medical mobile platform of choice for clinicians in exam rooms and on hospital rounds. Already, several EHR vendors offer their software on mobile platforms.

We can expect mobile technology to get smaller, faster, less expensive and be better integrated in the near future. Patient and clinician apps will continue to proliferate. Telemedicine will likely expand remote delivery of care using mobile technologies.

Key Points

- Handheld technology has moved quickly from personal digital assistants (PDAs) to smartphones and tablets

- We are entered an era of mHealth where mobile electronic devices will be employed to assist in healthcare worldwide

- Multiple medical software programs are available for mobile platforms that are free, shareware or fee-based

- We are starting to see enterprise level integration of mobile technology so interoperability is becoming less of an issue

Conclusion

Mobile technology continues to improve and gain popularity in the medical profession worldwide at an amazing pace. Mobile technology is being used for storing medical information, telephonic communication, patient monitoring and clinical decision support. In the not too distant future, it will likely be used commonly for geo-location and connectivity to electronic health records and other hospital networks. Smartphones have replaced most PDAs as processer speed, memory, network access and multimedia features continue to improve. Interest in smartphones will continue to increase due to more medical and non-medical applications developed, as well as evolving 4G networks. We believe that voice recognition has improved to the degree that it may become a prominent means of inputting for mobile devices in the not too distant future. Competition among the various operating systems is intense, driving functionality up and cost down. However, mobile technology has definite limitations and further research is needed to determine their actual impact and place in the armamentarium of most physicians. An excellent review *How Smartphones Are Changing Health Care for Consumers and Providers* by the California HealthCare Foundation appeared in 2010 that addresses the current and future state of smartphones and mobile technology. [43]

References

1. Torgan, C. The mHealth Summit: Local and Global Converge, 2010. www.caroltorgan.com (Accessed October 3 2011).

2. 3G Networks and future milestones. UMT. http://www.umtsworld.com/umts/history.htm (Accessed October 4 2011)

3. International Telecommunication Union www.itu.int (Accessed October 4 2011)

4. iPTVNews. http://www.iptv-news.com/iptv_news/2010/november_2010_2/almost_1bn_digital_tv_households_expected_globally_by_2015 (Accessed October 7 2011)

5. Gartner Research. June 2008. www.gartner.com (Accessed October 7 2011)

6. Infinite Research on Worldwide Tablets Market Research Report. January 18 2011. http://www.infiniteresearch.net/announcements/announcing-infinite-researchs-worldwide-tablet-market-forecast-report (Accessed June 20 2011)

7. Apple Newton. www.oldcomputers.net/apple-newton.html (Accessed October 7 2011)

8. Palm http://www.palm.com/us/ (Accessed April 10 2010)

9. Epocrates www.epocrates.com (Accessed April 10 2010)

10. Anderson P, Lindgaard AM, Prgomet M et al. Is selection of hardware device related to clinical task? A multi-method study of mobile and fixed computer use by doctors and nurses on hospital wards. J Med Internet Res 2009;11(3):e32

11. Smith A. 35% of American adults own a smartphone. Pew Internet. http://pewinternet.org/Reports/2011/Smartphones.aspx (Accessed September 2 2011)

12. Manhattan Research. Taking the Pulse US 11.0 www.manhattanresearch.com May 11 2011. (Accessed May 11 2011)

13. Gartner Research. September 22 2011. www.gartner.com (Accessed September 25 2011)

14. Free C, Phillips G, Felix L, et al. The effectiveness of M-health technologies for improving health and health services: a systematic review protocol. BMC Research Notes 2010;3:250 www.biomedcentral.com/1756-0500/3/250 (Accessed September 22 2011

15. Smith. A. Pew Internet Research. Americans and Text Messaging. September 19 2011. www.pewinternet.org (Accessed September 27 2011)

16. African-Americans, Women and Southerners Talk and Text The Most in the U.S. Nielsen Wire. August 24 2010. www.blog.nielsen.com (Accessed October 3 2011)

17. Downer SR, Meara JG, Da Costa AC Use of SMS text messaging to improve outpatient attendance. MJA. 183 (7):366-368

18. Da Costa TM, Salomao PL, Martha AS. The impact of short message service text messages sent as appointment reminders to patients' cell phones at outpatient clinics in Sao Paulo, Brazil. Int. J. Med. Inform. 2010;79:65-70

19. Text4Baby www.text4baby.org (Accessed October 2 2011)

20. Levine D, McCright J, Dobkin L,et al. SEXINFO: A Sexual Health Text Messaging Service for San Francisco Youth. Am J of Public Health 2008;98 (3):1-3

21. Yoon K, Kim H. A short message service by cellular phone in type 2 diabetic patients for 12 months. Diabetes research and clinical practice. 2008;79:256-261

22. Ostojic V, Cvoriscec B, Ostojic SB et al. Improving asthma control through telemedicine: A study of short message service. Telem J E Health 2005;11:28-35

23. Free C, Knight R, Robertson S et al. Smoking cessation support delivered via mobile phone text messaging (txt2stop): a single blind, randomized trial. Lancet 2011;378:49-55

24. Manhattan Research Strategic Insight v8.0 #12. www.manhattanresearch.com (Accessed October 2 2011)

25. Andrade AS, McGruder HF, Wu AW. A programmable prompting device improves adherence to highly active antiretroviral therapy in HIV-infected subjects with memory impairment. Clin Infect Dis 2005;41:875-882

26. Menon-Johansson AS, McNaught F, Madalia S, et al. Texting decreases the time to treatment for genital Chlamydia trachomatis infection. Sex Transm Infect 2006;82:49-51

27. Park H, Min W, Lee W, et al. Evaluating the short message service alerting system for critical value notification via PDA telephone. Ann Clin Lab Sci 2008; 38 (2): 149-156

28. Karasz H, Bogan S. What 2 know b4 u text: Short message service options for local health departments. Wash State J Pub Health Pract 2011;4 (1):20-27

29. CDC Mobile http://www.cdc.gov/mobile/ (Accessed October 2 2011)

30. Health Text Messaging Recommendations to the Secretary http://www.hhs.gov/open/initiatives/mhealth/recommendations.html (Accessed October 2 2011

31. mHealth for Development http://www.mobileactive.org/files/file_uploads/mHealth_for_Development_full.pdf (Accessed October 4 2011)

32. Feder FL. Cell-phone medicine brings care to patients in developing nations. Health Affairs. February 1 2010. http://www.ncpa.org/pdfs/Cell-Phone-Medicine-Brings-Care.pdf (Accessed February 3 2011)

33. Kiwanja. www.kiwanja.net (Accessed October 7 2011)

34. Mobile health www.wikipedia.org (Accessed October 7 2011)

35. Text To Change. www.texttochange.org (Accessed October 7 2011)

36. Apple iTunes App Store. www.apple.com/itunes (Accessed October 7 2011)

37. Android Market https://market.android.com (Accessed October 7 2011)

38. AirStrip www.airstriptech.com (Accessed October 7 2011)

39. IQMax http://www.iqmax.com/solutions/ (Accessed October 5 2011)

40. University of Cambridge and China Mobile. Mobile communications for medical care. 21 April 2011. http://mhealthinsight.com (Accessed October 14 2011)

41. eMocha www.emocha.org (Accessed October 1 2011)

42. SanaMobile www.sanamobile.org (Accessed October 1 2011)

43. Sarasohn-Kahn J. How Smartphones Are Changing Health Care for Consumers and Providers. California HealthCare Foundation. www.chcf.org April 2010 (Accessed April 10 2010)

13

Online Medical Resources

ROBERT E. HOYT

Learning Objectives

After reading this chapter the reader should be able to:

- State the challenges of staying current for the average clinician
- Describe the characteristics of an ideal educational resource
- Describe the evolution from the classic textbook based library to the online digital library
- Compare and contrast the different formats of digital libraries
- Describe the future of digital resources integrated with electronic health records
- Describe emerging Web 2.0 technologies in medicine
- Identify the most commonly used free and commercial online libraries

Introduction

Trying to keep up with the latest developments in medicine is very difficult, primarily due to the accelerated publication of medical information and the significant time constraints placed on busy clinicians. Dr. David Eddy said it best "The complexity of modern American medicine is exceeding the capacity of the unaided human mind." It is likely that clinicians are in fact so busy that they have no idea what new educational resources are available to them. They would like to move from the "information jungle" to the "information highway" but who will show them the way? This chapter is devoted to those clinicians who are seeking rapid retrieval of high quality medical information.

Challenges Faced by Clinicians

- Educational. More than 460,000 articles are added to Medline yearly.[1] The 2012 Physician's Desk Reference (PDR) is over thirty four hundred pages long making it exceedingly cumbersome to search for drug information.[2] It seems obvious that this is a disincentive to search for drug information and therefore is a patient safety issue. Standard medical textbooks are expensive and out of date shortly after publication. In addition, some argue that the descriptions of diseases are not always updated or evidence based.[3] Moreover, Shaneyfelt estimated that a general internist would need to read 20 articles every day just to maintain present knowledge.[4] There is a transition that is taking place in medical education where emphasis is no longer on developing physicians that know everything, but rather, practitioners who can find and use information when it is needed. Physicians find it difficult not to think of themselves as experts and it often shakes their confidence, but a new confidence can be found in knowing how to locate needed information.

- Diffusion of information. Recommendations from specialty organizations take time to trickle down to the generalists. There is no standard way to disseminate information that is either reliable or particularly effective. National guidelines, usually written by specialists, face the same challenges. Once there is a new standard of care for a disease such as diabetes, how do you get the word out, particularly to small or remote medical practices?

- Translational. Studies have shown that it may take up to ten or more years for research to be "translated" to the exam room (e.g., thrombolytics).[5] In a study by Antman, experts were also slow to make recommendations in textbooks even though high quality evidence was published many years prior.[6] On the other hand, many physicians are skeptical and wait for confirmatory studies. If they have been in practice for many years, they may have witnessed the pendulum sweep back and forth regarding, for example, the use of post-menopausal estrogens. Recent studies often contradict older studies due in part to better study design and larger subject populations.[7]

- Evolutionary. We can no longer teach "classic medicine," because diseases and their presentations change over time as demonstrated by new presentations for infectious diseases. Rocky Mountain Spotted Fever began to disappear as Lyme disease began to appear. Additionally, diseases were detected at more advanced stages in the older literature, because lab tests were lacking, making clinical presentations more dramatic. Currently we tend to diagnose diseases earlier, before the patient has advanced signs and symptoms due to better and earlier tests. Medical resources therefore must reflect new evidence.

- Retention. According to many studies there is an inverse relationship between current knowledge and the year of graduation from medical school. Ramsey compared board scores of Internists and the number of years elapsed since certification and demonstrated this inverse relationship.[8]

Patient-Related Questions and Answers

- Covell reported that on average internal medicine physicians had two questions for every three patients seen and found the answers for only 30%. [9]

- A study by Ely showed that Family Medicine physicians had 3.2 questions per 10 patients seen. The answer was pursued in only 36% of cases.[10]

- In a primary care survey Gorman noted 56% of physicians pursued answers where they thought an answer existed and 50% of answers dealt with an urgent issue. Most physicians turned to other physicians for answers and not the traditional medical library. Lack of time was the universal reason not to pursue answers in most studies.[11]

- In another study by Ely, the most common questions dealt with drugs, Ob-Gyn and adult infectious disease issues. Answers to 64% of questions were not pursued and physicians averaged less than two minutes per search. The most common resources used were books and colleagues and only two physicians performed literature searches.[12] It is important to point out that all of the above studies evaluated primary care physicians, so the needs of other physicians such as surgeons are less clear. Also, after these studies were published software programs such as Epocrates appeared and significantly changed how we seek drug information.

State of Medical Libraries Today

In the article, "Quiet in the Library," Lee speaks of the quiet created as a consequence of physicians no longer needing to go to the library to do research. Although today's medical libraries provide timely, pertinent and authoritative knowledge-based information in support of patient care, education and research, a significant percentage of the journal and textbook literature has migrated from print to online during the last decade. Physicians can research their clinical questions from their desktops without going to the library.[13, 14] Burrows in a paper reviewing electronic journal use at the Louis Calder Library at the University of Miami School of Medicine reports an 88% decrease in the use of print journals in the period from 1995 to 2004.[15] Libraries have moved their collections online and re-designed stack space into study areas, computer workstations and collaborative areas.

Evolution from Traditional Library to Online Resources

Within a very short time the internet has become the educational resource of choice due to the speed of retrieval and depth of information. A 2001 study by the American Medical Association showed that 75% of physician practices had internet access and 79% used it to research answers. Three out of ten medical practices had their own website.[16] These statistics continue to rise as does the availability of broadband access. Patients using the internet as an online library have closely mirrored the habits of physicians. The internet now hosts more than three billion web sites. As an indication of growth, a Google search for the words "medical education" in 1995 by one of the authors yielded 760 results,[17] whereas a search in 2010 yielded about 126 million citations. Although the 21st century searcher is at the center of a virtual library, he or she must cope with the quantity of easily retrieved information and be capable of evaluating that information for reliability, currency and authority.

Before the advent of the internet physicians used print resources for verifying factual information related to patient care, sought insights from their colleagues on difficult cases and performed library research on exceptional cases themselves or with the aid of medical librarians. The nature of physician information-seeking has not changed substantially in the last decade, but mode of access to resources has changed dramatically. To provide the best care for their patients, physicians still need to check drug information, differential diagnosis tools, current textbooks, or the journal literature, but instead of heading to the library they turn to the internet for answers. The resources available online are far more extensive than the personal libraries or hospital libraries that physicians used in a print world.

Reference materials in the twenty-first century have been migrated to online formats. Drug information compendia, laboratory references and textbooks have been converted to electronic formats, although only the major references and texts are available online. Two types of online journals have emerged: the electronic version of the print journal and the born-electronic journal. Either type can be open access (free to all users) or available by subscription only. Recently publishers are experimenting with hybrid journals that offer their most important content online, while still publishing print issues. Although many predict the demise of the print journal, the transition phase may last another decade. Today medical library collections are a mosaic of print and online content, but the mission of the medical library remains the same. Medical libraries today provide more extensive offerings than the print collections found in most hospital libraries a decade ago, but these resources are expensive and strictly controlled by site licenses. To have access to the most authoritative information a well-informed physician still needs to have an affiliation with a medical library or be willing to read premium medical content on a subscription or pay per view basis.

Journal indexes were the pioneers of research online. The Medline database that today has over 16 million citations from over 5,000 journals searchable online began in 1966 as an electronic archive of citations to the medical literature that was only searchable by highly trained librarians.

When MEDLINE introduced end-user searching in the pre-internet era citations previously accessed manually through the *Index Medicus* were available directly to clinicians at their desks. Although end-user searching revolutionized access to the journal literature, it was limited to titles and abstracts. With the advent of the internet came online journals and the ability to link the full-text journal articles to Medline citations. Despite these improvements Medline searching still is not an easy path to high quality, quick answers to clinical questions. Several studies have shown that finding an answer is difficult and takes too much time for a busy clinician.[18, 19] A pertinent abstract might be located, but it requires additional time to obtain the full-text whether it be from a free or fee-based online source or through the medical library. A Medline search should be reserved for rare medical problems, research, writing a paper or creating a clinical practice guideline (CPG). Medline will be discussed in more detail in the chapter on search engines.

In 1994 Shaughnessy stated that the usefulness of medical information is equal to the relevance times validity divided by the amount of work to access it.[20] A 2004 study in the journal *Pediatrics* comparing retrieval of information from online versus paper resources showed it took eight minutes for an answer via an online resource as compared to twenty minutes using traditional paper-based resources.[21] There is little doubt about the tremendous potential of online resources for speed of access, but the quest to find the precise, authoritative answer to a clinical question within the limitations of a patient visit remains elusive. Turning to resources of known quality appeared to be an efficient choice, so converting traditional resources to online formats was the logical first step.

Harrison's Online (the online version of *Harrison's Principles of Internal Medicine*, 18th edition) and Scientific American Medicine (now known as ACP Medicine) were among the first online full-text resources. The online versions of these popular textbooks are continually updated and are accessible from anywhere. Many libraries offer online access to these textbooks and individuals may purchase subscriptions to the online versions at about the same cost as copies of the print textbooks. Recent online versions offer a variety of subscription options and offer their content through several portals. The print edition of *Harrison's Principles of Internal Medicine*, 18th edition, published in 2011, offers supplementary material on DVD and is using RSS feeds and podcasts to disseminate its updates. Although these textbooks make valuable expert knowledge easily accessible, their main drawback is that they tend to cover only the basics about any subject and therefore lack depth. In spite of the fact that they have a search engine, like a standard textbook a reader may have to review multiple book chapters to find the answer.

More comprehensive aggregated resources followed the advent of online textbooks. MDConsult, Medscape, StatRef, and OVID were created to offer multiple resources such as books and journal articles, patient education materials, medical calculators and medical news in one product. Searches of these, otherwise excellent, resources yield multiple references to the full-text of various documents that must be analyzed to find the answer to the clinical question. You might have to read twenty books or journal pages to finally find the answer. This is not optimal if you are seeking an answer while the patient is still in the exam room or during hospital rounds. Ideal medical resources are those that are:

- Evidence based with references and level of evidence (explained in the chapter on evidence based medicine)

- Updated frequently

- Simple to access with a single sign-on

- Available at the point of care

- Capable of being embedded into an electronic health record

- Likely to produce an answer with only a few clicks

- Useful for primary care physicians <u>and</u> specialists

- Written and organized with the end user in mind

According to Richard Smith the "best information sources provide relevant, valid material that can be accessed quickly and with minimal effort."[22] The need for a synthesized resource that can easily provide evidence based answers to questions during the patient visit has given rise to several excellent, focused resources, often referred to as point of care resources or bedside information products. UpToDate, eMedicine, DynaMed, ACP-PIER and FirstConsult present their content, so that clinicians can answer clinical questions with current, comprehensive and rapid retrieval. These products focus on patient-oriented information and differ in the number of topics covered, the way the evidence is documented and the organization of the material. UpToDate, Essential Evidence Plus (formerly InfoRetriever), ACP Pier, Diseasedex, DynaMed and First Consult have been very well received, and clinicians develop their preferences among these offerings based on user interfaces and the ability of the database to answer questions. In an evaluation of five bedside information products, Campbell and Ash took a user-centered, task-oriented approach to testing the ability of these products to answer clinical questions. The study rated UpToDate the highest in ease of interaction, screen layout and overall satisfaction and found that users were able to answer significantly more questions quickly with UpToDate [23]; however, other researchers have found that users have preferred resources such as ACP Pier and Essential Evidence Plus, because of the way evidence levels are documented.[24] A 2004 study showed that 85% of medical students easily transitioned from traditional resources to primarily online medical resources (UpToDate and MDConsult).[25] In a report published in 2005, internal medicine residents were able to find answers 89% of the time and the information changed patient management 78% of the time. The most common resources accessed were UpToDate and PubMed.[26]

Use of the point-of-care tools discussed above begins with a diagnosis. In a controversial article in BMJ Tan and Ng[27] reported that Google could function as a useful diagnostic aid. Others would argue that specially designed tools such as the new generation of clinical decision support systems are more appropriately designed to improve medical diagnosis and reduce diagnosis errors by directing physicians to the correct diagnosis. Although clinical decision support systems have been around for years the new generation of tools as exemplified by *Isabel* (Isabel Healthcare) has been shown to suggest the correct diagnosis in approximately 96% of adult patients when tested with 50 consecutive internal medicine case records published in the New England Journal of Medicine.[28] Tools such as *Isabel* by assisting the physician in making the diagnosis provides an entry point into the literature and can link clinicians to resources such as UpToDate, PubMed and more in order to obtain information in depth on the case at hand.[29]

Several medical resource vendors are in the process of making the leap towards having the resource embedded into electronic health records. Examples would include iConsult, Dynamed, UpToDate and ACP-PIER to mention a few. Figure 13.1 demonstrates the evolution from the traditional library to the online library and integrated libraries into electronic health records.

Figure 13.1: The evolution from traditional to online medical libraries

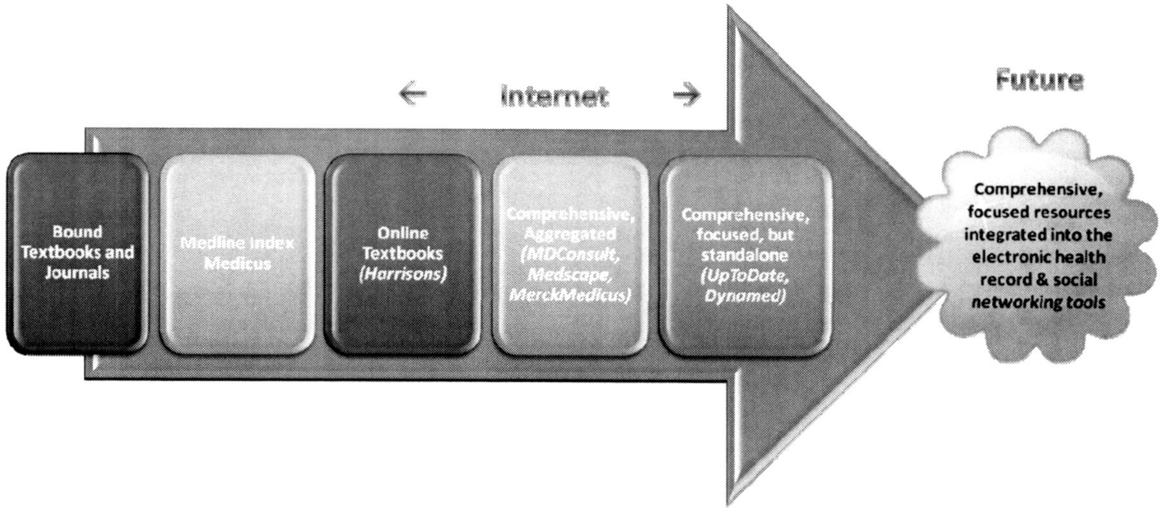

New Tools to Stay Abreast of Ever-Expanding Online Library

Many of the medical information resources described above mirror print resources and are written and designed to be read as questions arise. Lee mentions in his article "Quiet in the Library"[13] that "the flood of new information and the demands of simply getting through the day have become so overwhelming that many physicians no longer find the time for 'lifelong learning' through such activities as reading journals or attending grand rounds." To keep their medical practice current physicians need new tools. Interactive web technologies, known as Web 2.0, have emerged that allow knowledge sharing among users and allow customized content to be distributed to interested users. These tools can be harnessed to help physicians learn about the new developments that will improve their practice.[30, 31]

Weblogs or blogs, websites that build content through dated entries, have generated large repositories of focused medical content. Web 2.0 technologies hold the promise of an enriched learning environment enabled by collaboration on a large scale. Blogs such as *Kevin MD, Clinical Cases and Images* have developed large readership. Through the dialog between bloggers and readers physicians are discovering new ways to learn. British Medical Journal has launched a collaborative site called doc2doc to foster peer-to-peer communication.[32]

Wikis (the name comes from the Hawaiian word meaning quick) allows authors to collaborate to create peer-reviewed content that is written and edited by participating users. The most famous general example is Wikipedia. Several medical references have been developed using wiki architecture. WikiDoc[33] and Ganfyd[34] were created by physicians and are being developed by participating health professionals. WikiDoc, a collaborative online textbook, boasts over 155,000 textbook chapters and continuously updated medical news. Ganfyd, "a free medical knowledge base that anyone can read and any registered medical practitioner may edit," only allows credentialed individuals to provide content.

Keeping up with changes to important websites can be challenging. Really Simple Syndication (RSS) can make that process manageable. RSS is a format that delivers changes from multiple websites to one place. RSS allows physicians to request content from various websites and read that content in single place, known as an aggregator. Subscribe to an aggregator such as MedWorm or Bloglines, the look for the RSS icon () at a website. Subscribe to RSS feeds of interest and read them with the aggregator. Physicians are using RSS to receive textbook and website updates, Pub Med search results, journal tables of contents and medical news.

Audiocasts (podcasts) and videocasts offer educational programs and updates in multi-media formats. New England Journal of Medicine now offers weekly article summaries via podcasts and Johns Hopkins University offers weekly health news podcasts. To subscribe to multiple podcasts you will need to subscribe to an audio aggregator, known as a podcatcher. Choosing a podcatcher can be confusing. Podcatcher Matrix will assist you in choosing a podcatcher that is compatible with your operating system and mobile device.[35]

Sponsored Medical Web Sites

Multiple excellent web sites are available that are either sponsored or fee-based. Most of the sites discussed in this section have multiple features that continue to improve. Medical education traditionally has been based on reading journals or textbooks, but can also involve the presentation of interesting and unique cases. A thorough discussion of this alternative approach appeared in the February 2007 Mayo Clinic Proceedings.[36]

Medscape

- An all-purpose medical web site
- Covers 30+ medical specialties as well as sections for nurses, medical students and pharmacists

- Over 150 Resource Centers

- Provides updates, continuing medical education (CME), conference schedules, Medline, drug searches and multiple specialty articles and an eclectic selection of journal abstracts

- Weekly newsletters and updates (MedPulse) and Best Evidence; both are features unique to Medscape

- Drug and Device Digest providing the latest in alerts and approvals; helpful for patient safety concerns

- A free personal web site option

- Dermatology atlas

- Clinical practice guidelines and Cochrane connection

- Sponsored by advertising [37]

MerckMedicus

- Multipurpose site sponsored by Merck and Company, customizable for 20 specialties

- 60+ specialty textbooks

- 150+ full text journals

- Cochrane Reviews links

- Clinical podcasts

- Includes customized versions of MDConsult and OVID, DxPlain (differential diagnosis engine from Harvard), medical news and national meeting reports

- PDR Electronic Library

- Patient handouts

- Unique 3-D Atlas of the human body

- Professional development using CME, board reviews, medical meetings, medical school links and Braunwald's Atlas of Internal Medicine (1,500 slides you can copy). Also, a slide image bank of other slides that can be copied

- PDA portal, formatted for use with Palm or Pocket PC, includes news, the Merck Manual, Pocket Guide to Diagnostic Tests, and TheraDoc antibiotic assistant for PDA

- Journal abstracts and the ability to do a Medline search (If your PDA is not connected to wireless internet, searches will be done the next time you synchronize with your PC.)

- Sponsored by Merck and Company and state health professional license required for full access [38]

Amedeo

- This service will search major medical journals for a topic you select and then e-mail the results to you every week. It offers weekly webpage alerts displaying abstracts of selected journal tables of contents linked to PubMed

- Covers about 100 topics falling into 25 specialties

- Valuable if you are a subject expert and don't have the time to do a frequent journal search on your own

- Related websites include Free Books 4 Doctors www.freebooks4doctors.com and Free Medical Journals www.freejournals4doctors.com

- Similar tracking of articles also available through Google Alerts and NCBI (Pubmed)

- Self-supported non-profit site[39]

Sponsored and Non-Sponsored Resources

E-medicine

- 6,000 articles by 10,000 authors covering primary care and multiple sub-specialties

- Owned by *WebMD* and incorporated into Medscape

- Continually updated and peer reviewed Clinical Knowledge Base

- Articles are referenced and selectively cross-referenced

- References are presented at the end of each article with links to PubMed, but no footnotes in the text body

- Levels of evidence not given

- CME available

- Sponsored and institutional version available

- Institutional version is now known as Medscape Reference offers the following information: drugs, diseases, drug interactions, MEDLINE, anatomy, medical images and a healthcare directory.[40]

Online Epocrates

- Online Epocrates was the obvious next step after the successful PDA/mobile software program (see chapter on mobile technology)

- Both a free and fee-based online version is offered

- Program covers 3,300 drugs and 400 alternative medications

- Fee-based program includes local formulary information, pill identifier, MEDCALC 3000, alternative medications as well and an extensive drug library

- Free online program includes pill pictures and patient education

- The features online Epocrates offers over the mobile version: Ability to print or e-mail results, Medline search capability, pill pictures, MedCalc 3000 calculations and patient education sheets in English and Spanish. [41]

Government Medical Web Sites

National Library of Medicine

- PubMed (discussed in search engine chapter)

  o Provides free access to MEDLINE, NLM's database of citations and abstracts in the fields of medicine, nursing, dentistry, veterinary medicine, health care systems, and preclinical sciences

  o Links to many sites providing full text articles and other related resources

  o Provides a Clinical Queries search filters feature, as well as a Special Queries feature, which have recently combined in one interface

  o Links to related articles for a selected citation[42]

- NLM Gateway

  o Provide "one-stop shopping" for many of NLM's information resources

  o Offer citations, full text, video, audio, and images

  o Link within and across NLM databases[43]

- Toxnet

  o Cluster of databases covering toxicology, hazardous chemicals, environmental health and related areas[44]

National Guidelines Clearinghouse

- Comprehensive searchable database of evidence based clinical practice guidelines and related documents

- Structured abstracts (summaries) about the guideline and its development

- Links to full-text guidelines, where available, and/or ordering information for print copies

- Palm-based PDA downloads of the complete NGC Summary for all guidelines

- Guideline comparison utility for a side to side comparison of multiple guidelines [45]

MedlinePlus

- Premier online patient education site

- Important to have in exam room

- Service of the National Library of Medicine and the National Institutes of Health (NIH)

- Covers over 750 Health Topics in English and Spanish

- Drugs, Supplements, and Herbal Information

- Medical dictionary, encyclopedia and news

- 165 interactive video tutorials and surgical procedure videos

- Links to major patient education sites offered by health clinics, government and advocacy organizations such as, Mayo Clinic, National Institutes of Health (NIH), American Heart Association, etc.

- Links to Clinical Trials.gov to search research centers for specific diseases[46]
- In 2011, Medline Plus Connect was created that can link patient education to personal health records and medical health records[47]

Other Excellent Free Patient Education Sites

- Family Doctor http://www.familydoctor.org/
- Mayo Clinic http://www.mayoclinic.com
- Web MD http://www.webmd.com
- Kids Health http://kidshealth.org/

Free Medical Web Sites

HighWire Press

- Free site created by Stanford University to produce online peer-reviewed journals and scholarly content as open access or pay per view depending on the title
- Hosts 1,270 journals with over four million full text and two million free full text articles
- Capability to search HighWire and Medline at same time with access to both free and pay-for-view articles
- E-mail alerts and RSS feeds available
- Site hosts 37 free trials of journals, 43 free journals, 249 journals that offer back issues free and approximately 1,000 pay-for-view journals
- Offers e-books and a mobile access to journals for iPhone and iPad and Kindle reader [48]

Medical Algorithms

- Developed by the Institute for Algorithmic Medicine, a non-profit organization that develops online medical algorithms
- Currently includes 17,000+ scales, tools and assessments
- Algorithms are evidenced-based with multiple references
- Many algorithms are presented as an Excel spreadsheet so you can plug in actual patient numbers and get immediate results
- Covers many unusual calculations not found in MedCalc and other similar programs[49]

Medical Podcasts

- Information Network
- The Journal of the American Medical Association, New England Journal of Medicine and other journals now offer audio article summaries as podcasts
- Medical school library websites offer links to podcasts from a variety of journals

- CME providers are expanding Several medical organizations offer podcasts for medical education; mostly in audio format with some in video

- The American College of Cardiology posts "Heart Sounds" in a MP3 format as a download

- The Arizona Heart Institute and Hospital provides podcasts as part of the Cardiovascular Multimedia their use of podcasts[50]

Subscription (Fee-Based) Resources

MicroMedex

- Micromedex offers multiple drug databases of unbiased drug information searchable with a single query

- New interface organizes the database into a point-of-care tool

- Databases include Poisondex (toxicology), Diseasedex (Disease database), Lab advisor (laboratory information), DrugDex (drug interactions) , ReproRisk (human reproductive toxicology), CareNotes (patient education handouts in English and Spanish)

- Fully referenced drug database

- Handheld is available on all major platforms

- Unlike Epocrates it has:

  o Both renal and liver failure dosing

  o Drug-food interactions

  o Off label uses

  o Comparative efficacy

  o IV drug compatibility

  o Toxicology

  o Extensive references[51]

Lexi-Comp

- Comprehensive database of unbiased drug information

- Core pharmaceutical information includes population specific dosing, indication specific dosing, IV Compatibility, drug identification, drug interactions, toxicology and more

- Diseases and disorders via Harrison's Practice

- Laboratory and diagnostic medicine

- Formulary information

- Specific modules available for medicine, dentistry and oral surgery

- Patient handouts available in 18 languages.

- Handheld version includes the five most requested databases and Harrison's disease database

- Handheld is available on all major platforms including the Ipad[52]

OVID

- Several hundred textbooks in most specialties including drug references
- Approximately two thousand full text medical journals
- Access is to journal articles is available by institutional subscription or pay per view
- Search interface supports natural language and Boolean searching
- Medline search capability linked to online full-text of journal articles
- Cochrane Library is available under the title Evidence based Medicine Reviews
- Supports searching multiple databases simultaneously, i.e, Cochrane and Medline[53]

UpToDate

- Comprehensive resource containing over 97,000 pages of original, peer-reviewed text embedded with graphics and links to Medline abstracts
- Available online, on CD-ROMs or downloadable to handhelds
- Individual, educational and institutional subscriptions available
- Personal subscribers receive CME researching clinical questions
- Institutional subscribers may purchase online or single workstation CD-ROMs
- Covers 17 specialties
- Logically organized for fast answers
- 8,500 topics, written by 4,400 authors who review 440 journals
- Began grading recommendations for treatment and screening in 2006 and continues to expand that effort
- Continuously updated with about 40% of the content being edited each quarter
- Drug database includes drug-drug interactions
- Patient information topics in English
- Available for download to iPhone and iPad
- Integrated into GE Centricity EHR[54]

MDConsult

- 60+ textbooks
- Over 80 full text journals
- 35 Clinics of North America
- Comprehensive drug database
- 1,000 clinical practice guidelines
- 2,500 Patient education handouts
- 50,000 medical images

- Online CME and medical news

- Medline search capability

- Excellent Search engine for entire site

- Individual and institutional subscriptions available [55]

StatRef

- Offers about 200 textbooks and Medline online in a cross-searchable reference tool that includes textbooks and evidence based resources.

- ACP PIER, Journal Club & AHFS DI® Essentials™

- MedCalc3000

- Institutional subscriptions available [56]

Essential Evidence Plus (formerly InfoRetriever/InfoPOEMS)

- This is a program that was created by physicians for physicians. POEMS are "patient oriented evidence that matters." Specifically, this means the authors look for articles that are highly pertinent to patient care and patient outcomes.

- Consists of two products: DailyPOEMS and InfoRetriever

- DailyPOEMs are e-mailed to the subscriber Monday through Friday and are distilled from 100+ journals with only one in 40 accepted

- Site has 2000 POEMS

- POEM of the Week podcasts (RSS feeds available)

- Essential Evidence Plus (formerly InfoRetriever) available in online or for most smartphones

- Essential Evidence Plus tools: EBM guidelines (1,000 primary care practice guidelines, 3,000 evidence summaries and 1,000 photographs and images), Daily POEMS, Cochrane abstracts (2,193), selected practice guidelines (751), clinical decision rules (231).

- Number Needed to Treat (NNT) tool

- Derm Expert (photographic skin atlas)

- Diagnosis calculators (1,180)

- History and physical exam calculators (1,282)

- 5 Minute Clinical Consultant

- ICD-9 and E&M lookup tool

- Drug of Choice tool

- Searching results in a summary of resources on that topic categorized into typical quick reference categories like diagnosis, treatment, prognosis, etc. 5 Minute Clinical Consult monographs are listed first

- Individual and institutional subscriptions available [57]

ACP Medicine

- Publication of the American College of Physicians (ACP) and Web MD

- Previously known as Scientific American Medicine

- Evidence based and peer-reviewed

- Covers most subspecialties plus Psychiatry, Women's Health, Dermatology and Interdisciplinary medicine

- Available in binder, CD-ROMs and Online

- Up to 120 hours CME available

- Binder version is 2,800 pages

- Articles are dated and references are footnoted with PubMed links to the abstract

- Monthly updates (free) to be added to chapters

- Handheld point-of-care tool, *Best DX/Best Rx*

- Individual and institutional subscriptions available [58]

ACP PIER

- Organized into five topic types: diseases, screening and prevention, complementary and alternative medicine, ethical and legal issues and procedures

- Each of the 430 disease modules presents guidance statements and practice recommendations, supported by evidence of evidence

- PDA version available

- Drug resource, accessible from every module page

- Provides the medical resource content for Allscript's EHR

- What they cover they do well

- Like an online textbook; updated frequently

- Disease modules continue to be added

- Available directly from the ACP and through Stat!Ref by individual or institutional subscription [59]

FirstConsult

- Synthesizes evidence from journals and other sources into one database

- Offers concise, readable summaries of evidence that relate to patient care

- Organized into medical topics, differential diagnoses and procedures

- Updated weekly; major releases quarterly

- 475 topics

- 300 Patient education files in English and Spanish

- Procedure files and videos

- EHR ready
- Available for iPhone and iPad
- Lack of a drug database and limited topics are negatives
- Individual and institutional subscriptions available[60]

DynaMed

- Disease and condition reference
- Almost 3,200 clinical topics commonly seen in primary care
- Peer-reviewed and continually updated.
- Information presented based on validity, relevance and convenience
- All topics are organized in the same categories such as, general information, causes and risk factors, complications and associated conditions, history, physical, diagnosis, prognosis and treatment
- Bottom line recommendations are presented first, along with level of evidence. Links to articles will take you to the full text article if available and free online. Other links take you to PubMed where some are linked through medical libraries to full text articles
- Weekly e-mail of important articles; also available as podcast
- Handheld version available on popular platforms and is free with subscription
- Can be linked to an EHR with the EBSCOhost Integration Toolkit
- Individual and institutional subscriptions available[61]

Future Trends

Without question, we will continue to see a decline in print textbooks. The medical library at Johns Hopkins will close its doors, partly as a result of no longer needing a large facility to house medical literature.[62] Medical resources that cannot deliver current evidence based medicine at the point of care will lose relevance and market share. New medical information will arrive in the form of e-books, web resources, smartphone and tablet apps and electronic content that is integrated with electronic health records. Ultimately, medical resources will be context sensitive; as the user navigates through the electronic health record in different areas such as laboratory results clinical decision support tools will educate the user with the most current guidelines. We will likely see medical resources that are similar to UpToDate; web-based and updated frequently.

Key Points

- Clinicians are overwhelmed by the amount of new information and the lack of time
- We have shifted from traditional print textbooks in our medical libraries to online libraries
- Multiple resources exist that are both free and fee-based to serve as rapid high-quality references
- Ideal medical resources should be easy to access and fast to retrieve the most current information

Conclusion

Online resources are becoming the medical library of choice for healthcare workers due to depth of content and speed of retrieval. Furthermore, subject matter can be updated more rapidly compared to standard textbooks. Many excellent resources are free and the subscription resources are competitive with traditional textbooks. Resources vary from a low of about 400 topics to a high of 8,000+ topics. Prices tend to correlate with the scope of the content offered. There are many free resources that should be considered by all clinicians such as Epocrates Online, MedlinePlus and Medscape. The authors want to stress that very extensive resources such as UpToDate, eMedicine and DynaMed offer the greatest possibility of finding an answer in a few clicks. Other resources may point you to multiple book chapters and journal articles where you must sift through the data to find the answer. Clinicians are strongly encouraged to "test drive" these resources, adopt the ones that make the most sense and add them as desktop icons in each exam room.

Acknowledgement

We would like to thank Jane Pelligrino MSLS, AHIP Head of Library Services Department at the Naval Medical Center Portsmouth for her contributions

References

1. Medline Fact Sheet http://www.nlm.nih.gov/pubs/factsheets/medline.html (Accessed 2 July 2010)

2. Physician Desk Reference http://www.pdrbookstore.com/ (Accessed September 28 2011)

3. Richardson WS, Wilson MC Textbook descriptions of disease—where's the beef? ACP Journal Club July/August 2002: A-11-12

4. Shaneyfelt T. Building bridges to quality. JAMA 2001;286: 2600-01

5. Contopoulos-Ioannidis DG, Ntzani E, Ioannidis JP. Translation of highly promising basic science research into clinical applications. Am J Med. 2003 Apr 15; 114(6):477-84.

6. Antman EM et al. A comparison of results of meta analyses of randomized control trials and recommendations of clinical experts JAMA 1992; 268: 240-248

7. Ioannidis JP.A Contradicted and Initially Stronger Effects in Highly Cited Clinical Research JAMA 2005;294: 219-228

8. Ramsey PG et al. Changes over time in the knowledge base of practicing internists JAMA 1991; 266: 1103-1107

9. Covell, DG, Umann GC, Manning PR. Information needs in office practice: are they being met? Ann Int Med 1985; 103: 596-599

10. Ely J, Osheroff J, Ebel M et al. Analysis of questions asked by family doctors regarding patient care BMJ 1999; 319: 358-361

11. Gorman PN, Helfand M. How Physicians Choose Which Clinical Questions to Pursue and Which to Leave Unanswered. Med Decision Making 1995; 15: 113-119

12. Ely JW et al. Analysis of questions asked by Family doctors regarding patient care BMJ 1999;319:358-361

13. Lee T. Quiet in the Library NEJM 2005;352: 1068-70

14. Lindberg DAB, Humphreys BL. 2015-The Future of Medical Libraries NEJM 2005;352:1067-1070

15. Burrows S. A review of electronic journal acquisition, management and use in health sciences libraries JM:A 2006; 94(1):67-74

16. Technology usage in physician practice management AMA survey Dec 2001

17. Miccioli G. Researching Medical literature on the internet---2008 Update www.llrx.com/features/medical2008.htm (Accessed 3 July 2010)

18. Gorman PN. Can primary care physicians' questions be answered using the medical journal literature? Bull Med Libr Assoc 1994; 82:140-146

19. Chambliss ML. Answering clinical questions J Fam Pract 1996; 43:140-144.

20. Shaughessy A, Slawson D, Bennett J. Becoming an Information Master: A guidebook to the Medical Information Jungle J Fam Pract 1994; 39: 489-499

21. D'Alessandro DM, Kreiter CD and Petersen MW. An Evaluation of Information Seeking Behaviors of General Pediatricians Pediatrics 2004; 113: 64-69

22. Smith R. What Clinical Information do Doctors Need? BMJ 1996; 313: 1062-1068

23. Campbell R, Ash J. An evaluation of five bedside information products using a user-centered, task-oriented approach JMLA 2006; 94(4): 435-441

24. Trumble JM et al. A Systematic Evaluation of Evidence based Medicine Tools for Point-of-Care, presented at South Central Chapter, Medical Library Association Meeting, October, 2006, http://ils.mdacc.tmc.edu/papers.html (Accessed 3 July 2010)

25. Peterson MW et al. Medical student's use of information resources: is the digital age dawning? Acad Med 2004;79:89-95

26. Schilling LM et al. Residents' patient specific clinical questions: opportunities for evidence based learning Acad Med 2005; 80: 51-56

27. Tan H, Ng JHK. Googling for a diagnosis – use of Google as a d diagnostic aid: internet based study, BMJ 2006; 333: 1143-5

28. Graber ML. Performance of a Web-Based Clinical Diagnosis Support System for Internists, 2007; 23(Suppl 1) 37-40

29. Johnson C. What physicians don't know, Medicine on the Net 2008; 14(1): 1-5

30. Giustini D. How Web 2.0 is changing medicine, BMJ 2006; 333: 1283-4

31. Liesegang TJ. Web 2.0,Library 2.0, Physician 2.0, Am J Ophthalmology 2007; 114 (10):1801-3

32. Doc2Doc. http://doc2doc.bmj.com/ (Accessed September 28 2011)

33. WikiDoc http://www.wikidoc.org (Accessed September 28 2011)

34. Ganfyd http://ganfyd.org (Accessed September 28 2011)

35. Podcatcher Matrix http://www.podcatchermatrix.org/ (Accessed September 28 2011)

36. Pappas G, Falagas ME Free Internal Medicine Case-based education through the World Wide Web: How, Where and With What? Mayo Clin Proc 2007;82(2):203-207

37. Medscape http://www.medscape.com (Accessed September 28 2011)

38. Merck Medicus http://www.merckmedicus.com (Accessed September 28 2011)

39. Amedeo http://www.amedeo.com (Accessed September 28 2011)

40. E-Medicine http://www.emedicine.com (Accessed September 28 2011)

41. Epocrates http://www.epocrates.com (Accessed September 28 2011)

42. PubMed Fact Sheet http://www.nlm.nih.gov/pubs/factsheets/pubmed.html (Accessed September 28 2011)

43. NLM Gateway Fact Sheet http://www.nlm.nih.gov/pubs/factsheets/gateway.html (Accessed September 28 2011)

44. Toxnet http://toxnet.nlm.nih.gov (Accessed September 28 2011)

45. National Guidelines Clearinghouse http://guidelines.gov (Accessed September 28 2011)

46. MedlinePlus http://www.nlm.nih.gov/medlineplus/ (Accessed September 28 2011)

47. MedlinePlus Connect http://www.nlm.nih.gov/medlineplus/connect/overview.html (Accessed September 28 2011)

48. High Wire Press http://highwire.stanford.edu/ (Accessed September 28 2011)

49. Medical Algorithms. www.medal.org (Accessed 4 July 2010)

50. Collaborate CME Using Web 2.0 Technologies. Almanac. Alliance for CME, August 2008.http://www.med.upenn.edu/cme/Almanc%20ACME_08%20(2).pdf (Accessed 25 July 2010)

51. Micromedex http://www.thomsonhc.com/(Accessed September 28 2011)

52. Lexi-Comp http://www.lexi.com (Accessed September 28 2011)

53. OVID http://gateway.ovid.com (Accessed September 28 2011)

54. UpToDate http://www.uptodate.com (Accessed September 28 2011)

55. MDConsult http://www.mdconsult.com (Accessed September 28 2011)

56. StatRef http://www.statref.com (Accessed September 28 2011)

57. Essential Evidence Plus (formerly Inforetriever) http://www.essentialevidenceplus.com/ (Accessed September 28 2011)

58. ACP Medicine http://www.acpmedicine.com (Accessed September 28 2011)

59. ACP Pier http://pier.acponline.org (Accessed September 28 2011)

60. First Consult http://www.firstconsult.com (Accessed 4 September 28 2011)

61. DynaMed http://www.ebscohost.com/dynamed (Accessed September 28 2011)

62. Dawson C. Sign of the times: Johns Hopkins shuttering its medical library. ZD Net Education www.zdnet.com/blog/education (Accessed November 4 2011)

14

Search Engines

ROBERT E. HOYT

Learning Objectives

After reading this chapter the reader should be able to:

- State the significance of rapid high quality medical searches

- Define the role of Google and Google Scholar in healthcare

- Describe the role of PubMed and Medline searches

- Identify the variety of search filters essential to an excellent PubMed search

- Enhance PubMed searching with third party PubMed tools

- Use NLM Mobile

Introduction

The most rapid and comprehensive way to access information today from anywhere in the world is a search of the World Wide Web via the internet. If we assume that the internet is the new global library with more than three billion web sites, then it should come as no surprise that search engines are the gateway. Popular search engines such as Google provide successful searches for medical and non-medical issues. Although PubMed is the search engine of choice for formal searches of the medical literature, most inquiries are informal so searches need to yield primarily rapid and relevant results. Given the prevalence of web surfing for answers, multiple articles have been written about "search wars."[1, 2] It is unclear to what extent the use of search engines has changed human behavior and medical knowledge in last approximately 15 years. Previously, questions such as what is the difference between HDL and LDL cholesterol meant a trip to the library, the purchase of a book or a visit with a doctor. Now anyone can execute a search and have a reasonable likelihood that the search will be successful.

Just as important as selecting the search engine with which you are comfortable is learning to use all of the advanced search options. It is imperative to use filters to refine a search or you will become frustrated by the avalanche of information returned. In this chapter we will begin with a discussion of Google, followed by other less well-known search engines and finally a primer on PubMed searching.

Google

Google is by far the most widely used search engine in the world.[3] Its name is derived from the word *googol* which is the mathematical term for the number *1* followed by *100 zeroes*.[4] Google's success is based largely on its intuitiveness, retrieval speed and productive results. Google is listed as one of the ten forces to flatten the world in Thomas Friedman's book *The World is Flat*.[5]

Google has proven to be a fascinating company with a myriad of innovations on a regular basis. Google was developed by Larry Page and Sergey Brin in 1996 when they were graduate students at Stanford University. They created the "backrub" strategy which meant that a search would prioritize the results by ranking the page that is linked the most first (page ranking).[6] Some could argue that it used a popularity contest as a strategy. A shortcoming of this approach would be that new important web sites might take time to be linked. As the world's largest and fastest search engine it performs one billion searches daily by utilizing thousands of servers (server farm) running the Linux operating system.[7] "Googling" has deeply influenced the users' expectations about the answers to their questions and the way of searching the web."[8] Google can be criticized for being a shotgun and not a rifle in terms of returning too many results but this has not diminished its popularity. Because Google yields so many results in an average search, it is very important to learn about how to narrow or filter a search.

Google can be an acceptable medical search engine for common as well as rare conditions that are not likely to be found in journals or textbooks. Google provides a very global review, returning articles from the lay press, medical journals, magazines, etc. Dr. Robert Steinbrook noted that Google (56%) was the most common search engine used to refer someone to find a medical article at High Wire Press; compared to PubMed (8.7%).[9]

Google will cite Medline abstracts and occasionally full text articles; so for an informal search it is not unreasonable to start with Google to see if you find an answer in the first few citations listed. It is likely you will find an acceptable answer in less time than it takes to use PubMed, particularly if you narrow the search with additional descriptors and use an advanced search strategy. Meats et al. showed that clinicians searching for medical information prefer to use a simple strategy of the disease term and the population in question.[10] Google makes that type of searching possible, albeit inefficient, if advanced search techniques are not employed. If you search with the terms "type 2 diabetes foot checks frequency" you will likely retrieve clinical articles that describe how often foot checks should be performed in diabetics. In a Google search the most important term should be listed first.

Several recent articles in the medical literature have confirmed that Google has become a common medical search engine; even at academic centers.[11-13] Google is very aware that many patients use Google to search for medical answers to common and complex health questions.[14]

Successful searching depends on maximizing Google search options:

- Begin by setting preferences
  - Language you prefer e.g. English
  - Number of search results per page e.g. 10 or 20
  - Whether you want your search to launch in new window (recommended because when you exit the current page, you lose your search.)
- Select the *Advanced Search* option on the main page. As of 2011 the *Advanced Search* function is now located at the very bottom of the page.
  - Under *Find webpages that have...* you can search for a term or terms in the title only or in the body or both
  - Search for synonyms using "or" as the operator
  - Select search by *File type*: Word, Excel, PDF, PowerPoint, etc.
  - Put quotation marks around the words to search for an exact phrase, e.g. "University of West Florida" so you don't retrieve every citation with the words Florida, West or University

- Use *Advanced Search Tips* at the top right of the *Advanced Search* page to find *Operators* to refine the search

  - Type *define:* before a word or phrase to have Google serve as a dictionary

  - Enter an arithmetic string and Google will function as a calculator

- *Basic search help* provides helpful tips to improve the search process (http://support.google.com/websearch/bin/answer.py?hl=en&answer=134479):

  - Searches are not case sensitive

  - The word "and" is not necessary, because unlike PubMed, "and" is implied in Google. Use *Advanced Search* to use "or"

  - The top two or three paid advertisements pertaining to your search will appear at the top of the page, followed by the most popular web sites[15]

Google Scholar

Google Scholar is an offspring of Google that searches the full text of peer-reviewed scholarly journal articles at publishers' websites and the citations and abstracts provided online by the National Library of Medicine through PubMed. Google Scholar uses the same search technology as its parent. Because the search technology relies on an algorithm that weighs articles by their links to other relevant content, it is difficult to retrieve recent articles. Google Scholar delivers the quantity of retrieval, but not the quality, necessary to allow it to standalone. Because Google Scholar searches the full text of articles in contrast to PubMed that searches the title and abstract, Google Scholar enables the searcher to retrieve articles that contain words and phrases not found in the title or abstract. Google Scholar makes searching easier, but it should be used in conjunction with PubMed, not as a replacement for it. Google Scholar offers "cited by referencing" which is not available elsewhere for free. Google Scholar is a good tool for accessing the open source literature, it provides access to unique content not in other search tools and it is free and openly available.[16-20]

PubMed Search Engine

PubMed is a web-based retrieval system developed by the National Center for Biotechnology Information (NCBI) at the National Library of Medicine (NLM).[21] PubMed is one of twenty-three databases in NCBI's retrieval system, known as *Entrez,* that index information in toxicology, bioinformatics and genomics and include textbooks.

MEDLINE is the primary *Entrez* database, containing 19 million citations from the world's medical literature from the 1940s to the present, covering the fields of medicine, nursing, dentistry, veterinary medicine, health care administration, the pre-clinical sciences and some other areas of the life sciences.[22] NLM licenses its data to vendors to be used through proprietary interfaces, but PubMed search interface for MEDLINE is only available directly from the National Library of Medicine.

For simple answers to common problems PubMed may not be the place to begin a search, but it is the primary search engine for physicians seeking information on unusual cases and research topics. Although some would argue that the PubMed search process is too labor intensive, all healthcare workers who seek evidence based medical answers should learn to use PubMed. It is especially important in an academic or research environment. Without proper training PubMed searching can be challenging and frustrating. This section emphasizes the important features and shortcuts to make a search easier and more successful. Excellent tutorials exist on the PubMed site to teach you the basics of a good search. Also, several helpful review articles have been written that address PubMed tools and features. [23-25]

The query box in PubMed (Figure 14.1) allows keyword, Medical Subject Heading (MESH) and natural language (Google-type) entries. Search terms may be entered alone or connected by Boolean search operators, such as "AND" or "OR." The goal of the search is to find specific citations on the topic described by the search terms.

Figure 14.1: PubMed homepage [26]

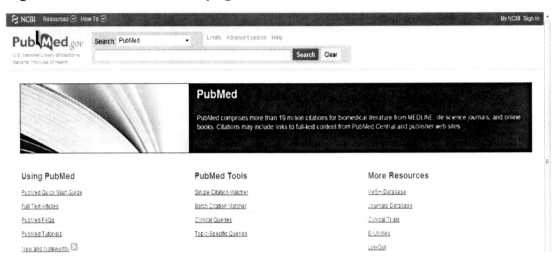

PubMed citations include the author, title, journal, publication date and PubMed identification number (PMID) as shown in Figure 14.2 and only 65% of Medline citations include an author abstract. PubMed does not search the full-text of cited articles.

Figure 14.2: Medline citation (Courtesy National Library of Medicine)

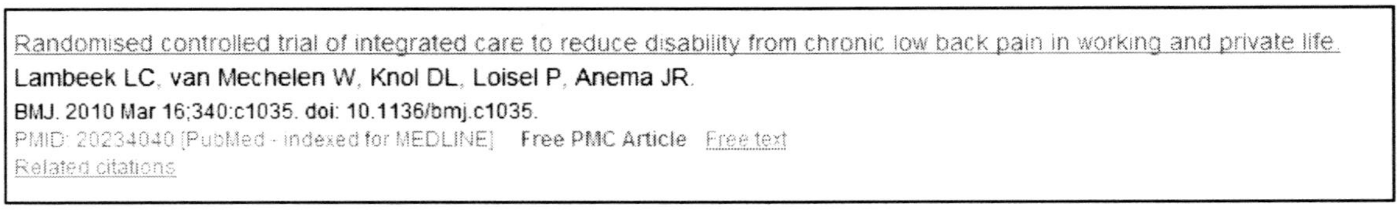

Randomised controlled trial of integrated care to reduce disability from chronic low back pain in working and private life.
Lambeek LC, van Mechelen W, Knol DL, Loisel P, Anema JR.
BMJ. 2010 Mar 16;340:c1035. doi: 10.1136/bmj.c1035.
PMID: 20234040 [PubMed - indexed for MEDLINE]   Free PMC Article   Free text
Related citations

Medical Subject Heading (MeSH): Journal articles are categorized by NLM indexers in order to facilitate searching. Articles are saved under two or more subject headings using a structured vocabulary called MeSH. Understanding what these terms are and how they can refine a search is an important first step in harnessing the power of PubMed. As you can imagine, terms such as low back pain could be labeled lumbar pain, osteoarthritis of the lumbar spine, etc. It will improve your search significantly, if you search with the preferred term, so take a moment to look at MeSH. You can access MeSH in the drop down menu in the search window or by choosing the MeSH Database in the menu on the left.

Figure 14.3 shows how the term "low back pain" is organized in MeSH. The MeSH entry shows a definition of the term and its synonyms and displays a set of subheadings with which to narrow a search on low back pain.

Figure 14.3: MESH term display (Courtesy National Library of Medicine)

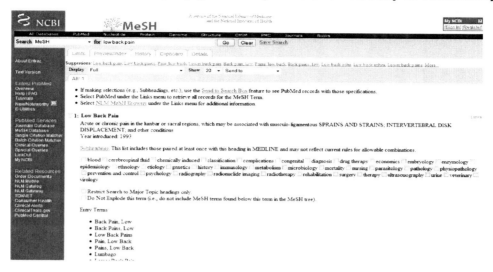

Figure 14.4 illustrates a search for sinusitis in MeSH. Different types of sinusitis are listed. At the bottom of each MeSH entry is the categorical display or "MeSH Tree" as shown in Figure 14.5. Searching the term, sinusitis, includes all the specific types listed under it, broadening the search. Conversely, reviewing sinusitis in MeSH allows you to discover the specific type of sinusitis available so that you may search the one that fits your query the best. MeSH is valuable in broadening or narrowing a search query.

Figure 14.4: MESH term search

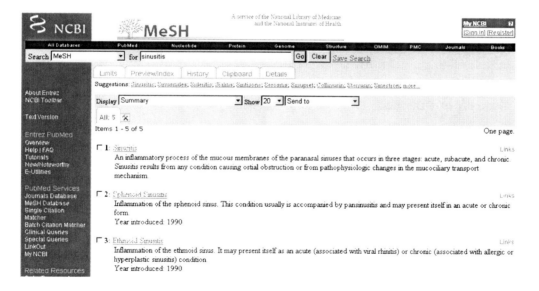

Figure 14.5: MeSH categories

All MeSH Categories
Diseases Category
Respiratory Tract Diseases
Nose Diseases
Paranasal Sinus Diseases
Sinusitis
Ethmoid Sinusitis
Frontal Sinusitis
Maxillary Sinusitis
Sphenoid Sinusitis

If you are struggling with your search terms in PubMed and not finding what you need, you may want to check your search terms in MeSH to see if the term is accepted by PubMed. Searching with the correct term can make all the difference. As is the case with Google searching, learning to use filters such as MeSH will result in more successful retrieval of information.

PubMed _Limits_ Option allows a search to be narrowed by date, age of subjects, gender, humans or animals, language, publication types, topics and field tags.

- You can also search for full text and free full text articles and abstracts. (Keep in mind that most articles before 1975 did not contain abstracts.)

- You can search by author or journal name

- Searchable main publication types include Clinical Trial, Editorial, Letter, Meta-Analysis, Practice Guideline, Randomized Controlled Trial, and Review

- Searchable topic subsets include AIDS, Bioethics, Cancer, Complementary Medicine, History of Medicine, Systematic Reviews and Toxicology

- Field tags. You can stipulate whether you want the search term in the title or body of the article. Multiple other choices are listed as well

Entering a Search in PubMed

PubMed is based on an architecture that uses indexed concepts (MeSH Headings) and Boolean logic to retrieve information. Search questions should be analyzed and broken down into concepts that are described using MESH Headings or text words. These search terms are then joined together by AND to retrieve articles that contain both concepts, or joined together by OR to retrieve articles that contain either concept. The Boolean operators should be capitalized. To search for articles on sinusitis caused by bacteria search bacterial infections AND sinusitis (see Figure 14.6).

Although the search box in PubMed looks very much like Google, the words entered in the search box are processed based on concept searching rather than natural language searching. Recognizing that most searchers are accustomed to Google searching PubMed is developing a natural language search engine that works together with the concept search engine to retrieve articles.

Figure 14.6: Combining MeSH terms with Boolean operators

Selecting limits: Once the concept search has been entered you may limit the search with the search parameters (filters). We are now going to limit the sinusitis search. In addition to searching for articles where sinusitis is the main topic (sinusitis [MAJR]) we will limit the search by age (Adult: 19 to 44), humans, Core clinical journals (all of which are in English), added to PubMed in past five years and those with links to free full text. We could have also selected clinical trial, random controlled trial or review, or checked the box for all four (see Figure 14.7).

Figure 14.7: Selecting multiple search limits (Courtesy National Library of Medicine)

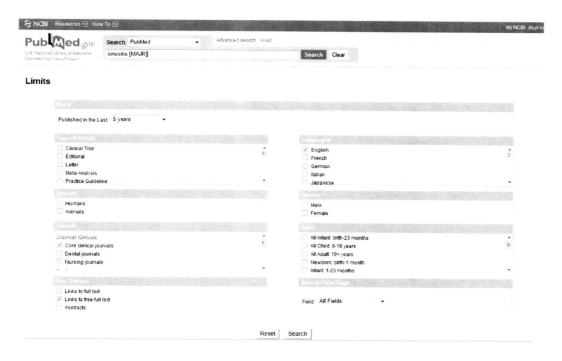

Figure 14.8: Search for sinusitis with multiple limits (Courtesy National Library of Medicine)

Our search with limits has greatly reduced the number of returned citations and improved the quality (Figure 14.8). Requesting free full text articles also reduces the search considerably. Changing the search to any abstract, instead of free full text articles, returns 79 citations.

- Note that the most current articles are listed first.

- Many articles are associated with an abstract that summarizes the article (Figure14.9)

- You must go to the full text article for more detail

Figure 14.9: Example of an abstract (Courtesy National Library of Medicine)

Arch Otolaryngol Head Neck Surg. 2006 Oct;132(10):1099-101.

Bacteriology of chronic sinusitis and acute exacerbation of chronic sinusitis.

Brook I.

Department of Pediatrics, Georgetown University School of Medicine, Washington, DC 20016, USA. ib6@georgetown.edu

Abstract

OBJECTIVE: To establish the microbiological characteristics of acute exacerbation of chronic sinusitis (AECS). SETTING: Academic medical center. PATIENTS: Thirty-two patients with chronic sinusitis and 30 patients with AECS. MAIN OUTCOME MEASURE: The aerobic and anaerobic microbiology of maxillary AECS and chronic maxillary sinusitis. RESULTS: A total of 81 isolates (33 aerobic and 48 anaerobic) were recovered from the 32 cases (2.5 per specimen) with chronic sinusitis. Aerobes alone were recovered in 8 specimens (25%), anaerobes only were isolated in 11 (34%), and mixed aerobes and anaerobes were recovered in 13 (41%). The predominant aerobic and facultative bacteria were Enterobacteriaceae and Staphylococcus aureus. The predominant anaerobic bacteria were Peptostreptococcus subspecies, Fusobacterium subspecies, anaerobic gram-negative bacilli, and Propionibacterium acnes. Twenty-one beta-lactamase-producing bacteria were recovered from 17 specimens (53%). A total of 89 isolates (40 aerobic and facultatives, and 49 anaerobic) were recovered from the 30 patients (3.0 per specimen) with AECS. Aerobes were recovered in 8 instances (27%), anaerobes only in 11 (37%), and mixed aerobes and anaerobes were recovered in 11 (37%). The predominant aerobes were Streptococcus pneumoniae, Enterobacteriaceae, and S aureus. The predominant anaerobes were Peptostreptococcus subspecies, Fusobacterium subspecies, anaerobic gram-negative bacilli, and P acnes. Thirty-six beta-lactamase-producing bacteria were recovered from 28 specimens (53%). CONCLUSIONS: This study demonstrates that the organisms isolated from patients with AECS were predominantly anaerobic and were similar to those generally recovered in patients with chronic sinusitis. However, aerobic bacteria that are usually found in acute infections (eg, S pneumoniae, Haemophilus influenzae, and Moraxella catarrhalis) can also emerge in some of the episodes of AECS.

PMID: 17043258 [PubMed - indexed for MEDLINE]   Free Article

⊕ MeSH Terms

⊕ LinkOut - more resources

Other options

Using the Advanced Search Screen: As you search, PubMed records your search statements. To review your previous searches and to combine search statements go to Advanced Search. Combine the statements by clicking search statement numbers and choosing the appropriate Boolean operator from the menu (see Figure 14.10).

Figure14.10: Example of Advanced Search Screen (Courtesy National Library of Medicine)

To see how the search you entered was executed by PubMed click on **Details.** On the Advanced Search Screen at the bottom of the page you will find navigation links for access to other PubMed modules.

Single Citation Matcher: When you are trying to locate a specific article and you only have fragments of the citation you may find Single Citation Matcher helpful. Type the information you know into the form to find the article of interest. You can search by author, journal, date, volume, issue, page or title words.

Clinical Queries: *Clinical Queries* provides another way to search for articles reporting the results of randomized controlled trials with the use of built-in filters. The research methodology behind the filters was created at McMaster University and the filters are built into the *Clinical Queries* interface, so you can search for randomized controlled trials by etiology (cause), diagnosis, therapy, prognosis or search for clinical prediction guides. Searches can be modified to be either broad/sensitive manner or narrow/specific.

Systematic reviews, a type of review that critically appraises multiple random controlled trials to give conclusions more strength (covered in more detail in the chapter on evidence based medicine), can be searched from the Clinical Queries page.

Display Options: Search results display automatically defaults to summary view, but you can also select abstract, Medline, XML and others

- Under Show tab you can elect to show up to 200 citations per page

- Under Sort by tab you can sort by author, journal or published date

Options for Saving Your Results

- *Clipboard* found under the *Send to* link allows you store up to 200 citations for up to eight hours

- *E-mail* lets you send your selected results to a colleague or yourself

- *Text* lets you save the results as a text file. (Sometimes useful for bringing the citations into MS Word.)

- *File* lets you put the results into a format that is suitable for a bibliographic software program

- *Order* allows you to send the citation to an affiliated library under the *Loansome Doc* program for document delivery

- *My NCBI*, located on upper right hand corner of the main screen, provides a valuable storage area for searches and collections of articles you have retrieved allowing you to

  - Save searches (otherwise gone in eight hours)

  - Set up e-mail alerts so you are notified when new articles are published on your topic of interest

  - Display links to online full-text of articles (*LinkOut*)

  - Choose filters that group search results

  - Registration is required for this free service

Other Features

Related articles and links:

To the right of each article you will see a hyperlink to similar or related articles. Select Links and it can link you to:

- PubMed books

- PubMed Central (www.pubmedcentral.nih.gov) journal articles links to articles in free and full text

- Link Out – links to external resources such as OVID or MDConsult to which the library with which you are affiliated subscribes

- Patient information from MedlinePlus

- PubMed® for Handhelds website offers several search options for MEDLINE® with the web browser of any mobile device. (Figure 14.11)

- PubMed PICO search, part of PubMed for Handhelds (Figure 14.12) – aids in the construction of a well thought out question prior to initiating a search. This tool it divides the question into sections defining the patient or problem, intervention, comparison and outcome (P.I.C.O). The URL or web address could be a desktop icon shortcut or a program on your handheld for fast searches.[27]

  - (P)atient or problem – how do you describe the patient group you are interested in? Elderly? Gender?

  - (I)ntervention, prognostic factor or exposure – Drug? Lab test? Tobacco?

  - (C)omparison – with another drug or placebo?

  - (O)utcome – what are you trying to measure? Mortality? Reduced heart attacks?

- PubMed Central—hosts multiple free full text articles. Unfortunately, many are located in minor journals of recent vintage. They are also more weighted towards a bioinformatics search.

Figure 14.11: PubMed for handhelds (Courtesy National Library of Medicine) [28]

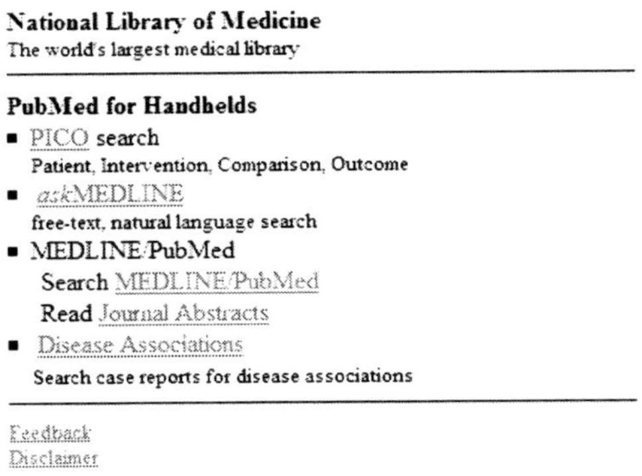

National Library of Medicine
The world's largest medical library

PubMed for Handhelds
- PICO search
  Patient, Intervention, Comparison, Outcome
- *ask*MEDLINE
  free-text, natural language search
- MEDLINE/PubMed
  Search MEDLINE/PubMed
  Read Journal Abstracts
- Disease Associations
  Search case reports for disease associations

Feedback
Disclaimer

Figure 14.12: PICO search (Courtesy National Library of Medicine)

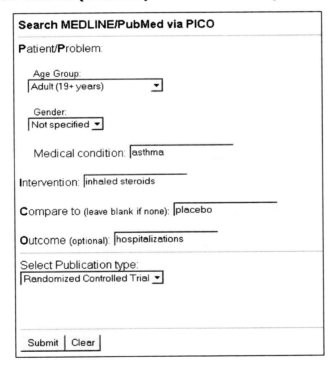

Third Party PubMed Tools

The National Library of Medicine (NLM) makes its database of citations available to the public for searching, and it also makes its data available through an application programming interface (API). The API allows interested users to write programs that mine the MEDLINE database in new ways. Several applications designed to optimize MEDLINE searching are available and others are emerging continually in an effort to exploit the MEDLINE data and may more accessible to the user.[29] Below are some examples of these third party PubMed tools that are noteworthy.

PubMed PubReMiner: Medical Subject Heading searching gives power to a PubMed searching. PubReMiner allows the searcher to enter keywords or PMIDs related to a query and then analyzes the relevant PubMed citations and their indexing to develop a list of terms with which to expand the search. PubReMiner is available directly from the website or through a web browser plug-in that is available for Mozilla Firefox or Internet Explorer 7.0 (see Figure 14.13).[30]

Figure 14.13: PubMed PubReMiner search on sinusitis and bacterial infections in title and abstract

PubMed EX: PubMed EX is a browser extension for Mozilla Firefox and Internet Explorer that marks up PubMed search results with additional information derived from data mining. PubMed EX provides background information that allows searchers to focus on key concepts in the retrieved abstracts. (see Figure 14.14) [31]

Figure 14.14: PubMed search on sinusitis and bacterial infections using a browser with the PubMed EX add-on

Gopubmed is a semantic search engine for searching PubMed enabling searchers to find articles easier and faster. Simply enter keywords or MeSH headings into the search box and the search engine will display the frequency of relevant terms with which to formulate your search. In Figure 14.15 the types of sinusitis as defined by MeSH is displayed on the left of the gopubmed screen. [32]

Figure 14.15: GoPubMed Basic Search on sinusitis and bacterial infections

Quertle: Quertle is a semantic search engine that helps users:

- Find true relationships between terms, rather than simple co-occurrences

- Use Power Terms to find appropriate categories of terms

- Use easily employed filters to limit a search by year or publication type

- See Figure 14.16 [33]

Figure 14.16: Quertle Search on sinusitis and bacterial infections in adults

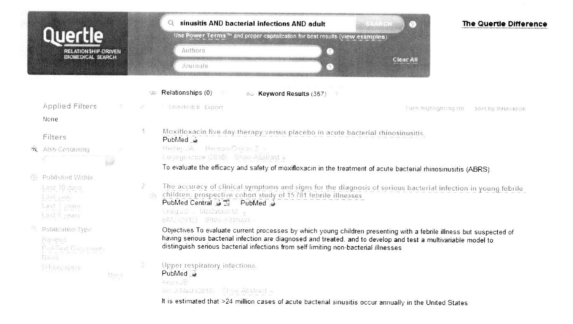

NLM Mobile: NLM Mobile includes several programs for Palm devices and Pocket PCs some of which can be used from the desktop These programs include AIDSinfo PDA Tools, WISER (Wireless System for First Responders, PubMed for Handhelds and the NCBI Bookshelf (see Figure 14.17).[34]

Figure 14.17: NLM Mobile (Courtesy National Library of Medicine)

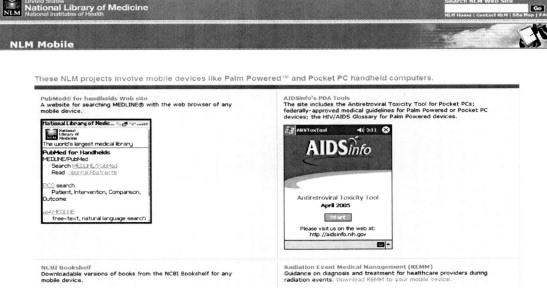

Future Trends

It is difficult today to conceive of a search engine more powerful or successful than Google. PubMed will continue to refine MESH and the filtering process. All search engines will improve as the internet evolves with more intelligent searches. Search engines will be part of most software programs and the search process will be faster and more focused due to artificial intelligence, faster networks and the Semantic Web. Computational knowledge search engines such as Wolfram Alpha will provide standalone answers to complex questions or be integrated with other programs. [35]

Key Points

- Search engines exist that can provide rapid high quality medical information
- Google has become a *defacto* initial medical search engine for many
- New search engines and meta-search engines continue to appear on the scene
- PubMed searches are important for formal searches of the medical literature
- All searches benefit from appropriate filters
- PubMed searching is enhanced by third party PubMed tools

Conclusion

At this time, Google is the premier search engine for non-medical and perhaps medical searches. With proper filtering and experience, Google and Google Scholar can be used with significant success. Today the average person can search for answers to a variety of medical questions. Although this may produce some "cyber-hypochondria" in a minority of searchers, it is likely to produce better informed patients in the majority. Better studies are needed to compare Google with other search engines and PubMed for quality and speed of retrieval. Familiarity with PubMed and its new features is important for healthcare workers who need to conduct formal searches of the medical literature. With knowledge and experience a PubMed search can result in relevant results in a timely fashion.

Acknowledgement

We would like to thank Jane Pelligrino MSLS, AHIP Head of Library Services Department at the Naval Medical Center Portsmouth for her contributions

References

1. Al-Ubaydli. Using search engines to find online medical information PLOS Medicine 2005 http://medicine.plosjournals.org/archive/1549-1676/2/9/pdf/10.1371_journal.pmed.0020228-S.pdf (Accessed 3 July 2010)

2. Goldman D. Search wars: Wolfram Alpha joins the battle. CNN Money.com. http://money.cnn.com/2009/05/27/technology/search_engines (Accessed 3 July 2010)

3. Google search basics: more search help http://www.google.com/support/bin/static.py?page=searchguides.html&ctx=advanced&hl=en (Accessed 3 July 2010)

4. Google.pedia. The ultimate Google resource. Michael Miller. Que publishing. 2007

5. Friedman, Thomas. The World is Flat. Farrar, Straus and Giroux. New York. 2006

6. The Anatomy of a large scale hypertextual web search engine http://infolab.stanford.edu/pub/papers/google.pdf (Accessed 3 July 2010)

7. Wikipedia: Google http://en.wikipedia.org/wiki/Google (Accessed 12 July 2009)

8. Giglia E. To Google or not to Google, that is the question Eur J Phys Rehabil Med 2008; 44:221-7

9. Steinbrook R. Searching for the Right Search—Reaching the Medical Literature NEJM 2006;354:4-7

10. Meats E et al. Using the Turning Research into Practice (TRIP) database: how do clinicians really search? JMLA 2007; 95(2): 156-163

11. Tang H, Ng JHK. Googling for a diagnosis-use of Google as a diagnostic aid: internet study BMJ Nov 11 2006. http://www.bmj.com/cgi/reprint/333/7579/1143 (Accessed 5 July 2010)

12. Correspondence. And a Diagnostic Test was Performed. NEJM 2005;353:2089-2090

13. Turner MJ. Accidental Epipen injection into a digit—the value of a Google search Ann R Coll Surg Engl.2004;86:218-9

14. Google Health Gains Partners. March 3 2010. http://news.cnet.com/8301-27083_3-10462961-247.html (Accessed March 10 2010)

15. Google search basics: basic search help http://www.google.com/help/basics.html (Accessed 3 July 2010)

16. Butler D. Science searches shift up a gear as Google Starts Scholar engine Nature 2004 Nov 25;432(7016):423

17. Jacso P. Google Scholar revisited. Online Information Review 2008; 32 (1): 102-114.

18. Giustini D. How Google is changing medicine BMJ. 2005 December 24; 331(7531): 1487–1488

19. Hazing AW, Vander Wal R. Google Scholar: the democratization of citation analysis. http://www.harzing.com/dlownload/gsdemo.pdf (Accessed 25 July 2010)

20. Sure, Google Scholar is ideal for some things The Search Principle blog http://blogs.ubc.ca/dean/2010/05/sure-google-scholar-is-ideal-for-some-things/ (Accessed 25 July 2010)

21. NML Databases http://www.nlm.nih.gov/databases/ (Accessed September 30 2011)

22. Fact Sheet MEDLINE http://www.nlm.nih.gov/pubs/factsheets/medline.html (Accessed September 30 2011)

23. Ebbert JO, Dupras DM, Erwin PJ. Searching the Medical Literature Using PubMed: A tutorial Mayo Clin Proc. 2003; 78:87-91

24. Sood A, Erwin PJ, Ebbert JO. Using Advanced Search Tools on PubMed for Citation Retrieval Mayo Clin Proc. 2004; 79:1295-1300

25. Haynes RB, Wilczynski N. Finding the gold in MEDLINE: Clinical Queries ACP Journal Club Jan/Feb 2005;142: A8-A9

26. Entrez PubMed http://www.pubmed.gov/ (Accessed September 30 2011)

27. NLM PICO http://askmedline.nlm.nih.gov/ask/pico.php (Accessed September 30 2011)

28. PubMed for Handhelds http://pubmedhh.nlm.nih.gov/nlmd/ (Accessed September 30 2011)

29. Rothman D. Archive for 3rd Party PubMed/MEDLINE Tools http://davidrothman.net/category/technology/3rd-party-pubmedmedline-tools/ (Accessed 25 July 2010)

30. PubReMiner http://hgserver2.amc.nl/cgi-bin/miner/miner2.cgi (Accessed September 30 2011)

31. PubMed EX http://bws.iis.sinica.edu.tw/PubMed-EX/index.html (Accessed September 30 2011)

32. gopubmed http://www.gopubmed.com (Accessed September 30 2011)

33. Quertle http://www.quertle.info/v2/ (Accessed September 30 2011)

34. NLM Mobile http://www.nlm.nih.gov/mobile/ (Accessed September 30 2011)

35. Wolfram Alpha www.wolframalpha.com (Accessed October 31 2011)

15

Evidence Based Medicine and Clinical Practice Guidelines

ROBERT E. HOYT

M. HASSAN MURAD

Learning Objectives

After reading this chapter the reader should be able to:

- State the definition and origin of evidence based medicine

- Define the benefits and limitations of evidence based medicine

- Describe the evidence pyramid and levels of evidence

- State the process of using evidence based medicine to answer a medical question

- Compare and contrast the most important online and smartphone evidence based medicine resources

- Describe the interrelationship between clinical practice guidelines, evidence based medicine, electronic health records and pay-for-performance

- Define the processes required to write and implement a clinical practice guideline

Introduction

"The great tragedy of Science - the slaying of a beautiful hypothesis by an ugly fact"

- Thomas Huxley (1825-1875)

Some might ask why Evidence Based Medicine (EBM) is included in a textbook on health informatics. The reason is that medical performance is based on quality and quality is based on the best available evidence. Clearly, information technology has the potential to improve decision making through online medical resources, electronic clinical practice guidelines, electronic health records (EHRs) with decision support, online literature searches, digital statistical analysis and online continuing medical education (CME).This chapter is devoted to finding the best available evidence and discussing one of its end products, clinical practice guidelines. Although one could argue that EBM is a buzz word like quality, in reality it means that clinicians should seek and apply the highest level of evidence available. According to the Center for EvidenceBased Medicine, EBM can be defined as:

"the conscientious, explicit and judicious use of current best evidence in making decisions about the care of individual patient"[1]

In *Crossing the Quality Chasm,* the Institute of Medicine (IOM) states:

> "Patients should receive care based on the best available scientific knowledge. Care should
> not vary illogically from clinician to clinician or from place to place" [2]

What the IOM is saying, is that every effort should be made to find the best answers and that these answers should be standardized and shared among clinicians. Such standardization implies that clinical practice should be consistent with the best available evidence that would apply to the majority of patients. This is easier said than done because so many clinicians are independent practitioners with little allegiance to any one healthcare organization. It is true that many questions cannot be answered by current evidence so clinicians may have to turn to subject matter experts. It is also true that the medical profession lacks the time and the tools to seek the best evidence. More than 1,800 citations are added to MEDLINE every day, making it impossible for a practicing clinician to stay up-to-date with the medical literature, not to mention that interpreting this evidence requires certain expertise and knowledge that not every clinician has. One does not have to look very far to see how evidence changes recommendations, e.g. bed rest is no longer recommended for low back pain[3] or following a spinal tap (lumbar puncture); routine activity is recommended instead. [4]

Until these older recommendations were challenged with high quality randomized controlled trials the medical profession had to rely on expert opinion, best guess or limited research studies.

Three pioneers are closely linked to the development of EBM. Gordon Guyatt coined the term EBM in 1991 in the American College of Physician (ACP) Journal Club.[5] The initial focus of EBM was on clinical epidemiology, methodology and detection of bias. This created the first fundamental principle of EBM: not all evidence is equal; there is a hierarchy of evidence that exists. In the mid-1990's, it was realized that patients' values and preferences are essential in the process of decision making, and addressing these values has become the second fundamental principle of EBM, after the hierarchy of evidence. Archie Cochrane, a British epidemiologist, was another early proponent of EBM. Cochrane Centers and the International Cochrane Collaboration were named after him as a tribute to his early work. The Cochrane Collaboration consists of review groups, centers, fields, methods groups and a consumer network. Review groups, located in 13 countries, look at randomized controlled trials. As of 2011 they have completed about 4600 systematic reviews, even though there have been 300,000 randomized controlled trials published.[6,7] The rigorous reviews are performed by volunteers, so efforts are slow. David Sackett is another EBM pioneer who has been hugely influential at The Centre for Evidence Based Medicine in Oxford, England and at McMaster University, Ontario, Canada. EBM has also been fostered at McMaster University by Brian Haynes who is the Chairman of the Department of Clinical Epidemiology and Biostatistics and the editor of the American College of Physician's (ACP) Journal Club. Although EBM is popular in the United Kingdom and Canada it has received mixed reviews in the United States. The primary criticisms are that EBM tends to be a very labor intensive process and in spite of the effort, frequently no answer is found.

The first randomized controlled trial was published in 1948.[8] For the first time subjects who received a drug were compared with similar subjects who would receive another drug or placebo and the outcomes were evaluated. Subsequently, studies became "double blinded" meaning that both the investigators and the subjects did not know whether they received an active medication or a placebo. Until the 1980s evidence was summarized in review articles written by experts. However, in the early 1990s, systematic reviews and meta-analyses became known as a better and more rigorous way to summarize the evidence and the preferred way to present the best available evidence to clinicians and policy makers. Since the late 1980s more emphasis has been placed on improved study design and true patient outcomes research. It is no longer adequate to show that a drug reduces blood pressure or cholesterol; it should demonstrate an improvement in patient-important outcomes such as reduced strokes or heart attacks.[9]

Importance of EBM

Learning EBM is like climbing a mountain to gain a better view. You might not make it to the top and find the perfect answer but you will undoubtedly have a better vantage point than those who choose to stay at sea level. Reasons for studying EBM resources and tools include:

- Current methods of keeping medically or educationally up-to-date do not work

- Translation of research into practice is often very slow

- Lack of time and the volume of published material results in information overload

- The pharmaceutical industry bombards clinicians and patients every day; often with misleading or biased information

- Much of what we consider as the "standard of care" in every day practice has yet to be challenged and could be wrong

Without proper EBM training we will not be able to appraise the best information resulting in poor clinical guidelines and wasted resources.

Traditional Methods for Gaining Medical Knowledge

- Continuing Medical Education (CME). Traditional CME is desired by many clinicians but the evidence shows it to be highly ineffective and does not lead to changes in practice. In general, busy clinicians are looking for a non-stressful evening away from their practice or hospital with food and drink provided.[10, 11] Much of CME is provided free by pharmaceutical companies with their inherent biases. Better educational methods must be developed. A recent study demonstrated that online CME was at least comparable, if not superior to traditional CME.[12]

- Clinical Practice Guidelines (CPGs). This will be covered in more detail, later in this chapter. Unfortunately, just publishing CPGs does not in and of itself change how medicine is practiced and the quality of CPGs is often variable and inconsistent.

- Expert Advice. Experts often approach a patient in a significantly different way compared to primary care clinicians because they deal with a highly selective patient population. Patients are often referred to specialists because they are not doing well and have failed treatment. For that reason, expert opinion needs to be evaluated with the knowledge that their recommendations may not be relevant to a primary care population. Expert opinion therefore should complement and not replace EBM.

- Reading. It is clear that most clinicians are unable to keep up with medical journals published in their specialty. Most clinicians can only devote a few hours each week to reading. All too often information comes from pharmaceutical representatives visiting the office. Moreover, recent studies may contradict similar prior studies, leaving clinicians confused as to the best course.

EBM Steps to Answering Clinical Questions

The following are the typical steps a clinician might take to answer a patient-related question:

- You see a patient and generate a question

- You formulate a well constructed question. Here is the PICO method, developed by the National Library of Medicine:

  o Patient or problem: how do you describe the patient group you are interested in? Elderly? Gender? Diabetic?

- Intervention: what is being introduced, a new drug or test?

- Comparison: with another drug or placebo?

- Outcome: what are you trying to measure? Mortality? Hospitalizations? A web-based PICO tool has been created by the National Library of Medicine to search Medline. This tool can be placed as a short cut on any computer.

- It has been recently suggested to add a T to PICO (i.e., PICOT) to indicate the Type of study that would best answer the PICO question.

- Seek the best evidence for that question via an EBM resource or PubMed.

- Appraise that evidence using tools mentioned in this chapter.

- Apply the evidence to your patient considering patient's values, preferences and circumstances[14]

Terminology Used in Answering Clinical Questions

- Evidence appraisal: When evaluating evidence, one needs to assess its validity, results and applicability.

- Validity: Validity means is the study believable? If apparent biases or errors in selecting patients, measuring outcomes, conducting the study, or analysis are present, then the study is less valid.

- Results: Results should be assessed in terms of the magnitude of treatment effect and precision (narrower confidence intervals or statistically significant results indicate higher precision).

- Applicability: Also called external validity, applicability indicates that the results reported in the study can be generalized to the patients of interest.

Most Common Types of Clinical Questions

- Therapy question. This is the most common area for medical questions and the only one we will discuss in this chapter

- Prognosis question

- Diagnosis question

- Harm question

- Cost question

The Evidence Pyramid

The pyramid in Figure 15.1 represents the different types of medical studies and their relative ranking. The starting point for research is often animal studies and the pinnacle of evidence is the meta-analysis of randomized trials. With each step up the pyramid our evidence is of higher quality associated with fewer articles published.[15] Although systematic reviews and meta-analyses are the most rigorous means to evaluate a medical question, they are expensive, labor intensive, and their inferences are limited by the quality of the evidence of the original studies.

- Case reports/case series. Consist of collections of reports on the treatment of individual patients without control groups, therefore they have much less scientific significance.

- Case control studies. Study patients with a specific condition (retrospective or after the fact) and compare with people who do not. These types of studies are often less reliable than randomized controlled trials and cohort studies because showing a statistical relationship does not mean that one factor necessarily caused the other.

- Cohort studies. Evaluate (prospectively or followed over time) and follow patients who have a specific exposure or receive a particular treatment over time and compare them with another group that is similar but has not been affected by the exposure being studied. Cohort studies are not as reliable as randomized controlled studies, since the two groups may differ in ways other than the variable under study.

- Randomized controlled trials (RCTs). Subjects are randomly assigned to a treatment or a control group that received placebo or no treatment. The randomization assures to a great extent that patients in the two groups are balanced in both known and unknown prognostic factors, and that the only difference between the two groups is the intervention being studied. RCTs are often "double blinded" meaning that both the investigators and the subjects do not know whether they received an active medication or a placebo. This assures that patients and clinicians are less likely to become biased during the conduct of a trial, and the randomization effect remains protected throughout the trial. RCTs are considered the gold standard design to test therapeutic interventions.

- Systematic reviews. Defined as protocol-driven comprehensive reproducible searches that aim at answering a focused question; thus, multiple RCTs are evaluated to answer a specific question. Extensive literature searches are conducted (usually by several different researchers to reduce selection bias of references) to identify studies with sound methodology; a very time consuming process. The benefit is that multiple RCTs are analyzed, not just one study.

- Meta-analyses. Defined as the quantitative summary of systematic reviews that take the systematic review a step further by using statistical techniques to combine the results of several studies as if they were one large single study.[15] Meta-analyses offer two advantages compared to individual studies. First, they include a larger number of events, leading to more precise (i.e., statistically significant) findings. Second, their results apply to a wider range of patients because the inclusion criteria of systematic reviews are inclusive of criteria of all the included studies.

Figure 15.1: The Evidence Pyramid

We will be dealing exclusively with therapy questions so note that randomized controlled trials are the suggested study of choice.[16]

Table 15.1: Suggested studies for questions asked

Type of Question	Suggested Best Type of Study
Therapy	RCT > cohort > case control > case series
Diagnosis	Prospective, blind comparison to a gold standard
Harm	RCT + cohort > case control > case series
Prognosis	Cohort study > case control > case series
Cost	Economic analysis and modeling

Studies that don't randomize patients or introduce a therapy along with a control group are referred to as observational studies (case control, case series and cohort). Most studies that have been reported on health information technology (HIT) are observational studies. This is important because cause and effect are difficult to prove, compared to a RCT. As an example, randomizing physicians to electronic prescribing (vs. paper prescribing) is difficult to implement and often impractical. In an observational study, physicians who volunteer to try electronic prescribing are likely "early adopters" and not representative of average physicians, which would skew the results. Alternate methods of randomization are feasible and desired, however. For example, "cluster randomization" would be a practical methodology in this situation. Here, several clinics or hospitals can be randomized as a whole practice to electronic prescribing whereas other clinics or hospitals can be randomized to paper prescribing.

Evidence of harm should be derived from both RCTs and cohort study designs. Cohort studies have certain advantages over RCTs when it comes to assessing harm: larger sample size, longer follow up duration, and more permissive inclusion criteria that allow a wide range of patients representing a real world utilization of the intervention to be included in the study.

Levels of Evidence (LOE)

Several methods have been suggested to grade the quality of evidence, which on occasion, can be confusing. The most up-to-date and acceptable framework is the GRADE (Grading of Recommendations, Assessment, Development and Evaluation).[17] The following is a description of the levels of evidence in this framework:

- Level 1: High quality evidence (usually derived from consistent and methodologically sound RCTs)

- Level 2: Moderate quality evidence (usually derived from inconsistent or less methodologically sound RCTs; or exceptionally strong observational evidence)

- Level 3: Low quality evidence (usually derived from observational studies)

- Level 4: Very low quality evidence (usually derived from flawed observational studies, indirect evidence or expert opinion

In this framework, RCTs start with a level 1 and observational studies start with a level 3. The rationale for this rating reflects the rigor of the RCTs and the strong inference they provide. For example, a recent systematic review and meta-analysis[18] reported that seven observational (non-randomized) studies demonstrated a beneficial association between chocolate consumption and the risk of cardiometabolic disorders. The highest levels of chocolate consumption were associated with significant reduction in cardiovascular disease and stroke compared with the lowest levels. Although these results seem impressive at face value, it is implausible that the effect of chocolate consumption is that profound (37% and 29%

reduction in the risk of cardiovascular disease and stroke). This magnitude of effect rivals the best available drugs and interventions used to prevent these diseases. Observational studies like these, have likely exaggerated the magnitude of benefit due to many factors (i.e., bias and confounding). It is possible that chocolate users are healthier, wealthier, more educated or have other characteristics that make them have lower incidence of disease. The opposite is also possible. Therefore, our confidence in estimates of effects generated from observational studies is lower than that of randomized trials. Hence, we derive evidence with different quality rating. Furthermore, it is important to recognize that the quality of evidence can be upgraded or downgraded if additional criteria based on study methodology and applicability are available.

Risk Measures and Terminology

Overall, therapy trials are the most common area of research and ask questions such as, is drug A better than drug B or placebo? In order to determine what the true effect of a study is, it is important to understand the concept of risk reduction and the number needed to treat. These concepts are used in studies that have dichotomous outcomes (i.e., only two possible answers such as dead or alive, improved or not improved); which are more commonly utilized outcomes. We will define these concepts and then present an example for illustration.

- Risk is defined as the rate of events during a specific period of time. It is calculated by dividing the number of patients suffering events by the total number of patients at risk for events.

- Odds are defined as the ratio of the number of patients with events to the number of patients without events.

    Notice that *Odds=1 / (1+risk)*

Example

Amazingstatin is a drug that lowers cholesterol. If we treat a 100 patients with this drug and five of them suffer a heart attack over a period of 12 months, the risk of having a heart attack in the treated group would be 5/100= 0.050 (or 5%). The odds of having a heart attack would be 5/95= 0.052. In the control group, if we treat 100 patients with placebo and seven suffer heart attacks, the risk in this group is 7/100=0.070 or 7% and the odds are 7/93=0.075.

Notice that the risk in the experimental group is called experimental event rate (EER) and the risk in the control group is called control event rate (CER). To compare risk in two groups, we use the following terms:

- Relative Risk (RR) is the ratio of two risks as defined above. Thus, it is the ratio of the event rate of the outcome in the experimental group (EER) to the event rate in the control group (CER).

    RR = EER/CER.

- Relative Risk Reduction (RRR) is the difference between the experimental event rate (EER) and the control event rate (CER), expressed as a percentage of the control event rate.

    RRR = (EER-CER)/CER.

- Absolute Risk Reduction (ARR) is the difference between the EER and the CER.

    ARR = EER-CER

    (Note that "difference" is not the same as subtracting CER from EER. For example if the EER is 1.5 amd the CER is 2.0, the difference is .5, not -.5)

- Number Needed to Treat (NNT) is the number of patients who have to receive the intervention to prevent one adverse outcome.[19]

$$NNT = 1/ARR$$

(or 100/ARR, if ARR is expressed as a percentage instead of a fraction)

- Odds Ratio (OR) is the ratio of odds (instead of risk) of the outcome occurring in the intervention group to the odds of the outcome in the control group.

On Amazingstatin, 5% (EER) of patients have a heart attack after 12 months of treatment. On placebo 7% (CER) of patients have a heart attack over 12 months

$$RR = 5\% / 7\% = 0.71$$

$$RRR = (7\% - 5\%) / 7\% = 29\%$$

$$ARR = 7\% - 5\% = 2\%$$

$$NNT = 100/2 = 50$$

As we calculated above, the odds for the intervention and control group respectively are 0.052 and 0.075; the odds ratio (OR) = 0.52/0.075 = 0.69

Comments

RR and OR are very similar concepts and as long as the event rate is low, their results are almost identical. These results show that this drug cuts the risk of heart attacks by 29% (almost by a third), which seems like an impressive effect. However, the absolute reduction in risk is only 2% and we need to treat 50 patients to prevent one adverse event. Although this NNT may be acceptable, using RRR seems to exaggerate our impression of risk reduction compared with ARR. Most of what we see written in the medical literature and the lay press will quote the RRR. Unfortunately, very few studies offer NNT data, but it is very easy to calculate if you know the ARR specific to your patient. Nuovo et al. noted that NNT data was infrequently reported by five of the top medical journals in spite of being recommended. [20] In another interesting article, Lacy and co-authors studied the willingness of US and UK physicians to treat a medical condition based on the way data was presented. Ironically, the data was actually the same but presented in three different ways. Table 15.2 suggests that US physicians may need more training in EBM.[21]

Table 15.2: Physician's Likelihood of Prescribing Medication Based on How Research Data is Presented

Physicians From	Relative Risk Reduction (RRR)	Absolute Risk Reduction (ARR)	Number Needed To Treat (NNT)
United States	54%	4%	10%
United Kingdom	24%	11%	22%

Examples of Using RRR, ARR and NNT

A full page article appeared in a December 2005 Washington Post newspaper touting the almost 50% reduction of strokes by a cholesterol lowering drug. This presented an opportunity to take a look at how drug companies usually advertise the benefits of their drugs. Firstly, in small print, you note that patients have to be diabetic with one other risk factor for heart disease to see benefit. Secondly, there are no references. The statistics are derived from the CARDS Study published in the Lancet in Aug 2004.[22] Stroke was reported to occur in 2.8% in patients on a placebo and 1.5% in patients taking the drug Lipitor. The NNT is therefore

100/1.3 or 77. So, you had to treat 77 patients for an average of 3.9 years (the average length of the trial) to prevent one stroke. This doesn't sound as good as "cuts the risk by nearly half." Now armed with these EBM tools, look further the next time you read about a miraculous drug effect.

Number Needed to Harm (NNH) is calculated similarly to the NNT. If, for example, Amazingstatin was associated with intestinal bleeding in 6% of patients compared to 3% on placebo, the NNH is calculated by dividing the ARR (%) into 100. For our example the calculation is 100/.03 = 33. In other words, the treatment of 33 patients with Amazingstatin for one year resulted, on average, in one case of intestinal bleeding as a result of the treatment. Unlike NNT, the higher the NNH, the better.

The Case of Continuous Variables. The results of studies (effect measures) described so far (i.e., RR, OR, ARR) are used when outcomes are dichotomous (such as dead or alive, having a heart attack or not, etc.). However, outcomes can also be continuous (e.g. blood cholesterol level). These outcomes are usually reported as a difference in the means of two study groups. This difference has a unit, which in the cholesterol example, is mg/dL. In addition to the mean difference, results would also include some measure that describes the spread or dispersion of measurements around the mean (i.e., standard deviation, range, interquartile range or a confidence interval).

If the metrics of continuous variables do not have intuitive intrinsic meaning (e.g., a score on a test or a scale), the *effect size* can be standardized (i.e., divided by the standard deviation; which makes the data measured in standard deviation units). This process allows the comparison of students taking different tests, or tests taken in different years, or comparing the results of studies that used different scales as their outcomes. This is possible because all these measurements are standardized (have the same unit, which is standard deviation unit). A commonly used effect size is Cohen's d, which is a standardized difference in means. It is interpreted arbitrarily as a small, moderate or large effect, if d was 0.2, 0.5 or 0.7; respectively. In addition to knowing that a result is statistically significant, calculating the effect size gives one an idea of how big the difference actually is.

Cost of Preventing an Event (COPE). Many people reviewing a medical article would want to know what the cost of the intervention is. A simple formula exists that sheds some light on the cost: COPE = NNT x number of years treated x 365 days x the daily cost of the treatment. Using our example of Amazingstatin = 40 x 1 x 365 x $2 or $29,200 to treat 40 patients for one year to prevent one heart attack. Now you can compare COPE scores with other similar treatments.[23]

Limitations of the Medical Literature and EBM

Because evidence is based on information published in the medical literature, it is important to point out some of the limitations researchers and clinicians must deal with on a regular basis:

- There is a low yield of clinically useful articles in general [24]

- Conclusions from randomized drug trials tend to be more positive if they are from for-profit organizations [25]

- Up to 16% of well publicized articles are contradicted in subsequent studies [26]

- Peer reviewers are "unpaid, anonymous and unaccountable" so it is often not known who reviewed an article and how rigorous the review was [27]

- Many medical studies are poorly designed: [28]
  - The recruitment process was not described [29]
  - Inadequate power (size) to make accurate conclusions. In other words, not enough subjects were studied [30]

o Studies with negative results (i.e., results that are not statistically significant) are not always published or take more time to be published, resulting in "publication bias." In an effort to prevent this type of bias the American Medical Association advocates mandatory registration of all clinical trials in public registries. Also, the International Committee of Medical Journal Editors requires registration as a condition to publish in one of their journals. However, they do not require publishing the results in the registry at this time. Registries could be a data warehouse for future mining and some of the well-known registries include:

- ClinicalTrials.gov

- WHO International Clinical Trials Registry

- Global Trial Bank of the American Medical Informatics Association

- Trial Bank Project of the University of California, San Francisco [31]

In spite of the fact that EBM is considered a highly academic process towards gaining medical truth, numerous problems exist:

- Different evidence rating systems by various medical organizations

- Different conclusions by experts evaluating the same study

- Time intensive exercise to evaluate existing evidence

- Systematic reviews are limited in the topics reviewed (3,000 in the Cochrane database) and are time intensive to complete (6 to 24 months). Often the conclusion is that current evidence is weak and further high quality studies are necessary

- Randomized controlled trials are expensive. Drug companies tend to fund only studies that help a current non-generic drug they would like to promote

- Results may not be applicable to every patient population

- Some view EBM as "cookbook medicine" [32]

- There is not good evidence that teaching EBM changes behavior [33]

Other Approaches

EBM has had both strong advocates and skeptics since its inception. One of its strongest proponents Dr. David Sackett published his experience with an "Evidence Cart" on inpatient rounds in 1998. The cart contained numerous EBM references but was so bulky that it could not be taken into patient rooms.[34] Since that article, multiple, more convenient EBM solutions exist. While there are those EBM advocates who would suggest we use solely EBM resources, many others feel that EBM "may have set standards that are untenable for practicing physicians."[35, 36]

Dr. Frank Davidoff believes that most clinicians are too busy to perform literature searches for the best evidence. He believes that we need "Informationists" who are experts at retrieving information.[37] To date, only clinical medical librarians (CMLs) have the formal training to take on this role. At large academic centers CMLs join the medical team on inpatient rounds and attach pertinent and filtered articles to the chart. As an example, Vanderbilt's Eskind Library has a Clinical Informatics Consult Service.[38, 39] The obvious drawback is that CMLs are only available at large medical centers and are unlikely to research outpatient questions.

According to Slawson and Shaughnessy you must become an "information master" to sort through the "information jungle." They define the usefulness of medical information as:

$$\text{Usefulness} = \frac{\textit{Validity x Relevance}}{\textit{Work}}$$

Only the clinician can determine if the article is relevant to his/her patient population and if the work to retrieve the information is worthwhile. Slawson and Shaughnessy also developed the notion of looking for "patient oriented evidence that matters" (POEM) and not "disease oriented evidence that matters" (DOEM). POEMS look at mortality, morbidity and quality of life whereas DOEMS tend to look at laboratory or experimental results. They point out that it is more important to know that a drug reduces heart attacks or deaths from heart attacks (POEM), rather than just reducing cholesterol levels (DOEM). [40] This school of thought also recommends that you not read medical articles blindly each week but should instead learn how to search for patient specific answers using EBM resources.[41] This also implies that you are highly motivated to pursue an answer, have adequate time and have the appropriate training. See case study below for example of EBM being applied to a clinical scenario.

Case Study

People with blockage of the carotid artery are at risk of stroke and death. They can be treated via surgery (called endarterectomy) or a less invasive procedure (putting a stent in the blocked area by going through the arteries, i.e., without surgery). The choice of procedure is controversial.

The evidence

A systematic review and meta-analysis appraised the quality of the totality of existing evidence in this area. They found 13 randomized controlled trials that enrolled a total of 7,484 patients. The methodological quality of the trials was moderate to high. Compared with carotid endarterectomy, stenting was associated with increased risk of stroke (relative risk [RR], 1.45; 95% confidence interval [CI], 1.06-1.99) and decreased risk of myocardial infarction (MI) caused by surgery (RR, 0.43; 95% CI, 0.26- 0.71). For every 1,000 patients opting for stenting rather than endarterectomy, 19 more patients would have strokes and 10 fewer would have MIs.

Patients values, preferences and context

Patients vary in their values such as aversion (fear) of stroke vs death and their fear of surgery and surgical complications such as scars in the neck and anesthesia. Patients also vary in their surgical risk (e.g., those with history of heart disease may prefer less invasive procedure to avoid prolonged anesthesia).

Guidelines

Due to the different impact of these procedures on the different outcomes, the guidelines were nuanced and stratified and allowed patients values and preferences, age, surgical and anatomical risk factors to be used in decision making. This example highlights the importance of patients' values and preferences as the second principle of EBM

References

Murad MH, Shahrour A, Shah ND, Montori VM, Ricotta JJ. A systematic review and meta-analysis of randomized trials of carotid endarterectomy vs stenting.J Vasc Surg. 2011 Mar;53(3):792-7. Epub 2011 Jan 8

Ricotta JJ, Aburahma A, Ascher E, et al. Updated Society for Vascular Surgery guidelines for management of extracranial carotid disease: executive summary. J Vasc Surg. 2011 Sep;54(3):832-6.

EBM Resources

There are many first-rate online medical resources that provide EBM type answers. They are all well referenced, current and written by subject matter experts. Several include the level of evidence (LOE). These resources can be classified as **filtered** (an expert has appraised and selected the best evidence, e.g., up-to-date or **unfiltered** (non-selected evidence, e.g., PubMed). For the EBM purist, the following are considered traditional or classic EBM resources:

- Clinical Evidence [42]
    - British Medical Journal product with two issues per year
    - Sections on EBM tools, links, training and articles
    - Evidence is oriented towards patient outcomes (POEMS)
    - Very evidence based with single page summaries and links to national guidelines
    - Available in paperback (Concise), CD-ROM, online or PDA format

- Cochrane Library [43]
    - Database of systematic reviews. Each review answers a clinical question
    - Database of review abstracts of effectiveness (DARE)
    - Controlled Trials Register
    - Methodology reviews and register
    - Fee-based

- Cochrane Summaries [44]
    - Part of the Cochrane Collaboration
    - Reviews can be accessed for a fee but abstracts are free. A search for low back pain in 2011, as an example, returned 393 reviews (abstracts)

- EvidenceUpdates [45]
    - Since 2002 BMJ Updates has been filtering all of the major medical literature. Articles are not posted until they has been reviewed for newsworthiness and relevance; not strict EBM guidelines
    - You can go to their site and do a search or you can choose to have article abstracts e-mailed to you on a regular basis
    - These same updates are available through www.Medscape.com

- ACP Journal Club [46]
    - Bimonthly journal that can be accessed from OVID or free if a member of the American College of Physicians (ACP)
    - Over 100 journals are reviewed but very few articles make the cut: in 1992 only 13% of articles from the NEJM made the Journal Club, all other journals were much lower
    - They have a searchable database and email alerting system

- Practical Pointers for Primary Care [47]

  - Free online review of articles from the New England Journal of Medicine, Journal of the American Medical Journal, British Medical Journal, the Lancet, the Annals of Internal Medicine and the Archives of Internal Medicine

  - Program can be accessed via the web or monthly reports e-mailed to those who subscribe

  - Editor dissects the study and makes summary comments that are very helpful to the average reader

- Evidence Based On-Call [48]

  - User friendly site intended for quick look-ups for clinicians on call

  - Has multiple critically appraised topics (CATs) that point out the most important clinical pearls, with level of evidence

- Others

  - TRIP Database [49]

  - OVID has the ability to search the Cochrane Database of Systematic Reviews, DARE, ACP Journal Club and Cochrane Controlled Trials Register at the same time. Also includes Evidence Based Medicine Reviews [50]

  - SUMSearch. Free site that searches Medline, National Guideline Clearing House and DARE [51]

  - Bandolier. Free online EBM journal; used mainly by primary care doctors in England. Provides simple summaries with NNTs. Resource also includes multiple monographs and books on EBM that are easy to read and understand [52]

  - Centre for Evidence Based Medicine is a comprehensive EBM site presented by Oxford University [53]

Clinical Practice Guidelines

The Institute of Medicine in 1990 defined clinical practice guidelines (CPGs) as:

> "systematically developed statements to assist practitioner and patient decisions about health care for specific clinical circumstances"[54]

Clinical practice guidelines (CPGs) take the very best evidence based medical information and formulate a game plan to treat a specific disease or condition. If one considers evidence as a continuum that starts by data generated from a single study, appraised and synthesized in a systematic review, CPGs would represent the next logical step in which evidence is transformed into a recommendation. Many medical organizations use CPGs with the intent to improve quality of care, patient safety and/or reduce costs. Information technology assists CPGs by expediting the search for the best evidence and linking the results to EHRs and smartphones for easy access. Two areas in which CPGs may be potentially beneficial include disease management and pay-for-performance, covered in other chapters. CPGs are important for several reasons such as 83% of Medicare beneficiaries have at least one chronic condition and 68% of Medicare's budget is devoted to the 23% who have five or more chronic conditions. [55] There is some evidence that guidelines that address multiple comorbidities (concurrent chronic diseases) actually do work. As an example, in one study of diabetics, there was a 50% decrease in cardiovascular and microvascular complications with intensive treatment of multiple risk factors. [56]

In spite of evidence to suggest benefit, several studies have shown poor CPG compliance by patients and physicians. The well publicized 2003 RAND study in the New England Journal of Medicine demonstrated that "overall, patients received 54% of recommended care."[57, 58] In another study of guidelines at a major teaching hospital there was overuse of statin therapy (cholesterol lowering drugs). Overuse occurred in 69% of primary prevention (to prevent a disease) and 47% of secondary prevention (to prevent disease recurrence or progression), compared to national recommendations.[59]

It should be emphasized that creating or importing a guideline is the easy part because hundreds have already been created by a variety of national organizations. Implementing CPGs and achieving buy-in by all healthcare workers, particularly physicians, is the hard part.

Developing Clinical Practice Guidelines

Ideally, the process starts with a panel of content and methodology experts commissioned by a professional organization. As an example, if the guideline is about preventing venous thrombosis and pulmonary embolism, multi-disciplinary content experts would be pulmonologists, hematologists, pharmacists and hospitalists.

Methodology experts are experts in evidence based medicine, epidemiology, statistics, cost analysis, etc. The panel refines the questions, usually in PICO format, that was discussed in the previous chapter. A systematic literature search and evidence synthesis takes place. Evidence is graded and recommendations are negotiated. Panel members have their own biases and conflicts of interest that should be declared to CPG users. Voting is often needed to build consensus since disagreement is a natural phenomenon in this context.

The Strength of Recommendations

Guideline panels usually associate their recommendations by a grading that describes how confident they are in their statement. Ideally, panels should separately describe their confidence in the evidence (the quality of evidence, described in previous chapter) from the strength of the recommendation. The reason for this separation is that there are factors other than evidence that may affect the strength of recommendation. These factors are: (1) how closely balanced are the benefits and harms of the recommended intervention, (2) patients' values and preferences, and (3) resource allocation.

For example, even if there is very high quality evidence from randomized trials showing that warfarin (a blood thinner) decreases the risk of stroke in some patients, the panel may issue a weak recommendation considering that the harms associated with this medicine are substantial. Similarly, if high quality evidence suggests that a treatment is very beneficial, but this treatment is very expensive and only available in very few large academic centers in the US, the panel may issue a weak recommendation because this treatment is not easily available or accessible.

Application to Individuals

A physician should consider a strong recommendation to be applicable to all patients who are able to receive it. Therefore, physicians should spend his/her time and effort on explaining to patients how to use the recommended intervention and integrate it in their daily routine.

On the other hand, a weak recommendation may only apply to certain patients. Physicians should spend more time discussing pros and cons of the intervention with patients, use risk calculators and tools designed to stratify patients' risk to better determine the balance of harms and benefit for the individual. Weak recommendations are the optimal condition to use decision aids, which are available in written, videographic and electronic formats and may help in the decisionmaking process by increasing knowledge acquisition by patients and reduce their anxiety and decisional conflicts.

Appraisal and Validity of Guidelines

There are several tools suggested to appraise CPGs and determine their validity. These tools assess the process of conducting CPGs, the quality and rigor of the recommendations and the clarity of their presentation. The following list includes some of the attributes that guidelines users (clinicians, patients, policy makers) should seek to determine if a particular CPG is valid and has acceptable quality:

- Evidence based, preferably linked to systematic reviews of the literature
- Considers all relevant patients groups and management options
- Considers patient-important outcomes (as opposed to surrogate outcomes)
- Updated frequently
- Clarity and transparency in describing the process of CPGs development (e.g., voting, etc.)
- Clarity and transparency in describing the conflicts of interests of the guideline panel
- Addresses patients' values and preferences
- Level of evidence and strength of recommendation are given
- Simple summary or algorithm that is easy to understand
- Available in multiple formats (print, online, PDA, etc.) and in multiple locations
- Compatibility with existing practices
- Simplifies, not complicates decision making [60]

Barriers to Clinical Practice Guidelines

Attempts to standardize medicine by applying evidence based medicine and clinical practice guidelines have been surprisingly difficult due to multiple barriers:

- Practice setting: inadequate incentives, inadequate time and fear of liability. A 2003 study estimated that it would require 7.4 hours/working day just to comply with all of the US Preventive Services Task Force recommendations for the average clinician's practice![61]

- Contrary opinions: local experts do not always agree with CPG or clinicians hear different messages from drug detail representatives

- Sparse data: there are several medical areas in which the evidence is of lower quality or sparse. Guideline panels in these areas would heavily depend on their expertise and should issue weak recommendations (e.g. suggestions) or no recommendations if they did not reach a consensus. These areas are problematic to patients and physicians and are clearly not ready for quality improvement projects or pay-for-performance incentives. For years, diabetologists advocated tight glycemic control of patients with type 2 diabetes; however, it turned out from results of recent large randomized trials that this strategy does not result in improved outcomes.[62]

  o We need more information about why clinicians don't follow CPGs. Persell et al. reported in a 2010 study that 94% of the time when clinicians chose an exception to the CPG it was appropriate. Three percent (3%) were inappropriate and 3% were unclear.[63]

- Knowledge and attitudes: there is a lack of confidence to either not perform a test (malpractice concern) or to order a new treatment (don't know enough yet). Information overload is always a problem.[64]

- CPGs can be too long, impractical or confusing. One study of Family Physicians stated CPGs should be no longer than two pages.[65-67] Most national CPGs are 50 to 150 pages long and don't always include a summary of recommendations.

- Where and how do you post CPGs? What should be the format?

- Less buy-in if data reported is not local since physicians tend to respond to data reported from their hospital or clinic.

- No uniform level of evidence (LOE) rating system

- Too many CPGs posted on the National Guideline Clearinghouse. For instance, a non-filtered search in November 2011 by one author for "diabetes" yielded 631 diabetes-related CPGs. The detailed search option helps filter the search significantly.[67]

- Lack of available local champions to promote CPGs

- Excessive influence by drug companies: A survey of 192 authors of 44 CPGs in the 1991 to 1999 time frame showed:

    o 87% had some tie to drug companies

    o 58% received financial support

    o 59% represented drugs mentioned in the CPG

    o 55% of respondents with ties to drug companies said they did not believe they had to disclose involvement[68]

- Quality of national guidelines: National guidelines are not necessarily of high quality. A 2009 review of CPGs from the American Heart Association and the American College of Cardiology (1984 to Sept 2008) concluded that many of the recommendations were based on a lower level of evidence or expert opinion, not high quality studies.[69]

- No patient input. At this point patients are not normally involved in any aspect of CPGs, even though they receive recommendations based on CPGs. In an interesting 2008 study, patients who received an electronic message about guidelines experienced a 12.8% increase in compliance. This study utilized claims data as well as a robust rules engine to analyze patient data. Patients received alerts (usually mail) about the need for screening, diagnostic and monitoring tests. The most common alerts were for adding a cholesterol lowering drug, screening women over age 65 for osteoporosis, doing eye exams in diabetics, adding an ACE inhibitor drug for diabetes and testing diabetics for urine microalbumin.[72] It makes good sense that patients should be knowledgeable about national recommendations and should have these guidelines written in plain language and available in multiple formats. Also, because many patients are highly "connected" they could receive messages via cell phones, social networking software, etc. to improve monitoring and treatment.

Initiating Clinical Practice Guidelines

Examples of Starting Points:

- High cost conditions: heart failure

- High volume conditions: diabetes

- Preventable admissions: asthma

- There is variation in care compared to national recommendations: deep vein thrombophlebitis (DVT) prevention

- High litigation areas: failure to diagnose or treat

- Patient safety areas: intravenous (IV) drug monitoring

The Strategy

- Leadership support is crucial

- Use process improvement tools such as the Plan-Do-Study-Act (PDSA) model

- Identify gaps in knowledge between national recommendations and local practice

- Locate a guideline champion who is a well-respected clinical expert.[71] A champion acts as an advocate for implementation based on his/her support of a new guideline

- Other potential team members:

  o Clinician selection based on the nature of the CPG

  o Administrative or support staff

  o Quality Management staff

- Develop action plans

- Educate all staff involved with CPGs, not just clinicians

- Pilot implementation

- Provide frequent feedback to clinicians and other staff regarding results

- Consider using the checklist for reporting clinical practice guidelines developed by the 2002 Conference on Guideline Standardization (COGS)[72]

Clinical Practice Guideline Examples

The CPG in Figure 15.2 was written for the treatment of uncomplicated bladder infections (cystitis) in women. The goal was to use less expensive antibiotics and treat for fewer days. The protocol or algorithm can be administered by a triage nurse when a patient telephones or walks in. Figure 15.3 demonstrates that the use of the first line drug (sulfa family) increased after the start of the CPG, whereas second line drug use decreased. Success of this program was based on educating all members of the healthcare team and reporting the results at medical staff meetings and other venues. It was also aided by an easy to follow guideline and full support by the nursing staff.

Figure 15.2: CPG for uncomplicated dysuria or urgency in women

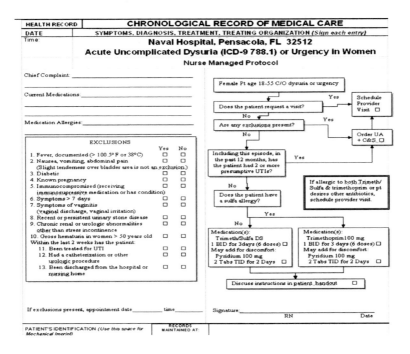

Figure 15.3: Results of CPG implementation (Courtesy Naval Hospital Pensacola)

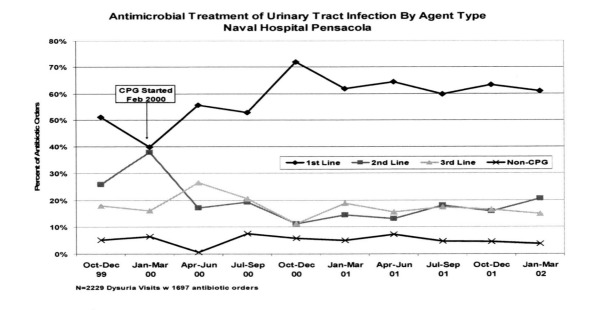

Electronic Clinical Practice Guidelines

CPGs have been traditionally paper-based and often accompanied by a flow diagram or algorithm. With time, more are being created in an electronic format and posted on the internet or Intranet for easy access. Zielstorff outlined the issues, obstacles and future prospects of online practice guidelines in an early review.[73] What has changed since then is the ability to integrate CPGs with electronic health records and smartphones.

CPGs on smartphones: These mobile platforms function well in this area as each step in an algorithm is simply a tap or touch of the screen. In Figures 15.4 and 15.5 programs are shown that are based on national guidelines for cardiac risk and cardiac clearance. Figure 15.4 depicts a calculator that determines the 10 year risk of heart disease based on serum cholesterol and other risk factors. A cardiac clearance program determines whether a patient needs further cardiac testing prior to an operation (Figure 15.5)[74] Many excellent guidelines for the smartphone exist that will be listed later in this chapter.

Figure 15.4: 10 year risk of heart disease

Figure 15.5: Cardiac clearance

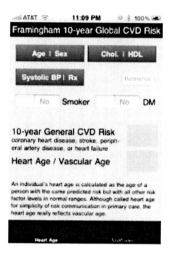

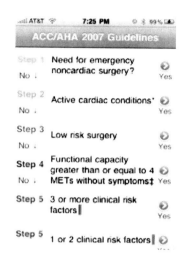

Web-Based Risk Calculators: Many of these are available on a mobile platform and are also available online. While these are not CPGs exactly, they are based on population studies and are felt to be part of EBM and can give direction to the clinician. As an example, some experts feel that aspirin has little benefit in preventing a heart attack unless your 10 year risk of one exceeds 20%. The following is a short list of some of the more popular online calculators:

- ATP III Cardiac risk calculator: estimates the 10 year risk of a heart attack or death based on your cholesterol, age, gender, etc.[75]

- FRAX fracture risk calculator: estimates the 10 year risk of a hip or other fracture based on all of the common risk factors for osteoporosis. Takes into account a patient's bone mineral density score, gender and ethnicity.[76]

- GAIL breast cancer risk assessment tool: estimates a patient's risk of breast cancer, again, based on known and accepted risk factors.[77]

- Stroke risk calculator: based on the Framingham study it predicts 10 year risk of a stroke based on known risk factors.[78]

- Risk of stroke or death for new onset atrial fibrillation: also based on the Framingham study, it calculates five year risk of stroke or death.[79]

EHR CPGs

Although not all electronic health records have embedded CPGs, there is definite interest in providing local or national CPGs at the point of care. CPGs embedded in the EHR are clearly a form of decision support. They can be linked to the diagnosis or the order entry process. In addition, they can be standalone resources available by clicking, for example, an "info-button." Clinical decision support provides treatment reminders for disease states that may include the use of more cost effective drugs. Institutions such as Vanderbilt University have integrated more than 750 CPGs into their EHR by linking the CPGs to ICD-9 codes.[80] The results of embedded CPGs appears to be mixed. In a study by Durieux using computerized decision support reminders, orthopedic surgeons showed improved compliance to guidelines to prevent deep vein thrombophlebitis.[81] On the other hand, three studies by Tierney, failed to demonstrate improved compliance to guidelines using computer reminders for hypertension, heart disease and asthma.[82-84] Clinical decision support, to include order sets is discussed in more detail in the chapters on electronic health records and patient safety.

There are other ways to use electronic tools to promulgate CPGs. In an interesting paper by Javitt, primary care clinicians were sent reminders on outpatient treatment guidelines based only on claims data. Outliers were located by using a rules engine (Care Engine) to compare a patient's care with national guidelines. They were able to show a decrease in hospitalizations and cost as a result of alerts that notified physicians by phone, fax or letter. This demonstrates one additional means of changing physician behavior using CPGs and information technology not linked to the electronic health record.[85] Critics might argue that claims data is not as accurate, robust or current as actual clinical results.

Software is now available (EBM Connect) that can compute compliance with guidelines automatically using administrative data. The program translates guidelines from text to algorithms for 20 disease conditions and therefore would be much more efficient than chart reviews. Keep in mind it will tell you if, for example, LDL cholesterol was ordered, not the actual results.[86]

Clinical Practice Guideline Resources

Web-based CPGs

- **National Guideline Clearinghouse.** This program is an initiative of the Department of Health and Human Services and is the largest and most comprehensive of all CPG resources. Features offered:
    - Includes about 2664 guidelines
    - There is extensive search engine filtering i.e. you can search by year, language, gender, specialty, level of evidence, etc.
    - Abstracts are available as well as links to full text guidelines where available
    - CPG comparison tool
    - Forum for discussion of guidelines
    - Annotated bibliography
    - They link to 17 international CPG resource sites[87]

- **National Institute for Health and Clinical Excellence (NICE)**
    - Service of the British National Health Service
    - Approximately 100 CPGs are posted and dated
    - A user-friendly short summary is available as well as a lengthy guideline, both in downloadable pdf format
    - Podcasts are available[88]

- Agency for Health Care Research and Quality (AHRQ)
  - 1 of 12 agencies within the Department of Health and Human Services (HHS)
  - AHRQ supports health services research initiatives that seek to improve the quality of health care in America
  - AHRQ's funds evidence practice centers that conduct evidence appraisal and reviews to support the development of clinical practice guidelines[89]
- Health Team Works (formerly Colorado Clinical Guidelines Collaborative)
  - Free downloads available for Colorado physicians and members of CCGC
  - As of October 2011 they have 14 CPGs available
  - Guidelines are in easy to read tables, written in a pdf format
  - References, resources and patient handouts are available[90]

Smartphone-based CPGs

Most CPGs can be downloaded for the iPhone or iPad through the iTunes Store or the Android Market. For further information about medical apps, we refer you to the chapter on mobile technology. The following are a sample of CPGs available for smartphones:

- NCCN Clinical Practice Guidelines in Oncology (NCCN Guidelines™) are available for iPhone and Android.

- Skyscape has multiple free CPGs available for download and also has 150+ fee-based CPGs. For example, Pediatric Clinical Practice Guidelines & Policies provides access to more than 30 clinical practice guidelines and more than 380 policy statements, clinical reports and technical reports.[91]

- mTBI Pocket Guide provides evidence based information about traumatic brain injury (TBI) and is available on the Android Market.

Future Trends

The field of EBM continues to evolve. Methodologists continue to identify opportunities to improve our understanding and interpretation of research findings. We anticipate more standardization of reporting and more transparency. Two studies published in 2010 help refine our knowledge base:

- Trials are often stopped early when extreme benefits are noted in the intervention group. The rationale for stopping enrollments of participants is that it is "unethical" to continue randomizing patients to the placebo arm because we are depriving them from the benefits of the intervention. However, it was found that stopping trials early for benefit exaggerates treatment effect by more than 30%; simply because we are stopping the trial at a point of extreme benefit that is clearly made extreme by chance. Such exaggeration leads to the wrong conclusions by patients and physicians embarking on comparing the pros and cons of a treatment and also leads to the wrong decisions by policymakers. In fact, stopping early may be unethical from a societal and individual point of view.[92]

- The second recent advancement in methodology relates to the finding that authors who have financial affiliation with the industry are three times more likely to make statements that are favorable to the sponsored interventions. It is very plausible that this bias is subconscious and unintentional; nevertheless, as readers of the literature, we should recognize the potential and implications of this bias.[93]

Key Points

- Evidence Based Medicine (EBM) is the academic pursuit of the best available answer to a clinical question

- The two fundamental principles of EBM are: (1) a hierarchy of evidence exists (i.e., not all evidence is equal) and (2) evidence alone is insufficient for medical decision making. It should rather be complemented by patient's values, preferences and circumstances.

- Health information technology will hopefully improve medical quality, which is primarily based on EBM.

- There are multiple limitations of both EBM and the medical literature.

- The average clinician should have a basic understanding of EBM and know how to find answers using EBM resources.

- Clinical Practice Guidelines (CPGs), based on evidence based medicine, are the roadmap to standardize medical care.

- CPGs are valuable for chronic disease management or as a means to measure quality of care.

Conclusion

Knowledge of evidence based medicine is important if you are involved with patient care, quality of care issues or research. Rapid access to a variety of online EBM resources has changed how we practice medicine. In spite of its shortcomings, an evidence based approach helps healthcare workers find the best possible answers. Busy clinicians are likely to choose commercial high quality resources, while academic clinicians are likely to select true EBM resources. Ultimately, EBM tools and resources will be integrated with electronic health records as part of clinical decision support.

The jury is out regarding the impact of CPGs on physician behavior or patient outcomes. Busy clinicians are slow to accept new information, including CPGs. Whether embedding CPGs into EHRs will result in significant changes in behavior that will consistently result in improved quality, patient safety or cost savings remains to be seen. It is also unknown if linking CPGs to better reimbursement (pay-for-performance) will result in a higher level of acceptance. While we are determining how to optimally improve healthcare with CPGs, most authorities agree that CPGs need to be concise, practical and accessible at the point of care. Every attempt should be made to make them electronic and integrated into the workflow of clinicians.

Acknowledgement

We thank Dr. Brian Haynes and Dr. Ramón Puchades for their contributions to this chapter.

References

1. Evidence Based Medicine: What it is, what it isn't. http://www.cebm.net/ebm_is_isnt.asp (Accessed September 3 2005)

2. Crossing the Quality Chasm: A new health system for the 21th century (2001) The National Academies Press http://www.nap.edu/books/0309072808/html/ (Accessed September 3 2005)

3.   MedlinePlus http://www.nlm.nih.gov/medlineplus/ency/article/003108.htm (Accessed September 3 2006)

4.   Teece I, Crawford I. Bed rest after spinal puncture. BMJ http://emj.bmjjournals.com/cgi/content/full/19/5/432 (Accessed Aug 24 2006)

5.   Guyatt GH. Evidence based medicine. ACP J Club 1991;114:A16

6.   Evidence Based Medicine. Wikipedia. http://en.wikipedia.org/wiki/Evidence_based-medicine (Accessed September 5 2005)

7.   Levin A. The Cochrane Collaboration Ann of Int Med 2001;135:309-312

8.   Medical Research Council. Streptomycin treatment of pulmonary tuberculosis. BMJ 1948;2:769-82

9.   Gandhi GY, Murad MH, Fujiyoshi A, et al. Patient-important outcomes in registered diabetes trials. JAMA. Jun 4 2008;299 (21):2543-2549.

10.   Davis DA et al. Changing physician performance. A systematic review of the effect of continuing medical education strategies. JAMA 1995; 274: 700-1.

11.   Sibley JC. A randomized trial of continuing medical education. N Engl J Med 1982; 306: 511-5.

12.   Fordis M et al. Comparison of the Instructional Efficacy of Internet-Based CME with Live Interactive CME Workshops. JAMA 2005;294:1043-1051

13.   National Library of Medicine PICO http://askmedline.nlm.nih.gov/ask/pico.php (Accessed September 7 2005)

14.   Centre for Evidence Based Medicine http://www.cebm.net/learning_ebm.asp (Accessed September 7 2007)

15.   Haynes RB. Of studies, syntheses, synopses and systems: the "4S evolution of services for finding the best evidence." ACP J Club 2001;134: A11-13

16.   The well built clinical question. University of North Carolina Library http://www.hsl.unc.edu/Services/Tutorials/EBM/Supplements/QuestionSupplement.htm (Accessed September 20 2005)

17.   Guyatt GH, Oxman AD, Vist G, Kunz R, Falck-Ytter Y, Alonso-Coello P, Schünemann HJ. The GRADE Working Group. Rating quality of evidence and strength of recommendations GRADE: an emerging consensus on rating quality of evidence and strength of recommendations. BMJ 2008;336:924-926

18.   Buitrago-Lopez A, Sanderson J, Johnson L, Warnakula S, Wood A, Di Angelantonio E, Franco OH. Chocolate consumption and cardiometabolic disorders: systematic review and meta-analysis. BMJ. 2011 Aug 26;343:d4488. doi: 10.1136/bmj.d4488.

19.   Henley E. Understanding the Risks of Medical Interventions Fam Pract Man May 2000;59-60

20.   Nouvo J, Melnikow J, Chang D. Reporting the Number Needed to Treat and Absolute Risk Reduction in Randomized Controlled Trials JAMA 2002;287:2813-2814

21.   Lacy CR et al. Impact of Presentation of Research Results on Likelihood of Prescribing Medications to Patients with Left Ventricular Dysfunction. Am J Card 2001;87:203-207

22.   Collaborative Atorvastatin Diabetes Study (CARDS) Lancet 2004;364:685-96

23.   Maharaj R. Adding cost to number needed to treat: the COPE statistic. Evidence Based Medicine 2007;12:101-102

24.   Haynes RB. Where's the Meat in Clinical Journals? ACP Journal Club Nov/Dec 1993: A-22-23

25. Als-Neilsen B, Chen W, Gluud C, Kjaergard LL. Association of Funding and Conclusions in Randomized Drug Trials. JAMA 2003; 290:921-928

26. Ioannidis JPA. Contradicted and Initially Stronger Effects in Highly Cited Clinical Research JAMA 2005;294:218-228

27. Kranish M. Flaws are found in validating medical studies The Boston Globe August 15 2005 http://www.boston.com/news/nation/articles/2005/08/15/flaws_are_found_in_validating_medic al_studies/ (Accessed June 12 2007)

28. Altman DG. Poor Quality Medical Research: What can journals do? JAMA 2002;287:2765-2767

29. Gross CP et al. Reporting the Recruitment Process in Clinical Trials: Who are these Patients and how did they get there? Ann of Int Med 2002;137:10-16

30. Moher D, Dulgerg CS, Wells GA. Statistical Power, sample size and their reporting in randomized controlled trials JAMA 1994;22:1220-1224

31. Evidence Based Medicine. Clinfowiki. www.informatics-review.com/wiki/index.php/EBM (Accessed June 19 2007)

32. Straus SE, McAlister FA Evidence Based Medicine: a commentary on common criticisms Can Med Assoc J 2000;163:837-841

33. Dobbie AE et al. What Evidence Supports Teaching Evidence Based Medicine? Acad Med 2000;75:1184-1185

34. Sackett DL, Staus SE. Finding and Applying Evidence During Clinical Rounds: The "Evidence Cart" JAMA 1998;280:1336-1338

35. Grandage K et al. When less is more: a practical approach to searching for evidence based answers. J Med Libr Assoc 90(3) July 2002

36. Schilling LM et al. Resident's Patient Specific Clinical Questions: Opportunities for Evidence Based Learning Acad Med 2005;80:51-56

37. Davidoff F, Florance V.The Informationist: A New Health Profession? Ann of Int Med 2000;132:996-999

38. Giuse NB et al. Clinical medical librarianship: the Vanderbilt experience Bull Med Libr Assoc 1998;86:412-416

39. Westberg EE, Randolph AM. The Basis for Using the Internet to Support the Information Needs of Primary Care JAMIA 1999;6:6-25

40. Slawson DC, Shaughnessy AF, Bennett JH. Becoming a Medical Information Master: Feeling Good About Not Knowing Everything J of Fam Pract 1994;38:505-513

41. Shaughnessy AF, Slawson DC and Bennett JH. Becoming an Information Master: A Guidebook to the Medical Information Jungle J of Fam Pract 1994;39:489-499

42. Clinical Evidence www.clinicalevidence.com (Accessed November 11 2011)

43. Cochrane Library http://www3.interscience.wiley.com/cgi-bin/mrwhome/106568753/HELP_Cochrane.html (Accessed November 11 2011)

44. Cochrane Review http://www.cochrane.org/reviews/index.htm (Accessed November 11 2011)

45. EvidenceUpdates http://plus.mcmaster.ca/evidenceupdates (Accessed November 11 2011)

46. ACP Journal Club. http://plus.mcmaster.ca/acpjc (Accessed November 12 2011)

47. Practical Pointers for Primary Care www.practicalpointers.org (Accessed November 12 2011)

48. Trip Database www.tripdatabase.com (Accessed November 12 2011)

49. Evidence Based On-call. www.eboncall.org (Accessed November 12 2011)

50. OVID http://gateway.ovid.com (Accessed November 12 2011)

51. SUMSearch http://sumsearch.uthscsa.edu (Accessed November 12 2011)

52. Bandolier http://www.medicine.ox.ac.uk/bandolier/ (Accessed November 12 2011)

53. Centre for Evidence Based Medicine www.cebm.net (Accessed November 12 2011)

54. Institute of Medicine (1990). Clinical Practice Guidelines: Directions for a New Program. Field MJ and Lohr KN (eds). Washington DC. National Academy Press. Page 38

55. O'Connor P. Adding Value to Evidence Based Clinical Guidelines JAMA 2005;294:741-743

56. Gaede P. Multifactorial intervention and cardiovascular disease in patients with type 2 diabetes NEJM 2003;348:383-393

57. McGlynn E . Quality of Health Care Delivered to Adults in the US RAND Health Study NEJM Jun 26 2003

58. Crossing the Quality Chasm: A new Health System for the 21th century 2001. IOM. http://darwin.nap.edu/books/0309072808/html/227.html (Accessed March 5 2006)

59. Abookire SA, Karson AS, Fiskio J, Bates DW. Use and monitoring of "statin" lipid-lowering drugs compared with guidelines Arch Int Med 2001;161:2626-7

60. Oxman A, Flottorp S. An overview of strategies to promote implementation of evidence based health care. In: Silagy C, Haines A, eds Evidence based practice in primary care, 2nd ed. London: BMJ books 2001

61. Yarnall KSH, Pollak KL, Østbye T et al. Primary Care: Is There Enough Time for Prevention? Am J Pub Health 2003;93 (4):635-641

62. Montori VM, Fernandez-Balsells M. Glycemic control in type 2 diabetes: time for an evidence based about face? Ann Intern Med 2009;150 (11):803-808

63. Persell SD, Dolan NC, Friesema EM et al. Frequency of Inappropriate Medical Exceptions to Quality Measures. Ann Intern Med 2010;152:225-231

64. Grol R, Grimshaw J. From Best evidence to best practice: effective implementation of change in patient's care Lancet 2003;362:1225-30

65. Wolff M, Bower DJ, Marabella AM, Casanova JE. US Family Physicians experiences with practice guidelines. Fam Med 1998;30:117-121

66. Zielstorff RD. Online Practice Guidelines JAMIA 1998;5:227-236

67. National Guideline Clearinghouse www.guideline.gov (Accessed November 11 2011)

68. Choudry NK et al. Relationships between authors of clinical practice guidelines and the pharmaceutical industry JAMA 2002;287:612-7

69. Tricoci P, Allen JM, Kramer JM et al. Scientific Evidence Underlying the ACC/AHA Clinical Practice Guidelines JAMA 2009;301(8):831-841

70. Rosenberg SN, Shnaiden TL, Wegh AA et al. Supporting the Patient's Role in Guideline Compliance: A Controlled Study. Am J Manag Care 2008;14 (11):737-744

71. Stross JK. The educationally influential physician – Journal of Continuing Education Health Professionals 1996; 16: 167-172)

72. Shiffman RN, Shekelle P, Overhage JM et al. Standardized Reporting of Clinical Practice Guidelines: A Proposal form the Conference on Guideline Standardization. Ann Intern Med 2003;139:493-498

73. Zielstorff, RD. Online Practice Guidelines. Issues, Obstacles, and Future Prospects. JAMIA 1998;5:227-236

74. Cardiac Clearance www.statcoder.com (Accessed October 29 2011)

75. ATP III Risk calculator http://hp2010.nhlbihin.net/atpiii/calculator.asp?usertype=prof

76. Frax Calculator http://www.shef.ac.uk/FRAX/ (Accessed October 29 2011)

77. Gail Breast Cancer Risk http://www.cancer.gov/bcrisktool/ (Accessed October 29 2011)

78. Stroke Risk Calculator http://www.stroke-education.com/calc/risk_calc.do (Accessed October 29 2011)

79. Stroke or death due to atrial fibrillation http://www.zunis.org/FHS%20Afib%20Risk%20Calculator.htm (Accessed October 29 2011)

80. Giuse N et al. Evolution of a Mature Clinical Informationist Model JAIMA 2005;12:249-255

81. Durieux P et al. A Clinical Decision Support System for Prevention of Venous Thromboembolism: Effect on Physician Behavior JAMA 2000;283:2816-2821

82. Tierney WM et al. Effects of Computerized Guidelines for Managing Heart Disease in Primary Care J Gen Int Med 2003;18:967-976

83. Murray et al. Failure of computerized treatment suggestions to improve health outcomes of outpatients with uncomplicated hypertension: results of a randomized controlled trial Pharmacotherapy 2004;3:324-37

84. Tierney et al. Can Computer Generated Evidence Based Care Suggestions Enhance Evidence Based Management of Asthma and Chronic Obstructive Pulmonary Disease? A Randomized Controlled Trial Health Serv Res 2005;40:477-97

85. Javitt JC et al. Using a Claims Data Based Sentinel System to Improve Compliance with Clinical Guidelines: Results of a Randomized Prospective Study Amer J of Man Care 2005;11:93-102

86. Welch, PW et al. Electronic Health Records in Four Community Physician Practices: Impact on Quality and Cost of Care. JAMIA 2007;14:320-328

87. National Guideline Clearing House www.guideline.gov (Accessed October 20 2011)

88. National Institute for Health and Clinical Excellence www.nice.org.uk (Accessed October 29 2011)

89. Agency for Health Care Research and Quality (AHRQ) http://www.effectivehealthcare.ahrq.gov/

90. Health Team Works www.healthteamworks.org (Accessed October 29 2011)

91. Skyscape www.skyscape.com (Accessed October 29 2011)

92. Bassler RD, Briel M, Murad MH et al. Stopping Randomized Trials Early for Benefit and Estimation of Treatment Effects: Systematic Review and Meta-Regression Analysis. JAMA 2010;303 (12):1180-7

93. Wang AT, McCoy CP, Murad MH. Association Between Affiliation and Position on Cardiovascular Risk with Rosiglitazone: Cross Sectional Systematic Review. BMJ 2010. March 18:340.c1344. doi:10.1136/bmj.c1344 (Accessed April 10 2010)

16

Disease Management and Disease Registries

ROBERT E. HOYT

ANN K. YOSHIHASHI

Learning Objectives

After reading this chapter the reader should be able to:

- Define the role of disease management in chronic disease

- Describe the need for rapid retrieval of patient and population statistics to manage patients with chronic diseases

- Compare and contrast the various disease registry formats including those that integrate with electronic health records

- Elaborate on how Meaningful Use objectives impact disease management and electronic health records

- Describe the interrelationships between disease registries, evidence based medicine and quality improvement programs

Introduction

Disease Management (DM) Programs are important for several reasons that will be pointed out in this chapter. First, we need enterprise-level approaches to disease management in order to evaluate, track and treat chronic diseases. The Institute of Medicine has stated that our strategy has been insufficient because "the current delivery system responds primarily to acute and urgent health care problems. Those with chronic conditions are better served by a systematic approach that emphasizes self-management, care planning with a multidisciplinary team and ongoing assessment and follow up."[1] A systematic approach implies the means to coordinate care and share information which requires information technology. Second, there is both a national and international rise in chronic diseases which is of great concern to governments trying to deal with rising healthcare costs. For this reason, disease management programs now exist in most developed countries. In the United States disease management is part of Meaningful Use (HITECH Act) and Accountable Care Organizations (Affordable Care Act) discussed in this and multiple other chapters.

In the next section we will define key terms that are important in understanding disease management, population health and public health.

Definitions

- Public Health: "the science and art of preventing disease, prolonging life and promoting health through the organized efforts and informed choices of society, organizations, public and private, communities and individuals."[2] Public health focuses on surveillance that includes tracking infectious disease epidemics, chronic diseases, bioterrorism and other events. For a more detailed discussion we refer readers to the chapter on public health informatics.

- Population Health: "the health outcomes of a group of individuals, including the distribution of such outcomes within the group."[3] Some authorities include disease, lifestyle, demand and condition management programs under population health.

- Disease Management (DM): "a systematic population based approach to identify persons at risk, intervene with a specific program of care and measure clinical and other outcomes."[4] DM focuses on specific diseases, e.g. diabetes.

- Lifestyle Management: focuses on personal risk factors (e.g. smoking)

- Demand Management: focuses on improved utilization (e.g. emergency room usage)

- Condition Management: focuses on temporary conditions (e.g. pregnancy)

- Patient Registry: "is an organized system that uses observational study methods to collect uniform data (clinical and other) to evaluate specified outcomes for a population defined by a particular disease, condition, or exposure, and that serves one or more predetermined scientific, clinical, or policy purposes". [5]

Disease Management Programs (DMPs)

The goal of all DMPs is to improve multiple patient outcomes: clinical, behavioral, financial, functional and quality of life outcomes.

Historically, DMPs were created in part because health maintenance organizations (HMOs) wanted to control the rising cost of chronic diseases. DMPs were established in the 1980s at Group Health of Puget Sound and Lovelace Health System in New Mexico and now are part of many large health care organizations. As an example, in a survey of over 1,000 healthcare organizations, disease registries were established with the following frequencies: diabetes (40.3%), asthma (31.2 %), heart failure (34.8 %) and depression (15.7 %).[6]

Chronic diseases affect about 20% of the general population yet account for 75% of health care spending. By the year 2030, 20% of the US population will be 65 or older. Chronic diseases are more likely to affect lower income populations who have limited access to medical care and limited insurance coverage. Figure 16.1 shows the predicted prevalence of chronic disease by year.[7]

The most common chronic diseases to be managed are heart failure, diabetes and asthma due to high prevalence and cost. Following close behind are obesity, hypertension, chronic renal failure and chronic obstructive lung disease (COPD).

Disease Management Program Participants

DMPs require a team approach as well as multiple internal and external partners interested in managing chronic diseases. The integration of multiple players is best demonstrated by the classic Chronic Care Model created by Dr. E.Wagner and the Macoll Institute for Healthcare Innovation. His model incorporates community resources, healthcare systems, information technology, patient participation and a disease management team.[8] The following are common examples of DMP participants:

- Quality Improvement Organizations (QIOs)

- State and Federal Governments (Medicare and Medicaid)

- Healthcare systems

- Physicians

- Employers

- Insurers

- Health Information Organizations (HIOs)

Figure16.1: Predicted chronic disease prevalence (millions) by year

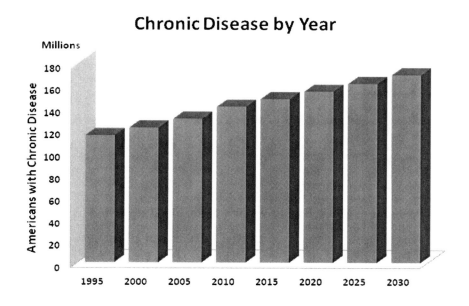

The Disease Management Approach

Establishment of a DM program usually involves the following questions and steps:

- Identifying a disease or condition and a target population e.g. type 2 diabetes in the uninsured.

- Determining if the problem is common enough or expensive enough to warrant a DM team. Is the disease or condition a high volume problem, a high cost problem or both?

- Defining the goal, e.g. decrease diabetic complications, decrease trips to the emergency room by asthmatics, etc.

- Determining if information systems already exist for the program. Data retrieval is easier if systems are already in place.

- Comparing local to national data, e.g. a local hospital has an annual readmission rate for heart failure of 55%; the national average is 40%.

- Reviewing existing clinical practice guidelines to see if they can be used or modified. In other words, don't re-invent the wheel.

- Determining if outcomes are clearly defined, measurable and meaningful.

- Evaluating patient self-management education; a very important aspect of disease management. Do web-based or mobile applications exist?

- Evaluating process and outcome measurements with eventual feedback to clinicians. One of the most effective ways to get buy-in by busy physicians is to show them how they are doing, compared to other similar physicians. The hospital management team also needs feedback.

- Emphasizing systems and populations, not individuals.

- Planning the necessary coordination among multiple services and agencies. [9]

For an example of how a university medical center improved compliance with discharge instructions for heart failure patients see the following case study on the next page.

The Role of Health Information Technology and DMPs

Disease management is discussed in a health informatics textbook because it is dependent on health information technology (HIT). Multiple interrelationships are also identified (see Figure 16.2). The following are examples of how HIT can assist DMPs:

- Automated data collection and analysis, e.g. using a clinical data warehouse (CDW).

- Clinical practice guidelines (CPGs) that are web-based or embedded into the electronic health records (EHRs).

- Disease registries that are part of EHRs.

- Telemonitoring of patients at home, e.g. recording weight, blood sugar and blood pressure.

- Patient tracking using a registry to track e.g. all type 2 diabetes or all patients with pacemakers.

- Using mobile technology so that patients can upload personal health data to a personal health record or patient portal using their smartphones.

- Using disease specific web sites so data can be uploaded and educational information acquired.

- Using health information exchanges to connect multiple healthcare workers on the DM team. This also permits aggregating data from an entire region or state and submitting quality reports to governmental agencies.

Figure 16.2: Disease Management Interrelationships (EBM - evidence based medicine, CPGs - clinical practice guidelines, EHRs - electronic health records, P4P - pay-for-performance)

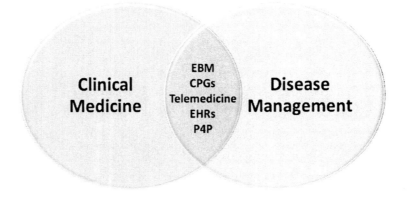

Case Study

Virginia Commonwealth University (VCU) Health System recognized that in order to reduce readmissions for heart failure (HF), the most common Medicare diagnosis-related group, it would need to improve compliance with six evidence based recommended discharge instructions. The instructions were in the areas of activity, diet, follow-up, medications, symptoms and weight monitoring. The challenge was to standardize and document compliance across the multiple hospital locations where HF patients might be discharged. They set the goal of 95% compliance with providing written HF instructions that included the six areas of patient education. The strategy was to embed clinical decision support (rules and alerts) as part of their enterprise electronic health record's (EHR) computerized physician order entry (CPOE). This required a multi-disciplinary approach that included clinicians, IT and the Office of Clinical Transformation. This effort complemented the existing 300 evidence based EHR order sets already in existence. The success rate from January 2006 to June 2010 is demonstrated in the graph below. http://www.himss.org/storiesofsuccess/docs/2011_submit/01_VCUHS-HF_Resubmit.pdf

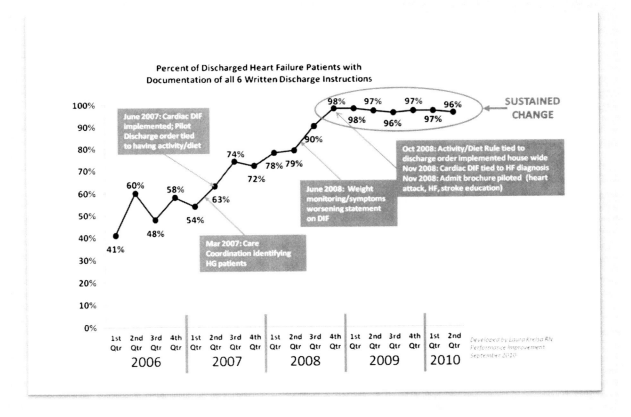

Disease Management and the US Federal Government

According to the Centers for Medicare and Medicaid (CMS) their costs account for about one-third of national health expenditures so those programs are constantly looking for ways to improve quality and reduce costs. A quote from the CMS web site: "About 14% of Medicare beneficiaries have congestive heart failure but they account for 43% of Medicare spending. About 18% of Medicare beneficiaries have diabetes, yet they account for 32% of Medicare spending. By better managing and coordinating the care of these beneficiaries, the new Medicare initiatives will help reduce health risks, improve quality of life, and provide savings to the program and the beneficiaries."[10]

CMS has created 10 pilot programs to see if disease management can save the government money over a three year period (phase I). The Chronic Care Improvement Program (part of the Medicare Modernization Act of 2003) is now known as the Medical Health Support Program. Companies involved will not get paid for disease management unless they can show a total savings of 5% compared to a control group. Companies that can demonstrate improved outcomes are asked to participate in phase II and will likely tackle diabetes or heart failure. The companies selected were: American Healthways, XL Health, Health Dialog Services, LifeMasters and McKesson Health Solutions. All participants will need robust information technology to succeed. As of November 2011 there has been no final report.[11]

The Affordable Care Act addressed the issue of chronic disease management by establishing several initiatives e.g., the accountable care organization model that is discussed in detail in the chapter on quality improvement strategies.

There is the expectation by the federal government that health IT, in particular EHRs, will result in better management and reporting of chronic diseases. For that reason, disease management reporting is part of Meaningful Use. There is a concern that that many EHRs are not capable of sending robust reports and government organizations such as Medicare/Medicaid are not ready to receive an avalanche of quality reports. To improve population health/disease management reporting, the Office of the National Coordinator released a free open source (Apache 2.0 license) population health reporting tool (popHealth) in early 2010. The goal of this tool is to allow for easier submission of quality reports to public health organizations. In addition, it will allow clinicians to create new ad hoc reports and perform their own population health analyses. Importantly, this tool integrates with EHRs because it complies with multiple data standards (CCD and CCR) and integrates with open source CONNECT, discussed in the chapters on health information exchange and data standards. The program was certified as a module for Meaningful Use. The popHealth application runs on JRuby or the Ruby programming language atop the Java Virtual Machine (JVM). Figure 16.3 shows how this application would run within the network of the user and generate quality reports. Figure 16.4 shows a screenshot of a typical quality report. On the left is a disease and condition menu demonstrating overall patient compliance with goals such as LDL cholesterol under 100. On the right the user can select gender or age to analyze the data further.[12]

Current Knowledge

The following are recent articles from the medical literature evaluating the impact of disease management:

- A study in the Journal of the American Medical Association (JAMA) demonstrated that 32 of 39 interventions showed improvement in at least one process or outcome measurement for diabetic patients; 18 of 27 studies involving three chronic conditions also demonstrated lower health care costs and/or lower utilization of services.[13]

- A comprehensive DM program for African-American diabetics showed large reductions in amputations, hospitalizations, emergency room visits and missed work days with an aggressive foot care program.[14]

Figure 16.3: popHealth Application (Courtesy Project popHealth)

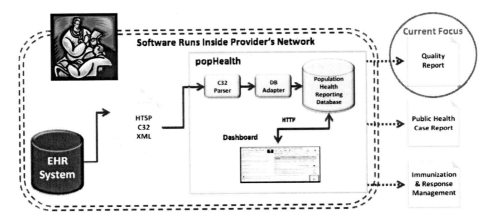

Figure 16.4: Patient Dashboard (Courtesy Project popHealth)

- HealthPartners Optimal Diabetes Care Impact: Program noted 400 fewer cases of retinopathy (eye damage) each year; 120 fewer amputations each year and 40 to 80 fewer myocardial infarctions (heart attacks) per year.[15]

- A systematic review/meta-analysis of DM programs for heart failure concluded that programs are effective in reducing admissions in elderly patients.[16]

- A DM program for myocardial infarctions reduced readmissions, emergency room visits and insurance claims.[17]

- A study of almost 800 chronically ill veterans using a web-based interactive disease dialogue telemedicine strategy at home was able to show a reduction in emergency room visits (40%), a reduction in hospital admissions (63%), a reduction in hospital bed days (60%), a reduction in nursing home admissions (64%) and a reduction in nursing home bed days (88%). Medication compliance improved as did compliance with national guidelines.[18]

- Grant et al. studied the effect of a specific diabetic web portal/personal health record that was integrated with an EHR. Although participants were more likely to have medications changed, their diabetic, blood pressure and cholesterol control was not better than a similar group of patients who had access to a standard web portal. One of the lessons learned was that patient participation in this trial was only 5% of their diabetic population. Also, poorly controlled diabetics were less likely to enroll in such a study.[19]

- Peikes et al. reviewed 15 disease management programs (Medicare Coordinated Care Demonstration) funded by Medicare. They studied 18,000 patients to determine if care coordination by nurses would improve chronic disease care or decrease costs. Only two of the 12 largest programs showed any statistically significant effects on hospital admissions. Expenditures were 8% to 41% higher in the intervention groups compared to controls. None of the programs generated net savings. They subsequently terminated all but two of the programs. They concluded that care coordinators (nurses) must interact with patients in person and not rely on telephones and technology. Also, coordinators must collaborate with the primary care clinicians to be successful. [20]

- Nephrologists (kidney specialists) working for Kaiser Permanente in Hawaii wanted to improve the number of referrals from generalists so they could intervene earlier for chronic kidney disease. Because they all used the same electronic health record, they were able to monitor kidney function in the entire population of 214,000 patients. Access to lab results, clinical notes and secure messaging allowed the specialists to contact the generalists with advice and schedule consultations with themselves rather than waiting for the generalists. The end result was the decrease in late referrals from 32% to 12%. This was a good example of using a disease registry to improve population health. Rather than rely on a computerized clinical decision support, the specialists provided the decision support. Actual patient outcomes such as whether kidney dialysis was delayed due to the specialists intervening early were not included. They outlined the key features of the EHR-based electronic population management database:

  o Access to comprehensive, current patient information

  o Database permitted risk stratification

  o Ability to annotate records to improve communication

  o Seamless integration of new data into the longitudinal record

  o Electronic messaging between specialists and generalists

- ○ Electronic alerts for deteriorating lab results
- ○ Generation of population level statistics
- ○ Ability to flag patient records by status [21]

- In another study from Kaiser Permanente they used their EHR to collect information on 650,000 individuals from 2002 to 2007. The EHR allowed them to easily note who had had a bone mineral density test (DEXA), who had a fracture and what meds the patients were on. Armed with this information they were able to show that hip fractures decreased 38%, DEXA testing increased 263% over the five years and the number of people on anti-osteoporosis drugs increased 153%. Again, population health is much easier with computable information obtained from robust EHR systems.[22]

In spite of some encouraging reports like those cited above, there are problems with the quality of the studies published thus far, such as lack of randomization or lack of a control group. In addition, many studies do not convincingly prove a reasonable return on investment.

Disease Registries

In the beginning of this chapter a patient registry is defined. Patient registries can track a variety of diseases and conditions so disease registries should be considered a type of patient registry. Patient registries serve several purposes: (1) Describing the natural history of disease (2) Determining the clinical impact or cost effectiveness of a program (3) Assessing the safety or harm of a treatment or approach (4) Measuring or improving the quality of care (5) Public health surveillance and (6) Disease control.[5]

Patient Registry Categories

Patient registries can be categorized as follows:

- Health Services Registries: used to track services such as hospitalizations, office visits, surgeries and infectious diseases.

- Disease/Condition Registries: used to track chronic diseases such as diabetes, heart failure and conditions such as pregnancy. Registries can also track rare diseases, e.g. alopecia areata or track resource intensive conditions, e.g. heart transplants.

- Product Registries: used to track patient safety-related concerns such as toxin exposure, certain medications, adverse drug events and devices, e.g. pacemakers.

- Combination Registries: quite possibly a patient might be in more than one registry such as coronary artery disease and coronary stent registries.[5]

An electronic disease registry is defined as " a software application for capturing, managing and providing access to condition specific information for a list of patients to support organized clinical care".[23] Stated another way; registries are tools that disease management programs use to track patients with chronic diseases or conditions, such as diabetes or smoking. As a result of this data DM programs can remind clinicians, nurses and patients to get lab work done and keep appointments. In addition, they can aggregate data to show, for example, the average hemoglobin A1c levels (blood test to measure blood sugar control) of a single patient or an entire clinic that could be useful for pay-for-performance (next chapter) programs (see Figure 16.5).

Figure 16.5: EHR-Disease registry that generates quality reports for reimbursement

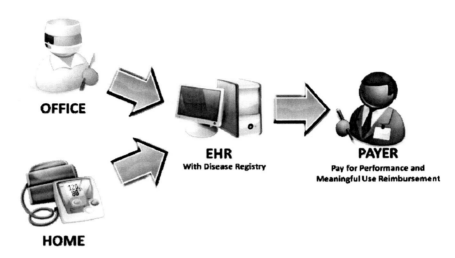

Disease registries can be populated through several mechanisms:

- Manual: data manually inputted onto paper or a computer database or spreadsheet or into a web-based program.

- Automatic: data automatically inputted into standalone software or web-based site using client-server software and integrated with, for example, a laboratory result program using LOINC and HL7 standards.

- Automated and integrated: data input, retrieval, tracking and graphing are all automatic and part of an electronic health record (EHR) or health information organization (HIO). This model is increasingly being adopted and is felt to have the greatest potential in DMP and pay-for-performance programs.

Large integrated delivery networks were the first healthcare organizations to develop sophisticated registries to measure and track diseases and other conditions. In the information box below we have presented the Kaiser Permanente total joint registry.[24] The Cleveland Clinic has a disease registry linked to their EHR, designed to collate, track and study patients who have chronic kidney disease (CKD). As of 2010 they had more than 50,000 patients enrolled.[25]

Total Joint Replacement Registry (TJRR)

In 2001 Kaiser-Permanente created a registry just for total joint replacement surgery, given the fact that Kaiser-Permanente surgeons perform 17,000 joint replacements each year. Electronic forms were created for data input that were integrated with the EHR. The registry is used for possible implant recalls and advisories, patient safety/quality improvement initiatives, to identify best practices and to conduct research. In addition to the TJRR, they have created four more orthopedic registries to monitor e.g. spine surgery. [23]

The ultimate solution will be to have universal adoption of EHRs that have robust disease registries. In this manner all fields are automatically populated with patient data, to include lab results, etc. (Figure 16.5). As discussed in chapter three, stage I Meaningful Use criteria included the requirement to generate patient lists for specific conditions to use for quality improvement, reduction of disparities and outreach. It also required outpatient quality reports and the ability to send reminders to patients for preventive care.[26]

Disease registries are not unique to the United States. A late 2011 article reported on the status of registries in Sweden, Australia, Denmark, United States and the United Kingdom. They identified many areas where improved patient care was associated with disease registries but causality can't be proven with this study design. It was clear that the registry was a valuable clinical tool but data transparency needed to take place as well as education of clinicians and patients, i.e. disease management. In the case of Sweden they had almost 90 government-supported registries established by medical sub-specialties so buy-in was not an issue. Importantly, it was the belief of the authors that disease registries are associated with significant cost savings globally. [27]

Disease Registry Limitations

Potential disease registry limitations were summarized in a 2010 monograph by the Robert Woods Johnson Foundation: (1) Standardizing data elements among disparate disease registries (2) Uniform method for patient identification (3) Assistance in linking registries with EHRs (4) Standardizing methodologies for statistical analysis (5) Ensuring high clinician participation (6) Guaranteeing registry sustainability (7) Clinical and administrative (claims) data should be combined in a registry (8) The need to manually input data for some registries (9) The need for accurate coding (10) The need for frequent updating and (11) The need for additional staff to maintain a registry.[28]

Disease Registry Cost

Approximately 50 disease registries exist that are free or fee-based. Cost is usually $500 to $600 annually per user for commercial registries. In general, free public registries have less functionality than commercial registries.

Disease Registry Resources

For an excellent in-depth review of 16 registries see Chronic Disease Registries: A Product Review by the California HealthCare Foundation.[29] They also review the IT tools used for chronic disease management. Five California foundations have combined resources to support a $4.5 million project known as "Tools for Quality" to test disease registries for the low income and underserved populations in their state. They have recruited 33 clinics thus far that will be paid on average about $40,000 to acquire and maintain disease registries.[30] The California HealthCare Foundation also has a 2008 monograph that compares electronic health records with disease management systems but does not offer specific examples or vendors.[31] A paper by Khan et al. discusses the current and future status of diabetic registries that has implications for other diseases.[32]

Disease Registry Examples

Chronic Disease Electronic Management Systems (CDEMS): This popular program is Microsoft Access-based and tracks diabetes and adult preventive health. The program is customizable and includes lab reminders for clinicians. The reports generated are also customizable and users have access to a web forum to discuss issues. A free add-on program inputs data automatically from several laboratory information systems (Quest, Labcorp, Dynacare and PAML). Shortcomings include the need to manually input data and access is limited to ten concurrent users (see Figure 16.6). [33]

Figure16.6: CDEMS Disease Registry (Courtesy Washington State Department of Health, Diabetes Prevention & Control Program and Centers for Disease Control, Diabetes Translation Division)

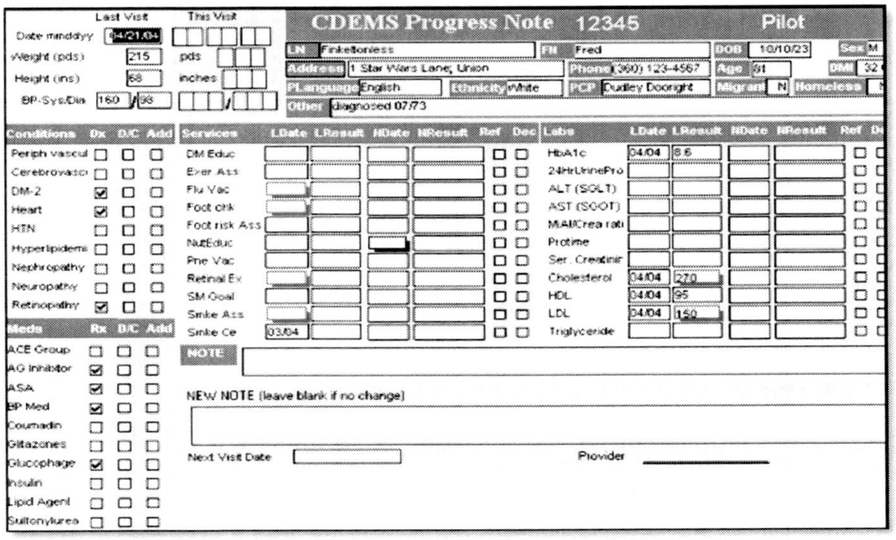

Population Health Navigator (PHN): Population Health Navigator is a program used by the Department of Defense (DOD) to track asthma, beta-blocker use following myocardial infarction, cardiovascular risk factors, breast cancer screening, cervical cancer screening, depression, diabetes, hypertension, COPD, hyperlipidemia, low back pain and high utilizers. Data can be analyzed by physician, clinic or hospital system. Data can be exported to MS Excel for data manipulation. Drawbacks include that it is not integrated into the electronic health record (AHLTA) and data is not available real time (about a 60 day delay). The site is secure and only available to DOD personnel with proper authority.[34]

Covisint DocSite Registry: One of the best known web-based commercial registries is Patient Registry by DocSite that will track multiple common diseases. It can be integrated with practice management software, EHRs and e-prescribing systems. Figure 16.7 displays a typical clinician report with lab results and due dates. Clinical practice guidelines can be embedded in the registry with the ability to make local modifications. Other features include HL7 links to input lab data, patient education, patient letter generation, and the ability to host data locally or on the DocSite server. They integrated the ACP-PIER resource into disease registries in 2008. The charge for this basic registry is $50/clinician/month. They also offer DocSite Enterprise for larger organizations. They continue to evolve and offer integration with pay-for-performance and medical home models.[35]

Figure 16.7: DocSite Patient Planner registry (Courtesy DocSite)

DocSite PatientPlanner

Service Activity: Provider Summary

Site: DocSite - Health Care Center
Provider: Dennis Smith

Measure Name	Patient Count	Panel Avg Value	% Met Pt Goal	% Met Pop Goal	% Overdue	Avg Times Checked	Panel Max	Panel Min
HbA1c	68	6.93	56%	56%	0%	5.13	13.00	5.00
HDL	68	45.40	76%	76%	0%	4.49	92.00	24.00
LDL	68	95.45	54%	54%	0%	4.49	251.00	25.00

Provider: Sam Admas

Measure Name	Patient Count	Panel Avg Value	% Met Pt Goal	% Met Pop Goal	% Overdue	Avg Times Checked	Panel Max	Panel Min
HbA1c	83	7.06	49%	49%	0%	5.16	13.00	4.00
HDL	70	41.97	69%	69%	0%	3.80	95.00	16.00
LDL	70	108.63	46%	46%	0%	3.80	207.00	24.00

Provider: Lucy Jones

Measure Name	Patient Count	Panel Avg Value	% Met Pt Goal	% Met Pop Goal	% Overdue	Avg Times Checked	Panel Max	Panel Min
HbA1c	44	7.16	50%	50%	0%	4.57	13.00	4.00
HDL	44	48.84	80%	80%	0%	3.59	104.00	21.00
LDL	44	104.77	45%	45%	0%	3.59	223.00	23.00

Remedy MD: This web-based site has more than 100 disease registries for clinicians and researchers. Application can capture, aggregate and analyze data from administrative, clinical and genetic information from EHRs as well as imaging applications and portals. The built-in OntologyManager™ supports all of the major standards such as LOINC, CPT, ICD, SnoMed, and UMLS. Registries are customizable.[36]

MAVIQ: This is the first open source for profit disease management application. Its CarePlus web-based module does the following: creates patient lists, manages chronic diseases, recalls patients for preventive health services, tracks patient care plans, has automatic reminders, patient portal for self-management and generates quality improvement and compliance reports. Data can be imported from EMR/PM and integrated with HL7 feed. Cost for about 100 patient recalls per day is $179.90 per month.[37]

Patient Electronic Care System (PECYS): This is a disease registry based on Wagner's Chronic Disease Model. It is used frequently by community health centers to manage chronic diseases. Clinical practice guidelines are embedded for decision support.[38]

The preceding section was an overview of the topic of patient registries. For additional reading we recommend an extensive 2010 monograph by AHRQ.[5]

Future Trends

Developed and developing nations are faced with escalating chronic diseases that are associated with high healthcare expenditures. Not only will there have to be healthcare reform to change the payment strategy, there will need to be more disease management programs. Health information technology will support coordinated care, patient tracking, data retrieval and outcome analysis. A myriad of technologies will need to be interoperable such as electronic health records, patient portals, health information exchanges, home telemedicine devices and mobile devices to provide coordinated disease management programs.

As larger organizations develop comprehensive disease management programs with their own data warehouses we can expect higher quality outcome studies. The goal which will be better data generating better medical practices, resulting in better patient outcomes.

Newer programs such as the Hospital Readmission Reduction Program will begin financially penalizing hospitals with higher than normal readmission rates for heart attacks, heart failure and pneumonia in FY 2013. Look for both a carrot and stick approach to disease management by the federal government. [39]

Future Meaningful Use requirements may force EHR vendors to have comprehensive and interoperable disease registries that include automated reporting.

Key Points

- Chronic diseases are on the rise in the USA and worldwide
- Chronic diseases are costly so disease management programs are commonplace, but benefits are controversial
- Disease management programs benefit from information technology by creating electronic disease registries
- Most EHRs have electronic disease management programs in order to meet Meaningful Use

Conclusion

For Disease Management programs to succeed there needs to be a mandate to improve the treatment of chronic disease coupled with financial support. Due to the rising costs of chronic diseases, CMS and managed care organizations are interested in new pilot programs. What must be shown is that DM programs improve patient outcomes and save money. It is much easier to show that programs improve processes such as lab tests drawn than improved patient outcomes, such as fewer heart attacks or strokes. The Congressional Budget Office in 2004 concluded that there was inadequate evidence that DM programs reduced healthcare spending and little has changed since then.[40] Bringing in more patients for preventive care will clearly increase medical costs, at least in the short run. The hope is that the costs will fall long term with preventive care.

Ultimately, all electronic health records will have comprehensive disease management features that will be customizable for clinicians and administrators. Data will be easier to retrieve and analyze in a real time mode and will be linked to reimbursement. Until that happens, however, we will rely on a variety of disease registries and disease management systems. Even with ARRA reimbursement of EHRs that have disease registries, it will be many years before we understand the true impact of electronic disease management and Meaningful Use reporting.

At this time, models that integrate human (nurse, physician, pharmacist, etc) involvement with technology seem to work better than purely technical solutions for disease management.

References

1. Crossing the Quality Chasm: A new health system for the 21th century. 2001. National Academies Press http://www.nap.edu/books/0309072808/html (Accessed March 5 2006)

2. Winslow, Charles-Edward Amory (1920 Jan 9). "The Untilled Fields of Public Health." Science 51 (1306): 23–33. doi:10.1126/science.51.1306.23. PMID 17838891. http://www.sciencemag.org/content/51/1306/23.long (Accessed October 4 2011)

3. Kindig D, Stoddart G. What is Population Health? Amer J Pub Health 2003;93(3):380-383

4. Epstein RS, Sherwood LM. 1996. From outcomes research to disease management: a guide for the perplexed. Ann Intern Med 124: 832-837

5. Registries for Evaluating Patient Outcomes: A User's Guide. AHRQ. 2010. Rockville, MD. AHRQ Pub. No.10-EHC049 http://www.effectivehealthcare.ahrq.gov/ehc/products/74/531/Registries%202nd%20ed%20final%20to%20Eisenberg%209-15-10.pdf (Accessed December 20 2011)

6. Casolino L, Gillies RR, Shortell SM, et al. External incentives, information technology, and organized processes to improve health care quality for patients with chronic diseases. JAMA. 2003;289: 434-41.

7. Wu, Shin-Yi and Green, Anthony. Projection of Chronic Illness Prevalence and Cost Inflation. RAND Corporation, October 2000

8. Chronic Care Model http://www.improvingchroniccare.org/change/model/components.html (Accessed March 2 2006)

9. Disease Management. Care Continuum. http://www.carecontinuum.org/dm_definition.asp (Accessed October 1 2011)

10. Xu S. Advancing Return on Investment Analysis for Electronic Health Investment. JHIM 2007;21:32-39

11. Centers for Medicare and Medicaid Services http://www3.cms.hhs.gov/apps/media/press/release.asp?Counter=1274 (Accessed March 2 2006)

12. Project popHealth http://projectpophealth.org (Accessed October 30 2011)

13. Bodenheimer T, Wagner, E H, Grumbach K. Improving Primary Care for Patients With Chronic Illness: The Chronic Care Model, Part 2 JAMA 2002;288:1909-1914

14. Patout CA et al. Effectiveness of a comprehensive diabetes lower extremity amputation prevention program in a predominately low income African-American population Diabetes Care 2000;23:1339-1342

15. HealthPartners. Dr Gail Amundsen (personal communication, August 2006)

16. Gonseth J et al. The effectiveness of disease management programmes in reducing hospital admissions in older patients with heart failure: a systematic review and meta-analysis of published reports Eur Heart Journal 2004;25:150-95

17. Young W et al. A disease management program reduced hospital readmission days after myocardial infarction CMAJ 2003;169:905-10

18. Meyer, M, Kobb R, Ryan P. Virtually Healthy: Chronic Disease Management in the Home. Disease Management 2002;5 (2):87-94

19. Grant RW, Wald JS, Schnipper JL et al. Practice-Linked Online Personal Health Records for Type 2 Diabetes Mellitus. Arch Intern Med 2008;168(16):1776-1782

20. Peikes D, Chen A, Schore J et al. Effects of Care Coordination on Hospitalization, Quality of Care, and Health Care Expenditures Among Medicare Beneficiaries. JAMA 2009;301(6):603-618

21. Lee BJ, Forbes K. The role of specialists in managing the health of populations with chronic illness: the example of chronic kidney disease. BMJ 2009;339:b2395

22. Dell RM, Greene D, Anderson D et al. Osteoporosis Disease Management: What Every Orthopedic Surgeon Should Know. J Bone Joint Surg Am 2009;91Suppl 6:79-86

23. Using Computerized Registries in Chronic Disease http://stage.chcf.org/documents/chronicdisease/ComputerizedRegistriesInChronicDisease.pdf (Accessed December 15 2011)

24. Paxton EW, Inacio MC, Khatod M et al. Kaiser Permanente National Total Joint Replacement Registry: aligning operations with information technology. Clin Orthop Relat Res. 2010 Oct;468(10):2646-2663

25. Navaneethan SD, Jolly SE, Schold JD et al. Development and Validation of an Electronic Health Record-Based Chronic Kidney Disease Registry. 2010. Clin J Am Soc Nephrol 5:2010. Doi:10.2215/CJN.04230510 (Accessed September 30 2011)

26. Proposed Rules. Federal Register. Vol 75, No. 8. January 13 2010 (Accessed April 11 2010)

27. Larsson S, Lawyer P, Garellick G et al. Use of 13 Disease Registries in 5 Countries Demonstrates The Potential to Use Outcome Data to Improve Health Care's Value. 2011. DOI: 10.1377/hithaff.2011.0762 (Accessed December 8 2011)

28. How Registries Can Help Performance Measurement Improve Care. White Paper. June 2010. www.hospitalqualityalliance.org/.../files/Final%20Registries%20paper.pdf (Accessed July 1 2010)

29. Chronic Disease Registries: A Product Review May 2004 www.chcf.org (Accessed March 5 2006)

30. Better Chronic Disease Care Through Technology: Health Care Foundations Unveil $4.5 Million Program June 11 2008 www.chcf.org (Accessed June 18 2009)

31. Electronic Health Records versus Chronic Disease Management Systems: A Quick Comparison. California HealthCare Foundation. March 2008. www.chcf.org (Accessed April 11 2010)

32. Khan L, Mincemoyer S, Gabbay RA. Diabetes Registries: Where We Are and Where Are We Headed? Diab Tech & Ther 2009;11 (4): 255-262

33. Chronic Disease Electronic Management Systems www.cdems.com (Accessed October 30 2011)

34. Navy & Marine Corps Public Health Center. http://www-nmcphc.med.navy.mil/Data_Statistics/Clinical_Epidemiology/pophealthnav.aspx (Accessed October 30 2011)

35. Covisint DocSite http://www.covisint.com/web/guest/healthcare/docsite (Accessed October 30 2011)

36. RemedyMD www.remedymd.com (Accessed October 30 2011)

37. MAVIQ www.maviq.com (Accessed October 30 2011)

38. PECYS. Aristos Group. www.aristos.com (Accessed October 30 2011)

39. Hospital Readmissions Reduction Program. www.aamc.org (Accessed January 1 2012)

40. Congressional Budget Office http://www.cbo.gov/showdoc.cfm?index=5909&sequence=0 (Accessed March 5 2006)

17

Quality Improvement Strategies

ROBERT E. HOYT

RONALD G. GIMBEL

Learning Objectives

After reading this chapter the reader should be able to:

- Define quality medical care and how it relates to patient safety

- State the goals of quality improvement (QI) programs

- List the components of the Quality Improvement Roadmap and National Quality Strategy

- Describe how health information technology (HIT) can support quality improvement

- List several quality improvement programs sponsored by the Centers for Medicare and Medicaid Services (CMS)

- Compare and contrast the patient centered medical home and accountable care organization models and how they are supported by HIT

- List the concerns and limitations of current QI programs for the average clinician

Introduction

When compared to other developed countries medical care in the United States is expensive, accounting for about 16% of gross national product (GNP), and is not associated with improved longevity.[1] A 2011 study by Nolte and McKee looked at preventable mortality in 16 developed countries and found that there was improvement in all countries, but the least improvement occurred in the United States.[2] Another study comparing medical quality in developed countries noted that the United States physicians reported: the highest percent (58%) of patients claiming difficulty affording medication, the lowest after-hours support (29%) for patients, last place in use of electronic health records (EHRs) and one of the lowest rates in use of teams to treat chronic diseases.[3] However, as result of reimbursement for Meaningful Use of electronic health records by Medicare/Medicaid, EHR adoption is catching up (see chapter 3 on EHRs).

While it is beyond the scope of this book to discuss all the factors that impact the quality of healthcare in the United States we will outline some of the more important factors:

- The US health care system is fragmented and poorly organized for improvement. Unlike an integrated delivery network such as Kaiser Permanente, most of the country is based on small independent medical practices that receive reimbursement based on fee-for-service.

- According to the Institute of Medicine (IOM), health care in the U.S. has experienced the growing complexity of science and technology yet has not been able to fully exploit the revolution in health information technology.

- There has been an increase in chronic conditions, e.g. obesity, diabetes and heart failure, and health care has poorly designed delivery systems that are not organized around quality and patient safety. [4]

- The Agency for Healthcare Research and Quality (AHRQ) has demonstrated that there is too much variation in health care when comparing states e.g. coronary angiography rates. [5]

- A well publicized 2003 Rand study suggested that only 55% of Americans received recommended care.[6] It should be pointed out, however, that the methodological approach for the study has been challenged and according to a follow-on study it would take, on average, 7.4 hours daily for the average physician to comply with all recommendations for preventive care.[7]

The bottom line is the federal agencies involved with health care have been extremely concerned about the high cost of health care, the less than optimal health care delivered, and sub-optimal patient safety. As a result, agencies seek new non-traditional health care delivery and reimbursement models aimed at incentivizing quality and reducing variation in outcomes; select strategies will be addressed in the remainder of the chapter. An example is the Centers for Medicare and Medicaid Services' (CMS) *Quality Improvement Roadmap* where the agency espouses a simple vision "The right care for every person every time". The Roadmap lists six criteria of the right health care, adopted from the Institute of Medicine's *Crossing the Quality Chasm*:

- Safe: care does not harm patients

- Effective: care prevents disease and complications and minimizes suffering, disability and death

- Efficient: patients receive care without waste

- Patient centered: care is coordinated and continuous; patients are informed and educated and involved in decision making

- Timely: patients and staff do not experience unwanted delay

- Equitable: care is equal, regardless of race, language, personal resources, diagnosis or condition

The core strategies of the Quality Improvement Roadmap can be summarized as follows:

- Publish quality measurements and information: Use the same performance measures among all health care organizations and select those that are the most evidence based

- Pay-for-performance: Principles are explained later in this chapter

- Promote health information technology: Includes the adoption of electronic health records, e-prescribing and health information exchanges

- Work through partnerships: Select national, federal, and civilian quality-oriented partners (e.g. Agency for Healthcare Research and Quality, National Quality Forum, American Health Quality Association and National Committee on Quality Assurance)

- Improve access to better treatments: Accelerate the availability and effective use of the best treatments [8]

Although this vision derives from the Institute of Medicine (IOM), it has been incorporated by most federal, state and civilian healthcare organizations. To accomplish this vision organizations have developed multiple quality improvement strategies (e.g. pay-for-performance, care coordination, patient safety initiatives, e-prescribing, electronic health records, quality performance reporting and clinical practice guidelines). All of these are discussed in detail in other chapters.

In early 2011 the Department of Health and Human Services announced the National Strategy for Quality Improvement in Health Care (National Quality Strategy) that was mandated by the Affordable Care Act.

Private and public partners (e.g. AHRQ) will carry out the strategy. More than 300 organizations provided input into the creation of the Strategy. Further Strategy details such as pilot initiatives are available. The three major goals of the National Quality Strategy:

- Better care by improving quality and making healthcare more patient-centered, reliable, accessible and safe

- Healthy people and communities by improving interventions that address behavioral, social and environmental health determinants

- Affordable care by reducing the cost of healthcare for individuals, families, employers and government

Six priorities will help achieve these aims:

- Making care safer by reducing harm

- Ensuring that individuals and families are engaged in their care

- Promoting effective communication and coordination of care

- Promoting effective prevention and treatment practices for the leading causes of mortality, starting with cardiovascular disease

- Working with communities to promote wide use of best practices to enable healthy living

- Making quality care more affordable by spreading new health care deliver models [9]

Several of these priorities will be facilitated by health information technologies such as electronic health records, clinical decision support, personal health records, health information exchanges, disease registries; all discussed in detail in other chapters.

Quality Improvement Strategies

Pay-for-Performance

We are in the process of seeing newer payment and delivery models but we have little data to analyze at this point. One strategy known as pay-for-performance (P4P) has captured attention and funding. The Centers for Medicare and Medicaid Services define P4P as a "quality improvement and reimbursement methodology aimed at changing current payment structure which primarily reimburses based on the number of services provided regardless of outcome. P4P attempts to introduce market forces and competition to promote payment for quality, access, efficiency and successful outcomes".[8]

There have been numerous studies since the IOM classic *Crossing the Quality Chasm* report that confirm we are not getting our money's worth from American medicine. As an example, a study by the Commonwealth Fund demonstrated that the quality of care delivered to Medicare recipients was not related to the amount of money spent.[10] The IOM has been consistently critical of the variation in care delivery and outcomes as well as serious patient safety issues (see patient safety chapter). As a result, they have repeatedly called for an increase in payments to clinicians who offer higher quality care. These concerns about "value-based care" are further aggravated by the fact that the United States has an annual $2.3 trillion dollar health care price tag that continues to rise each year. The IOM released *Rewarding Provider Performance: Aligning Incentives in Medicare* report in 2006 that called for a change in reimbursement that would result in higher quality of care delivered.[11]

Statements by organizations such as the IOM have helped support the notion that we need major changes in the field of medicine, to include how we determine reimbursement for care. P4P (also known as value-based purchasing) has gained traction in the United States in a surprisingly short period of time. The momentum may in part be due to the 2004 statement made by Mark McClellan, administrator for the Centers for Medicare and Medicaid Services in the Wall Street Journal:

> "In the next five to ten years, pay-for-performance based compensation could account for 20-30% of what the federal programs pay providers."[12]

As a further example of the rise of P4P programs, Rosenthal et al. in a 2006 article examined the incidence of P4P programs in 252 Health Maintenance Organizations (HMOs). They determined that over half had P4P programs; 90% of programs were for physicians and 38% were for hospitals.[13]

Table 17.1 shows the types of data, clinical scenarios and examples of information technology used in P4P programs.

Table 17.1: Types of data, clinical scenarios and IT support for P4P programs (EHR = Electronic health record, HIE = health information exchange)

Types of data	Clinical Scenarios	IT Support
Utilization data	Emergency room visits	Data repositories, EHRs, HIE
Clinical quality	Women who have had mammograms	Patient lists, disease registries, EHRs, HIE
Patient satisfaction	Percent of patients who would recommend their primary care manager	Online surveys
Patient safety	Percent of patients questioned about allergic reactions	EHRs, e-prescribing module

In spite of the potential of information technology to improve quality, numerous issues exist. Most EHRs are not ready for generating P4P type reports. Ideally, data would be automatically generated from the EHR if the data were inputted into data fields via structured templates rather than free text. Unfortunately, most clinical notes are not written using structured templates and problem summary lists are not updated often enough to be a reliable data source. Perhaps natural language processing (NLP) will eventually be able to scan a dictated patient encounter and automatically submit a P4P report as well as a coding level. Lab results are often easier to report because they are coded by data standards such as Logical Observation Identifiers Names and Codes (LOINC). Similarly, the federal government is not yet ready to receive voluminous quality reports from EHRs as part of Meaningful Use requirements. A 2007 article by Baker on automated review of quality measures for heart failure using an EHR concluded that the current system was insufficient, e.g. it lacked the ability to tell why a drug was not started or why it was stopped. Chart reviews were the only way to tell why recommended medications were not used or were discontinued.[14] Furthermore, there is a need to identify acute versus chronic problems and active versus inactive problems in EHRs. Until EHRs are universal, organizations must consider a transitional plan like disease registries and disease flow sheets. Health care systems may benefit from health information exchange that includes a central data repository (CDR) or data warehouse with a rules engine.[15] Data could be pushed or pulled from the CDR for monthly reports. Further information about the role of HIT in quality improvement can be found in the chapters on EHRs, medical data, disease management and HIE.

In order for P4P to be well received there needed to be a set of outpatient clinical performance measures that would be accepted by clinicians.[16, 17] Many of the early P4P projects were actually pay-for-reporting that looked at processes and not pay-for-performance that focused on clinical outcomes. Process measurements check to see if a test was done and not the actual result. This typically allows for easy retrieval of data using administrative or insurance claims data. Organizations such as the National Quality Forum are developing medical quality measures that will be used in all quality improvement programs.

CMS has a game plan over the next three to five years to transition from a passive fee-for-service reimbursement plan to a proactive value-based purchasing model. Much of the innovation in health care delivery will likely be realized through the new CMS Innovation initiative.[18] The Innovation Center was created under the Affordable Care Act, in order to "test innovative payment and service delivery models to reduce program expenditures, while preserving or enhancing the quality of care" for those who get Medicare, Medicaid or CHIP benefits. The Center received $10 billion in direct funding in fiscal years 2011 through 2019 to support this mission. Through the Innovation Center, CMS is working to transform from a claims payer in a fragmented care system into a partner that helps achieve better value for our health care dollars.

Meaningful Use (MU)

MU was discussed in detail in the chapter on EHRs with the core and menu objectives posted in the appendix at the end of the chapter. MU is mentioned in this section because EHR adoption is pivotal to health care reform and quality improvement in many areas. Health care data must be digital and discrete in order to be shared and analyzed so this is impossible with paper records. In order to be reimbursed for using a certified EHR, an eligible physician would have to demonstrate (and prove) MU. Three out of five of the overarching goals of MU have these implications: (1) improve quality, safety, efficiency and reduce health disparities, (2) improve care coordination, and (3) improve population and public health. These goals are achieved through EHR tools such as e-prescribing, disease registries, CPOE, clinical decision support, quality reports, HIE and electronic patient summaries.

The Patient-Centered Medical Home (PCMH) Model

This model is intended to be an improved model for health care quality and delivery, payment reform, chronic disease management and practice innovation. It is based on the relationship between the patient and their primary care physician (PCP). It is up to the PCP and his/her team to manage and coordinate chronic diseases with the goal of keeping the patient healthy and at home. Although the concept has been around since 1967, it was promoted by major medical associations in 2007. Since then the concept has been embraced by private insurers [19], Medicare [20], and the Department of Defense.[21] Part of the concept for PCMH is technology support using disease registries, EHRs, personal health records, e-prescribing, patient portals, secure messaging, e-visits, HIEs and tele-home care. In this model, practices would have to handle more walk-ins and same-day appointments. CMS has demonstration projects in eight states and early evidence suggests a positive effect on cost and quality. Projects will eventually include Federally Qualified Health Centers (FQHCs).[22]

Bates and Bitton maintain that EHRs are pivotal for this model but frequently lack the desired functionality. [23] Rittenhouse et al. suggest that small to medium sized medical practices (the majority of US primary care) use very few PCMH processes, most likely due to limited IT support. [24] For a review of the topic and more detail we refer readers to two recent articles.[25, 26]

Accountable Care Organizations (ACOs)

An ACO is a type of healthcare payment and delivery reform model that associates reimbursement with quality measures and cost reduction for a defined patient population. The Patient Protection and Affordable Care Act (ACA) created the Medicare Shared Savings program, allowing ACOs to contract with Medicare by January 2012. A minimum of five thousand patients must be enrolled and participation (voluntary) must be for a minimum of three years. If organizations can show cost savings and adherence to quality metrics, then they are eligible for "shared savings." If they fail to do so, they would be penalized. Participants will need to report on 33 quality measures in four domains: (1) patient experience domain with seven measures, (2) care coordination/patient safety domain with six measures, (3) preventive medicine domain with eight measures, and (4) at risk populations domain measuring care for diabetes, heart failure, hypertension and coronary disease with 12 measures. EHR use is voluntary but counts double as a quality measure. Year one will be pay-for-reporting and by year three pay-for-performance. These measures are aligned with other CMS quality programs. ACOs will have baseline performance recorded July 2011 to March 2012, measuring all Medicare part A&B payments as the economic benchmark. A scoring system has been developed that will substantiate payment for achievement or improvement in performance, compared to baseline. Practices will need to be innovative, evidence based, patient centric and care coordination will be essential. Additionally, ACOs must be more of a team effort and exchange data more readily to succeed. A variety of ACO pilot projects are underway in the United States by civilian insurers and Medicaid. [27]

HIT will be an integral part of ACOs to promote evidence based medicine (EHR order sets and clinical practice guidelines) and patient engagement (patient portals, PHRs and secure messaging), quality and cost reports (EHR generated), and care coordination (HIE, continuity of care documents, telehealth and remote patient monitoring). Figure 17.1 demonstrates how HIT is integral to ACOs.

Figure 17.1: ACO technology infrastructure (adapted from Battani) [28]

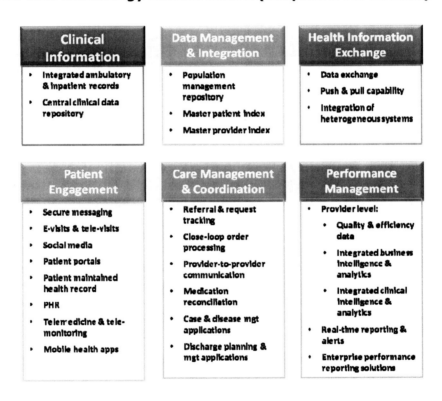

Quality Improvement Projects

It is estimated that more than 100 organizations have P4P programs in place, in spite of the paucity of studies to prove efficacy or return on investment. Many of the programs are really pay-for-reporting programs, in that, clinicians are being reimbursed for submitting evidence that they checked on an important test, not that the test was optimal or met national recommendations.

Physician Quality Reporting Initiative (PQRI)

PQRI is a Medicare program that began in 2007 for the purpose of reimbursing for reporting quality measures. It is a pay-for-reporting initiative where participants can report individual quality measures, disease/condition-specific measures or reports through disease registries (that could be part of an EHR). In

2010 the PQRI bonus was approximately 2% (1.5% prior years) of the total Medicare Part B allowable charges during the reporting period. This amounted to about $4,000 per primary care physician. Data submitted on 30 consecutive patients is a program requirement. [29]

Clinicians are paid bonuses the following year and physician survey data suggests that this time-based delay has not been well received. In addition, clinicians have been slow to receive feedback on their progress from CMS.[30] A commercial disease registry DocSite will submit reports to Medicare for a fee.[31] The PQRI experience of one medical group was reported in 2008. In spite of this group having an EHR, they had to develop new software in order to create a PQRI report which drove up their cost.[32] For 2010 there were 175 measures and four reporting options: claims-based, registry-based, EHR-based and a new group practice option. The EHR option requires three out of 10 measures submission for a bonus incentive payment. Information about how to submit patient data with a disease registry or EHR is available at www.QualityNet.org.

CMS Premier Hospital Quality Incentive Demonstration Project

The project began in 2003 with 270 hospitals participating in a three year demonstration period. Hospitals were paid based on compliance with 34 quality indicators in five common areas (heart attack, heart failure, pneumonia, coronary artery bypass surgery and hip/knee disease). The program leverages the Premier Perspective database which is the largest of its kind in the nation. After review of first year data, hospitals scoring in the top 10% received a 2% bonus in Medicare payments. Hospitals scoring in the second 10% received 1% and those below received no bonus. Currently, it is possible for hospitals to have a 1% to 2% decrease in Medicare payments if at year three of participation they have not improved beyond the baseline.

A three year report was available on the Premier web site (June 2008) demonstrating a 15.8% improvement over the previous three years. As a result, the project was extended beyond initial demonstration period.[33, 34]

Surgical Care Improvement Project (SCIP)

The SCIP is a similar CMS P4P project for surgical care in hospitals. The overarching goal at the onset was to reduce surgical infections by 25% by 2010. Each of the SCIP target areas are advised by a technical expert panel (TEP). Beginning in July 2005,[35] the project explored post-surgery complications such as: site infections, adverse cardiac events, deep vein thrombosis (blood clots in the legs) and pneumonia. The current surgical quality measures are part of the National Hospital Inpatient Quality Measures.[36] Participation is voluntary and results are eventually posted on www.hospitalcompare.hhs.gov . A 2010 JAMA article reported adherence with individual SCIP measures, that are publicly reported, was not associated with a significantly lower probability of infection.[37]

Hospital Value-Based Purchasing Program

The Hospital Value-based Purchasing Program, authorized through the Affordable Care Act, begins in October 2012 (FY 2013) and may impact 3500 hospitals in the U.S. A primary program goal is to improve care and reduce costs and inpatient care; the largest part of Medicare payments. Reimbursement will be linked to quality improvement and patient satisfaction. Examples include (1) how fast hospitals provide balloon angioplasty to those needing it, (2) how often patients receive blood thinners to prevent blood clots, and (3) how often do patients with heart failure receive appropriate discharge instructions. Program quality measures were endorsed by the NQF and measures are posted on the Hospital Compare web site www.healthcare.gov. By 2015 hospitals will receive reduced payments if they are not using appropriate HIT to improve delivery of coordinated care.[38]

Comprehensive Primary Care Initiative

The Comprehensive Primary Care Initiative, launched in 2011 reimburses physicians a monthly care-management fee in addition to the usual Medicare fee-for-service. This initiative is part of the patient centered medical home model previously discussed. Primary care physicians will be reimbursed for care coordination, prevention and improved communication with care givers. The monthly care-management fee will decrease in year three and four. CMS is inviting private payers to join the program with the goal of providing better reimbursement so clinicians can hire more staff and health information technology to deliver better care. [39]

We now have a track record of projects conducted by purchasers, payers, QI organizations, Medicaid and other countries from which to draw conclusions. [40-52]

Quality Improvement Concerns and Limitations

The following are some of the concerns about quality improvement programs expressed primarily by physicians and their organizations:

- Do QI programs discriminate against practices without EHRs?

- Are EHRs sophisticated enough to provide accurate measures of quality?

- Should data be public?

- Will QI programs cause clinicians to "dump" non-compliant or sicker patients?

- Will QI programs result in higher quality care or long term return on investment?

- Will QI programs adjust for sicker, poorer and more elderly patients?

- Much of the practice of medicine does not have identified quality measures, so improvement may be spotty.

- Will the motive behind change be financial and not really improving quality?

- What is an "adequate financial bonus" to support QI programs?

- Is the extra work to report actually worth the relatively small incentive payment?

- Will the number of QI measurements for multiple government programs be excessive?

- Should bonuses be paid for improvement even if results do not meet national goals?

- At this time, the majority of QI reimbursement goes to primary care physicians and not specialists or hospitals. Is this likely to change with newer models?

- Waiting on performance "report cards" occasionally takes a long time and impedes next year's improvement [53-59]

Future Trends

Overall, the most important trend is the shift from paying for the volume of health care delivered to paying for the actual quality of care. The presumption by health care experts and the federal government is that high quality evidence based health care in the long run will be less expensive and be associated with improved patient safety. Certainly, hospitals that are highly rated by organizations such as HealthGrades have confirmed this association. How to transition smaller, rural and poorer healthcare organizations to higher quality remains to be seen and is a future challenge. We anticipate extensive lessons learned from the ACO model implementation.

Key Points

- U.S. health care is the most expensive in the world, yet many important quality outcomes demonstrate worse results than other countries who invest less in health care
- Civilian and federal insurers are looking at reimbursing for quality in lieu of just quantity of service
- Measuring quality is difficult and controversial but will likely benefit from new health information technology (particularly the electronic health record)
- Multiple new QI demonstration projects are underway
- It is unclear whether newer quality improvement strategies will really improve medical quality or reduce cost

---

Conclusion

The federal government, mandated and supported by the Affordable Care Act, developed a National Quality Strategy with the goals of improving quality and reducing cost. Newer quality improvement strategies such as the patient-centered medical home and accountable care organization models will require robust information technology to record and transmit quality measures. There is a legitimate concern that we might not be ready to generate, receive and analyze an avalanche of new quality measures, even with widespread EHR adoption. We will likely benefit from pilot projects that produce lessons learned and public-private collaborations that produce innovation.

We are still in the infancy stage of optimally employing HIT to improve quality and reduce cost. Readers should be on the look out for interesting examples of quality improvement and lessons learned. AHRQ produced a 2010 monograph "Using Health IT: Eight Quality Improvement Stories that discusses real world examples of HIT implementations that lead to quality improvement.[60]

References

1. World Health Organization http://www.who.int (Accessed November 4 2011)

2. Nolte E, McKee M. Variations in Amenable Mortality—Trends in 16 High-Income Nations. Health Pol 2011; 103(1): 47-52.

3. Davis K, Schoen C, Schoenbaum SC et al. Mirror, Mirror on the Wall: An International Update on the Comparative Performance of American Health Care. May 2007. www.commonwealthfund.org (Accessed November 11 2011)

4. Institute of Medicine. http://www.iom.edu/Activities/Quality/PatientSafetyHIT.aspx (Accessed November 4 2011)

5. Agency for Healthcare Quality and Research. www.ahrq.gov (Accessed October 16 2009)

6. McGlynn EA et al. The Quality of Health Care Delivered to Adults in the United States NEJM 2003:2635-2645

7. Yarnell KSH, Pollak KI, Ostbye T et al. Primary Care: Is there enough time for prevention? Am J Pub Health 2003. 93(4):635-641

8. CMS Quality Improvement Roadmap. Executive summary https://www.cms.gov/CouncilonTechInnov/downloads/qualityroadmap.pdf (Accessed July 16 2010)

9. National Quality Strategy http://www.healthcare.gov/law/resources/reports/quality03212011a.html (Accessed September 28 2011)

10. Leatherman S, McCarthy D. Quality of Health Care for Medicare Beneficiaries: A Chartbook 2005. The Commonwealth Fund http://www.cmwf.org/publications/publications_show.htm?doc_id=275195 (Accessed March 2 2006)

11. Rewarding Provider Performance: Aligning Incentives in Medicare IOM September 2006 www.iom.edu (Accessed October 22 2006)

12. Landro, L. Pay for Performance Rewards Preventive, Follow-Up Care. Wall Street Journal September 17, 2004. http://www.asa.siuc.edu/isberner/HCM385/readings/WSJ_Landro.htm (Accessed October 15 2005)

13. Rosenthal MB et al. Pay for performance in commercial HMOs. NEJM 2006;355:1895-902

14. Baker, DW, Persell SD, Thompson JA et al. Automated Review of Electronic Health Records to Assess Quality of Care for Outpatients with Heart Failure. Ann Intern Med 2007;146:270-7

15. White paper: Pay for performance Information Technology Implications for Providers. First Consulting Group Feb 2005 www.fcg.com (Accessed November 20 2007)

16. Agency for Healthcare Research and Quality. Recommended Starter Set. http://www.ahrq.gov/qual/aqastart.htm (Accessed November 7 2007)

17. Medicare Physician Group Practice Demonstration http://www.cms.hhs.gov/DemoProjectsEvalRpts/downloads/PGP_Fact_Sheet.pdf (Accessed June 19 2009)

18. Centers for Medicare and Medicaid Innovation. http://innovations.cms.gov (Accessed September 28 2011)

19. North Dakota Health Plan to Use Health IT in Medical Home Initiative. October 16 2008. http://www.ihealthbeat.org (Accessed October 16 2008)

20. Medicare Medical Home Demonstration Project http://www.cms.hhs.gov/DemoProjectsEvalRpts/MD/itemdetail.asp?itemID=CMS1199247 (Accessed June 18 2009)

21. The Military Health System Blog. February 24 2009. http://www.health.mil (Accessed March 2 2009)

22. Reid RF, Coleman K, Johnson EA et al. The Group Health medical home at year two: cost savings, higher patient satisfaction and less burnout for providers. Health Aff (Millwood). 2010;29(5):835-843

23. Bates DW, Bitton A. The Future of Health Information Technology in the Patient-Centered Medical Home. Health Affairs 2010 29 (4):614-62

24. Rittenhouse DR, Casolino LP, Shortell SM et al. Small and Medium-Size Physician Practices Use Few Patient-Centered Medical Home Processes. Health Aff (Millwood) August 2011;30(8) Online http://content.healthaffairs.org/content/30/8/1575.full?ijkey=Cq.8ITSlIscIM&keytype=ref&siteid=healthaff (Accessed September 15 2011)

25. Rosenthal TC. The Medical Home: Growing Evidence to Support a New Approach to Primary Care. J Am Board Fam Med. 2008; 21(5):427-440

26. Meaningful Connections: a resource guide for using health IT to support patient centered medical home. Patient Centered Primary Care Collaborative. www.pcpcc.ent (Accessed April 10 2010)

27. Accountable Care Organizations. Centers for Medicare and Medicaid. https://www.cms.gov/ACO/ (Accessed October 3 2011)

28. Battani J. Preparing for Accountable Care: The Role of Health IT in Building Capability. 2011 http://www.csc.com (Accessed November 20 2011)

29. Physician Quality Reporting Initiative. www.cms.hhs.gov/pqri (Accessed April 15 2010)

30. Survey: Medicare PQRI Data Not Useful. September 8 2008) www.healthdatamanagement.com (Accessed June 19 2009)

31. PQRI. DocSite www.docsite.com (Accessed April 15 2010)

32. Wintz R, Rosenthal B, Zadem SZ. The Physician Quality Reporting Initiative: A Practical Approach to Implementing Quality Reporting. Advances in Chronic Kidney Disease 2008;15(1):56-63

33. CMS/Premier Hospital Quality Incentive Demonstration Project http://www.premierinc.com/quality-safety/tools-services/p4p/hqi/index.jsp (Accessed June 20 2008)

34. Lindenauer PK et al. Public Reporting and Pay for Performance in Hospital Quality Improvement. NEJM 2007;356:486-496

35. Martin CB. Medicare's Pay for Performance Legislation: A newsmaker interview with Thomas Russell MD www.Medscape.com September 15 2005 (Accessed February 20 2006)

36. Brennan KC, Spitz G. SCIP Compliance and the role of concurrent documentation. www.psqh.com January/February 2008 (Accessed February 28 2008)

37. Stulberg JJ, Delaney CP, Neuhauser DV et al. Adherence to Surgical Care Improvement Project Measures and the Association with Post Operative Infections. JAMA 2010;303(24):2479-2485

38. Healthcare.gov www.healthcare.gov (Accessed September 28 2011)

39. Centers for Medicare and Medicaid Innovation http://innovations.cms.gov/areas-of-focus/seamless-and-coordinated-care-models/cpci/ (Accessed October 1 2011)

40. Bridges to Excellence www.bridgestoexcellence.org/ (Accessed June 19 2009)

41. de Brantes PS, D' Andrea BG. Physician's Respond to Pay for Performance Incentives: Larger Incentives Yield Greater Participation. Am J of Man Care 2009;15(5):305-310

42. Endrado P. Pay for performance tools evolve as market shifts www.healthcareitnews.com September 26 2005 (Accessed February 9 2006)

43. Pay for performance www.acponline.org/weekly/2005/4/5/index.html (Accessed February 22 2006)

44. Merx K. Win-win program for docs, patients. Detroit free press. April 25 2005 www.freep.com (Accessed February 1 2006)

45. California Pay for Performance Collaboration http://www.pbgh.org/programs/documents/PBGH_ProjSummary_P4P_03_2005.pdf (Accessed February 11 2006)

46. Damberg CL, Raube K, Teleki SS et al. Taking Stock of Pay for Performance: A Candid Assessment From the Front Lines. Health Affairs 2009;28(2):517-525

47. Apland BA, Amundson GM. Financial Incentives, an indispensable element for quality improvement. Patient Safety & Quality Healthcare. Sept/Oct 2005. www.psqh.com (Accessed February 2 2006)

48. Primus + University of Nottingham http://www.primis.nhs.uk/pages/default.asp (Accessed November 20 2007)

49. Roland M. Linking Physicians' Pay to the Quality of Care — A Major Experiment in the United Kingdom NEJM 2004;351:1448-1454 (Accessed February 7 2006)

50. Campbell SM, Reeves D, Kontopantelis E et al. Effects of Pay for Performance on the Quality of Primary Care in England. NEJM 2009;361:368-78

51. Patmas MA. A Novel Pay for Performance Program www.uptodate.com/p4p.ppt (Accessed February 20 2007)

52. Majority of State Medicaid Programs Plan for Pay for Performance Standards. HealthDailyNews. April 12 2007 www.healthfinder.gov/news (Accessed February 21 2008)

53. Rohack JJ. The Role of Confounding Factors in Physician Pay for Performance Programs. Johns Hopkins Advanced Studies in Medicine 2005;5:174-75

54. Audet AM et al. Measure, Learn and Improve: Physicians' Involvement in Quality Improvement Health Affairs 2005;24:843-53

55. Raths D. Pay for Performance. Healthcare Informatics Feb. 2006; 48-50

56. Colwell J. Market forces push pay for performance. ACP Observer. May 2005

57. Shaw G. What can go wrong with pay for performance incentives ACP Observer March 2006

58. Colwell J. Pay for performance takes off in California ACP Observer Jan/Feb 2005

59. Bodenheimer T et al. Can money buy Quality? Physician response to pay for performance. http://hschange.org/CONTENT/807/ (Accessed February 23 2006)

60. Using Health IT: Eight Quality Improvement Stories www.healthit.ahrq.gov (Accessed September 28 2011)

18

Patient Safety and Health Information Technology

ROBERT E. HOYT

RONALD G. GIMBEL

Learning Objectives

After reading this chapter the reader should be able to:

- Identify why patient safety is a national concern

- Define medical errors, adverse events and preventable adverse events

- Compare and contrast how information technology can potentially improve or worsen patient safety

- Compare and contrast the governmental and non-governmental patient safety programs

- List the various technologies that are likely to improve medication error rates

- Identify the obstacles to widespread implementation of patient safety initiatives

Introduction

In this chapter we will discuss the role of health information technology and patient safety, with emphasis on how technology has the potential to improve patient safety. We will also discuss the limitation of HIT in improving patient safety and why. Medical errors will be broadly discussed but we will focus the majority of our attention on preventing medication errors. This chapter will also describe the current status of national and international patient safety programs. First we would like to begin with a set of definitions.

Patient Safety-Related Definitions [1-3]

- Patient Safety: "freedom from accidental injury"

- Medical Error: "failure of a planned action to be completed as intended or the use of a wrong plan to achieve an aim"

- Adverse Events: "an injury resulting from a medical intervention"

- Preventable Adverse Events: errors that result in an adverse event that are preventable

- Overuse: "the delivery of care of little or no value" (e.g. widespread use of antibiotics for viral infections)

- Underuse: "the failure to deliver appropriate care" (e.g. vaccines and cancer screening)
- Misuse: "the use of certain services in situations where they are not clinically indicated" (e.g. magnetic resonance imaging for routine low back pain)

In the classic report *To Err is Human* published in 2000, the Institute of Medicine (IOM) estimated that at least 98,000 inpatients die every year and 1,000,000 are injured due to preventable errors.[1] The mortality and morbidity rate may have been actually higher as many outpatient adverse events were not reported. While McDonald and others argue that the methodology used to report these statistics was flawed, most agree that American medicine is not as safe as it should be.[4] The 2001 Institute of Medicine report *Crossing the Quality Chasm* emphasized the importance of medical quality leading to improved patient safety.[2] The current medical system was described as an "era of Brownian motion in health care."

The IOM has long been an advocate of using information technology to improve healthcare quality and patient safety. They clearly state that safety is the first domain of medical quality. The 2001 IOM report recommended that we "improve access to clinical information and support clinical decision making" and "create a national information infrastructure to improve health care delivery and research." Also, one of their goals was to eliminate handwritten notes in the following decade.[2]

Errors can involve different aspects of medical care such as diagnosis, treatment and preventive care. Furthermore, medical errors can be errors of commission or omission and fortunately not all errors result in an injury and not all medical errors are preventable. Until the past two decades, there has been a paucity of articles written about patient safety and most articles have dealt specifically with medication errors and not errors occurring in other areas of medical practice. A 2003 article ranked the most common types of medical errors made by American family physicians: prescribing medications, getting the correct laboratory test for the correct patient at the correct time, filing system errors, dispensing medications and responding to abnormal test results.[5] Health information technology has the potential to improve these types of medical errors.

According to Dr. Leape, an early patient safety advocate, the first specialty to experience dramatic advances in patient safety was anesthesiology, with less than one death in 200,000 patients undergoing anesthesia.[6]

Most authorities believe that errors occur more often due to inadequate systems and not inadequate individuals. Most of these errors arise because our system of medical care including training, staffing, financial incentives, as well as local and federal policies, was not designed to prevent errors or mitigate their effects. Some authorities believe that about 50% of medical errors are preventable with better systems.[7] Also, our fee-for-service system did not traditionally reimburse based on quality or patient safety. This has changed greatly in recent years and will be discussed in more depth in the chapter on quality improvement strategies.

Other industries such as the airlines have dramatically reduced mishaps thru initiatives such as "crew resource management" (CRM).[8] CRM training focuses on interpersonal communication, situational awareness, leadership and decision making. This technique has been so successful hospitals often incorporate CRM as part of management training. In particular, some operating rooms employ a CRM-based check list prior to initiating surgery.

An estimate from the Society of Actuaries Health Section estimated that medical errors in the United States cost $19.5 billion in 2008.[9] In addition to the obvious increased cost, mortality and morbidity that results from medical errors there is a resulting increase in litigation. It was estimated in 2003 that US malpractice costs totaled $27 billion.[10]

This chapter will discuss how health information technology (HIT) may improve patient safety, largely through improving the quality of care delivered. Medical quality is discussed in more detail in the chapter on quality improvement strategies. It is important to stress, however, that HIT can also create new types of medical errors as discussed in this chapter and the chapters on electronic health records and electronic prescribing. In the next section we will list several well-known reports on patient safety and the quality of medical care in the United States.

Patient Safety Reports

Institute of Medicine (IOM) Reports

Earlier in this chapter we commented on the important IOM patient safety reports *To Err is Human, Building a Safer Health System* (2000) and *Crossing the Quality Chasm, A New Health System for the 21st Century* (2001). Their recommendations are listed in the executive summary:

- Congress should create a Center for Patient Safety within the Agency for Healthcare Research and Quality

- A nationwide reporting system for medical errors should be established

- Volunteer reporting should be encouraged

- Congress should create legislation to protect internal peer review of medical errors

- Performance standards and expectations by healthcare organizations should include patient safety

- FDA should focus more attention on drug safety

- Healthcare organizations and providers should make patient safety a priority goal

- Healthcare organizations should implement known medication safety policies [1, 2]

The IOM's *Patient Safety: Achieving a New Standard for Care* (2003) expanded on the prior two sentinel monographs with the following statements and recommendations: [11]

- Patient safety must be linked to medical quality

- A new healthcare system must be developed that will prevent medical errors in the first place

- New methods must be developed to acquire, study and share error prevention among physicians, particularly at the point of care

- The IOM recommended specific data standards so patient safety-related information can be recorded, shared and analyzed

In 2011 the IOM released a report titled *Health IT and Patient Safety: Building Safer Systems for Better Care*. Unlike its previous reports, this report focused exclusively on health IT as it relates to patient safety and quality. Somewhat alarming was a primary finding that the evidence about the impact of health IT on patient safety, as opposed to quality, is mixed but shows the challenges involve people and clinical implementation as much as the technology.[12] While published evidence suggests improvement in patient safety have been realized with HIT, a finding of "no effect" or actual "associated harm" were also realized. The report cites a lack of health IT-related safety data and suggested contributing causes which include the absence of measures and a central repository for analysis as well as contractual barriers preventing sharing of information. The report issues 10 recommendations to encourage:

- HHS to publish an "action and surveillance plan" within 12 months (i.e. 2012)

- HHS to push health IT vendors to support the free exchange of information about health IT experiences and issues

- ONC to work with public and private sectors to make comparative user experiences publicly available

- HHS should fund a Health IT Safety Council to assess and monitor the safe use of health IT and its use to enhance patient safety

- Health IT vendors to publicly register and list their products with ONC

- HHS to specify the quality and risk management processes that health IT vendors must adopt

- HHS should establish a mechanism for vendors and users of health IT-related deaths, serious injuries, or unsafe conditions

- HHS recommend to Congress establishing an independent federal entity to investigate patient safety deaths, serious injuries, or potentially unsafe conditions associated with health IT

- HHS should monitor and report progress of health IT safety annually and FDA begin developing framework for regulation

- HHS, in collaboration with others, should support cross-disciplinary research toward the use of health IT as part of a learning system [12]

HealthGrades 2011 Patient Safety Excellence Awards

This organization reviews the data from 40 million inpatient Medicare and Medicaid cases each year and rates hospitals, in terms of patient safety. Five thousand hospitals and 800,000 physicians are rated. Specifically, they rate risk-adjusted mortality and complications for 27 hospital procedures and complications. Ratings range from one star (lowest) to five star (highest). The most recent report covered data from 2008 to 2010. They estimate that the top ranking hospitals represent, on average, a 43% lower risk of a patient safety adverse event compared to the lowest ranking hospitals. They awarded 238 hospitals with the excellence award and this represented about 5% of the nation's hospitals. The awardees are listed on their web site. Samantha Collier MD of HealthGrades believes that the hospitals that traditionally have excellent safety scores have a "culture of safety" and they are the ones that have all of the mechanisms including technology in place to prevent and track patient safety issues. At this time HealthGrades has 11 million visitors accessing their web site each month.[13]

Patient Safety Culture

Virginia Mason is an integrated healthcare system located in Seattle WA. For the past decade they have reanalyzed multiple health care processes, based on the Toyota Production System. The result has been fewer inpatient falls, bed sores, etc. They post several independent quality reports on their web site that compares them to the state and national averages. They stress patient involvement in care, to include rapid response teams if a family member believes the patient is deteriorating. Their nurses work in teams so they can now spend 90% of their time with patients, instead of the national average of 35%.

www.virginiamason.org

Governmental and Non-Governmental Organizations and Programs Supporting Patient Safety

US Federal Agencies

The Department of Health and Human Services, as authorized by legislation, plays a key leadership role in patient safety and quality in the United States, delivered via several of its agencies.

Agency for Healthcare Research and Quality (AHRQ): This agency is the designated lead federal agency for patient safety. The agency maintains an active patient safety research portfolio and numerous grant and contract funding mechanisms to deliver on its mission. The patient safety portfolio is broad in scope including the: Patient Safety Organizations Network (as authorized by Public Law 109-41), preventing medication error, reducing unavoidable inpatient readmissions, preventing healthcare-associated infections, TeamSTEPPS™ training and an associated Health IT portfolio. Relatively new to the health IT portfolio is their Ambulatory Safety & Quality Program. This program accentuates the role of health IT through funding opportunity announcements (FOAs); specifically:

- Enabling quality measurement through health IT (includes patient safety focus)
- Improving quality through clinician use of health IT
- Enabling patient-centered care through health IT
- Improving management of individuals with complex health care needs through health IT [14]

Centers for Medicare and Medicaid Services (CMS): Effective October 1, 2008 Medicare stopped reimbursing hospitals for complications they deemed preventable. At this time, this new policy does not affect physicians. The list of non-reimbursable complications included:

- Objects left in a patient during surgery and blood incompatibility
- Catheter-associated urinary tract infections
- Pressure ulcers (bed sores)
- Vascular catheter-associated infections
- Surgical site infections
- Serious trauma while hospitalized
- Extreme blood sugar derangement
- Blood clots in legs or lungs [15]

CMS launched its new Partnership for Patients initiative, funded by up to $1 billion through the Patient Protection and Affordable Care Act (PL111-148), in 2011 with two core goals of keeping patients from getting sicker or injured in the health care system and helping patients heal without complications by improving transitions in care from hospitals to alternative settings.[16] As a result, hospitals with higher than anticipated 30 day readmission rates for heart attacks, heart failure and pneumonia will receive decreased reimbursement. This will occur in FY 2013 and the Secretary of HHS has the opportunity to increase the number of conditions in FY 2015. Specific program goals are to decrease hospital readmission rates by 20% and hospital-acquired conditions by 40% by 2013. To date there are over 4,000 partners (including 2,000+ hospitals) nationally. There is a funding mechanism associated with hospital membership where the institution received financial incentives for participating and demonstrating improvement in 10 areas associated with harm. They include:

- Adverse drug events

- Catheter-associated urinary tract infections

- Central line associated blood stream infections

- Injuries from falls and immobility

- Obstetrical adverse events

- Pressure ulcers

- Surgical site infections

- Venous thromboembolism

- Ventilator-associated pneumonia

- Other hospital-acquired conditions [16]

Noteworthy is an understanding of the key importance of health informatics and health IT in the Partnership for Patients initiative. Hospitals and other partners must have the ability to assemble, analyze and trend clinical and administrative data to capture baseline data and measure improvement over time. Health IT-based interventions, several mentioned below, are expected to assist the partners in realizing improvement.

As a component of their agency mission CMS manages the Quality Care Finder (www.hospitalcompare.hhs.gov) that allows consumers to review quality metrics e.g. morbidity (complications) and mortality (death), when making decisions about hospitals, physicians, nursing homes, home health, and dialysis centers in terms of their own care. Users can search by medical condition, surgical procedure or patient satisfaction measures. Figure 18.1 compares two large hospitals located in the same city, in terms of mortality rates from heart attacks.

Figure 18.1: Heart attack mortality comparison

(Source: www.hospitalcompare.hhs.gov**)**

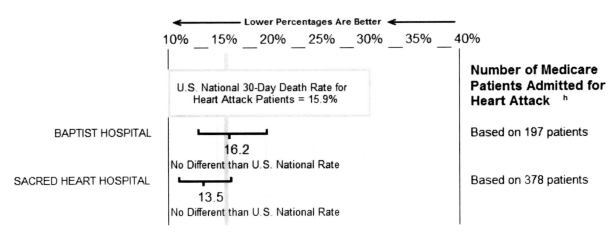

Health Resources and Service Administration: This agency is the primary Federal agency for improving access to health care services for people who are uninsured, isolated, ore medically vulnerable.[17] With respect to patient safety, HRSA manages a "Patient Safety and Clinical Pharmacy Services Collaborative (PSPC)." Currently there are over 450 community organizations of community-based health care providers participating in the PSPC. The initiative seeks to improve the quality of health care by integrating evidence based clinical pharmacy services into the care and management of high-risk, high-cost, complex patients. By embracing contemporary quality improvement methodology and sharing lessons through collaborative learning, the PSPC has realized a 54% gain in getting complex patients "under control" with respect to optimal medical management and a reduction of 49% in adverse drug events for this high risk patient population.[17]

The Food and Drug Administration: This agency plays a pivotal role in regulating, among other key areas, drugs, medical devices and radiation-emitting products. The agency manages a host of programs and initiatives to achieve its goal. Examples include:

- MedWatchPlus: The FDA is extending its MedWatch program to include safety reports on all FDA regulated products (drugs, devices, biologics, dietary supplements, human food, animal feed and veterinary products). The new program known as MedWatchPlus will partner with the National Institutes of Health (NIH) and will include a public web site and data collection tools. This will offer a single reporting portal and repository for data analytics. The program will be released in phases and slated to be complete by 2011.[18]

- The Center for Devices and Radiological Health (CDRH) is part of the FDA is responsible for the pre-market approval of all medical devices and radiation –emitting products. In 2011, the CDRH announced for public comment its intent to regulate mobile medical applications designed for use on smartphones and other mobile medical computing devices.[19] The move is broad and expansive, capturing the interest and concern of industry[20] and others including the American Medical Informatics Association.[21] At debate are issues surrounding what should be regulated and what should be exempt. The AMIA position suggests that primary FDA attention should center on applications (clinical decision support) where there is no mediation by a human being (e.g. clinician). The Association cautioned to FDA to address the rapidly emerging and converging devices along with evolving forms of care delivery such as medical homes and accountable care organizations.[21]

- Recalls, Market Withdrawals, and Safety Alerts are managed by the FDA and apply to drugs, medical devices, radiation-emitting products, food, and other key areas. Health systems, hospitals, pharmacies, and other care delivery organizations rely on their data systems and warehouses to identify patients at risk and respond appropriately to recall actions. The data systems and ability to mine and analyze the data rely on medical informatics skills and tools.

State Patient Safety Programs

States are in a position to positively influence patient safety as they are involved with purchasers and payers of healthcare and they benefit from improved patient safety, in terms of money saved. By early 2010, 27 states and the District of Columbia passed legislation or regulation related to hospital reporting of adverse events to a state agency.[22] The regulations will involve event report analysis and valuable lessons learned.

The National Academy for Health Policy (NASHP) was established in 2000 to assist states in delivering safe patient care. They offer a patient safety toolbox to the states that have been most involved that includes information (policies, practices, forms, reports, methods, and contracts) intended to improve states' reporting systems.[22, 23]

Non-Governmental Patient Safety Organizations and Programs

While governmental agencies and programs have a substantial voice in patient safety and quality, there are several key non-governmental organizations that are prominent actors in the patient safety and quality community.

National Patient Safety Foundation (NPSF): The foundation is an independent not-for-profit 501(C)(3) organization founded in 1997 with the aim of improving the safety of care provided to patients. To deliver on their mission, the NPSF structures their energy around five fundamental action steps that include: (1) identifying and creating a core body of knowledge, (2) identifying pathways to apply the knowledge, (3) developing and enhancing the culture of receptivity to patient safety, (4) raising public awareness and fostering communication around patient safety, and (5) improving the status of the Foundation and its ability to meet its goals. The NPSF is a funder of patient safety-related research and hosts a large annual patient safety congress where patient safety advances, quality improvement strategies, and research are presented. Technology-related projects, similar to those identified below, are representative of those promoted at the NPSF congress.[24]

The National Quality Forum (NQF): This public-private collaborative group was organized in 1999 for the purpose of quality measure development and public reporting. They establish national standards to improve the quality of medical care and by so doing improve patient safety. Currently they have posted 741 standards (over 100 dealing with patient safety) on their web site with a new search engine and analytic tools. They have a health information technology advisory committee (HITAC) comprised of national experts. One important aspect of the work they are doing relates to the Quality Data Model (QDM). This model describes clinical concepts in a standardized format for electronic quality measures that can be generated and communicated by electronic health records.[25]

The Joint Commission: Beginning in 2002, the Joint Commission began publishing National Patient Safety Goals (NPSGs) that are updated annually. The purpose of the national goals is to highlight attention to key patient safety areas ripe for improvement. These national goals are developed and updated by a widely recognized group of patient safety experts. While a legitimate argument can be made that all of the NPSGs can be addressed (or partially addressed) through health IT, three of the hospital 2011 NPSGs have direct association: identify patients correctly, improve staff communication and use medicines safely.[26] In 2008 the Joint Commission recommended that there be one national infrastructure to measure and track quality improvement data.[27] In addition they warned healthcare organizations, through their Sentinel Event Alert #42, that implementing health information and converging technologies can create or perpetuate patient safety risk and preventable adverse events. The alert identifies potential contributing factors of health information and converging technologies on patient safety and outlines action steps to prevent harm in this area. Some examples include assessing pre-implementation need and clinical workflow, actively engaging clinicians, ongoing monitoring, establishing training and refresher training for clinicians and staff, establishing related organizational processes, developing and testing order sets before automating, building in checks and balances to mitigate potential harm, and others.[28]

Institute for Healthcare Improvement (IHI): The IHI instituted a plan in December 2004 to save 100,000 lives from medical errors by getting hospitals to incorporate at least one of six safety measures. A report on June 14th 2006 estimated that 122,300 deaths have been prevented through the adoption of new safety measures by more than 3,000 participating hospitals over an 18th month period.[29] Currently the IHI, and its member, are engaged in numerous improvement projects at the nexus of patient safety, clinical quality and medical informatics.

LeapFrog Group: LeapFrog is a consortium of healthcare purchasers that demand better quality. One of the four areas they promote is the adoption of inpatient computerized physician order entry (CPOE). They maintain survey safety data from over 1,000 hospitals who volunteered to submit data, as well as a calculator to determine return on investment (ROI) for hospital pay-for-performance programs. A consumer can search hospital overall patient safety and safety related to specific procedures via a search engine on their site.[30]

HealthGrades: HealthGrades is an organization that rates different aspects of medical care. On their web site is a search engine for physicians, dentists and hospitals. Hospital reports compare a variety of surgical procedures or diagnoses by state. Physician reports compare disciplinary action, board certification and patient opinions. Quality awards are posted and patient safety indicators are described as average, worse than average or better than average.[31]

American Medical Informatics Association (AMIA): While not a patient safety organization, AMIA and its members are playing a pivotal in enhancing patient safety and quality through medical and health informatics research. This is not surprising given the association's commitment to promoting research on electronic health records, CPOE systems, medication management systems, clinical decision support, mobile technologies, electronic clinical documentation capture, and like projects that explicitly address patient safety and quality outcomes.[32]

Health Information Technology and Patient Safety

Medication error reduction is a prominent patient safety focus area impacted by healthcare IT. The Institute for Safe Medication Practices promotes the "five rights" of medication safety: right drug, right patient, right dosage, right route and right time.[33] It has been shown that adverse drug events (ADEs) account for up to 3.3 % of hospital admissions.[34] To compound the issue, serious ADEs reported to the FDA increased about 2.6 fold from 1998 to 2005, as did fatalities due to medications.[35, 36] While the IOM cited a study claiming that 7,000 deaths occurred in 1993 due to medication errors, one author maintains that 31% of deaths cited were actually due to drug overdoses.[37] Fortunately, 99% of medication errors do not result in injury. About 30% of ADEs are felt to be preventable and of those about 50% are preventable at the ordering stage.[38] It is worth noting that CPOE does not prevent errors of administration (e.g. wrong patient) or timing (e.g. wrong time).[39]

In spite of the fact that more drugs are prescribed for outpatients, inpatient drug use is very dangerous. Intravenous (IV) medications are associated with 54% of ADEs and 61% of serious or life threatening errors.[34] A 2007 monograph by the Institute of Medicine, *Preventing Medication Errors,* made several salient points:

- On average, a hospital patient is subject to one medication error per day

- About 1.5 million preventable ADEs occur yearly with about 400,000 preventable ADEs occurring in inpatients

- Estimated cost of $5,857 per inpatient error resulted in about $3.5 billion in 2006 dollars due to longer length of stay and additional services (figure excludes litigation)

- Estimates are probably low, based on how statistics were collected[40]

Technology has great potential in reducing medication errors but there are many unanswered questions. Several studies of health information technology and medication errors concluded that well-controlled studies are lacking, tend to be reported only at a select number of universities, and patient outcomes are lacking.[39, 40]

An article in Health Affairs in 2008 reported on adoption of medication safety related HIT by 4,561 non-federal hospitals in 2006. The IT applications studied were: electronic medical records, clinical decision support, CPOE, bar coding medication dispensing (BarD), medication dispensing robot, automated dispensing machine, electronic medication administration records (eMAR) and bar coding at medication administration (BarA). They concluded the following:

- Larger and urban hospitals had much higher adoption rates

- On average, only 2.24 of eight applications were adopted per hospital

- One-fourth of hospitals had not adopted any of the eight technologies

- Teaching hospitals had higher rates of adoption

- The most widely adopted application was the automated dispensing machine and least adopted was bar coding for medication administration (BarA) [43]

A 2007 survey by the American Society of Health System Pharmacists evaluated the adoption of pharmacy IT in the United States. They reported the following conclusions:

- 50% of respondents had at least one component of an EHR; 6% were paperless; 12% had CPOE with clinical decision support; 40% had digital documentation; it was a challenge for EHRs to connect to pharmacy IT systems and be available on all hospital units

- 24% had barcode medication administration systems

- 44% used smart pumps

- 83% used automated dispensing cabinets for drugs

- 10% used pharmacy robots

- 21% used e-prescribing

- 10% had a completely electronic medication reconciliation application

- 8.5% had electronic medication administration records (eMar)

- 54% use a paper-based system

- 20% had an eMar bundled with barcoding and electronic nursing documentation

- 36% had pharmacy IT personnel[44]

For the sake of completeness we will mention that diagnostic errors are also a concern in regards to patient safety. A Harvard Medical Practice Study suggested that 17% of preventable errors were due to a misdiagnosis.[43] Older studies, utilizing autopsies for inpatients that died, revealed about 35-40% of patients died due to the wrong diagnosis.[44] Unfortunately, we currently request autopsies infrequently so we can no longer correlate pre and post-mortem diagnoses. Although there is limited evidence at this point that technology improves diagnostic accuracy, there are ongoing research and development efforts in diagnosis tools that span clinical settings (e.g. primary care, radiology), platforms (e.g. EHR, CPOE), and medical education.

Technologies with the Potential to Decrease Medication Errors

CPOE systems and EHRs

- Inpatient CPOE. This functionality was recommended by the IOM in 1991. A 1998 study by Bates and colleagues demonstrated that CPOE can decrease serious inpatient medication errors by 55% (relative risk reduction).[45] In recent years there has been a substantial growth in researcher attention to measuring the impact of inpatient CPOE on reduction of prescribing errors and ADEs. This has led to several reviews[46] and systematic reviews of the literature, comparing the outcome from published studies.[47-50] While CPOEs were found to be effective in reducing medication errors[46,50] and in some studies ADEs,[49, 50] researchers concluded that the cumulative research is rather modest and many studies are flawed by design.[47,49,50] Issues related to organizational, technical, and design factors were also cited.[46-48] Inpatient CPOE is covered in more detail in the chapter on EHRs.

- Outpatient CPOE. There is a greater chance for a medication error written for outpatients because of the number of outpatient prescriptions written. Inpatient prescriptions, however, are more dangerous, particularly intravenous blood thinners, opiates and chemotherapy. Kuo et al. reported medication errors from primary care settings. Seventy percent (70%) of medication errors were related to prescribing, 10% were administration errors, 10% were documentation errors, 7% dispensing errors, and 3% were monitoring errors. ADEs resulted from 16% of medication errors with 3% hospitalizations and no deaths. In their judgment, 57% of errors might have been prevented by electronic prescribing.[51]

- Clinical Decision Support. It is important to note that clinical decision support may be embedded within CPOE (inpatient or outpatient), within the EHR, within mobile and converging technologies, or standalone support. It is clinical decision support that serves as the backbone for translating research into practice aiding clinicians in better diagnosis and treatment of patients. Computerized drug alerts have obvious potential in decreasing medication errors but they continue to be refined. Kuperman divided drug alerts into basic and advanced as demonstrated in Table 18.1.[52]

Table 18.1: Basic and Advanced Drug alerts

Basic	Advanced
Drug allergy	Dose adjustment for renal disease
Dosage guidance	Geriatric dosing
Formulary decision support	Medication-related laboratory testing
Duplicate drug orders	Drug-disease contraindications
Drug-drug interactions	Drug-pregnancy checking

Clinical decision support has many other safety implications than medication safety. Casolino et al. reported on how often patients fail to hear about results such as mammograms, Pap smears and stool specimens for blood. They concluded that about one in 14 abnormal tests are not adequately reported to patients and/or not documented in the chart. This study reinforces the concept that safety processes and work flow must be worked out ahead of time and apparent to all clinicians or problems will occur, regardless as to whether you use a paper-based or electronic system.[53] More information about clinical decision support and medication alerts can be found in the chapters on EHRs and e-prescribing.

Health Information Exchange (HIE)

As pointed out by John Halamka, HIE has the potential to improve patient safety by better communication between disparate healthcare participants. Meaningful Use that is accomplished with HIE should provide valuable information: during the transitions of care, by populating immunization registries and personal health records and by reporting syndromic surveillance-related data to public health.[54] Very few hospitals have the IT support and system sophistication to accomplish widespread sharing of information similar to Brigham & Women's Hospital in Boston, but contemporary advances are being realized as a result of HITECH-funded programs. Substantial gains in the use of HIEs in enhancing patient safety should spread nationally within the next three to five years.

Automated Inpatient Medication Dispensing Devices

Devices are like ATM machines, are kept on nursing units and communicate with pharmacy computers and dispense medications stocked by the pharmacy. Password protected devices keep medication records but unfortunately, there is limited evidence that these systems reduce errors or affect outcomes.[55,41] There is some evidence to suggest, perhaps due to clinical workflow and contextual issues, that these devices could introduce patient safety challenges.[56-58]

Home Electronic Medication Management Systems

At least one company is in the process of developing an ATM-like machine to administer medications to the elderly at home. Medications are loaded into the machine as a 6x9 inch blister pack with storage for up to 10 medications for one month. The device is connected to the pharmacy via the internet so they can monitor compliance and adjust doses. The device (EMMA) gives a visual and audible alert when it is time to take a medication.[59] It is unknown whether the device will be subjected to the proposed FDA regulatory process for mobile medical applications (identified above).

Pharmacy Dispensing Robots

Studies suggest that robotic systems save space, decrease manpower, increase the speed to fill a prescription and decrease errors. Robots are very helpful when there is a shortage of pharmacists or staff. Technology allows pharmacists to have more of a supervisory role. Ideally, systems would receive electronic prescriptions from outpatient and inpatient areas, then be checked by both the EHR and the pharmacist, then labels are printed and the prescription filled.[60]Robots are available in different models that handle a variety of drugs (50 to 200), giving pharmacies financial flexibility.[61]

Electronic Medication Administration Record (eMAR)

This technology eliminates legibility issues as there is no need to rewrite the MAR when medications are changed or discontinued. Provides ready access to the patient's chart to see what medications the patient is on and provides allergy and timing alerts. Application is available to nurses and physicians who usually make separate rounds. Program can be web-based and can be wireless.[62]

"Smart" Intravenous (IV) Infusion Pumps

Intravenous sedatives, insulin, anticoagulants and narcotics pose the highest risk of harm from medication errors.[63] Early IV pumps allowed for constant infusion rates without programmable alerts. Newer smart pumps can be programmed to deliver the correct amount of IV drugs and are associated with drug libraries and alerts that the dose differs from hospital guidelines. This feature is known as a "dose error reduction system" (DERS) and is particularly important if there were a decimal point error or the units of administration such as mg/hour were incorrect. The end result is that the infusion will not begin until the discrepancy is corrected. As an added benefit some pumps also wirelessly transmit data so that specific events can be captured and studied.

Smart pumps will eventually link to eMars, CPOE and pharmacy IT systems. Evidence thus far indicates that smart infusion pumps avert serious IV medication errors. It is important to realize that even a small reduction in errors that involve dangerous IV drugs is an important advance.[64] A 2005 study found that serious medication errors were unchanged compared to a control group. This was thought to be due to the fact that the default data entry interface bypassed the error reduction system, leading many nurses to not consult the drug library. Also, alert overrides were common and there were many undocumented verbal orders. It would be important for hospitals to set the drug library as the default for the program. An unanticipated bonus of this program was the fact that the memory system of the infusion pump was a treasure trove of information, pointing out future areas of training and changes in nursing protocols.[65] Smart pumps with built in bar coding are available.[66]

Calculators

Johns Hopkins University created a web-based pediatric total parenteral (IV) nutrition (TPN) calculator and as a result reduced medication errors in half with an annual projected saving of $60,000 to $80,000. The infusion calculator was associated with 83% fewer errors.[67] Other web-based and handheld medical calculators are available but little is known regarding their impact on patient safety. These types of devices are targeted in the proposed FDA regulation of mobile medical applications due to the potential harm they can introduced if not properly designed, tested, and implemented.[19]

Bar Coded Medication Administration (BCMA)

BCMA involves a variety of elements: bar code printers, scanners, a network (wired or wireless) to connect to a server, server with bar coding software and integration with the pharmacy information system and any CPOE system. A typical linear bar code is most common but newer two dimensional bar codes exist that encode more information in a smaller space and can be read from different angles (Figure 18.2 and 18.3).

Figure 18.2: Bar codes linear and 2-D   **Figure 18.3: Bar code bracelets**

How does a BCMA system work? A standard scenario would be for a nurse to scan his/her ID bar code, the patient's bar code and the medication's bar code. This information could be sent wirelessly to the program server with software that determines that the correct medication is going to the correct patient at the correct time. In general, the system will generate a warning or an approval. Studies have shown that about 35% of medication errors occur at the administration stage. Further breakdown of errors that might be prevented by BCMA include: dose omission (21%), wrong patient (4%), wrong time (4%), wrong route (1%).[68]

Most healthcare organizations use three linear barcodes: codes 128, 39 and Reduced Space Symbology (RSS). Two dimensional barcodes are available that can store 3,000 characters of patient information. Bar codes can be placed on patient ID bands, medications, vials of blood and transfusion bags. FDA mandated that drug companies apply bar codes on unit dose medications and blood components. Barcodes must contain the national drug code (NDC) that can be used to identify medications. The price tag is likely to be $300,000 to $1 million for hospitals to adopt barcode technology.

Expect more innovations with 2-D (QR) codes in the future. One EHR vendor has partnered to have QR codes on vaccine bottles such that when scanned, the vaccine supply can be monitored, information can be sent to the EHR and the state immunization registry and clinical decision support can confirm that the correct patient received the correct immunization.[69]

There are very few studies looking at patient outcomes with this technology. Poon et al. studied dispensing errors before and after implementation of BCMA at the Brigham and Women's Hospital Pharmacy. They demonstrated that the target dispensing error rate dropped by 0.25% to 0.018% (93 % relative risk reduction).[70] A 2010 follow-up study by the same author from the same hospital concluded that bar coding coupled with an eMAR reduced medication errors at the transcription and administration stages. In addition there was a reduction in potential adverse drug events; with true adverse drug events (documented harm to patients) not reported.[71] BCMA in an adult medical intensive care unit was reported in 2009. The system had the potential to improve several areas of medication management but showed an improvement in only administration time errors after implementation of a BCMA system.[72] Veteran's Affair hospitals have had bar coding since 1999 in their 161 hospitals. Once scanned, the software confirms that the correct medication in the correct dose and frequency has been given to the correct patient. It also updates the electronic medication record. As a result of this technology one VA hospital was able to decrease medication errors by 66% over five years.[73]

Bar coding is also used for laboratory specimen labeling. As an example, an inpatient's ID bracelet is scanned and it confirms that this patient requires a certain blood test. A mobile printer prints labels that are attached to the tube of blood at the bedside.[74] A study from a pediatric oncology hospital demonstrated a decrease from 0.03% to 0.005% in mislabeling errors after one year of implementation. The incidence of unlabeled specimens continued to be the same, after implementation. There were a few misreads due to the curvature of the wrist band, that will likely be prevented with a two-dimensional (2-D) bar code band. They estimated that the cost of the system added $1.75 to each specimen processed.[75] AHRQ funded pilot programs in multiples states but in spite of some successes they concluded that implementation is not easy.[76]

Problems with BCMA include: high cost, nurse work flow issues, some meds need to be re-packaged in order to be read and scanners are not interoperable so institutions may have to buy different scanners.[77] It is known that nurses often have to create "workarounds" to solve BCMA shortcomings.[78] A 2010 study concluded this technology inconsistently decreased ADEs and created several new types of medication errors that were not part of the "5 rights" approach.[79]

Radio Frequency Identification (RFID)

Radio frequency identification is a relatively new technology that that has some similarities to barcoding but important differences. Unlike bar coding, RFID can be read-only or read-write capable and RFID tags can be read if wet or thru clothing; they are therefore better for blood and IV bags. Tags can be active (needs battery, larger, more memory, longer range and more expensive) or passive (smaller, cheaper, short range and no battery) (Figure 18.4) RFID tags can be low, medium or high frequency. A scanner must interface with an established database to identify the object with the RFID tag. The tags are cheap but transceivers (scanners) are expensive.

Figure 18.4: Passive RFID tag on back of a drug label (Courtesy CPTTM)

RFID is used in healthcare primarily to locate and track patients and inventory, but with a few new wrinkles.

RFID systems can track patients within a hospital with an active tag that works like a transmitter and gives location and time. RFID tracking will also allow for better business and time analysis. In the info box is an example of RF-scanning of the surgical patient before the wound is closed to be sure no surgical sponges are left in.[80]

In 2007 the Mayo clinic began using passive RFID tags attached to specimen bottles, used to hold biopsies. The RFID system was provided by 3M and over 30,000 specimens have been processed. The RFID holds a unique patient number stored in a database that must match. The error rate prior to RFID was 9.2%/100 bottles and .55%/100 bottles after transition.[81] A 2008 article raised serious concerns about RFID. When an active or passive RFID tag is read by the scanner it emits electromagnetic inference (EMI). They reported frequent potentially hazardous incidents in a non-clinical scenario when devices like pacemakers and ventilators were exposed to EMI, even at distances greater than 12 inches. Although they tested only RFID tags produced by two vendors, there should be a note of caution with all RFID devices around critical equipment.[82]

Surgical Sponge RFID Detection System

A retained surgical sponge after an operation is a significant patient safety issue because it usually means infection and repeat surgery. Sponges embedded with RFID technology means the abdomen can be scanned prior to closing to be sure no sponges were left in. One study of 1,600 patients showed no false positive or negative results.[80]

Medication Reconciliation

It is well known that when patients transition from hospital-to-hospital, from physician-to-physician or from floor-to-floor, medication errors are more likely to occur. Home medications are occasionally forgotten or incompletely recorded. The Joint Commission mandated hospitals must reconcile a list of patient medications on admission, transfer and discharge. Medication reconciliation is now part stage I Meaningful Use criteria. A report of "errors of transition" concluded the following: 66% occurred at transition to another level of care e.g. ICU, 22% occurred on admission and 12% occurred on discharge.[83] If all medical offices, pharmacies and hospitals had the same EHR or were connected to a shared health information organization, then the answer would be simpler and electronic. Instead, we find completely disparate systems that are not interoperable. Patients can compound the issue by using multiple pharmacies, taking alternative drugs and not keeping records. Multiple IT solutions are available but none are comprehensive because of the disparate process. The following are vendors or initiatives related to medication reconciliation.[84-87]

The significance of having prior prescribing information available at the time a prescription is written should not be underestimated. Researchers reported a study in which clinicians were given six months of prescription claims data compared to a control group with no such information. Those with the additional information were more likely to change dosages (21% vs. 7%); add drugs (42% vs. 14%) and discontinue drugs (15% vs. 4%). Also, physicians with prior drug histories detected non-compliance in about one-third of patients versus none in the control group.[88] Another important issue concerning medication error reduction is the ability to reconcile all outpatient medications when a patient is admitted to a hospital. In many instances the information given by the patient is not correct. Lau reported that 61% of patients had at least one drug missing and 33% had two or more drugs missing on initial admission interview.[89]EHRs, HIOs and pharmacy claims data all offer the opportunity to provide additional patient drug history.

While pharmacy claims data derived from pharmacy benefits managers makes sense, it will not help the uninsured who do not have records. Also, many patients take herbal medications they fail to report and are not retrievable electronically.

Electronic Prescribing

Electronic prescribing will be covered in the next chapter

Barriers to Improving Patient Safety through Technology

Organizational

Medicine, as structured within the United States, is primarily a decentralized system with no unifying philosophy. Many small physician groups have limited loyalty to hospitals or other healthcare organizations. They may not interact frequently with other physician offices or healthcare organizations and may not share data. In summary, the U.S. healthcare system was not optimally designed for quality and hence information technology may not solve existing problems. With respect to patient safety, organizational barriers are often substantial and broad in scope. A critical organizational challenge to patient safety realized in hospitals and health systems is the absence of a "culture of safety."[90-93] Other organizational barriers cited in the literature include clinical workflow issues, [47,49,91,94] communication challenges and a lack of teamwork among the health care team, to name just a few. [90,93]

Financial

Who will pay for what? It is estimated that it will cost $500 to $700 billion dollars over the next 10 years to have a full-fledged interoperable electronic health record nationwide. This is 3% to 4% of the total health care budget which is a lower percentage than what other industries spend on technology. In 1996 the healthcare industry spent about $543 per worker as compared to $12,666 per worker spent by security brokers and other industries for information technology.[94] Patient safety is organizationally represented as a cost for healthcare organizations as there is no income stream associated with keeping patients safe. Accordingly, financial constraints have been cited as a barrier to adoption of patient safety-oriented health IT. There is, however, the expectation that patient safety activities may at least in part pay for themselves through reduced litigation, and nurse time utilization.[95]

Error reporting

Reporting continues to be voluntary and inadequate at best. In a recent study of over 90,000 voluntary electronic error reports from 26 hospitals, most were from nurses and only 2% were reported by physicians.[96] A survey of over 1,000 physicians revealed that 45% did not know if their institution had an error reporting system. Seventy percent (70%) thought the current reporting systems were inadequate. Physicians believed reporting would improve if information was kept confidential, nondiscoverable, it was quick to input and it was nonpunitive.[97] Clearly, there are legal and licensure issues associated with error reporting. Currently, there is no universal method to standardize error reporting in the US. A new FDA portal is discussed in a prior section.[18] Previously, alerts to physicians about defective devices and drug alerts were mailed. To improve the situation the Health Care Notification Network was created that will e-mail alerts as well as public health emergencies and bioterrorism events.[98, 99]

Classen et al. reported the results of using three medical error reporting tools in an evaluation of admissions to three large tertiary hospitals with robust patient safety programs. The tools were the "global trigger tool" developed by the Institute for Healthcare Improvement, the Quality Patient Safety Indicator developed by AHRQ and the Utah/Missouri Adverse Event Classification. It should be noted that they evaluated adverse events, regardless whether harm was preventable or not. The global trigger tool utilized a non-physician primary team of chart reviewers as well as a team of physicians or secondary review. They examined discharge notes and codes, medications, operation records, all progress notes and any other note to indicate a trigger, such as antidote administered. The major findings were that the global trigger tool detected far more adverse events than the other tools and had a sensitivity of 95% and a specificity of 100%. As a result, adverse events occurred in 33% of hospital admissions. They believed that only real time reporting would be more accurate.[100]

Error reporting is important but is "after the fact" so it fails to prevent morbidity and mortality. This may change as more organizations have robust IT systems coupled with artificial intelligence and rules engines. For example, the Cleveland Clinic is analyzing its patient charts with a cycle of analytics to look for potential complications while the patient is still in the hospital.[101]

Future Trends

We anticipate the eventual development of standardized patient safety parameters and triggers as well as a universal reporting system. Once that is in place there will be a movement from retrospective to real time data analysis to detect and/or prevent adverse patient events. This will require a very robust information technology system with a data warehouse that is assisted by evidence based rules engines and artificial intelligence to pick up issues real time. We also anticipate more penalties from payers who are becoming less tolerant to healthcare systems associated with greater than average adverse events.

Jha and Classen are of the opinion that patient safety reporting should be part of Meaningful Use such that EHR vendors must include patient safety features to record, track and trend adverse events.[102] This would likely require mandated inpatient and outpatient adverse event capture in the EHR but how often would this require human inputting?

Key Points

- Patient safety is a major issue facing U.S. medicine today. Far too many people die from medical errors each year.

- Both governmental and non-governmental organizations and programs are tackling patient safety challenges in healthcare organizations; health IT projects are broadly supported by these organizations as a strategy to improve current conditions.

- There is great hope that information technology, particularly clinical decision support as part of the electronic health record, will improve patient care and safety.

- There is some evidence that clinical decision support and alerts may reduce medication errors.

- Bar code medical administration also appears to reduce some medication related errors but is expensive and complicated.

- A dedicated and focused patient safety strategy and culture should accompany any deployment of health information technology.

Conclusion

Better studies are needed to clearly demonstrate health information technology consistently improves patient safety. Until then, we will have to rely on anecdotal and limited studies. Somewhat surprising, there is not a national database or method to store and analyze medical errors.[97] Moreover, CEOs and CIOs will be looking for a reasonable return on investment. However, if improved patient safety means a larger market share, fewer law suits or a better hospital ranking by the state or federal government, then adoption will likely occur. According to HealthGrades, there is evidence that the highest ranked hospitals for quality have lower mortality rates.[103] Additionally, it appears that the most wired hospitals also have lower mortality rates but it is too early to establish clear-cut cause and effect.[104] One could also draw on the experience of the Veterans Affairs hospitals to show how their electronic health record has markedly improved the quality of care and efficiency.[105] Is their dramatic systemic improvement solely due to their EHR or is it due to the visionary Dr. Kiser who saw the need for modernization and the establishment of a culture of quality and safety? A study by Menachemi et al. evaluated 98 Florida hospitals' IT adoption and patient outcome measures and concluded there was a definite correlation. They felt that IT systems for clinicians provided up-to-date guidelines at the point of care.[106] The relationship between HIT and patient outcomes is likely to be more complicated and involves more than just technology, such as the effects of better leadership, training, etc.

References

1. To Err is Human: building a safer health system. 2000. IOM http://www.nap.edu/books/0309068371/html/ (Accessed October 10 2009)

2. Crossing the Quality Chasm: A New Health System for the 21st Century (2001) Institute of Medicine (IOM) http://lab.nap.edu/books/0309072808/html/3.html (Accessed October 4 2009)

3. Agency for Health Care Research and Quality. Patient Safety Glossary. http://psnet.ahrq.gov/glossary.aspx (Accessed November 1 2011)

4. McDonald CJ, Weiner M, Hui, SL. Deaths Due to Medical Errors are Exaggerated in Institute of Medicine Report JAMA 2000;284(1): 93-95

5. Types of medical errors commonly reported by family physicians. American Family Physician 2003 67(4):697

6. Leape LL. Preventing Medical Injury. Quality Review Bull. 1993;19(5):144-149

7. Vilamovska AM, Conklin A. Improving Patient Safety: Addressing Patient Harm Arising form Medical Errors. RAND Policy Insight. April 2009 Vol 3. Issue 2

8. Airline industry since 1970 http://www.airsafe.com/airline.htm (Accessed October 4 2005)

9. Shreve J, Van Den Bos J, Gray T et al. The Economic Measurement of Medical Errors. Milliman 2010. http://www.crosstelecom.com/assets/pdf/research-econ-measurement.pdf (Accessed October 10 2010)

10. Langreth R .Fixing Hospitals. Forbes, June 20, 2005. http://www.forbes.com/forbes/2005/0620/068.html (Accessed January 13, 2012)

11. Patient Safety: Achieving a new standard of care http://www.nap.edu/catalog/10863.html (Accessed April 19 2006)

12. Health IT and Patient Safety: Building Safer Systems for Better Care (2011) Institute of Medicine. www.iom.org/reports (accessed November 23 2011)

13. HealthGrades Excellence Awards 2011. http://cdn.mm-health.com/7e/652a00f8d011e08a3c12313d033e31/file/NEW%20CONSUMERISM%20PDF.pdf (Accessed November 6 2011)

14. Agency for Healthcare Research and Quality www.ahrq.gov (Accessed October 16 2009)

15. Baker B. Hospitals works with admitting doctors on documentation. ACP Internist. October 2008, p. 9.

16. Partnership for Patients: A Common Commitment (2011). HealthCare.Gov. http://www.healthcare.gov/compare/partnership-for-patients/about/index.html (Accessed November 2 20111)

17. About HRSA. (2011). Health Resources and Services Administration. http://www.hrsa.gov/about/index.html (Accessed November 23 2011)

18. FDA Developing Web Portal To Ease Adverse Event Reporting. October 24 2008. www.ihealthbeat.org (Accessed October 28 2008)

19. FDA outlines oversight of mobile medical applications. (2011). U.S. Food and Drug Administration http://www.fda.gov/newsevents/newsroom/pressannouncements/ucm263340.htm (Accessed November 23 2011)

20. Cerrato, P. Mobile medical apps meet the FDA, part 2. Information Week Healthcare. October 31, 2011. http://www.informationweek.com/news/healthcare/mobile-wireless/231901916 (Accessed November 23 2011)

21. AMIA advises FDA on clinical decision support via mobile medical applications. American Medical Informatics Association. November 14th, 2011. http://www.amia.org/news-and-publications/press-release/amia-advises-fda (Accessed November 23 2011)

22. NASHP http://www.nashp.org/pst-welcome (Accessed September 22 2011)

23. Weinberg J, Hilborne LH, Nguyen QT. Regulation of Health Policy: Patient Safety and the States. www.ahrq.gov/downloads/pub/advances/vol1/Weinberg.pdf (Accessed September 26 2011)

24. National Patient Safety Foundation: about us. (2011). http://www.npsf.org/au/mission_vision.php (Accessed November 25 2011)

25. National Quality Forum. www.qualityforum.org (Accessed September 27 2011)

26. The Joint Commission. www.jointcommission.org (Accessed November 7 2011)

27. Joint Commission Alert: Prevent Technology-Related Health Care Errors. Patient Safety & Quality Healthcare December 17 2008 www.psqh.com (Accessed December 17 2008)

28. Facts about the National Patient Safety Goals. (2011). The Joint Commission. http://jointcommission.org/facts_about_the_national_patient_safety_goals (Accessed November 25 2011)

29. Institute for Healthcare Improvement www.ihi.org (Accessed November 7 2011)

30. Leapfrog www.leapfroggroup.org (Accessed November 8 2011)

31. HealthGrades. www.healthgrades.com (Accessed November 8 2011)

32. About AMIA. (2011). The American Medical Informatics Association. http://www.amia.org/about-amia (Accessed November 25 2011)

33. Institute for Safe Medication Practices http://www.ismp.org/faq.asp (Accessed November 24 2011)

34. Averting Highest Risk Errors Is First Priority. Patient Safety & Quality Healthcare May/June 2005

35. Fattinger K et al. Epidemiology of drug exposure and adverse drug reactions in two Swiss departments of internal medicine. Br J Clin Pharm 2000;49(2):158-167

36. Moore TJ et al. Serious Adverse Drug Events Reported to the Food and Drug Administration, 1998-2005. Arch Intern Med. 2007;167:1752-1759

37. Rooney Cl. Increase in US medication-error deaths. Letter to the Editor Lancet 1998;351:1656-1657.

38. Bates DW et al. Incidence of Adverse Drug Events and Potential Adverse Drug events: Implications for Prevention. JAMA 1995;274:29-34

39. Fitzhenry F et al. Medication Administration Discrepancies Persist Despite Electronic Ordering. JAMIA 2007;14:756-764

40. Preventing Medication Errors. Committee on Identifying and Preventing Medication Errors, Aspden P, Wolcott A, Bootman KL, Cronenwett, LR (eds). Institute of Medicine. The National Academies Press. Washington, DC 2007

41. Oren E, Shaffer ER and Guglielmo JB. Impact of emerging technologies on medication errors and adverse drug events. Am J Health Syst. Pharm 2003;60:1447-1458

42. Chaudhry B et al. Systematic Review: Impact of Health Information Technology on Quality, Efficiency and Costs of Medical Care. Ann of Int Med 2006;144:E-12-E-22

43. Furukawa MF, Raghu TS, Spaulding TJ, Vinze A. Adoption of Health Information Technology for Medication Safety in US Hospitals, 2006. Health Affairs 2008;27:865-875

44. Pedersen CA, Gumpper KF. ASHP national survey on informatics: Assessment of the adoption and use of pharmacy informatics in US Hospitals—2007. Am J Health-Syst Pharm 2008;65:2244-2264

45. Bates DW et al. Effect of computerized physician order entry and a team intervention on prevention of serious medication errors. JAMA 1998;280:1311-1316

46. Masslove, DM, Rizk, N., Lowe, H.J. Computerized physician order entry in the critical care environment: a review of current literature. J Intensive Care Med. 2011; 26(3): 165-71

47. Reckmann, MH, Westbrook, J.I., Koh, Y, et al. Does computerized provider order entry reduce prescribing errors for hospital inpatients? A systematic review. J Am Med Inform Assoc, 2009; 16(5): 613-23

48. Khajouei, R, Jaspers, M.W. The impact of CPOE medication systems' design aspects on usability, workflow and medication orders: a systematic review. Methods Inf Med. 2010; 49(1): 3-19

49. Wolfstadt, J.L., Gurwitz, J.H., Fields, T.S. et al. The effect of computerized physician order entry with clinical decision support on the rates of adverse drug events: a systematic review. J Gen Intern Med. 2008; 23(4): 451-8

50. Ammenwerth, E., Schnell-Inderst, P., Machan, C. et al. The effect of electronic prescribing on medication errors and adverse drug events: a systematic review. 2008; 15(5): 585-600

51. Kuo GM, Phillips RL, Graham D. et al. Medication errors reported by US family physicians and their office staff. Quality and Safety In Health Care 2008;17(4):286-290

52. Kuperman GJ, Bobb A, Payne TH et al. Medication-related clinical decision support in computerized provider order entry systems: A review. J Am Med Inform Assoc 2007;14 (1):29-40

53. Casolino LP, Dunham D, Chin MH et al. Frequency of Failure to Inform Patients of Clinically Significant Outpatient Test Results. Arch Intern Med 2009;169(12):1123-1129

54. Halamka, J The Safety of HIT-Assisted Care. February 1 2011. Life as a Healthcare CIO. http://www.geekdoctor.blogspot.com(Accessed February 4 2011)

55. Murray M. Automated Medication Dispensing Devices http://www.ahrq.gov/clinic/ptsafety/chap11.htm (Accessed April 23 2006)

56. Nebeker, J.R., Hoffman, J.M., Weir, C.R. et al. High rates of adverse drug events in a highly computerized hospital. Arch Intern Med. 2005; 165(10): 1111-6

57. Woehick, H.J., McQueen, A.M., & Connolly, L.A. Off-hours unavailability of drugs during emergency situations with automated drug dispensing machines. Can J Anaesth. 2007; 54(5): 403-4

58. Balka, E., Kahnamoui, N. & Nutland, K. Who is in charge of patient safety? Work practices, work processes and utopian views of automatic drug dispensing systems. Int J Med Inform. 2007; 76(Suppl. 1): S48-57

59. In Range Systems www.inrangesystems.com) (Accessed November 24 2007)

60. Hospital Pharmacist http://www.pjonline.com/pdf/papers/pj_20050618_automateddispensing.pdf (Accessed April 10 2006)

61. ScriptPro www.scriptpro.com (Accessed November 15 2007)

62. Ascend eMAR www.hosinc.com (Accessed April 23 2006)

63. Winterstein AG, Hatton RC, Gonzalez-Rothi R. Identifying clinically significant preventable adverse drug events through a hospital's database of adverse drug reaction reports. Am J of Health Sys Pharm 2002;59:1742-1749

64. Vanderveen T. Smart Pumps: Advanced and continuous capability. Patient Safety & Quality Healthcare. Jan/Feb 2007. p40-48

65. Rothschild JM, Keohane CA, Cook EF. A Controlled Trial of Smart Infusion Pumps to Improve Medication Safety in Critically Ill Patients. Critical Care Medicine 2005;33 (3):533-540

66. Vanderveen T. IVs First, a New Barcode Implementation Strategy. Patient Safety & Quality Healthcare May/June 2006

67. Ball MJ, Merryman T, Lehmann CU. Patient Safety: A tale of two institutions. J of Health Info Man 2006;20:26-34

68. Cummings J, Bush P, Smith D, Matuszewski K. Bar-coding medication administration overview and consensus recommendations. Am J Health-Syst Pharm 2005;62:2626-262

69. Miliard M. Cook Children's pioneers 2D barcode system for vaccines. Healthcare IT News. October 13 2011. http://www.healthcareitnews.com(Accessed October 13 2011)

70. Poon EG et al. Medication Dispensing Errors and Potential Adverse Drug Events before and after Implementing Bar Code Technology in the Pharmacy. Ann of Int Med 2006;145:426-434

71. Poon EG, Keohane CA, Yoon CS et al. Effect of Bar Code Technology on the Safety of Medication Administration. NEJM 2010;362 (18):1698-1707

72. DeYoung JL, Vanderkooi ME, Barlfletta JF. Effect of bar code assisted medication administration on medication error rates in an adult medical intensive care unit. Am J Health Syst Pharm 2009;66(12):1110-5

73. Coyle GA, Heinen M. Evolution of BCMA within the Department of Veterans Affairs. Nurs Admin Q 2005;29:32-38

74. Murphy D. Barcode basics. Patient Safety & Quality Healthcare. July/August 2007:40-44

75. Hayden RT et al. Computer-Assisted Bar-Coding System Significantly Reduces Clinical Laboratory Specimen Identification Errors in a Pediatric Oncology Hospital. J Pediatr 2008;152:219-24

76. Decisionmaker Brief: Bar-Coded Medication Administration (BCMA) http://healthit.ahrq.gov (Accessed September 24 2008)

77. Gee T. Bar coding: Implementation Challenges. Patient Safety & Quality Healthcare. March/April 2009 pp 22-26

78. Koppel R, Wetterneck T, Telles JL et al. Workarounds to Barcode Medication Administration Systems: Their Occurrences, Causes and Threats to Patient Safety. JAMIA 2008;15:408-423

79. Young J, Slebodnik M, Sands L. Bar Code Technology and Medication Administration Error. J Patient Saf 2010;6:115-120

80. RF Surgical Systems. http://www.rfsurg.com/benefits.htm (Accessed January 4 2008)

81. Study:RFID Technology Can Reduce Errors During Biopsy Analysis. October 8 2008 www.ihealthbeat.org (Accessed October 9 2008)

82. van der Togt R et al. Electromagnetic Interference From Radio Frequency Identification Inducing Potentially Hazardous Incidents in Critical Care Medical Equipment. JAMA 2008;299(24):2884-2890

83. Clancy, C. Medication Reconciliation: Progress Realized, Challenge Ahead. Patient Safety & Quality Healthcare. July/August 2006. www.psqh.com (Accessed July 20 2007)

84. RelayHealth. www.relayhealth.com (Accessed July 20 2007)

85. MedsTracker. www.designclinicals.com/meds-tracker.html (Accessed July 20 2007)

86. DrFirst and MEDITECH to deliver sophisticated mediation reconciliation, a strategic alliance to integrate. EHR Consultant. January 7 2008 www.emrconsultant.com (Accessed January 12 2008)

87. Hamann C et al. Designing a medication reconciliation system. AMIA 2005 Proceedings p 976 (Accessed January 12 2008)

88. Bieszk N et al. Detection of medication non-adherence through review of pharmacy claims data. Am J Health Syst Pharm 2003; 60:360-366

89. Lau HS et a.l The completeness of medication histories in hospital medical records of patients admitted to general internal medicine wards. Br J Clin Pharm 2000;49:597-603

90. Patient safety primers: safety culture (2011). Agency for Healthcare Research & Quality. http://psnet.ahrq.gov/primer.aspx?primerID=5 (Accessed November 25 2011)

91. Pronovost, P.J., Weast, B., Holzmueller, C.G., et al. (2003). Evaluation of the culture of safety: survey of clinicians and managers in an academic medical center. Qual Saf Health Care. 2003; 12:405-10

92. Huang, D.T., Clermont, G., Sexton, J.B. et al. Perceptions of safety culture vary across the intensive care units of a single institution. Crit Care Med. 2007; 35:165-76

93. White, D.E., Straus, S.E., Stelfox, H.T. et al. What is the value and impact of quality and safety teams? A scoping review. Implem Sci. 2011; 6:97

94. Health Professions Education: A Bridge to Quality (2003) Board on Health Care Services (HCS) Institute of Medicine (IOM) http://darwin.nap.edu/books/0309087236/html/29.html (Accessed February 11 2006)

95. Kaushal, R., Jha, A.K., Franz, C. et al. Return on investment for a computerized physician order system. J Am Med Inform Assoc. 2006; 13(3): 261-6

96. Milch CE et al. Voluntary Electronic Reporting of Medical Errors and Adverse Events. J of Gen Int Med 2006

97. Garbutt J, Waterman AD, Kapp JM et al. Lost Opportunities: How Physicians Communicate About Medical Errors. Health Affairs 2008;27(1):246-255

98. Health Care Notification Network www.hcnn.net (Accessed June 20 2009)

99. Pizzi R. Cardiologists and OB-GYNs join online national drug alerts network. Healthcare IT News July 22 2008. www.healthcareitnews.com (Accessed June 20 2009)

100. Classen DC, Resar R, Griffin F et al. Global Trigger Tool Shows that Adverse Events in Hospitals May be Ten Times Greater than Previously Measured. Health Affairs. 2011;30(4):581-589

101. Goedert J. Cleveland Clinic Puts Charts to the Test. October 4 2011. www.healthdatamanagement.com (Accessed October 5 2011)

102. Jha AK, Classen DC. Getting Moving On Patient Safety—Harnessing Electronic Data for Safer Care. NEJM 2011;365:1756-1758

103. Study: Hospitals rated in top 5% have mortality rates 27% lower. Patient Safety & Quality Healthcare March/April 2006: 57

104. Annual list of most-wired hospitals released. 7/12/2005. www.ihealthbeat.org (Accessed April 5 2006)

105. Stires D. Technology has transformed the VA. CNN Money http://money.cnn.com . May 5 2006 (Accessed May 5 2006)

106. Menachemi N et al. Hospital Adoption of Information Technologies: A Study of 98 Hospitals in Florida. J of Healthcare Man Nov/Dec 2007;52(6)

19

Electronic Prescribing

ROBERT E. HOYT

KENNETH G. ADLER

Learning Objectives

After reading this chapter the reader should be able to:

- Identify the problems and limitations of handwritten prescriptions

- List the confirmed and potential benefits of electronic prescribing

- Describe the SureScripts network and how e-prescribing works

- Enumerate the steps towards e-prescribing adoption and implementation

- List the obstacles to widespread e-prescribing

Introduction

E-prescribing is simply the generation of a digital prescription that can be transmitted to a pharmacy over a secure network. The reason we are moving in this direction can be demonstrated by the prescription shown in Figure 19.1 that resulted in a patient death as the pharmacist interpreted the drug prescribed as Plendil and not Isordil. This was the first medical malpractice case successfully prosecuted due to illegible handwriting. The jury awarded $450,000 in damages with 50% of responsibility on the cardiologist and 50% on the pharmacist.[1]

Cases such as this led the Institute for Safe Medication Practices (ISMP) to push to "eliminate handwritten prescriptions within three years." The ISMP points out that up to 7,000 Americans die each year due to medication errors resulting in a cost of about $77 billion annually.[2] A majority of these errors are due to illegible handwriting, wrong dosing and missed drug-drug or drug-allergy reactions. E-prescribing has been promoted as a solution to improve patient safety, decrease costs and streamline the prescription process.

Figure 19.1: Illegible prescription

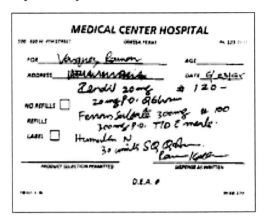

Approximately three billion prescriptions are written annually in the United States but the majority are still paper-based. This trend is changing, due to increased EHR adoption and the multiple advantages of e-prescribing:

- Legible and complete prescriptions that help eliminate handwriting errors and decrease pharmacy "callbacks" (150 million yearly) and rejected scripts (30%) [3]

- Abbreviations and unclear decimal points are avoided

- The wait to pick up prescriptions would be reduced

- Fewer duplicated prescriptions

- Better compliance with fewer drugs not filled or picked up

- Should reduce workload for pharmacists

- Timely notification of drug alerts and updates

- Better use of generic or preferred drugs

- The ability to check plan-level and patient-level formulary status and patient copays

- E-prescribing can interface with practice and drug management software

- The process is secure and HIPAA compliant

- It is the HIT platform for future clinical decision support, alerts and reminders. It could integrate decision support related to both disease states and medications

- Digital records improve data analysis of prescribing habits

- Programs offer the ability to look up drug history, drug-drug interactions, allergies and compliance

- While entering an e-script is slower than writing a paper script, clinicians have options to speed up the process like batch refills and choosing from lists of drugs most commonly prescribed in a practice

- Provides a single view of prescriptions from multiple clinicians

- Applications have the ability to check eligibility, co-pays and it can file drug insurance claims

- Reduced cost. A 2005 study suggested that e-prescribing reduced labor costs $0.97 for a new prescription and $0.37 for a renewed prescription [4]

The Medical Group Management Association published an important report documenting the time and money spent by physicians and staff to refill medications, verify proper formulary choices, etc. It demonstrated that non-electronic prescribing can be time intensive and expensive if you factor in the calls back and forth to a pharmacy. For an average 10 physician medical group there were an average of seven phone calls per day, 63% for refills at an annual price tag of $157,000 for the time spent by the office staff and physicians to handle pharmacy related calls. [5]

The Center for Information Technology Leadership's 2003 *Report on the Value of Computerized Physician Order Entry in Ambulatory Settings* estimated that e-prescribing would save $29 billion annually from fewer medication errors; reduced overuse, misuse and adverse drug event related hospitalizations and more cost effective selection of generic or less expensive medications.[6]

The concept of e-prescribing is gathering momentum in the United States and all states now approve eRx. The following are summary data about the status of e-prescribing in the United States as of the end of 2010, from the National Progress Report on E-prescribing and Interoperable Healthcare:

- Prescription Benefit: Electronic responses to requests for prescription benefit information grew 125% from 188 million in 2009 to 423 million in 2010

- Medication History: Prescription histories delivered to prescribers grew 184% from 81 million in 2009 to 230 million in 2010

- Prescription Routing: Prescriptions routed electronically grew 72% from 191 million in 2009 to 326 million in 2010

- EMR vs. Standalone E-Prescribing Software: About 79% of e-prescribers used EMRs in 2010, up from 70% in 2009

- Prescribers: The number of prescribers routing prescriptions electronically grew from 156,000 at the end of 2009 to 234,000 by the end of 2010, representing about 34% of all office-based prescribers

- Payers: At the end of 2010, Surescripts could provide access to prescription benefit and history information for more than 66% of patients in the U.S

- Community and Mail Order Pharmacies: At the end of 2010, approximately 91% of community pharmacies in the U.S. were connected for prescription routing and six of the largest mail order pharmacies were able to receive prescriptions electronically[7]

E-prescribing is available in two modes: (1) a standalone software program installed on a PDA/smartphone, computer or available as a web-based application and (2) eRx integrated into an electronic health record. The standalone choice is simpler, less expensive and easier to learn but the EHR mode offers greater patient information at the time of prescribing. With the web-based application, patient data resides on a remote server and not on your computer. Another point worth mentioning is, although an increase in e-prescribing is occurring, many clinicians choose to print out an electronic script and then fax it to a pharmacy, thus negating several of the benefits of electronic prescribing. In order for standalone e-prescribing to function well, patient lists need to be uploaded into the system to eliminate manually inputting of information for new patients at the time of e-prescribing. Patient lists can derive from practice management systems or a practice can pay for an interface to be built to upload practice patient demographic.

The federal government has supported e-prescribing for many years. The first incentive program was authorized under the Medicare Improvements for Patients and Providers Act of 2008 (MIPPA) and was implemented in 2009 and 2010. Changes were made in 2011: the incentive decreased from 2% to 1%; a penalty was introduced for eligible professionals that have not implemented a qualified e-prescribing system by November 1, 2011. The penalty is 1% of total allowed charges in 2012 and a 1.5% penalty in 2013; physicians eligible for reimbursement by Medicare/Medicaid for EHR use as a result of HITECH will not qualify for MIPPA reimbursement.

As pointed out in chapters 1 and 2, the HITECH Act, as part of the American Recovery and Reinvestment Act, will include reimbursement by Medicare and Medicaid for EHRs, starting in 2011. To achieve Meaningful Use eligible physicians will have to utilize e-prescribing. All EHR systems certified in 2008 to 2011 by CCHIT meet these requirements. SureScripts® lists the standalone and EHR vendors who have certified products and the exact functionality that was certified on their web site. The following are ways that Stage 1 Meaningful Use affects e-prescribing:

- 40% of prescriptions must be transmitted electronically
- Drug-drug and drug-allergy interaction checking must be performed
- Active medication lists must be maintained
- Allergy list must be maintained
- Medication reconciliation must be performed
- Formulary checking (menu option in Stage 1, expected to be required in Stage 2)

On March 31st 2010 the DEA published a final rule that allowed for controlled drugs (narcotics, etc.) to be prescribed electronically. As the rule stands, users will have to provide two-factor authentication (i.e. two means of identification): something you know (e.g. password), something you have (e.g. hard token or smart card) or something you are (e.g. finger print). It will be up to EHR and e-prescribing vendors to possibly modify their software to comply with all DEA regulations. Their software will need to be certified for E-prescribing of Controlled Substances (EPCS). In addition, clinicians will need to apply to a private entity that verifies identities. As of 2011 not all states have aligned their prescribing regulations with the latest DEA rules.[8]

Surescripts Network

SureScripts was founded in 2001 by the National Association of Chain Drugs Stores and the National Community Pharmacists Association to improve the quality, safety and efficiency of the overall prescribing process. One of the strongest motivating forces behind this collaboration was the need to reduce the number of physician call backs. Figure 19.2 demonstrates how the network ties together the multiple parties related to e-prescribing.

The Surescripts Network is the largest network to link electronic communications between pharmacies and physicians, allowing the electronic exchange of prescription information. In 2009 about 90% of major pharmacy chains and about 50% of independent pharmacies were certified to connect to the network. Surescripts works with software companies that supply electronic health records (EHRs) and electronic prescribing applications to physician practices and pharmacy technology vendors to connect their solutions to the network. Although the network is free to physicians, pharmacies pay a small amount (21.5 cents per transaction) to the software vendor like DrFirst and they in turn pay Surescripts. Vendors must complete a certification process that establishes rules that safeguard the prescribing process, to include patient choice of pharmacy and physician choice of therapy.

SureScripts created the Center for Improving Medication Management to promote research into improved prescribing through technology. In 2008 Surescripts merged with RxHub, a pharmacy benefits managing operation to create one organization.[9]

Figure 19.2: Surescripts Network connections (Courtesy Surescripts)

How E-Prescribing Works

Typically, a patient arrives at a physician's office with a new medical problem. As part of the check-in process or during the visit with the physician the patient's prescription benefits and potential co-pays can be reviewed. If a medication is prescribed, the prior prescription history is available, known drug allergies are posted and the formulary choices offered by the insurance company can be reviewed. Checking eligibility and formulary choices will decrease pharmacy call-backs. Accessing the medication history may point out past compliance problems. The electronic prescription is forwarded electronically to the patient's pharmacy of choice where it is queued until it is filled by the pharmacist.

In the case of the renewal or refill of a chronic medication the patient usually contacts their pharmacy and the pharmacist sends a secure message to the appropriate physician. The clinician can generally renew or refill a single or multiple medications with a mouse click. The electronic approval is then sent back to the pharmacy for processing.

E-Prescribing Clinical Decision Support

It is not thought that simply switching from paper to electronic prescriptions will improve patient safety; it will require clinical decision support (CDS) discussed in the chapter on electronic health records.

Perhaps the most important CDS is the reminder that a patient has a confirmed allergy to a drug, thus preventing a potential serious reaction. It is most helpful if the actual details of the allergy are listed. The next important CDS feature is drug-drug interaction determination. In elderly patients on multiple medications it is particularly important to understand the effect of one drug on another. Notification of an interaction will usually cause the prescribing physician to reduce the dose of one drug or make another safer choice. There are many other types of CDS that might be important associated with e-prescribing. Drug-condition/disease alerts might remind a physician that drug A is not safe in a pregnant woman. Reminders about dosages out of range (too high or too low), age or BMI extremes would be very valuable, particularly with toxic drugs such as chemotherapy medications. Reminders about duplicate drugs and drugs prescribed by other physicians are also very important.

As electronic health records become smarter by using rules engines and artificial intelligence we can expect alerts about potential prescribing problems based on liver or kidney problems and other considerations. Eventually, there may be summary alerts based on age, gender, BMI, liver/kidney function, etc. such as "This patient is at risk of drug side effects, recommend lisinopril dose reduction by 50%."

Current Knowledge

This section will discuss several interesting articles about e-prescribing. The chapters on EHRs and patient safety also discuss aspects of computerized physician order entry (inpatient and ambulatory) and clinical decision support. More detail about drug alerts and patient safety can be found there. The following is a summary of the articles:

- A 2008 systematic review reported that a majority of studies showed a reduction in adverse drug events (ADEs) and potential ADEs.[10]

- A handwritten versus computerized prescription study was conducted in an emergency room setting. Computer written scripts were three times less likely to include errors and five times less likely to require a pharmacist's clarification. This resulted in a decrease in wait time for patients and call back time by pharmacists to physicians.[11]

- A 2009 studied examined drug alerts from e-prescribing and found that physicians tended to find them more of a nuisance than an educational tool.[12]

- One study evaluated the pharmacist's perspective and disclosed unique new e-prescribing issues: incorrect drugs, doses and patient instructions continue to occur; in spite of an electronic process prescribing delays persisted. They recommended that only clinicians forward e-prescriptions, clinical decision support should be used, scripts should be sent together (bundled); software standardization would be helpful and there should be a mechanism to message physicians about issues.[13]

- DesRoches et al. reported the results of a large survey of physicians who had either an integrated (part of EHR) eRx system or standalone system. The physicians who used the integrated system tended to be younger and practice in larger systems. These same physicians used their eRx system more extensively, in more depth and in a manner that was more educational.[14]

- A study of 3850 outpatient electronic prescriptions reported in 2011 revealed an error rate of 11.7%, with about a third having the potential to cause adverse drug events (ADEs). Two thirds of the prescribing errors were due to omissions of drug dose, instructions, etc. Actual ADEs were not reported.[15]

- A qualitative study of e-prescribing was reported in 2011 and recorded some of the existing issues physicians and pharmacists are facing[16]:

  o The refill process had more problems and errors than the initial new prescription process and resulted in workarounds for both physicians and pharmacies

  o Some pharmacies don't accept electronic scripts because they don't want to pay Surescripts fees

  o Mail order pharmacies still lack consistent e-prescribing capabilities. Most of their refills are still done by fax o Physicians write sigs (instructions) that aren't patient friendly and pharmacists have to rewrite them

o Physicians often receive duplicate requests from pharmacies for a variety of reasons

Preparing for E-Prescribing

An excellent starting point to prepare for eRx would be to review the *Clinician's Guide to Electronic Prescribing*, cited previously. This comprehensive resource provides guidance in multiple areas:

- Section 1: E-Prescribing as an Enabler to Transform Care

- Section 2: E-Prescribing, Health Information Technology for Economic and Clinical Health Act (HITECH) and "Meaningful Use"

- Section 3: Understanding Medicare Improvements for Patients and Providers Act (MIPPA)

- Section 4: Becoming a Successful, Meaningful User of E-Prescribing

- Section 5: Electronic Prescribing of Controlled Substances (EPCS)

- Appendices: Buyer's Guide, FAQs, Resources, Meaningful Use Criteria and Vision, and Registration for EPCS[8]

E-Prescribing Obstacles and Issues

In spite of increasing e-prescribing adoption challenges remain:

- Initial e-prescribing is slower than paper scripts, but not when you factor in time spent calling back pharmacists or playing "phone tag."[17] Refilling previously e-prescribed prescriptions is often faster than having to write out paper prescriptions.

- New errors have appeared due to the newness of electronic prescribing, in spite of improved legibility and certification. [15,18]

- E-prescribing systems should have data validation to not allow submission if there is missing information

- The graphical user interface (GUI) should be standardized

- EHR certification by ONC-ATCB should incorporate lessons learned and add features that would decrease medication errors

Future Trends

We anticipate continued increase in e-prescribing as it becomes the standard of care. Most eRx will be integrated with an electronic health record and will benefit from steady improvement in clinical decision support. Drug alerts will become more intelligent so that we see a decrease in adverse events and a decline in alert fatigue. The process will integrate better into the physician's, patient's and pharmacist's workflow so that eRx will be shown to save everyone's time.

Key Points

- Electronic prescribing adoption is rapidly increasing
- 79% of electronic prescriptions in 2010 were generated from EHRs
- Electronic prescribing has multiple advantages over paper prescriptions, but for impact on patient safety clinical decision support must be used

Conclusion

Electronic prescribing adoption has steadily increased over the last decade. Adoption has been positively impacted by EHR reimbursement by Medicare and Medicaid and e-prescribing as part of an EHR is now the most common mode. While an initial electronic prescription is slower to generate, compared to a paper prescription, the overall process is time neutral or even improved due to fewer pharmacy call backs and a faster refill process. E-prescribing errors are still seen but tend to be minor and related to omission of data which should be remedied with more stringent software certification. The greatest impact on patient safety is likely to occur when robust clinical decision support is used routinely with eRx.

More research is needed to evaluate barriers to adoption. One research obstacle is the delay between when e-prescribing is studied and the eventual publication. Studies reported in 2011 usually reflect studies performed in 2007 to 2008 and therefore don't take into account newer improved technologies and incentives.[19] As lessons learned are incorporated into newer e-prescribing software, we can anticipate an improved experience for clinicians, patients and pharmacists.

References

1. Poison in Prescription. Allscripts 3/19/2001 http://www.allscripts.com/ahsArticle.aspx?id=297&type=News%20Article (Accessed November 2 2005)
2. A Call to Action: Eliminate Handwritten Prescriptions within 3 years! White Paper. Institute for Safe Medication Practices www.ismp.org (Accessed November 3 2005)
3. RxHub www.rxhub.net (Accessed February 22 2007)
4. Rupp, MT. E-Prescribing: The Value Proposition. America's Pharmacist. April 2005. www.surescripts.com (Accessed November 5 2007)
5. The Medical Group Management Association www.mgma.com/about/default.aspx?id=280 (Accessed November 5 2007)
6. The Value of Computerized Provider Order Entry in Ambulatory Settings 2003 www.citl.org/research/ACPOE.htm (Accessed February 22 2007)
7. The National Progress Report on E-prescribing and Interoperable Healthcare. SureScripts www.surescripts.com (Accessed October 15 2010)
8. Clinician's Guide to E-Prescribing http://www.surescripts.com/media/800052/cliniciansguidee-prescribing_2011.pdf (Accessed October 13 2011)
9. Surescripts. www.surescripts.com (Accessed October 13 2011)
10. Ammenwerth E et al. The Effect of Electronic Prescribing on Medication Errors and Adverse Drug Events: A Systematic Review. JAMIA PrePrint June 25 2008 doi:10.1197/jamia.M2667 (Accessed July 2 2008)
11. Bizovi KE et al. The Effect of Computer-Assisted Prescription Writing on Emergency Department Prescription Errors. Acad Emerg Med 2002;9:1168-1175

12. Issac T, Weissman JS, Davis RB et al. Overrides of Medication Alerts in Ambulatory Care. Arch Intern Med 2009;169(3):305-311

13. Rupp MT, Warholak TL. Evaluation of e-prescribing in chain community pharmacy: Best-practice recommendations. J Am Pharm Assoc 2008;48:364-370

14. DesRoches CM, Agarwal R, Angst CM et al. Differences Between Integrated and Stand-Alone E-prescribing Systems Have Implications for Future Use. Health Affairs 2010;29(12):2268-2276

15. Nanji KC, Rothschild JM, Salzberg C et al. Errors associated with outpatient computerized prescribing systems. JAMIA 2011 doi:10.1136/amiajnl-2011-000205 (Accessed July 13 2011)

16. Grossman JM, Cross DA, Boukus ER et al. Transmitting and Processing Electronic Prescriptions: Experiences of Physician Practices and Pharmacies. JAMIA 2011;doi:10.1136/amiajnl-2011-000515

17. Hollingsworth W et al. The Impact of e-Prescribing on Prescriber and Staff Time in Ambulatory Care Clinics: A Time-Motion Study. JAMIA 2007;14:722-730

18. Pharmacists Newsletter www.pharmacistsletter.com December 8 2009 (Accessed March 1 2009)

19. Wang JC, Mihir HP, Schueth AJ et al. Perceptions of Standards-Based Electronic Prescribing Systems as Implemented in Outpatient Primary Care: A Physician Survey. J Am Med Inform Assoc 2009;16:493-5

20

Telemedicine

ROBERT E. HOYT

REYNALD FLEURY

Learning Objectives

After reading this chapter the reader should be able to:

- State the difference between telehealth and telemedicine

- List the various types of telemedicine such as teleradiology and teleneurology

- List the potential benefits of telemedicine to patients and clinicians

- Identify the different means of transferring information with telemedicine such as store and forward

- Enumerate the most significant ongoing telemedicine projects

Introduction

According to the Office for the Advancement of Telehealth, Telehealth is defined as:

> "the use of electronic information and telecommunications technologies to support long-distance clinical health care, patient and professional health-related education, public health and health administration"[1]

Similar to the term e-health, telehealth is an extremely broad term. A review by Oh et al. found 51 definitions for e-health, suggesting that the term is too general to be useful and the same is probably true regarding telehealth.[2] One could argue that Health Information Organizations (HIOs), Picture Archiving and Communication Systems (PACS) and e-prescribing are also examples of telehealth if they exchange healthcare information between distant sites. Clearly, telehealth is the broader term that incorporates clinical and administrative transfer of information, whereas telemedicine relates to remote transmission or exchange of only clinical information. In this chapter we will use the term telemedicine instead of telehealth and define it as follows:

> "the use of medical information exchanged from one site to another via electronic communications to improve patients' health status"[3]

Telemedicine was postulated in the 1920s when an author from Radio News magazine demonstrated how a doctor might examine a patient remotely using radio and television. Ironically this was proposed before television was even available.[4] (Figure 20.1) The first instance of remote monitoring has been attributed to monitoring the health of astronauts in space in the 1960's.[5] Very rudimentary telemedicine has been conducted using telephone communication for the past fifty years or more. With the advent of the internet and video conferencing many new modes of communication are now available.

Figure 20.1: Early Telemedicine (Courtesy Radio News)

The goal of telemedicine ultimately is to provide timely and high quality medical care remotely. Telemedicine is becoming increasingly popular for the following reasons: (1) With the rising cost of healthcare worldwide, newer delivery models are appearing that will include telemedicine. In the case of the United States where Medicare will not reimburse for readmission for certain diseases we need new strategies to prevent readmissions, to include telemedicine. (2) There is a shortage of primary care and intensivist physicians. Moreover, they are maldistributed to urban and not rural areas. Remote delivery of medical care with telemedicine is a partial fix. (3) We need additional means to deliver medical care, given the rise in chronic diseases and our graying population. (4) Telemedicine results in improved collaboration among physicians and disparate healthcare organizations. (5) Telemedicine raises patient satisfaction when it results in better access to specialty care, less time lost from work and/or fewer long distant trips to tertiary medical centers.

Like many of the other topics in health informatics covered in this textbook there are multiple interrelationships. Telemedicine can be employed for disease management and as a strategy for improved patient care and communication, thus being part of consumer health informatics. Telemedicine is slowly being integrated with a variety of technologies and platforms such as electronic health records, PACS, health information organizations, mobile and picture archiving and communication systems. Due to the pervasive nature of mobile technology, it is also a player in telemedicine.

Telemedicine is part of healthcare reform internationally, in that it aims to improve access to high quality care and education remotely. It can be used for populations at risk, such as rural patients and the elderly. As medical care becomes more patient-centric telemedicine will become part of the patient centered medical home and accountable care organization models.

Telemedicine Communication Modes

In this chapter we will mention multiple ways patients can receive remote care, starting from simple e-mail to complex video teleconferencing. In the past several years we have seen new telemedicine technologies and business models appear with more on the way. Table 20.1 shows several of the communication modes used in telemedicine, along with pros and cons.

Table 20.1: Telemedicine Communication Modes

Communication Mode	Pros	Cons
Patient-Portal secure-messaging	Asynchronous. Able to attach photos. Response can be formatted with template. Could use VoIP. Audit trail is available	Not as personal as live visit. Usually not connected to EHR or other information
Telephone	Widely available, simple and inexpensive. Real-time	Not asynchronous. Unstructured. No audit trail. Only real-time
Audio-Video	Maximal input to clinician. Can include review of x-rays, etc. Perhaps more personal than just messaging	Currently, most expensive in terms of networks and hardware but that is changing

Telemedicine Transmission Modes

There are three telemedicine transmission modes:

- Store-and-forward. Images or videos are saved and sent later. As an example, a primary care physician takes a picture of a rash with a digital camera and forwards it to a dermatologist to view when time permits. This method is commonly used for specialties such as dermatology and radiology. This could also be referred to as asynchronous communication.

- Real time. A specialist at a medical center views video images transmitted from a remote site and discusses the case with a physician. This requires more sophisticated equipment to send images real time and often involves two way interactive telemonitors. Telemedicine also enables the sharing of images from peripheral devices such as electronic stethoscopes, otoscopes, etc. This would be an example of synchronous communication.

- Remote monitoring. A technique to monitor patients at home, in a nursing home or in a hospital for personal health information or disease management.

Telemedicine Categories

We are going to divide Telemedicine into the categories noted below based on current knowledge and initiatives. It should be pointed out that virtual patient visits (televisits or e-visits) could be part of telemedicine or consumer health informatics. We have elected to discuss this topic in the chapter on consumer health informatics.

- Televisits: see chapter on consumer health informatics
- Teleconsultations: teleradiology, teledermatology, etc.
- Telemonitoring:
  - Telerounding: hospital inpatients
  - Telehomecare: monitoring physiological parameters, activity, diet, etc. at home

Teleconsultations

Teleconsultation is a worldwide phenomenon because specialists tend to practice in large metropolitan areas, and not in rural areas. Most programs consist of a central medical hub and several rural spokes. Programs attempt to improve access to services in rural and underserved areas, to include prisons. This reduces travel time and lowers the cost for specialists and patients alike. Programs have the potential to raise the quality of care delivered and help educate remote rural patients and physicians. The most commonly delivered services are mental health, dermatology, cardiology and orthopedics. Currently, in the United States there are over 200 telemedicine programs that are operational in forty eight states. [5]

Teleradiology

The military has taken the lead in this area partly due to the high attrition rate of radiologists and the desire to enhance radiology support for military deployments. By 2007 most Army x-rays became digital, which helped the storage, transmission and interpretation of images. With this newer technology a computerized tomography (CT) scan performed in Afghanistan can be read at the Army medical center in Landstuhl, Germany. Another example of military teleradiology can be found on the Navy hospital ships Mercy and Comfort where digital images can be transmitted to shore based medical centers for interpretation or consultation.

In the civilian sector, vRad (formerly NightHawk Radiology Services) helps smaller hospitals by supplying radiology services. All are board certified; most trained in the United States and carry multiple state licenses. They list a staff of 400+ radiologists and interpret seven million studies per year. They offer conventional radiology as well as CT, MRI, Ultrasound and Nuclear Medicine interpretation.[6]

Another more common but important example of teleradiology is the practice of radiologists reading films after-hours at home. They must have high resolution monitors and high speed connections to the internet but with this set up and voice recognition software; they can be highly productive at home. This is becoming the standard practice for radiologists. Instead of driving in or staying at the hospital at night to interpret images, they can deliver interpretations while at home.

Teleneurology

Many regions lack neurologists to see patients with stroke-like symptoms to determine if they need clot-busting drugs (thrombolytics) or need to be transferred to a higher level of care. This is, in part, due to the increased malpractice risk and decreased reimbursement situation of treating emergency patients. With the advent of telemedicine, the case can be discussed real time and the patient and their x-rays can be viewed remotely by a stroke specialist. One company, REACH Call Inc. developed a web-based solution that includes a complete audio-visual package so neurologists can view the patient and their head CT (CAT scan). REACH Call Inc. was developed by neurologists at the Medical College of Georgia. Because the program is web-based, the physician can access the images from home or from the office. Likewise, the referring hospital only has to have an off-the-shelf web camera, a computer and broadband internet connection.[7] Specialists-on-Call is a Massachusetts based organization that has 40 part time or full time neurologists on board to handle emergency consults via telemedicine for about 60 private community hospitals. In 2011 they added telepsychiatry. Their capabilities include the ability to transfer head CT images and bidirectional audio and video conferencing with remote physicians and families. To accomplish this they have an infrastructure that consists of a PACS, a call center, an electronic health record and videoconferencing equipment. The cost for this service is not inexpensive; for a 200 bed hospital it would cost $400 per day and $40,000 for initial installation fees. It is unknown if third party payers will eventually reimburse for this service.[8, 9] A teleneurology study is reported later in this chapter.

Telepharmacy

Like teleradiology, this field arose because of the shortage of pharmacists to review prescriptions. Vendors now sell systems with video cameras to allow pharmacists to approve prescriptions from a remote location. This is very important at small medical facilities or after-hours when there is not a pharmacist on location.[10]

The North Dakota Telepharmacy Project operates 36 remote sites where pharmacy technicians receive approval for a drug by distant pharmacists via teleconferencing. In this manner a full drug inventory is possible even in small rural communities and the pharmacists still perform utilization reviews and other services remotely.[11]

Telepsychiatry

Several studies have indicated that telepsychiatry is equivalent to face-to-face psychiatry for most patients.[12] The American Psychiatric Association promotes telepsychiatry, primarily for remote or underserved areas, using live video teleconferencing. During a telesession, there can be individual or group therapy, second opinions and medication reconciliation. In general, virtual visits help team medicine and patient satisfaction has been good. On the American Psychiatric Association web site, there are valuable resources.[13] Another telepsychiatry trend that is appearing is the use of free commercial-off-the-shelf (COTS) audiovisual programs such as Skype. Voyager Telepsychiatry uses this popular program to hold virtual telepsychiatry sessions.[14] One of the most important areas for telepsychiatry will involve military members who return from war with Posttraumatic Stress Disorder (PTSD) and Traumatic Brain Injury (TBI). About 40% of veterans live in rural areas, where transportation may be an issue. The VA has opened three Veterans Rural Health Resource Centers in Iowa, Utah and Vermont to help develop and evaluate telemedicine programs.[15]

Teledermatology

With the advent of good quality digital cameras and cell phones with medium quality cameras, the concept of teledermatology was born. The Teledermatology Project, created in 2002, has the goal of providing free worldwide dermatology expertise, particularly for third world countries and the underserved. Physicians can easily obtain a teleconsultation and diagnostic and therapeutic advice using the store and forward mode. A 2003 survey indicated that there were 62 teledermatology programs in the United States.[16, 17] iDoc24 is a Teledermatology project that began in Sweden for the European Union. It was designed for those patients who were traveling or did not have access to their physician and had a new skin condition. Patients can take a picture of their skin lesion with a digital camera or cell phone and forward it (can be anonymous) as an attachment to a text message and it would be followed by a response by a dermatologist within 24 hours. The image can also be integrated with the regional personal health portal that is part of the Swedish National Health Service. The goal is to provide better service, answer anonymous requests and decrease overall face-to-face visits to dermatologists.[18]

For more details on teledermatology, we refer readers to an e-Medicine article[19] and a 2009 review by the California HealthCare Foundation.[20]

E-Mail Teleconsultation

Audio and video teleconferencing is not the only way to communicate remotely. The Army has established a teleconsultation service for deployed military clinicians, based on e-mail communication. The service is available 24/7 for all branches of the military with most responses completed in less than six hours. Almost every specialty is available to the military physician while on ships, the battlefield or part of humanitarian or disaster relief operations. The most common specialty consult requested is dermatology and the most common location requesting assistance has been Iraq. The program is administered as part of the Office of the Surgeon General Teleconsultation Program. [21]

Telemonitoring

Telerounding

This is a new concept developed to help address the shortage of physicians and nurses. Telerounding is being rolled out in facilities with reasonably good reviews, in spite of obvious criticisms that it further compromises the already strained doctor-patient relationship.

Robot Rounds. A study in 2005 in the Journal of the American Medical Association showed that surgeons could make a second set of rounds using a video camera at the patient's bedside (InTouch Robots). A physician assistant makes the actual rounds, backed up by the attending physician remotely via the robot. Robot units are five and a half feet tall, weigh 220 lbs. and have a computer monitor as a head. The cost is more than $100,000 each or they can be leased for $5,000 monthly plus $5,000 per viewing station. At this time they are being used in 20 plus hospital systems in the United States. They can move around and can project x-ray results to the patient. Ellison et al. reported on urological patients who either received face-to-face rounds post-operatively or robotic telerounding. They concluded that robotic rounding was safe and well received by patients. Two-thirds of patients stated they would rather see their own physician remotely than a stranger making rounds in person. [22-24]

E-ICU Rounding. In the United States it is predicted that we need approximately 35,000 intensivists (physicians who specialize in ICU care), but we only have 6,000. Moreover, in spite of the fact that hospital beds are not increasing, ICU beds are. Therefore, remote monitoring makes sense particularly during nighttime hours when physicians might not be present. The Leapfrog Group has advocated care delivered by intensivists for all ICUs as one of its four patient safety recommendations; but this goal remains elusive.[25] Hospitals that use e-ICUs believe there are patient safety and financial benefits but both need to be proven. An e-ICU service may be less expensive than recruiting full time intensivists. Also, because ICU care can cost $2,500 daily, any cost saving modality that positively affects length of stay or mortality will gain market attention. Avoiding law suits in the ICU also means cost savings. It is estimated that over 100 hospitals now have e-ICU programs, even though there is no reimbursement by insurers.[26]

A few hospital systems have created their own eICU systems but most have used the VISICU platform. It was founded by two intensivists from Johns Hopkins in 1998 and later purchased by Phillips Electronics Healthcare division. VISICU extended support of care outside the ICU in 2007. Their plan is to use the eCareMobile™ unit (Figure 20.2) to monitor sick patients on medical surgical floors, emergency departments, step-down units and post anesthesia units. [27]

Figure 20.2: eCareMobile™ unit (Courtesy VISICU)

The cost for e-ICUs is significant in light of the uncertain benefits. The University of Massachusetts Memorial Health Care network spent $8 million to create a virtual ICU network to connect eight intensive care units. Specialists can now remotely view electronic health records, nursing notes, test results and video images of patients as well as access the latest clinical practice guidelines. They will add two-way video feeds so patients' families can communicate with specialists.[28] Sutter Health paid more than $25 million to establish its VISICU e-ICU system. Based on their analysis they have saved about $2.6 million in treatment costs by preventing deaths due to sepsis. In addition, they estimate that if sepsis is treated early, the ICU stay is shortened by four days.[29] It is unfortunate that many understaffed rural hospitals will not be able to afford intensivists or these services unless they are part of a larger network or there is reimbursement by insurers. eICU studies are reported later in this chapter. The bottom line is that further research is needed to provide the kind of detail necessary to determine the benefit of this type of telemedicine. For example, is the benefit greater for a small hospital with limited ICU expertise compared to a large integrated ICU system with an abundance of intensivists?

Telehomecare

Telehomecare is remote monitoring of the patient at home. One healthcare expert has stated that "home is the new hub of health" which implies the home needs to be interoperable with the rest of the healthcare system. [30] It usually involves monitoring vital signs, weights, blood sugars, etc. that can be sent via a wired or wireless mode from homes to physicians' offices, health information exchanges, etc. While home telemonitoring can also include fitness programs and "aging in place" technologies, we will focus on chronic disease management and post-acute care monitoring. The goal is to better educate and monitor patients at home in an effort to provide better patient-centric healthcare, while reducing readmissions and unnecessary emergency room visits, thus saving money. There are multiple reasons telemonitoring is burgeoning:

- Chronic diseases are on the rise that will likely increase hospitalizations, readmissions and unnecessary emergency room visits. Measures like home monitoring might decrease this trend. The goal is to intervene immediately, rather than wait till the next appointment.

- Medicare changed reimbursement to home health agencies from the number of visits to a diagnoses based system, leading to decreased reimbursement for visiting nurses.

- Telemonitoring programs potentially support audio and visual communication with patients at home and therefore can reduce home visits by a nurse or physician. Nurses can make visits only if there is a problem, such as a change in symptoms or vital signs.

- One consulting organization predicts a nursing shortage of 800,000 and a physician shortage of 85,000 to 200,000 by the year 2020. [31]

- Baby boomers are tech savvy and more likely to demand services like telemonitoring.

- Monitoring may be possible using the ubiquitous cell phone and new microsensors.

- Linking home monitoring devices to EHRs with decision support and health information exchanges will increase the functionality of this new technology. The potential to save costs is attractive but elusive and will require high quality confirmatory studies.

- CMS has administered Medicare Medical Home Demonstration projects to test the "medical home" and "hospital at home" concepts. Medical groups will be paid for coordination of care, health information technology, secure e-mail and telephone consultation and remote monitoring. Details are preliminary and available on the CMS site.[32] Accountable Care Organizations (ACOs) may also incorporate telemedicine. For additional information on the patient-centered medical home model and ACOs see the chapter on quality improvement strategies.

- The Affordable Care Act will reduce payments to hospitals deemed to have excessive readmission rates for heart failure, acute myocardial infarction (heart attack) and pneumonia, beginning in 2012. [33]

Many health IT vendors are developing home monitors and sensors that will transmit information to a physician's office or other healthcare organization. Programs will be interactive and include patient education for issues such as drug compliance. This data may interface with an electronic health record, health information organization (HIO) or web site for others to evaluate. Some predict that houses (smart homes) will be wired with multiple small sensors known as "motes" that will monitor daily activities such as taking medications and leaving the house. The information would be transmitted to a central organization that would notify the patient and/or family if there was non-compliance or a worrisome trend.

Telemonitoring is actually a process with multiple steps depicted in Figure 20.3.

Figure 20.3: Telemonitoring cyclical process

Home Telemonitoring System Examples

More than 50+ companies offer technology to monitor patients at home and the list continues to grow and include large companies such as Intel and General Electric. Devices can be standalone or be integrated with another system such as an electronic health record or personal health record. Devices connect externally using USB, Bluetooth, telephony (POTS), WiFi and 3G/4G telecommunication networks.

Health Buddy is a FDA approved device that is certified by the National Committee for Quality Assurance and used by the Veterans home telemedicine programs. Health Buddy is used by over 12,000 patients and has been shown in one study (of limited design) to increase medication compliance and reduce outpatient visits. [34, 35] Additional studies are reported in another section. The Centers for Medicare and Medicaid Services tested the system with about 2,000 patients with chronic diseases and the results are in the section on Telemedicine Studies. Features include:

- Data is sent via phone lines
- Device comes with desktop decision support software
- Program covers 45 disease protocols
- Device connects to a glucometer, BP machine, weight scales and peak flow meter for asthmatics
- Program is interactive with patients; it asks questions daily

HoneyWell HomMed has over 15,000 monitors currently in use and more than 300,000 patients have been monitored. Features include: [36]

- Voice messages to patients in multiple languages

- Digital weight scale, blood pressure, oximetry, glucometer, peak flow meter, blood tests (PT/INR), temperature and EKG

- Data is transmitted via phone lines

- LifeStream Connect™ is a new option that interfaces with EHRs

More Telemonitoring Systems

- MyCareTeam: a fee for service diabetic portal developed in cooperation with Georgetown University. Now integrates with Allscripts EHR. [37]

- Voluntis: patient relationship management platform based on the web-based medpassport and mobile infrastructure. They specialize in asthma, heart failure and diabetes management (see info box).[38, 39]

- MediCompass: patient portal for diabetes, asthma and cardiovascular disease. Provides a dashboard view, a registry, an executive view and messaging.[40]

- Intel Health Guide: Intel along with General Electric has entered the telehomecare market with a comprehensive program called Care Innovations. One product, *Connect* offers wellness surveys, brain games, medication reminders and messaging. Another product *Guide* connects the patient to their physician and allows vital sign uploads, two way video messaging, wellness surveys and secure messaging.[41]

- ViTelCare T400: this system includes a touch screen monitor that can measure and store blood pressure, blood glucose, weights and oximetry data. Includes disease specific alerts, interactive programs and patient education for heart failure, hypertension, diabetes, COPD, depression, PTSD and substance abuse. Data is transmitted over phone lines, WAN/LAN or broadband connections.[42]

Diabeo

Telemedicine system is part of Voluntis disease management programs and consists of smartphone software and a web portal diabetic management teams can access. Software has insulin calculators based on blood glucose, diet and activity.

Six month study (TeleDiabi 1 Study) of poorly controlled insulin dependent diabetics showed improvement in HbA1C levels using Diabeo, compared to control patients. Those patients who used the system and received feedback from the diabetic team experienced the most benefit.[38, 39]

Telemedicine Projects

The following section provides a sampling of some interesting telemedicine projects:

Informatics for Diabetes Education and Telemedicine (IDEATel): The largest government sponsored telemedicine program in the US. The project evaluated approximately 1,650 computer illiterate patients living in urban and rural New York State. Patients received a home telemedicine unit that consisted of a computer with video conferencing capability, access to a web portal for secure messaging and education and the ability to upload glucose and blood pressure data. These same subjects were assigned a case manager who was under the supervision of a diabetic specialist. They used the Veterans Affairs clinical practice guidelines on diabetes. They were compared to a control group that didn't receive the home monitoring system. The results of this project are reported in the next section.[43]

Georgia Partnership for Telehealth (GPT): Georgia has 159 counties, many at the poverty level. This network is the first statewide effort to link 36 rural hospitals and clinics with specialists at eleven large urban hospitals. Project created partnerships among Wellpoint (Blue Cross/Blue Shield) and the state government. Importantly, telemedicine consults were reimbursed as office visits due to a new Georgia law and 20 specialties were felt to be appropriate for telemedicine. In 2009 the top categories for encounters were wound care and telepsychiatry. GPT will be the lead agent for the new Southeastern Telehealth Resource Center.[44-46]

University of Texas Medical Branch at Galveston: Program is the largest telemedicine system in the world with 300 locations and 60,000 annual telemedicine sessions. Sixty per cent of visits deal with a prison population. They also offer specialty services in neurology, addiction medicine and psychiatry.[47]

VA Rocky Mountain Healthcare Network: In Colorado veterans with heart failure, diabetes and emphysema were enrolled in a telemedicine program. The VA reported a 53% reduction in hospital stays resulting in a $508,000 savings overall. Outpatient visits dropped 52% and overall estimated savings of the program was $1.2 million.[48]

Teleburn Project: University of Utah Burn Center used telemedicine to treat burn patients in three states. Specialists can view videos or digital photos of burn patients for initial determination or follow up. The demonstration project was funded by the Department of Commerce.[49]

TeleKidcare: Urban project operated by the Kansas University Medical Center to deliver care at school so that children do not have to leave school to receive medical care. Reasons for televisits include: 47% ear, nose and throat problems, 31% for behavioral problems and 10% for eye related problems. In spite of private and government support they do not have a long term business plan for sustainability.[50]

TelePediatrics and TeleDentistry: The University of Rochester created the Health-e-Access program in 2001. The program was initially set up to connect pediatricians to inner city child care centers and elementary schools using telemedicine and two-way video conferencing via the internet. The program has allowed the children and their parents to not leave the centers or their jobs. The program was started by grants but insurers have been willing to cover this initiative, presumably because it cuts down on emergency room visits. The director of the project has stated that he believes about 28% of pediatric visits to the emergency room in upstate New York could have been treated with telemedicine.[51]

Veteran's Teleretinal Program: Since 2000 the Veterans Health Affairs has operated a VHA Teleretinal Imaging project at 104 sites. Because 20% of veterans have diabetes they felt they had to conduct retinal screening for diabetic damage even though they might not have retinal experts (ophthalmologists) at each clinic. They took high resolution retinal images and "stored and forwarded" them to eye specialists who would later made a determination.[52]

California Central Valley Teleretinal Program: Using a non-proprietary, open source web-based program (EyePACS) images can be forwarded to an ophthalmologist for interpretation. Images are stored on a SQL Server and images are viewed with a web browser. A simple software program on the PC allows for uploading images to the server. There is e-mail notification to the consultant and back to the individual who sent the images. California will expand the program from the Central Valley to the entire state to serve 100 clinics and approximately 100,000 patients. [53, 54]

Northwest Telehealth: This initiative has 65 sites using about 100 telemedicine devices. They offer the following services: clinical care (15 specialties), teleER, telepharmacy, distance education, administrative and operational planning/coordination.[55]

Federal Communications Commission (FCC): In 2006 they announced a $400 million budget for pilot projects to promote broadband networks in rural areas. The goal is to create networks for public healthcare organizations and non-profit clinicians that will eventually connect to a national backbone. The network could be used for telemedicine or other medical functions in rural areas. In 2007 the FCC created a $417 million fund that would support pilot projects to connect more than 6,000 hospitals, research centers, universities and clinics. The FCC paid up to 85% of the cost to design, engineer and construct the networks. Internet2 or the LambdaRail Network will be used. Many of the projects will involve multi-state areas and most will enhance telemedicine. Much of the funding will come from the Universal Service Fund that derives from a fee added to consumers and telecommunication companies.[56] The New England Telehealth Consortium announced in January 2008 that it will use the $24.7 million in FCC grant money to link 555 clinics, physician offices, hospitals, public health offices and universities in Maine, Vermont and New Hampshire. The network will act like a second internet to allow the transmission of records and x-rays and the creation of videoconferences.[57] In early 2009 Congress directed the FCC to develop a National Broadband Plan with goal of providing broadband access to every American and funded it as part of the American Recover and Reinvestment Act (ARRA). In 2010 the FCC posted the Plan on a new web site. [58]

Case Study

United Kingdom Department of Health Whole System Demonstrator Program

This is the largest (6000 patients) randomized telemedicine trial underway to study the impact of technology on diabetes, coronary heart disease and chronic lung disease. The study looked at outcomes, use of services, user and professional experiences, etc. Preliminary results published in December 2011 indicate: 20% reduction in emergency admissions, 15% reduction in emergency department visits, 14% reduction in elective admission, 14% reduction in bed days and a 45% reduction in mortality.

These results are interesting but until the final results are published and scrutinized by outside reviewers we will not know the actual impact of this project.

http://www.dh.gov.uk/en/Publicationsandstatistics/Publications/PublicationsPolicyAndGuidance/DH_131684

Current Knowledge

The following are a sample of some of the more interesting and recent telemedicine articles to appear in the medical literature:

- A 2001 systematic review of telemedicine by Roine et al. looked at reported patient outcomes, administrative changes or economic assessments. Of 1,124 potential articles, 50 were felt to fit criteria for review. Most of the studies reviewed pilot projects and were of low quality. They felt that teleradiology, teleneurosurgery (looking at head CT scans before transfer), telepsychiatry, the transmission of echocardiograms, the use of electronic referrals to enable e-mail consultations and video teleconferencing between primary and secondary clinicians had merit. They also felt that it was impossible to state the economic value of telemedicine based on current evidence.[59]

- The US Department of Veterans Affairs operates perhaps the largest telehomecare networks in the world (37,500 patients as of 2009). This is partly due to the fact that the VA has transitioned from inpatient to outpatient and home care. Also, with so many active duty members returning injured from the war zone they will eventually need telehomecare. Their Care Coordination /

Home Telehealth program is also a disease management program. The VA currently runs three programs: telehomecare, teleretinal and a video teleconferencing services that link 110 hospitals and 380 clinics. Data from home devices inputs into the VA's EHR.[60] A study of 17,000 VA home telehealth patients was reported in late 2008. Although the cost per patient averaged $1,600, it was considerably less expensive than in-home care. They utilized individual care coordinators who each managed a panel of 100-150 general medical patients or 90 patients with mental health related issues. They promoted self-management, aided by secure messaging systems and a major goal was early detection of a problem to prevent an unnecessary visit to the clinic or emergency room. 48% were monitored for diabetes, 40% for hypertension, 25% for heart failure, 12% for emphysema and 1% for PTSD. Patient satisfaction was very high. This study showed a 19% reduction in hospitalizations and a 25% reduction in the average number of days hospitalized.[61]

- The Electronic Communications and Home Blood Pressure Monitoring study compared home blood pressure (BP) monitoring along with a BP web portal, with and without the assistance of a pharmacist. The web portal was integrated with an enterprise EHR. In the group that received assistance from the online pharmacist, they showed significantly more patients achieving control than those who were monitored and had web portal access but no interaction with a pharmacist. Results might not pertain to other diseases and requires patients to have internet access and pharmacists to be able to have EHR access.[62]

- Web-based care for diabetes was evaluated by the same group (Group Health) who evaluated hypertension control in the above paragraph. They compared a group of Type 2 diabetes who received "usual care" with another group who had access to a web portal linked to an EHR. The web-based program included secure e-mail messaging with clinicians, feedback on blood sugar results, educational web resources and an interactive online diary to record diet, etc. After one year the control of diabetes, based on a glycated hemoglobin was marginally better (decrease of .7%) but there was no difference in blood pressure or cholesterol control between the two groups. There was no correlation between improvement and the number of times the web portal was accessed. They only used one care manager so it is unknown if their results would have been different with multiple care managers.[63]

- Group Health conducted another study of 1,500+ diabetics aged >18 years old to determine if those who used secure messaging with their clinician had better blood sugar, blood pressure and cholesterol control. Only 19% of patients chose to message their physician. Those that did had better blood sugar control, but not better control of blood pressure or cholesterol but had a higher rate of outpatient visits. Patients were not randomized for this study and the study was not prospective, so results are more difficult to interpret.[64]

- The one year results of the IDEA TEL were published in late 2007 and showed mild improvement in blood sugars, cholesterol and blood pressure compared to the control project. Patient and physician satisfaction were positive but detailed cost data was lacking. Ironically, Medicare claims were higher in the study patients than in the control group, for unclear reasons.[65] The five year results were published in 2009 and although they showed some statistically significant improvement in blood sugar, cholesterol and blood pressure control, they were of doubtful clinical significance. Importantly, users of this technology had a dropout rate greater than 50%.[66] In 2010 a final report from this group concluded that "telemedicine case management was not associated with a reduction in Medicare claims."[67]

- Teleneurology or telestroke care was evaluated by a study by Meyer in 2008. They compared the outcomes of patients with a possible impending stroke and consultation by telephone, versus full video teleconferencing. Correct treatment decisions were made more frequently (98% versus 82%) for the teleconferencing sessions, but patient outcomes were the same. There was no difference in death rates or hemorrhaging after the clot busting drugs (thrombolytics) were administered.[68] An excellent review article on stroke telemedicine was published by Demaerschalle et al. in the Mayo Clinic Proceedings.[69] The jury is out whether stroke telemedicine is cost effective or a reasonable choice, compared to telephonic consultation.[70]

- A study was reported at the 2009 Society of Critical Care Medicine meeting by Avera Health from a 30 month implementation of VISICU reported multiple benefits, to include cost savings.[26] In spite of the many potential virtues of the e-ICU, a 2009 article by Berenson et al. expressed the opinion that the actual value of e-ICUs was far from proven and there was a major interoperability issue between the e-ICU software and critical ICU systems like IV fluids and mechanical ventilation.[71] Another article in 2009 by Thomas et al. evaluated the medical care in six ICUs before and after the implementation of an e-ICU system. They concluded that there was not an overall improvement in mortality or length of stay.[72] A 2011 meta-analysis by Young et al. showed a decreased mortality and length of stay in the ICU but not the overall hospital mortality or length of stay.[73] Another 2011 article from a single academic medical center reported a lower hospital mortality and length of stay, improved guideline adherence and reduced preventable complications.[74] However, as pointed out in an editorial, we still don't know if eICUs can improve care in rural/remote settings and whether hired intensivists on site would be more valuable.[75]

- A study reported in JAMA in 2009 looked at whether telephone delivered care for post cardiac bypass depression by nurses would be equivalent to usual care. In this randomized controlled trial telephonic collaborative care was superior in terms of mental health-related quality of life, physical functioning and mood symptoms at eight-month follow up.[76]

- An international meta-analysis of 10 randomized controlled trials looked at remote patient monitoring (RPM) of heart failure patients. They concluded RPM reduced the risk for all-cause mortality and hospitalization for heart failure. The number needed to treat (discussed in evidence based medicine chapter) was 50 for all-cause mortality and 14 for heart failure hospitalization.[77] Another Cochrane meta-analysis also showed benefit of telemedicine[78] while a 2010 telephone monitoring study reported no benefit, thus again, no consensus.[79]

- The telemonitoring *HealthBuddy* was evaluated by CMS and another research group using claims data to evaluate the economic impact of this tool in patients with COPD, heart failure or diabetes. Importantly, only 37% of eligible patients opted to join the program. CMS was unable to show cost savings compared to a control group whereas the research group showed small savings but did not factor in the cost of the monitoring devices.[80, 81]

Barriers to Telemedicine

The barriers to telemedicine are similar to the barriers to all health information technology we have covered in several other chapters. The most significant barriers are as follows:

- Limited reimbursement. Most telemedicine networks are created with federal grants. Medicare will reimburse if there is a formal consultation linked by live two-way video teleconferencing and the patient resides in a professional shortage area. Medicaid at the federal level does not reimburse for telemedicine. Medicare will reimburse physicians, nurse practitioners, physician assistants, nurse midwives, clinical nurse specialists, clinical psychologists and clinical social workers. The originating sites can be offices, hospitals, skilled nursing homes, rural health clinics and community mental health centers. Medicare reimburse for telemedicine services for initial inpatient care, outpatient care, pharmacologic management, end stage renal disease-related visit and psychiatric diagnostic interviews. Clinicians at the remote site submit claims using the correct CPT or HCPCS codes as well as the telemedicine modifier GT.[82] In 2010 Medicare added three new outpatient and inpatient HCPCS codes for Telemedicine.[83] Many private insurers don't cover telemedicine, but a few provide the same coverage as face-to-face visits. In 2010 Virginia became the 12th state to require insurers to cover telemedicine for interactive audio or video visits.[84]

- Limited research showing reasonable benefit and return on investment. A systematic review of telehealth economics concluded that standard economic evaluation methods were not used therefore the results were not generalizable.[85] A review of 80 systematic reviews of telehealth effectiveness reported 21 were positive, 18 found the evidence promising but limited and 41 reported the evidence is limited and inconsistent.[86] In summary, the studies on telemedicine are mixed and are of low quality. Most studies are based on a small patient population as large randomized controlled trials are expensive. Therefore results can't be generalized to every population. Similarly, many studies are not conducted over a long period of time so attrition rates might not be accurately reported. Moreover, there are many flavors of technology (telephone, smartphone, internet, interactive device, etc.) used in telemedicine making comparisons more difficult. It does seem like the addition of a skilled healthcare worker, such as a nurse or pharmacist, is necessary to experience benefits from a telemedicine program. Healthcare organizations that have excellent health IT support as well as disease management teams are the most likely to benefit from telemedicine. In this early stage of telemedicine, the technology by itself does not seem to produce significant benefit.

- High cost or the limited availability of high speed telecommunications.

- Bandwidth issues, particularly in rural areas where telemedicine is most needed. VPN connections slow the process further.

- High resolution images or video require significant bandwidth, particularly if x-rays or images or pills have to be read by remote clinician. Telepsychiatry may require lower resolution. The following are average file sizes (megabytes): Xray 10 MB, MRI 45 MB, Mammogram 160 MB and 64 slice CT 3,000 MB.[87]

- State licensure laws when telemedicine crosses state borders. Some states require participating physicians to have the same state license. In 2011 CMS loosened the requirements for telehealth. The new rule will allow hospitals receiving telehealth services to be privileged and credentialed from the hospital providing telehealth services.[88]

- Lack of standards.

- Lack of evaluation by a certifying organization.

- Fear of malpractice as a result of telemedicine. Who is going to evaluate telemonitoring data 24/7?

- Sustainability is a concern due to an inadequate long term business

- Lack of sophistication on the part of the patient, particularly in the elderly and under-educated.

UVA Center for Telehealth

The University of Virginia created this network in 1994 to provide care for rural underserved patients. This network already has more than 40 subspecialists in 85 locations in Virginia. They will now host the new Mid-Atlantic Telehealth Resource Center that will support Delaware, Kentucky, Maryland, North Carolina, the District of Columbia and West Virginia. They are one of 12 Telehealth Resource Centers in the US.[89]

Telemedicine Organizations and Resources

Organizations

- Office for the Advancement of Telehealth (OAT): falls under Health Resources and Services Administration (HRSA) that is an agency of the Department of Health and Human Services. Its goal is to promote telemedicine in rural/underserved populations, provide grants, technical assistance and "best practices."[1]

- Regional Telehealth Resource Centers: The United States now has 12 regional telehealth resource centers created to set up a national telehealth network. The last three funded in 2011 by HRSA were Virginia (see info box), Maine and Indiana.

- American Telemedicine Association (ATA): a non-profit international organization with paid membership that began in 1993. Individual state telemedicine policies are included on their web site. ATA has created a set of telemedicine standards and guidelines covering telemental health, diabetic retinopathy, teleradiology, telemedicine operations and telepathology. Goals of the ATA are as follows:

    o "Educating government about telemedicine as an essential component in the delivery of modern medical care

    o Serving as a clearinghouse for telemedicine information and services

    o Fostering networking and collaboration among interests in medicine and technology

    o Promoting research and education including the sponsorship of scientific educational meetings and the Telemedicine and e-Health Journal

    o Spearheading the development of appropriate clinical and industry policies and standards"[3]

- USDA Rural Development Telecommunications Program: The USDA has a program to finance the rural telecommunications infrastructure. In 2007 there were grants and loans totaling $128 million to achieve the goals of broadband access for distant learning and remote medical care. The USDA Rural Development agency has funded several e-ICU programs in the US, including the study by Avera Health noted in an above section.[90]

- The Agency for Healthcare Research and Quality (AHRQ): AHRQ has funded a number of telemedicine projects looking at virtual ICUs, telewound projects, cancer management, medication management, heart failure management and others.[91]

Resources

- Sarashon-Kahn J. *The Connected Patient*. The California Health Care Foundation. February 2011 www.chcf.org

- Lykke F, Holzworth M, Rosager M et al. Telemedicine: An Essential Technology for Reformed Healthcare. www.csc.com

- For an international perspective: International Society for Telemedicine & eHealth http://www.isfteh.org/

Future Trends

Telemedicine is a relatively new field created because of the misdistribution of physicians, the need for remote delivery of medical care and the emergence of nascent technologies. Televisits will likely increase if found to be helpful to patients and clinicians for minor illnesses, even if reimbursement lags. Teleconsultation is on the rise worldwide to address access problems for populations at risk: rural, poor, incarcerated, elderly and those with multiple chronic diseases. Telemonitoring is complex because it traditionally required sophisticated and expensive technology as well as skilled human intervention to deliver virtual ICU care or home telemedicine. With cell phone cameras, web cams and simple programs such as Skype ™ the technology is maturing and more affordable. It is unknown whether newer healthcare delivery models such as accountable care organizations will adopt and promote telemedicine as a means to increase access to expert care, while saving money.

Key Points

- Telehealth is a neologism that relates to long distance clinical care, education and administration
- Telemedicine refers to the remote delivery of medical care using technology
- Almost all specialties now have telemedicine initiatives
- In spite of the lack of reimbursement, virtual ICUs have gained in popularity because they have perceived benefits
- Telehomecare is a new telehealth initiative that has appeared due to the graying of the US population and the increase in chronic diseases
- Lack of uniform reimbursement, lack of standards and lack of high quality outcome studies have impacted the adoption of telemedicine

Conclusion

Telemedicine is still in its infancy in most areas of the world. New organizations such as the Middle East Society of Telemedicine (MESOTEL) have emerged to cover the Middle East and North Africa.[92] The barriers are largely financial due to the high cost to set up the system and the lack of reimbursement in many cases. With the price of telemedicine systems dropping, telemedicine for rural patients is likely more cost-effective than referral to distant urban specialists. If the FCC and ARRA initiatives are successful and/or HIOs flourish, we may have the infrastructure required for telemedicine throughout the United States. Transmission and storage of large images and the ability to compare old and new imaging studies will be greatly aided by Internet2, LambdaRail and modern web PACS. If future studies prove there is substantial return on investment then it is a matter of time before more payers support telemedicine. At this time, successful telemedicine programs require an engaged patient and physician, a supportive infrastructure, disease managers and payer reimbursement.

References

1.	Office for the Advancement of Telehealth http://www.hrsa.gov/telehealth/ (Accessed November 12 2011)

2.	Oh H, Rizo C et al. What is eHealth?: a systematic review of published definitions. J Med Inter Res 2005; 7 (1) e1

3. American Telemedicine Association http://www.atmeda.org (Accessed November 1 2011)

4. Telemedicine: A guide to assessing telecommunications in health care. Marilyn Field ed. National Academies Press 1996. http://www.nap.edu/catalog/5296.html (Accessed September 25 2006)

5. Puskin DS HHS Perspective on US Telehealth www.ieeeusa.org/volunteers/committees/mtpc/Saint2001puskin.ppt (Accessed December 6 2006)

6. vRad www.virtualrad.com (Accessed October 25 2011)

7. REACH www.reachcall.com (Accessed November 12 2011)

8. Teleneurology Helps Combat Specialist Shortage, Wait Times. July 17 2007 www.ihealthbeat.org. (Accessed July 18 2007)

9. Specialists oncall www.specialistsoncall.com (Accessed October 25 2011)

10. Envision Telepharmacy www.envision-rx.com (Accessed October 25 2011)

11. North Dakota Telepharmacy Project. http://telepharmacy.ndsu.nodak.edu (Accessed November 12 2011)

12. O'Reilly R, Bishop J, Maddox K et al. Is telepsychiatry equivalent to face to face psychiatry? Results from a randomized equivalence trial. Psychiatr Serv 2007;58(6):836-843

13. American Psychiatry Association http://www.psych.org (Accessed October 25 2011)

14. Telepsychiatry www.telepsychiatry.com (Accessed October 25 2011)

15. Joch A. Tele-therapy. Government Health IT November 2008 p.30-31

16. Telederm Project www.telederm.org (Accessed October 25 2011)

17. Soyer HPk Hofmann-Wellenhof R, Massone C et al. Telederm.org: Freely Available Online Consultations in Dermatology www.Plosmedicine.org 2005 2(4) (Accessed June 10 2009)

18. iDoc24. www.epractice.eu/en/cases/idoc24 (Accessed October 20 2011)

19. Teledermatology. E-Medicine http://emedicine.medscape.com/article/1130654-overview (Accessed October 25 2011)

20. Armstrong AW, Lin SW, Liu et al. Store-and-Forward Teledermatology Applications. December 2009. California HealthCare Foundation www.chcf.org (Accessed February 10 2010)

21. Army Telecommunications Program. www.cs.amedd.army.mil/teleconsultation.aspx. (Accessed May 24 2009)

22. Ellison LM, Nguyen M, Fabrizio MD, Soha A. Postoperative Robotic Telerounding Arch Surg 2007;142(12):1177-1181

23. Roberts R. Robots on Rounds. Kansas Business Journal Sept 5 2005 (Accessed September 10 2005)

24. Robotic Doctor Makes Rounds in Baltimore. www.Ihealthbeat.org Feb 27 2006 (Accessed March 10 2006)

25. Leapfrog. www.leapfroggroup.org (Accessed November 12 2011)

26. Breslow MJ. Effect of a multiple-site intensive care unit telemedicine program on clinical and economic outcomes: An alternative paradigm for intensivist staffing. Crit Care Med 2004;32:31-38

27. Phillips VISICU www.healthcare.phillips.com (Accessed October 25 2011)

28. Massachusetts Hospital System Taps eICU To Offset Staff Shortages. www.ihealthbeat.org November 19 2007. (Accessed November 24, 2007)

29. Sutter Health Taps eICU System to Combat Sepsis-Related Deaths. June 26 2007. www.ihealthbeat.org. (Accessed June 26 2007)

30. Sarashon-Kahn, J. The Connected Patient. California Healthcare Foundation. February 2011. www.chcf.org (Accessed September 10 2011)

31. Healthcare Staffing Growth Assessment. Staffing Industry Strategic Research. June 2005 http://media.monster.com/a/i/intelligence/pdf (Accessed September 25 2006)

32. Medicare Medical Home Demonstration https://www.cms.gov/demoprojectsevalrpts/md/itemdetail.asp?itemid=cms1199247 (Accessed October 25 2011)

33. Affordable Care Act http://www.healthcare.gov/law/index.html (Accessed October 25 2011)

34. Health Buddy® http://www.bosch-telehealth.com/content/language1/html/55_ENU_XHTML.aspx (Accessed October 25 2011)

35. Cherry JC, Moffatt TP, Rogriquez C and Dryden K. Diabetes Disease Management Program for an Indigent Population Empowered by Telemedicine Technology. Diab Tech Ther 2002;4:783-791

36. Honeywell HomMed www.HomMed.com (October 25 2011)

37. MyCareTeam www.mycareteam.com (Accessed October 25 2011)

38. Voluntis www.voluntis.com (Accessed October 25 2011)

39. Charpentier G, Benhamou P, Dardari D et al. The Diabeo Software Enabling Individualized Insulin Dose Adjustments Combined with Telemedicine Support Improves HbA_{1c} in Poorly Controlled Type 1 Diabetic Patients. Diabetes Care 2011; 34: 533-539

40. MediCompass www.medicompass.com (Accessed October 25 2011)

41. Care Innovations www.careinnovations.com (Accessed October 26 2011)

42. Vitelcare http://www.vitelcare.com/monitor.htm (Accessed October 26 2011)

43. Informatics for Diabetes Education and Telemedicine http://www.ideatel.org/ (Accessed September 10 2007)

44. Rural Georgia Hospitals To Receive Telemedicine Funds www.ihealthbeat.org December 2 2004 (Accessed September 10 2006)

45. Georgia Taps Telemedicine for Rural Health care. www.ihealthbeat.org. March 26 2007. (Accessed March 27 2007)

46. Brewer R, Goble G, Guy P. A Peach of a Telehealth Program: Georgia Connects Rural Communities to Better Healthcare. Winter 2011. Perspectives in Health Information Management. www.perspectives.ahima.org (Accessed October 20 2011)

47. Texas Telemedicine Used To Treat Rural Patients, Prisoners. November 30 2006. www.ihealthbeat.org. (Accessed December 1 2006)

48. Austin M. Telehealth a virtual success. Denver Post October 24 2005 (Accessed November 10 2006)

49. Telemedicine Network Facilitates Burn Care in Three States www.ihealthbeat.org February 17 2006 (Accessed February 18 2006)

50. Egan, C. School-Based Telemedicine Program Holds Promise for Student Health. September 14 2007 www.ihealthbeat.org. (Accessed September 14 2007)

51. University of Rochester Medical Center www.urmc.edu (Accessed May 12 2008)

52. VHA Teleretinal Imaging Program www.va.gov/occ/Teleret.asp (Accessed February 7 2007)

53. CHCF Expands Project to Prevent Diabetes-Related Blindness through Telemedicine. www.chcf.org (Accessed December 18 2007)

54. EyePACS. www.eyepacs.org. (Accessed November 12 2011)

55. Northwest Telehealth www.nwtelehealth.org (Accessed November 12 2011)

56. Federal Communications Commission www.fcc.gov (Accessed November 12 2011)

57. Network Aims to Link Health Care Sites Across New England. January 08 2008. www.ihealthbeat.org. (Accessed January 8 2008)

58. National Broadband Plan. www.broadband.gov (Accessed March 31 2010)

59. Roine R, Ohinmaa, Hailey D. Assessing telemedicine: a systematic review of the literature. CMAJ 2001;165(6):765-71.

60. Buxbaum P. True Believer. Government Health IT. March/April 2009 p.13

61. Darkins A, Ryan P, Kobb R et al. Care Coordination/Home Telehealth: the Systematic Implementation of Health Informatics, Home Telehealth and Disease Management to Support the Care of Veteran Patients with Chronic Conditions. Telemedicine and e-Health. 2008;14(10):1118-1126

62. Green BB, Cook AJ, Ralston JD et al. Effectiveness of Home Blood Pressure Monitoring, Web Communication, and Pharmacist Care on Hypertension Control: A Randomized Controlled Trial. JAMA 2008;299(24):2857-2867

63. Ralston JD, Hirsch IB, Hoath J et al. Web-Based Collaborative Care for Type 2 Diabetes. Diabetes Care 2009;32(2):234-239

64. Harris LT, Haneuse SJ, Martin DP et al. Diabetes Quality of Care and Outpatient Utilization Associated with Electronic Patient-Provider Messaging: A Cross-Sectional Analysis. Diabetes Care 2009;32:1182-1187

65. Shea S. The Informatics for Diabetes and Education Telemedicine (IDEATEL) Project. Trans Amer Clin Clim Assoc 2007;118:289-300

66. Shea S, Weinstock RS, Teresi JA et al. A Randomized Trial Comparing Telemedicine Case Management with Usual Care in Older, Ethnically Diverse, Medically Underserved Patients with Diabetes Mellitus: 5 Year Results of the IDEATel Study. JAMIA 2009;16:446-456

67. Palmas W, Shea S, Starren J et al. Medicare payments, healthcare service use and telemedicine implementation costs in a randomized trial comparing telemedicine case management with usual care in medical underserved participants with diabetes. J Am Med Inform Assoc 2010;17:196-202

68. Meyer BC, Raman R, Hemmen T et al. Efficacy of site-independent telemedicine in the STRokEDOC trial: a randomized, blinded, prospective study. August 3 2008 www.thelancet.com/neurology e-publication (Accessed August 10 2008)

69. Demaerschalk BM, Miley ML, Kiernan TJ et al. Stroke Telemedicine. Mayo Clin Proc 2009;84(1):53-64

70. Berthoid J. Help from afar: telemedicine vs. telephone advice for stroke. ACP Internist April 2009 p. 19

71. Berenson RA, Grossman JM, November EA. Does Telemonitoring of Patients—The eICU—Improve Intensive Care? Health Affairs 2009;28(5):w937-947

72. Thomas EJ, Lucke JF, Wueste L et al. Association of Telemedicine for Remote Monitoring of Intensive Care Patients With Mortality, Complications and Length of Stay. JAMA 2009;302(24):2671-2678

73. Young LB, Chan PS, Lu X. Impact of telemedicine intensive care unit coverage on patient outcomes. Arch Intern Med 2011;171(6):498-506

74. Lilly CM, Cody S, Zhao H et al. Hospital Mortality, Length of Stay and Preventable Complications Among Critically Ill Patients Before and After Tele-ICU Reengineering of Critical Care Processes. JAMA 2011;305(1):2175-2228

75. Kahn JM. The Use and Misuse of ICU Telemedicine. JAMA 2011;305(21):227-228

76. Rollman BL, Belnap BH, Hum DB et al. Telephone-Delivered Collaborative Care for Treating Post-CABG Depression. JAMA 2009;302(19):2095-2103

77. Klersy C, De Silvestri A, Gabutti G et al. A meta-analysis of remote monitoring of heart failure patients. J Am Coll Cardiol 2009;54:1683-94

78. Inglis S. Structured Telephone Support or Telemonitoring Programmes for Patients with Chronic Heart Failure. Cochrane Database of Systematic Reviews 2010; no.8, doi 10.1002/14651858.CD007228.pub 2 (Accessed September 20 2011)

79. Chaudhry SI, Mattera JA, Curtis JP et al. Telemonitoring in Patients with Heart Failure. NEJM November 16 2010. Doi 10.1056/NEJMoa1010029 (Accessed December 4 2010)

80. McCall N, Cromwell J, Urato C. Evaluation of Medicare Care Management for High-Cost Beneficiaries Demonstration: the Health Buddy Consortium. RTI International April 2011 Report.

81. Baker LC, Johnson SJ, Macaulay D et al. Integrated Telehealth and Care Management Program for Medicare Beneficiaries With Chronic Disease Linked to Savings. Health Affairs. 2011;30(9):1689-1697

82. Majerowicz A, Tracy S. Telemedicine. Bridging Gaps in Healthcare Delivery. Journal of AHIMA. May 10 2010. 52-56

83. Centers for Medicare and Medicaid Services www.cms.hhs.gov/telmedicine (Accessed June 20 2009)

84. Virginia Gov. Signs Bill Requiring Insurers to Cover Telemedicine. April 6 2010. www.ihealthbeat.org (Accessed April 12 2010)

85. Bergmo TS. Can economic evaluation in telemedicine be trusted? A systematic review of the literature. BiomedCentral October 2009 (Accessed November 22 2010)

86. Ekeland AG, Bowes A, Flottorp S. Effectiveness of Telemedicine: A Systematic Review of Reviews. Int J Med Inform 2010;79:736-771

87. Health Care Broadband in America. www.broadband.gov (Accessed January 3 2011)

88. AAMI. CMS telemedicine restrictions eased. http://www.aami.org/news/2011/051211.cms.html (Accessed June 4 2011)

89. University of Virginia Center for Telehealth. http://www.healthsystem.virginia.edu/pub/office-of-telemedicine/center-for-telehealth.html (Accessed October 15 2011)

90. USDA Rural Development http://www.rurdev.usda.gov/UTP_Programs.html (Accessed October 26 2011)

91. Agency for Healthcare Quality and Research. www.ahrq.gov (Accessed October 26 2011)

92. Mesotel. www.mesotel.org (Accessed October 20 2011)

21

Picture Archiving and Communication Systems

ROBERT E. HOYT

LAJUANA EHLERS

Learning Objectives

After reading this chapter the reader should be able to:

- Describe the history behind digital radiology and the creation of picture archiving and communication systems

- Enumerate the benefits of digital radiology to clinicians, patients and hospitals

- List the challenges facing the adoption of picture archiving and communication systems

- Describe the difference between computed and digital radiology

- Understand the new possibilities with web-based PACS and mobile technology

Introduction

The following is a detailed definition of PACS:

> "Systems that facilitate image viewing at diagnostic, reporting, consultation, and remote computer workstations, as well as archiving of pictures on magnetic or optical media using short or long-term storage devices. PACS allow communication using local or wide-area networks, public communications services, systems that include modality interfaces, and gateways to healthcare facility and departmental information systems."[1]

Digital imaging appeared in the early 1970's by pioneers such as Dr. Sol Nudelman and Dr. Paul Capp. The first reference to PACS occurred in 1979 when Dr. Lemke in Berlin published an article describing the functional concept. In 1983 a team lead by Dr Steven Horii at the University of Pennsylvania began working on the data standard Digital Imaging and Communications in Medicine (DICOM) (see chapter on data standards) that would facilitate image sharing. The US Army Medical Research and Materiel Command installed the first large scale PACS in the US in 1992.[2] The University of Maryland hospital system was the first to go "filmless" in 1999. While PACS had many early contributors the father of PACS in the United States is felt to be Andre Duerinckx MD PhD. [3]

Many hospitals and radiology groups have made the transition from analog to digital radiography. To their credit, radiologists have pushed for this change for years but have had to wait for better technology and financial support from their healthcare organizations. Early pioneers understood that a digital system would mean no more bulky film jackets, frequently lost films and slow retrieval. We are now at a point where the technology is mature and widely accepted but cost is still an issue at smaller healthcare organizations.

Initially, hospitals purchased film digitizers so routine x-rays could be converted to the digital format and this was followed by scanning the digital image directly into the PACS.

Importantly, with the increasing use of electronic health records (EHRs), there is a need to integrate PACS with EHRs, hospital information systems (HISs) and radiology information systems (RISs). The Veterans Health Administration launched a nationwide teleradiology network in 2009 that interfaces with its EHR (VistA). All images will be sent to a server in California that all VA radiologists can access. The Department of Defense is planning for a similar multi-facility PACS solution in the near future.

PACS is now a mainstream tool among most large healthcare systems because of ease of use, popularity among physicians, speed of image retrieval, and flexibility of the imaging platform. It is estimated that about 90% of large teaching hospitals have PACS but usage by small community hospitals is lower.[4, 5] According to a June 2010 report on PACS adoption, replacement is now the most common reason for purchase of a system, with only 15% of purchases occurring for the first time. In this report 84% of hospitals of 100+ beds surveyed had implemented PACS in multiple locations outside the hospital, with the remainder of hospitals having single implementations. Additionally, 59% of hospitals also had Cardiac PACS that could store and display cardiology studies such as cardiac catheterization.[6]

One of the future challenges of teleradiology will be to share PACS images among disparate healthcare organizations.[7] SuperPACS is a new concept that would allow a radiology group that serves multiple sites with different PACS, radiology information systems (RISs) and hospital information systems (HISs) to view the sites as a single entity.[8] PACS was initially associated with expensive work stations ($50,000) using thick-client technology. Now the trend is for thin or smart clients that permit clinicians to access PACS via a web browser from the office or home.[9] Health Information Organizations (see chapter on health information exchange) are beginning to link to web-based PACSs so images from different organizations can be viewed and shared.

PACS is made possible by faster processors, higher capacity disk drives, higher resolution monitors, more robust hospital information systems, better servers and faster network speeds. PACS is also frequently integrated with voice recognition systems to expedite report turnaround. PACS usually has a central server that serves as the image repository and multiple client computers linked with a local or wide area network. Images are stored using the DICOM data standard. Input into PACS can also occur from a DICOM compliant CD or DVD brought from another facility or teleradiology site via satellite. Most diagnostic monitors are still grayscale as they have better resolution (three to five megapixels), compared to color. Newer "medical monitors" have 2,048 x 2,560 pixel resolution and can display 1,000+ shades of grey instead of the 250 shades of grey seen on a standard desktop monitor.

PACS Key Components (see Figure 21.1)

- Digital acquisition devices: the devices that are the sources of the images. Digital angiography, fluoroscopy and mammography are the newcomers to PACS

- The Network: ties the PACS components together

- Database server: high speed and robust central computer to process information

- Archival server: responsible for storing images. A server enables short term (fast retrieval) and long term (slower retrieval) storage. HIPAA requires separate back up

- Radiology Information system (RIS): system that maintains patient demographics, scheduling, billing information and interpretations

- Workstation or soft copy display: contains the software and hardware to access the PACS. Replaces the standard light box or view box

- Teleradiology: the ability to remotely view images[10]

Figure 21.1: PACS Components

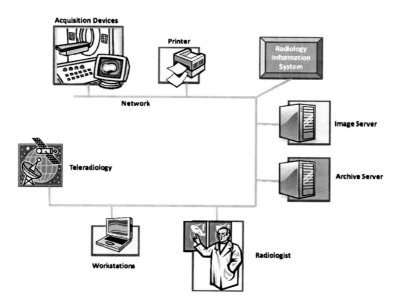

Types of Digital Detectors

- Computed radiography (CR): after x-ray exposure to a special cassette, a laser reader scans the image and converts it to a digital image. The image is erased on the cassette so it can be used repeatedly.[10] (Figure 21.2)

- Digital radiography (DR): does not require an intermediate step of laser scanning.[10]

Figure 21.2: Computed Radiography

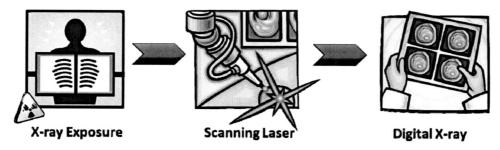

X-ray Exposure　　　**Scanning Laser**　　　**Digital X-ray**

As noted, a PACS should interface with both the HIS and RIS. Typically, the patient is identified in the HIS and an order created that is sent to the RIS via an HL7 protocol. Orders will go to the imaging device via the DICOM protocol and the image is created and sent to the PACS server. A diagnostic report is generated by the radiologist and stored on the PACS server. Diagnostic reports are sent back to the HIS via HL7 messages so they can be viewed on a DICOM viewer.

It is important to point out that many facilities with digital systems or PACS still print hard copies or have some non-digital services. This could be due to physician resistance, lack of resources or the fact that it has taken longer for certain imaging services such as mammography to go digital. *Full PACS* means that images are processed from ultrasonography (US), magnetic resonance imaging (MRI), positron emission tomography (PET), computed tomography (CT), routine radiography and endoscopy. *Mini-PACS*, on the other hand, is more limited and processes images from only one modality.[11]

Web PACS Integration with Health Information Exchanges

Diagnostic imaging plays a significant role in the medical care a patient receives. Movement of paper-based patient records across geographic and institutional borders can decrease the ability for care providers to have immediate access to the patient's entire medical record and imaging history without the implementation of a health information exchange system. Erin Chesson also states that "the power and reach of the web is empowering the health imaging world – completing the loop from radiology to specialist and back to the referring physician and even the patient." Furthermore, the benefits of web-based technology provide on-demand, online access to electronic images regardless of the location of patient records, reports and images.[12]

What does *web PACS* mean? Web PACS operates through the web environment just like the ASP model electronic health record, discussed in the second chapter. There is no client software residing on the viewer's computer. Instead, the web server and database are all available via the internet and identifiable with a unique URL. This is clearly different from "legacy PACS" systems where the client software was located on each user's computer. Table 21.1 compares the legacy PACS with web PACS. According to PACS marketing manager, Al Dryer of Agfa Healthcare, web-based PACS "is an application that uses different web technologies in a very open manner, regardless if the user is on a PC or Mac, using Linux or Windows for the operating system."[13]

Table 21.1: Legacy PACS compared to web PACS[14]

Legacy PACS	Web PACS
Only available on computers with proper software installed	Available anywhere with internet access
Upgrades must be manually installed	Upgrades are done centrally
Multiple user interfaces	One user interface
Difficult to integrate with health information exchanges	Easy to integrate with health information exchanges
Difficult to link to multiple EHRs	Easier to link to EHRs
Labor intensive for PACS administrator for maintenance and training	Much less labor intensive for maintenance and training
Could involve multiple operating systems	One operating system
Less likely to be standards-based	Utilizes JPEG compression, DICOM, HL7 and IHE profiles

Its goal is to offer seamless availability to radiologists, referring physicians, clinicians and nursing staff wherever they need images, i.e. at their office facilities, in the electronic health record, at their homes or wherever there is access to a remote, secure computer. To the patient it means that their physician has access to all of the medical information required to make informed decisions regarding their medical care: recent and previous images and reports, lab results, medication history, and other pertinent information.[14]

For example, a patient with a fractured lumbar spine can enter the emergency department at a medical facility located 90 miles away from Spokane, WA. The Emergency department (ED) physician there may be undecided about transporting the patient via helicopter to Spokane for neurosurgery, says Jon Copeland, CIO of Inland Imaging. The availability of web PACS affords medical personnel the technology to contact

Spokane and request the physicians there to view the patient's images. At the same time, the ED physician can contact a neurosurgeon at his home, who can log in from a home system using the web viewer to analyze the patient's back images for his recommendations. "Without shared, single image environment, this would not be possible," says Copeland.[12] For physician comments about web PACS, see case study in the side bar.

Other medical facilities such as Frederick Memorial Hospital and Peninsula Regional Hospital in Maryland are moving forward in planning for a health information exchange network. Marylanders will finally have the ability to retrieve their medical data regarding health care not just locally but regionally as well. Frederick Memorial also offers about 165 physicians a cost-effective means to log into its system for information exchange. There is also a physician portal available to physicians to view lab and PACS.[15]

The Consolidated Imaging Initiative (CI-PACS) in southern Maine, developed by MaineHealth and the Maine Medical Center, implemented a regional health information exchange system for radiology for rural hospitals. The system offers a shared, standards-based, interoperable PACS in two hospitals, Franklin Memorial Hospital and Miles Memorial Hospital. The last phase, web access rollout, provided for digital images to be web-based accessible, soft copy review, to the additional clinical areas. The system also provided access to remote sites and physicians' offices using the link into the hospital's CI-PACS connection.[16]

The Ochsner Health System in New Orleans integrates seven hospitals and 35 clinics. The hospital system had already implemented a widespread electronic clinical documentation process, "and had evolved its EMR platform forward" explains Dr. Lynn Witherspoon, CIO of the Ochsner Health System. The next goal was to use PACS in order to allow referring and ordering physicians access to patient images. According to Dr. Witherspoon, integrating all patient information for electronic accessibility to physicians is where PACS came in. He said, "PACS allowed us to put a web portal in the EMR. So now they'll open a folder on a patient record in the EMR, and they can open the PACS – PACS-EMR integration."[17]

Merge Healthcare announced in late 2011 it would introduce a free cloud-based service called Merge Honeycomb. This new service will allow users to upload, download, view, and share medical images at no cost. When a physician needs access to patient images, they can log into the image sharing network using any web browser. The network is permission-based, therefore images can only be seen by those who have been granted privilege. The goal of Merge Honeycomb is to help to decrease the need for exam duplication and patients' unnecessary exposure to radiation.[18]

Case Study

Why install a web-based PACS? According to Dr. Yaakov Applbaum, chairman of radiology, "Films were being lost." He proceeded to say, "My favorite stories were those of the trauma surgeons where hip films would be done in the trauma bay and the patient would be moved to CT. By the time the patient was at CT, the hip films were lost. Internally we needed PACS to survive. Once we decided to get a PACS, we wanted to be able to compete favorably with other institutions in the area and make it easier for referring physicians to see patient images. To facilitate that, we went with a web-based system that has 100% functionality outside the hospital equal to inside the hospital." This has been probably the most rewarding project of my life because we are actually improving patient care so dramatically and getting hospitals to cooperate in ways that have been unheard of in the past. The web viewer has broken down the barriers because it is just so effective"

PACS and Mobile Technology

Until recently, the U.S. Food and Drug Administration (FDA) had prohibited physicians from using radiology images displayed on mobile devices to make an official diagnoses. In February 2011 the FDA approved the first primary diagnostic radiology application for mobile devices. Performance evaluation reviewed by the

FDA consisted of tests for measured luminance, image quality (resolution), and noise referenced by international standards and guidelines. This new mobile radiology application will provide physicians access to view medical images on the Apple iPhone, iPad, and iPod. Keith Dreyer, D.O., Ph.D., vice-chair of radiology for informatics at Massachusetts General Hospital and serves as an associate professor of radiology at Harvard Medical School, stated in the May, 2011 RSNA publication news, "I see these devices being a mainstay for radiologists on call away from a clinical workstation." He further states, "The devices may currently be too limited in functionality and screen size to provide adequate throughput for a heavy case load, but for answering an immediate question, they will be quite adequate for many examination types." This new mobile application is named the Mobile MIM and includes a VueMe version for patients. While the viewers are free, there is a charge for storing and viewing images on the company's servers. The FDA emphasized that this application should only be utilized when there isn't access to a PACS workstation to view images and to make medical diagnosis of CT, MRI imaging and PET examinations.[19, 20]

Evidence is building in favor of mobile device usage in medicine. These systems have a high potential to improve efficiency and communication in medical imaging. Polomar Pomerado Health (PPH) in San Diego began a mobility initiative in July 2011 with Cisco Cius tablets. These tablets are internally built platforms named "MIAA (Medical Information Anytime Anywhere)." According to Orlando Portale, chief innovation officer at PPH, this mobile medical device makes available electronic health information to incorporate radiology reports and images. Another example of mobile use is at John Hopkins' department of radiology and radiological sciences. The radiology department is providing iPads for all residents. Carl Miller, MD, chief resident, states "We think [the iPad] has tremendous potential to transform clinical education in radiology." Furthermore, according to Paul Nagy, PhD, visiting associate professor at John Hopkins University radiology department in Baltimore, "we've seen PACS vendors respond, migrating their platforms onto the iPad." [21]

A study performed in 2009 by R.J. Toomey et al., compared a Dell Axim PDA and an Apple iPod Touch device. The researchers concluded that handheld devices show promise in the area "emergency teleconsultation for detection of basic orthopedic injuries and intracranial hemorrhage." However, further investigation is necessary.[22] Regarding the image quality of handheld devices, it is unclear at this point whether or not the image quality is adequate "for primary diagnosis and secondary consultation." The limitations of small screen size, intrinsic low resolution of display and minimal memory, security problems, are a few concerns about display of radiologic images on PDAs.

Currently there are several vendors that provide handheld viewing of PACS images. Among the most common manufacturers are: Imco Technologies' IMCO-STAT software functions with PACS to permit the radiologists and physicians to share images, video, voice and text on mobile devices such as smartphones and tablets. Phillips Medical Systems has iSite PACS using the iSyntax core transfer protocol which allows handheld viewing of images. Siemens Medical Solutions has created the Syngo Suite and Soarian HIS clinical information system. This system makes available web-based, real-time access to patient's EHR, and PACS images on an assortment of mobile device solutions. [23]

Image Resolution

Unfortunately, the field of radiology has had a gradual approach to mobile device technology for interpreting images mainly due to the concern of image resolution. Randall Stenoien, President, Innovative Radiology, PA, and CEO of Houston Medical Imaging, LLC, says "As radiologists, we adhere to FDA or American College of Radiology (ACR) criteria in terms of the resolution of the studies and the quality of the monitors. We have to test our monitors where we are going to read cases." Dr. Stenoien is optimistic regarding the direction towards making diagnosis on mobile devices. Dr. Stenoien continues "The mobile app is a web-based interface using PACS, which is going to be browser agnostic - whether you're using Firefox, Safari, or Internet Explorer. That, for me, is going to make a huge difference in my practice for referring docs. Using an iPad, the referring doctor can log-in to see their patients' images without having to push the images at all."[24]

According to Elliot K. Fishman, MD, FACR, John Hopkins University Department of Radiology, Baltimore, MD, mobile devices do have drawbacks. The screens are smaller compared to workstation monitors, and presently users are not able to dictate reports or view comparison films side by side. On the up side, there are features on the iPad and iPhone that place these devices on the same level, or above, the vigorous PACS. Lawrence White, Senior Marketing Manager for GE Healthcare Imaging Solutions-PACS Mobility, confirms that the technology is available. The biggest adjustment is in the user experience. According to White, the experience from the standpoint of navigating an iPad or Android or Tablet is going to be similar, thus the reason GE has developed an application on a "native graphic user interface rather than trying to port it on an existing application." The most recent "iteration of the iPad screen supports a 9.7-inch (diagonal) LED-backlit glossy widescreen at 1024 x 768-pixel resolution at 132 pixels per inch (ppi). Although the iPhone 4's screen spans just 3.5-in diagonally, at 960 x 640-pixel resolution at 326 ppi, it is equipped with the latest Apple Retina display for sharper images, videos, and text." Google's Android operating system allows the radiologist to select screen size and resolution that is compatible with his/her interpretation needs and/or user preference.

Dr. Stenoien concludes, "the trend in radiology and in medicine in general is to have the radiologist available all the time. If we have an FDA-cleared way of providing some of these services, it is really going to make us a lot more mobile." [24]

ResolutionMD Mobile

In September 2011 Calgary Scientific received FDA approval to market ResolutionMD Mobile as a medical imaging diagnostic application. The new mobile device supports several mobile devices and operating systems. Calgary Scientific's conducted multiple hands-on trials, performed using patient data that evaluated reading performance among mobile devices and standard PACS workstations, results showed equivalent diagnostic performance. On the primary diagnosis, the radiologists unanimously concluded that in office lighting conditions, there was no change in switching from the mobile device in dim lighting to the PACS workstation.

ResolutionMD mobile's server-based software application allows physicians immediate access to the display, reports, and analysis of patient images such as CT and MR, stored within any healthcare facility, and to submit a clinical diagnosis via their medical devices. Images are not permanently stored on the mobile devices. ResolutionMD mobile performs on 3/4G wireless, and "ensures that no highly sensitive or confidential patient information is retained on the mobile device."[25-27]

Osirix Mobile DICOM Viewer

OsiriX is a DICOM PACS open source viewer for the MAC OS. In addition to viewing images on an Apple computer, they can be viewed on an iPhone and iPad.[28]

PACS for a Hospital Desktop Computer

The AGFA IMPAX 6.3 PACS is an example of a client-server based system used by the US Navy.[29] The PACS receives HL7 messages from the hospital information system (HIS) and provides diagnostic reports and other clinical notes along with the patient's images. Although resolution is slightly better with special monitors, the quality of the images on the standard desktop monitor is very acceptable for non-diagnostic viewing (see Figure 21.3). Any physician on the network can rapidly retrieve and view standard radiographs, CT scans and ultrasounds. The desktop program is intuitive with the following features:

- Zoom-in feature for close-up detail

- Ability to rotate images in any direction

- Text button to see the report
- Mark-up tool that does the following to the image:
  - Adds text
  - Has a caliper to measure the size of an object
  - Has a caliper to measure the ratio of objects: such as the heart width compared to the thorax width
  - Measures the angle: angle of a fracture
  - Measures the square area of a mass or region
  - Adds an arrow
  - Right click on the image and short cut tools appear
- Export an image to any of the following destinations:
  - Teaching file
  - CD-ROM
  - Hard drive, USB drive or save on clipboard
  - Create an AVI movie

The following are two scenarios that point out how practical PACS can be for the average primary care physician:

- Scenario #1: An elderly man is seen in the emergency room at the medical center over the weekend for congestive heart failure and is now in your office on a Monday morning requesting follow up. Your practice is part of the Wonderful Medicine Health Organization, so you pull up his chest x-ray on your office PC.
- Scenario #2: You are seeing a patient visiting your area with a cough and on his chest x-ray you note the patient has a mass in his left lung. You download this image on a CD (or USB drive) for the patient to take to his distant PCM where he will receive a further work up.

Figure 21.3: Chest X-ray viewed in PACS

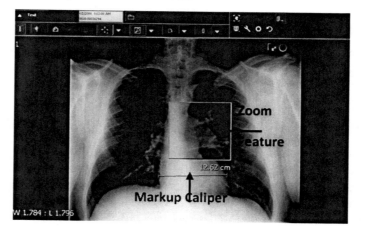

Open Source PACS

The commercial web-PACS vendor Heart Imaging Technologies uses the open source platform known as ClearCanvas. They offer a fee-based clinical (FDA approved) version and a free community version. Modules that can be downloaded include:

- Workstation DICOM PACS viewer software

- Radiology Information System for scheduled workflow, etc.

- Image Server software

- Platform (based on Microsoft.NET) and software developers kit (SDK) so plug-ins can be developed. [30]

PACS Advantages and Disadvantages

PACS Advantages

- Replaces a standard x-ray film archive which means a much smaller x-ray storage space; space can be converted into revenue generating services and it reduces the need for file clerks

- Allows for remote viewing and reporting; to also include teleradiology

- Expedites the incorporation of medical images into an electronic health record

- Images can be archived and transported on portable media, e.g. USB drive and Apple's iPhone

- Other specialties that generate images may join PACS such as cardiologists, ophthalmologists, gastroenterologists and dermatologists

- PACS can be web-based and use "service oriented architecture" such that each image has its own URL. This would allow access to images from multiple hospitals in a network

- Unlike conventional x-rays, digital films have a zoom feature and can be manipulated in innumerable ways

- Improves productivity by allowing multiple clinicians to view the same image from different locations

- Rapid retrieval of digital images for interpretation and comparison with previous studies

- Fewer "lost films"

- Reports are more likely to accompany the digital image

- Radiologists can view an image back and forth like a movie, known as "stack mode"

- Quicker reporting back to the requesting clinician

- Digital imaging allows for computer aided detection (CAD) [31]

- Increased productivity. Several studies have shown increased efficiency after converting to an enterprise PACS. In a study by Reiner, inpatient radiology utilization increased by 82% and outpatient utilization by 21% after transition to a film-less operation, due to greater efficiency.[32] In another study conducted at the University of California Davis Health System, transition to digital radiology resulted in: a decrease in the average image search time from 16 to two minutes (equivalent to more than $1 million savings annually in physician's time); a decrease in film printing by 73% and file clerk full time equivalents (FTEs) dropped by 50% (equivalent to more than $2 million savings annually).[33] The Health Alliance Plan implemented PACS at Henry Ford Health Systems in 2003. Results indicate: turnaround time for film retrieval dropped from 96 hours to 36 minutes; net savings of $15 per film and key players noted significant time savings. [34]

PACS Disadvantages

- Cost is the greatest barrier, although innovations such as open source and "rental PACS" are alternatives

- New legislation cutting reimbursement rates for certain radiology procedures, thus decreasing capital that could be used to purchase a PACS [35]

- Expense and complexity to integrate with hospital and radiology information systems and EHRs

- Lack of interoperability with other PACSs

- Bandwidth limits may require network upgrades

- Different vendors may use different DICOMS tags to label films

- Viewing digital images a little slower than routine x-ray films

- Workstations may require upgrades if high resolution monitors are necessary

Future Trends

In spite of its expense PACS has become the de facto standard of care for medical imaging. Making digital images available to all medical staff in a user friendly manner has been a quantum leap forward. Now there is a trend towards web PACS because it is more capable and is a better fit for large healthcare organizations, health information organizations and newer delivery models such as accountable care organizations. This is being supported and facilitated by faster networks, better monitor resolution and more digital imaging. Similarly, there will be better mobile platforms (smartphones and tablets) for viewing images by primary care and specialty physicians, patients and radiologists.

Key Points

- PACS is the logical result of digitizing x-rays, developing better monitors and medical networks

- PACS is well accepted by radiologists and non-radiology physicians because of the ease of retrieval, quality of the images and flexibility of the platform

- PACS is a type of teleradiology, in that, images can be viewed remotely by multiple clinicians on the same network

- Cost and integration are the most significant barriers to the widespread adoption of PACS

- WebPACS will promote better interoperability and sharing

- Mobile devices such as smartphones and tablet PCs offer a new viewing platform

Conclusion

PACS and digital imaging result from a predictable technological evolution beyond traditional film. For that reason, PACS has become a mainstream technology for moderate to large healthcare organizations. Like electronic health records (EHRs) PACS is an expensive technology to implement, but unlike EHRs, there is greater acceptance by clinicians. EHRs and Health Information Organizations will benefit by being interoperable with web PACS. Healthcare organizations will be looking for ways to interpret and distribute a wide range of images to the entire organization. The technology is moving closer to thin client or zero client web-based PACS for maximum flexibility and interoperability for the enterprise.

References

1. Vidar corp. http://www.filmdigitizer.com/about/news/glossary.htm (Accessed April 14 2006)

2. Hood MN, Scott H. Introduction to Picture Archive and Communication Systems. J Radiol Nurs. 2006;25:69-74

3. Wiley G. The Prophet Motive: How PACS was Developed and Sold http://www.imagingeconomics.com/library/tools/printengine.asp?printArticleID=200505-01 (Accessed April 14 2006)

4. Oosterwijk HT. PACS Fundamentals 2004 Aubrey Tx, Otech, Inc http://www.psqh.com/janfeb05/pacs.html (Accessed February 20 2006)

5. Matthews M. The PACS Picture. March 2009. Imaging Economics www.imaging economics.com (Accessed October 1 2009)

6. Miliard M. PACS adoption has reached the mature stage. June 25 2010. www.healthcareitnews.com (Accessed June 28 2010)

7. Moore J. Imaging meets the network. Government Health IT November 2008. pp 26-28

8. Benjamin M, Aradi Y, Shreiber R. From shared data to sharing workflow: Merging PACS and teleradiology. Eur J Rad 2010;73:3-9

9. Valenza T. Thin Wins. Imaging Economics March 2009 www.imagingeconomics.com (Accessed October 1 2009)

10. Samei, E et al. Tutorial on Equipment selection: PACS Equipment overview. Radiographics 2004; 24:313-34

11. Bucsko JK. Navigating Mini-PACS Options. Set sail with Confidence. Radiology Today. http://www.radiologytoday.net/archive/rt_071904p8.shtml (Accessed January 11 2007)

12. Chesson E. Choosing Web-based PACS. Health Imaging.com June 2006 http://www.dominator.com/assets/002/5131.pdf (Accessed October 5, 2010)

13. Massat M B. Will Web-Based PACS Take Over? Imaging Technology News January/February 2009 http://www.itnonline.net/node/30965 (Accessed October 5, 2010)

14. PACS Vendor Questions The Importance of Web-based vs Non Web-based Solutions. Dynamic Imaging http://www.dynamic-imaging.com/pdf/DI-Web-based-WhitePpr.pdf (Accessed October 9, 2010)

15. Task Force to Study Electronic Health Records. Infrastructure Management & Policy Development Workgroup. Health Information Exchange in Maryland. iHealth & Technology March 2005 http://mhcc.maryland.gov/electronichealth/shared/taskforce/february/hinfoexmd021307.pdf (Accessed October 10, 2010)

16. Loux, Stephenie et al. Consolidated Imaging: Implementing a Regional Health Information Exchange System for Radiology in Southern Maine http://www.ahrq.gov/downloads/pub/advances2/vol4/Advances-Loux_36.pdf (Accessed October 16, 2010)

17. PACS Grows Up. Healthcare Informatics November 2007 http://www.healthcare-informatics.com/ME2/Segments/Publications/Print.asp?Module=Publications::Article&id=6D7809D1B8D846F6974230A2F146BD32 (Accessed October 17, 2010)

18. Riedel, Catherine. Merge Healthcare Announces Nation's Largest Medical Image Sharing Network. News Release. (2011) http://www.merge.com/MergeHealthcare/media/company/Project-Honeycomb-PR-10_5_11-FINAL.pdf (Accessed October 22, 2011)

19. U.S. Department of Health & Human Services. FDA News Release. FDA clears first diagnostic radiology application for mobile devices. February 2011 http://www.fda.gov/NewsEvents/Newsroom/PressAnnouncements/ucm242295.htm (Accessed September 25 2011)

20. Radiological Society of North America Technology Forum. First FDA-approved Mobile Radiology App Poised for Daily Use. May 2011 http://www.rsna.org/Publications/rsnanews/May-2011/mobile_app_feature.cfm (Accessed September 25 2011)

21. Stevens, Mary & Beyers, Jeff. Mobile Devices Expand Radiologists' Horizons. HealthImaging. August 2011 http://www.healthimaging.com/index.php?option=com_articles&article=29182 (Accessed September 14 2011)

22. Toomey, Rachel J. et al. Diagnostic Efficacy of Handheld Devices for Emergency Radiologic Consultation (2010) February 2010 http://www.ncbi.nlm.nih.gov/pmc/articles/PMC2826276/ (Accessed September 13, 2011)

23. Fratt, Lisa. Images...Coming Soon to a Handheld Near You HealthImaging. May 2006 http://www.healthimaging.com/index.php?option=com_articles&view=article&id=4236(Accessed September 14 2011)

24. Bolan, Cristen. Apps mobilize radiology. Applied Radiology, volume 40, Number 05 May 2011 http://www.appliedradiology.com/Issues/2011/05/Tech-Trends/Apps-mobilize-radiology.aspx (Accessed September 28 2011)

25. Calgary Scientific. http://www.calgaryscientific.com/assets/files/resmd-mobile/ResolutionMDMobile_Features_001.pdf September 2011 (Accessed September 6, 2011)

26. Fratt, Lisa. FDA oks mobile app for iPhone, iPad viewing. Healthcare Imaging Regulatory News. September 2011 http://www.healthcaretechnologymanagement.com/_article/29693:fda-oks-mobile-app-for-iphone-ipad-viewing (Accessed October 4 2011)

27. Holmes, Colin. Networks Realities for "Anywhere, Anytime" Medical Imaging. Calgary Scientific. September 2011 http://www.calgaryscientific.com/assets/files/common/NetworkRealitiesForMedicalImaging_20110926.pdf (Accessed September 6 2011)

28. Osirix http://www.osirix-viewer.com (Accessed October 26 2011)

29. AGFA Healthcare. IMPAX 6. http://www.agfa.com/en/he/products_services/all_products/impax_60.jsp (Accessed June 4 2006)

30. ClearCanvas www.clearcanvas.ca/dnn (Accessed June 1 2010)

31. Ulissey MJ. Mammography-Computer Aided Detection. E-medicine. January 26 2005 www.emedicine.com (Accessed January 7 2007)

32. Reiner BI et al. Effect of Film less Imaging on the Utilization of Radiologic Services. Radiology 2000;215:163-167

33. Srinivasan M et al. Saving Time, Improving Satisfaction: The Impact of a Digital Radiology System on Physician Workflow and System Efficiency. J Health Info Man 2006;21:123-131

34. Innovations in Health Information Technology. AHIP. November 2005. www.ahipresearch.org (Accessed January 10 2007)

35. Phillips J et al. Will the DRA Diminish Radiology's Assets? http://new.reillycomm.com/imaging/article_detail.php?id=380 (Accessed June 18 2011)

Bioinformatics

ROBERT E. HOYT

INDRA NEIL SARKAR

Learning Objectives

After reading this chapter the reader should be able to:

- Define bioinformatics, translational bioinformatics and other bioinformatics-related terms

- State the importance of bioinformatics in future medical treatments and prevention

- Describe the Human Genome Project and its many important implications

- List private and governmental bioinformatics databases and projects

- Enumerate several bioinformatics projects that involve electronic health records

- Describe the application of bioinformatics in genetic profiling of individuals and large populations

Introduction

In this chapter we will discuss bioinformatics, the biomedical informatics sub-discipline that has gained increasing prominence in recent years thanks to initiatives such as the Human Genome Project, discussed in a later section. Bioinformatics can trace its formal beginning to about 30 years ago. However, in many ways bioinformatics has evolved independent of health informatics and thus has its own sets of definitions and background information.

Definitions

We begin with some common definitions and in the next section provide a short genomics primer.

- Bioinformatics, often times referred to as Computational Biology, is a general description of "the field of science in which biology, computer science and information technology merge to form a single discipline"[1] Bioinformatics makes use of fundamental aspects of computer science (such as databases and artificial intelligence) to develop algorithms for facilitating the development and testing of biological hypotheses, such as: finding the genes of various organisms, predicting the structure and/or function of newly developed proteins, developing protein models and examine evolutionary relationships.[2,3]

- Translational bioinformatics focuses on the "development of storage, analytic and interpretive methods to optimize the transformation of increasingly voluminous biomedical data into proactive, predictive, preventive and participatory health."4 Simply put, translational bioinformatics is the specialization of bioinformatics for human health.

- Genomics is the field that analyzes genetic material from a species

- Proteomics is the study at the of level of proteins (e.g., through gene expression)

- Pharmacogenomics is the study of genetic material in relationship with drug targets

- Metabolomics is the study of genes, proteins or metabolites

- Metagenomics is the analysis of genetic material derived from complete microbial communities harvested from natural environments 5

- Phenotype is the observable characteristic, structure, function and behavior of a living organism. Size and hair color could be examples. Phenotype is largely determined by the genotype

- Genotype is the genetic information that is often associated with phenotypes or regulation of biological function6

Genomic Primer

The human body has about 100 trillion cells and each one contains a complete set of genetic information (chromosomes) in the nucleus; exceptions are eggs, sperm and red blood cells. Humans have a pair of 23 chromosomes in each cell that includes an X and Y chromosome for males and two Xs for females. Offspring inherit one pair from each parent. Chromosomes are listed approximately by size with chromosome 1 being the largest and chromosome 22 the smallest. Organisms have differing numbers of chromosomes (e.g., our closest extant primate relatives, chimpanzees, have 24 pairs). Chromosomes consist of double twisted helices of deoxyribonucleic acid (DNA). DNA is composed of four sugar-based building blocks ("nucleotides": adenine [A], thymine [T], cytosine [C], and guanine [G]) that are generally found in pairs ("Watson-Crick" pairing: A-T, C-G). Genes are regions on chromosomes that encode instructions, which may result in proteins that then in turn enable biological functions. The process of decoding genes involves transcribing the DNA into ribonucleic acid (RNA) and then translation into amino acids that make up proteins (Figure 22.1). Collectively, the complete set of genes is referred to as a "genome." It is estimated that humans have between 20,000 and 30,000 genes and that genomes are about 99.9% the same between individuals. Variations in genomes between individuals are known as single nucleotide polymorphisms (SNPs) (pronounced "snips"). There are three types of alterations: single base-pair changes, insertions or deletions of nucleotides, and reshuffled DNA sequences. Although SNPs are common, their significance is complex and unpredictable.7-9

Importance of Bioinformatics

Besides diagnosing the 3,000 to 4,000 hereditary diseases that exist today, bioinformatics may be helpful to discover more targets for future drugs, develop personalized drugs based on genetic profiles and develop gene therapies to treat diseases with a strong genomic component, such as cancer. The most common way to achieve this is to use genetically altered viruses that carry human DNA. This approach, however, has not been definitely shown to work and has not been for general use by the FDA. Manipulation of genomes in other organisms, such as microbes, has shown promise for energy production ("bio-fuels"), environmental cleanup, industrial processing and waste reduction. Genetically engineered plants could also be made to be drought or disease resistant.

Figure 21.1: Genes (Courtesy of Nat. Inst. of General Medical Sciences)

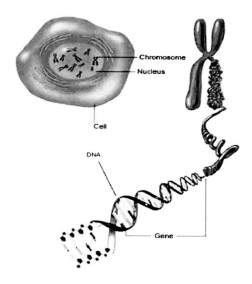

This chapter will deal primarily with transformational bioinformatics (TBI), a relatively newly identified area of focus in bioinformatics that is largely focused on the study of data contained within exponentially growing genetic and clinical databases. A significant goal of TBI is to enable bi-directional crossing of the translational barrier between the research bench and the bed in the medical clinic. With growing genome-wide and population-based research data sets we are uncovering more genotype-phenotype associations that potentially can detect and treat diseases with a genetic component earlier. Such associations may also help create tailor made drugs for higher efficacy. Figure 22.2 demonstrates the bidirectional nature of data and information flow between bioinformatics and health informatics. We have seen the emergence of translational bioinformatics primarily due to the rapid advances in technology on both sides. In other words, a variety of advances in bioinformatics, such as faster and cheaper DNA sequencing, and more widespread adoption of electronic health records have made this possible.

Figure 22.2: Translational bioinformatics (Adapted from Sarkar et al[10])

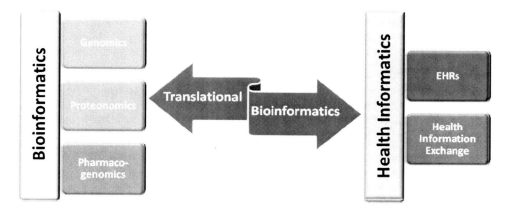

Pharmacogenomics is an excellent example of how translational bioinformatics can be used within the context of pharmaceutical development to utilize genomic information for better drug discovery and utilization. Drug companies are faced with the huge expense of drug development, the long road to producing a new drug and expiring patents. Drug failures are common and can be due to lack of clinical efficacy, side effects and commercial issues. Unfortunately, animal models are often times not adequate for the development and evaluation of drugs for treating human conditions. It is thus the goal to use genetic information for:

- New indications for an old drug (drug repurposing)

- New targets for existing drugs (e.g., treatment of tongue cancer using RET inhibitors)

- Drugs to work better in certain patient groups (gender, age, race, ethnicity, etc.) with possible genetic variants

- Knowing ahead of time what drugs to avoid due to higher incidence of side effects that are genetically modulated

- Develop clinical decision support in electronic health records based on pharmacogenomics[11,12]

Multiple projects are underway to integrate genetic and clinical data that will be discussed later in the chapter. We want to emphasize that burgeoning electronic health records (EHRs) and health information exchanges (HIEs), which are rapidly becoming ubiquitous, will contribute massive amounts of patient information (including demographic, laboratory, and clinical data). It is important to also note that in addition to genomic and clinical data, environmental data may offer valuable insights into the understanding and eventual treatment of disease.

Bioinformatics Projects and Centers

The Human Genome Project (HGP)

One of the greatest accomplishments in medicine in the current era of science was the Human Genome Project. This international collaborative project, sponsored by the US Department of Energy and the National Institutes of Health, was started in 1990 and finished in 2003. In the process of acquiring the human genome (as a complete set of DNA sequence, encompassing all 23 chromosomes), genome sequences for a number of other key organisms ("model" organisms) were also acquired. These included the *Escherichia coli* bacterium, fruit fly (*Drosophila melanogaster*), and house mouse (*Mus musculus)*. By mid-2007 about three million differences (SNPs) had been identified in human genomes. Appreciating the potential significant societal impact, the HGP also addressed the ethical, legal and social issues associated with the project. Since the completion of the HGP, attention is now more focused on the development of approaches to analyze and learn from volumes of data representing increasing numbers of individuals.[13-15]. These analyses include the annotation of information associated with disease onto chromosomes. Figure 22.3 displays the DNA sequencing of just chromosome number 12. Huge relational databases are necessary to store and retrieve this information. New technologies continue to emerge that reduce the necessity to sequence an entire human genome, such as DNA arrays (gene chips) that help speed the analysis and comparison of DNA fragments.[16] The cost of the HGP was close to $3 trillion; by 2010, a single gene chip can detect over a million variations in the base-pairs in a genome, in a few hours, costing only several hundred dollars.[7] Even more exciting is the prospect that within a decade it is expected that the cost of an entire human genome will cost around $1,000.

National Human Genome Research Institute (NHGRI)

NHGRI is an NIH institute that has many educational resources on their web site. Like other NIH institutes, they conduct and fund research within their intramural division, as well as support extramural research with external partners. Their health section has multiple resources for patients and healthcare professionals with particular emphasis on the Human Genome Project. The "Issues in Genetics" section covers important controversies in policy, legal and ethical issues in genetic research. They include a large glossary (200+) of genetics-related definitions, also available as a software app for the iPhone and iPad.[17]

Figure 22.3: Chromosome 12 (Courtesy of the National Library of Medicine)

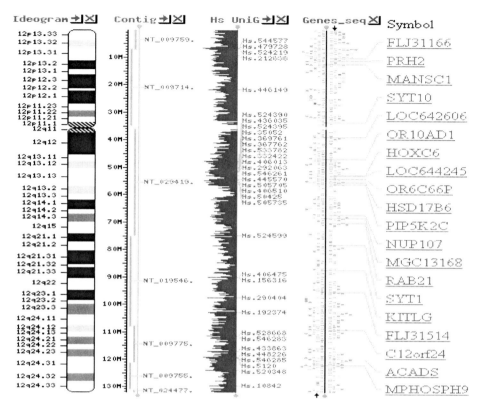

Human Microbiome Project (HMP)

It is estimated that less than 0.01% of microbes on Earth have been cultured, characterized, and sequenced. As an exception, the complete genome for the common human parasite *Trichomonas vaginalis* was reported in 2007 in the journal Science.[18] HMP is a NIH sponsored initiative that will study the myriad of organisms (oral, nasal, skin, gastrointestinal flora, etc.) that co-exist with humans and heretofore have been rarely studied. It will utilize metagenomics, as explained in the definitions section. As detailed on the HMP web site their goals are as follows:

- Determine whether individuals share a core human microbiome

- Understand whether changes in the human microbiome can be correlated with changes in human health

- Develop new technological and bioinformatic tools needed to support these goals

- Address the ethical, legal and social implications raised by human microbiome research[5]

Human Variome Project

This Australian initiative began in 2006 with the goal to create systems and standards for storage, transmission and use of genetic variations to improve health. Rather than catalogue "normal" genomes they focus on the abnormalities that cause disease. Another aspect of their vision is to provide free public access to their databases.[19]

National Center for Biotechnology Information (NCBI)

The NCBI was created in 1988 and is part of the National Library of Medicine at the National Institutes of Health. It hosts thousands of databases associated with biomedicine (including the popular MEDLINE and GenBank databases) and thereby is considered one of the world's largest biomedical research centers. The NCBI provides access to sequences from over 100,000 organisms (via GenBank), including the complete genomes of over 1,000 (via NCBI Genome). Genomes represent both completely sequenced organisms and those for which sequencing is still in progress. Popular NCBI databases, which are linked by a common interface (Entrez), are listed in Figure 22.4.

Figure 22.4: NCBI Databases (Courtesy National Library of Medicine)

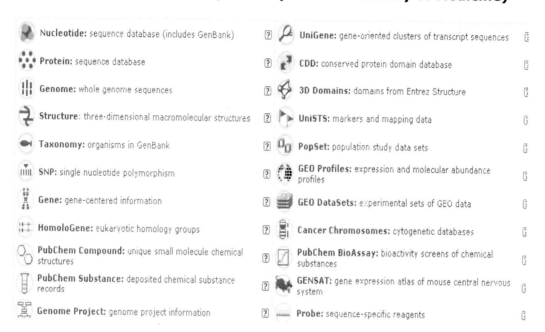

If you access the Genome project you can do a search for specific genes or proteins from different species. Figure 22.5 demonstrates the result of an Entrez Gene search for a tumor protein (TP53).

The NCBI site also provides access to BLAST (Basic Local Alignment Search Tool) that enables the identification of significantly related (based on a "expectation" value or "e-value") nucleotide or protein sequences from within the protein and nucleotide databases.[20]

GenBank

This database was established in 1982 and is the NIH sequence database that is a collection of all publicly available DNA sequences. Along with EMBL (Europe) and DDBJ (Asia), GenBank is a member of the International Nucleotide Sequence Database Consortium (INSDC), which provides free access to sequence data from nearly anywhere with an internet connection. As of this writing, there are approximately 126,551,501,141 bases in 135,440,924 sequence records in the traditional GenBank divisions. Interestingly, many biological and medical journals now require submission of sequences to a database prior to publication, which can be done with NCBI tools such as BankIt.[21]

Figure 22.5: Entrez search for tumor protein (Courtesy National Library of Medicine)

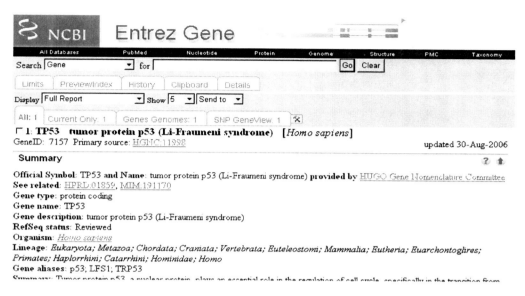

The Online Mendelian Inheritance in Man (OMIM)

This is another NCBI database of genetic data and human genetic disorders. It was originally developed and sponsored by Johns Hopkins University and Dr. Victor McKusick, a pioneer in genetic metabolic abnormalities. It includes an extensive reference section linked to PubMed that is continuously updated.[22]

World Community Grid

This project was launched by IBM in 2004 and simply asked people to donate idle computer time. By 2007 over 500,000 computers were involved in creating a super-computer used in bioinformatics. Projects include Help defeat Cancer, Fight AIDS@Home, Genome Comparison and Human Proteome Folding projects. This grid promises to greatly expedite biomedical research by analyzing complex databases more rapidly as a result of this grid.[23]

Pharmacogenomics Knowledge Base (PharmGKB)

This Stanford University based resource catalogues the relationships between genes, disease and drugs. There are sections on drugs, medical literature, variant genes, pathways, diseases and phenotypes that are searchable.[24]

Framingham Heart Study SHARe Genome-Wide Association Study

In 2007, the Framingham Heart Study began a new phase by genotyping 17,000+ subjects as part of the FHS SHARe (SNP Health Association Resource) project. The SHARe database is located at NCBI's dbGaP and will contain 550,000 SNPs and a vast array of phenotypical (combined characteristics of the genome and environment) information available in all three generations of FHS subjects. These will include measures of the major risk factors such as systolic blood pressure, total, LDL and HDL cholesterol, fasting glucose, and cigarette use, as well as anthropomorphic measures such as body mass index, biomarkers such as fibrinogen and C-reactive protein (CRP) and electrocardiography (EKG) measures such as the QT interval.[25]

The Mayo Clinic Bipolar Disorder Biobank

Researchers at the Mayo clinic and other institutions are analyzing the genetic and clinical information on 2000 patients in their biobank to determine genetic aspects of bipolar disorder. It is hoped that data generated from this project will lead to earlier and better treatment of this mental health disorder.[26]

Informatics for Integrating Biology and the Bedside (i2b2)

i2b2 is a National Institutes of Health National Center for Biomedical Computing located at Harvard Medical School. The Center has developed open source software that will enable investigators to mine existing clinical data for research. At this time there are 72 member institutions, including 12 that are international. The project was designed to allow users to query a system-wide de-identified repository for a set of patients meeting certain inclusion or exclusion criteria. On the web site, users can download client-software, client-server software and the source code.[27] The i2b2 infrastructure has been shown to be generalizable to multiple sites for a range of clinical conditions.[28]

Cancer Biomedical Informatics Grid (CaBIG)

CaBIG is sponsored by the National Cancer Institute at the National Institutes of Health. The architecture is known as CaGrid and is an open source service oriented architecture (SOA). The infrastructure is designed to support the collection and analysis of data from disparate systems to promote biomedical research. The core software and associated tools can be downloaded from their web site. For cancer biologists and researchers the site offers the following:

- Download genomic and clinical data from a wide variety of cancer types
- Query the database of animal models for human cancers
- Prepare microarray data for analysis
- Analyze proteomics data
- Query and share diverse data types via a web portal
- Organize and design clinical trials[29, 30]

For more information on translational bioinformatics and related databases in the context of biomedicine, we refer you to the textbook edited by Shortliffe and Cimino.[31]

Future Trends

Two major themes are appearing in the field of translational bioinformatics: (1) the potential for personal genetic ("direct to consumer") services and (2) integration of genomic information into electronic health records.

Personal Genomics

These trends are largely possible because of the availability of population-based genetic data and the decreasing cost for human genome determination.

- Population Studies: There are a number of ongoing initiatives that will leverage genomic data in the context of population studies. For instance, Oracle Corporation will partner with the government of Thailand to develop a database to store medical and genetic records. This initiative was undertaken to offer individualized "tailor made" medications and to offer bio-surveillance for future outbreaks of infectious diseases such as avian influenza.[32] Not all such initiatives have been successful. Perhaps the best known is DeCODE Genetics Corporation, which aimed to collect disease, genetic and genealogical data for the entire population of Iceland; however, it filed for chapter 11 bankruptcy in 2009.[33] Nonetheless, DeCODE continues some operations and the development of personal genomics based solutions, largely in partnership with organizations like Pfizer.

- Decreasing Cost of Human Genome Determination: Coinciding with the completion of the HGP, the NHGRI has kept track of the cost to perform DNA sequencing of an entire human genome over the past decade. As Figure 22.6 indicates, the cost has dropped from an initial cost of $100,000,000 to a current cost of about $10,000 per genome. Notably, the decrease in cost of genome sequence is exceeding Moore's Law (attributed to Intel co-founder Gordon Moore, and states that the cost of computing power will be halved every 18 months based on advances in technology).[17]

Figure 22.6: Cost per Genome over time (Courtesy National Human Genome Research Institute)

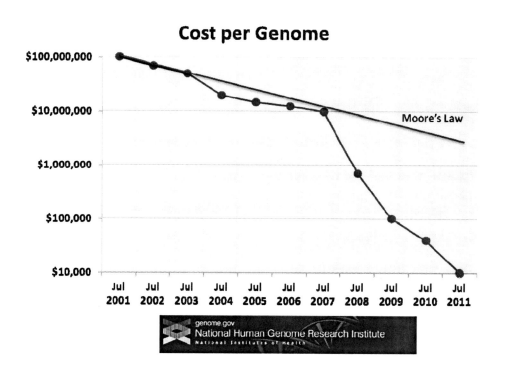

- Personal Genetics Testing. Many patients will want to know their own genetic profile, even if the consequences are uncertain. The following are examples of personal genetics companies ("direct to consumer genomics"):

  o Celera Genomics will take advantage of the genomics project to offer genetic mapping services and pharmacogenomics. They offer a cystic fibrosis genotyping assay.[34]

  o DNA Direct is another company that offers online genetic testing and counseling. They do offer both patient and physician education and have staff genetic counselors.[35]

  o deCODE Genetics offers whole genomic sequencing as well as deCODEme, an analysis for 47 diseases, traits and ancestry. They also can run risk profiles for type 2 diabetes, prostate cancer, atrial fibrillation, myocardial infarction, glaucoma and breast cancer. A simple mouth wash provides the DNA needed for analysis.[33]

○ 23andMe is a direct to consumer online genetic testing company. For $99 they will send a testing kit to homes based on analyzing saliva with a turnaround time of four to six weeks. Currently, they look for 97 diseases, carrier states and drug response conditions. They also offer an analysis of ancestry based on the genetic profile.[36] In 2010 a genome wide association study (GWAS) was published that used this technology and showed that patient questionnaire results correlated well with genetic results. Additionally, they were able to describe five new genotype-phenotype associations: freckling, photic sneeze reflex, hair curl and failure to smell asparagus.[37] Google's co-founder Sergey Brin has funded a project through this company to study the genetic inheritance of Parkinson's disease. They hope to recruit 10,000 subjects from various organizations and offer a discount price for complete analysis.[38]

However, as pointed out by Dr. Harold Varmus, personal genetics "is not regulated, lacks external standards for accuracy, has not demonstrated economic viability or clinical benefit and has the potential to mislead customers."[39]

In order for genetics to enter the mainstream, new technologies and specialties will need to be developed and numerous ethical questions will arise. Just finding the abnormal gene is the starting point. Genetic tests will have to be highly sensitive and specific to be accepted. In general, patients will not be willing to undergo major procedures (e.g., a prophylactic mastectomy or prostatectomy to prevent cancer) unless the genetic testing is nearly perfect. It is also important that genetic counseling be available to help patients understand the implication of genetic susceptibility tests (versus genetic guarantee of disease, such as the mutations associated with Huntington's disease).

Additionally, the Genetic Information Nondiscrimination Act of 2008 was passed to protect patients against discrimination by employers and healthcare insurers based on genetic information. Specifically, the Act prohibits health insurers from denying coverage to a healthy individual or charging that person higher premiums based solely on genetic information and bars employers from using individuals' genetic information when making decisions related to hiring, firing, job placement, or promotion.[40]

Many obstacles face the routine ordering of genetic tests by the average patient. Ioannidis et al. points out that in order for genetic testing to be reasonable several facts must be true. The disease you are interested in must be common. Even with breast cancer, when you evaluate seven established genetic variants, they only explain about 5% of the risk for the cancer. If the disease (e.g., Crohn's disease) is rare, then the test must be highly predictive. In order for genetic testing to be relevant you should have an effective treatment to offer, otherwise there is little benefit. The test must be cost effective, as many currently are too expensive. As an example, screening for sensitivity to the blood thinner warfarin (Coumadin) makes little sense at this time due to cost.[41]

A 2010 Lancet journal commentary also warned of additional concerns. Whole-genome sequencing will generate a tremendous amount of information that the average physician and patient will not understand without extensive training. At this point, we lack adequate numbers of geneticists and genetic counselors that understand the implications of data being made available thanks to continued advances in biotechnology. Patients will need to sign an informed consent to confirm that many of the findings will have unclear meaning. They will have to deal with the fact that they may be found to be carriers of certain diseases that *may* have impact on childbearing, etc. Genetic testing may cause many further tests to be ordered, thus leading to increased healthcare expenditures. As we gain more information about whole-genome sequencing, more patients will desire it but who will pay for it? And can the costs be justified?[42]

Two other recent articles drive home additional practical points. When the risk of cardiovascular disease based on the chromosome 9p21.3 abnormality was evaluated in white women, it only slightly improved the ability to predict cardiovascular disease above standard, well-accepted risk factors.[43] Meigs et al. looked at whether multiple genetic abnormalities associated with Type 2 diabetes would be predictive of the disease.

They found that the score based on 18 genetic abnormalities only slightly improved the ability to predict diabetes, compared to commonly accepted risk factors.[44]

For more information regarding future bioinformatics trends we refer to the review paper by Altman and Miller.[45]

Integration with Electronic Health Records

Eventually, the patient's genetic profile will be one more data field in the electronic health record. Recently, gene variants have been identified for diabetes, Crohn's disease, rheumatoid arthritis, bipolar disorder, coronary artery disease and multiple other diseases.[46] There are a number of forward-looking initiatives that have started on the path to integrate genomic data with traditional clinical data, for example:

- In late 2006 the Veterans Affairs healthcare system began collecting blood to generate genetic data that it will link to its EHR. The goal is to bank 100,000 specimens as a pilot project and link this information to new drug trials. The new voluntary program was officially launched in 2011 and is known as the Million Veteran Program (MVP). MVP will link genetic, military exposure, health and lifestyle into a single database.[47]

- Kaiser Permanente created the Research Program on Genes, Environment and Health and in the first phase two million members will be surveyed to determine their medical history, exercise and eating habits. As of mid-2011 genetic, medical and environmental information had been collected on 100,000 of its members. Kaiser plans correlative studies with its 15 years of digital health information, collected through its electronic health record system. Because the average age of participants is 65 it is anticipated that excellent information about aging will be generated. For example, they are measuring telomere length (the tips of chromosomes) that is thought to correlate with aging. This NIH funded initiative was completed in 15 months, thanks to newer technologies. It is anticipated that data will be analyzed and available to other researchers by 2012.[48, 49]

- The Electronic Medical Records and Genomics (eMERGE) Network is a consortium of biomedical informatics researchers across the United States. The National Human Genome Research Institute organizes this network, with additional funding from the National Institute of General Medical Sciences. An important theme is whether electronic health records are a vital resource for complex genomic analysis of disease susceptibility and patient outcomes in diverse patient populations.[50]

- Vanderbilt University recently published a strong correlation between their genetic biorepository known as BioVU (genotype) with clinical information (phenotype) obtained from their electronic health record. The diseases studied were rheumatoid arthritis, multiple sclerosis, Crohn's disease and type 2 diabetes.[51]

SNOMED CT is making changes to its codes to include genetic information and the National eHealth Initiative is developing "use cases" for family history and genetics so standards can be created by organizations like the Health Information Technology Standards Panel (HITSP). Organizations such as Partners HealthCare, IBM, Cerner and data mining vendors are all gearing up to add genetic information to what we currently know about patients and integrate that with electronic health records.[52]

The Agency for Healthcare Research and Quality (AHRQ) is developing computer-based clinical decision support tools to help clinicians use genetic information to treat conditions with a strong genetic component, such as breast cancer. Such tools that could be integrated into EHRs are: whether women with a family history of breast cancer need BRCA1/BRCA2 testing and which women who already have breast cancer may benefit from additional genetic testing.[53]

It is surprising that family history is often overlooked by clinicians and that it usually does not exist as computable data for analysis. To our knowledge, no electronic health record collects this information in a common computable format and uses it for clinical decision support; family history data are generally entered as unstructured text that can be of varying quality (based on provider-patient interviews). Data standards have been developed so family history can be part of EHRs and PHRs, in order to be shared.[54] There is a government sponsored free web tool available for the public to record their family history using the newest data standards. In this way, the results can be saved as a XML file and shared by EHRs and PHRs. The site, *My Family Health Portrait,* is available for English or Spanish speaking patients, is easy to use and does not store any patient information on the site. Instead, patients can store the XML file on their personal computers.[55] The program is open source and downloadable from this site.[56]

For further information about the role of EHRs and genomics we refer you to these citations.[57, 58]

Key Points

- Traditionally, bioinformatics has been a field remote from clinical medicine, but translational bioinformatics will likely bridge this gap
- Advances in biotechnology (such as genome sequencing) will likely introduce a treasure trove of genetic information that will enable deeper understandings of the manifestation of disease as well as the development of a new cadre of therapeutics over the next decade
- The inclusion of genetic profiles is being contemplated for electronic health records
- At this time, direct to consumer genetic testing is still in its early stages, and cannot be used as a replacement for traditional clinical tests (but may be used in complement)

Conclusion

The Human Genome Project and bioinformatics may seem foreign to many clinicians. The promise of translational bioinformatics is to transform biological knowledge (such as can be inferred from genomic data) into clinically actionable items. The success of translational bioinformatics will not be realized until clinicians can access and clinically interpret data that tells them who should be screened for certain conditions and which drugs are effective in which patients as part of day-to-day practice. In the meantime, biomedical scientists and companies will continue to add to the many genetic databases, develop genetic screening tools and get ready for one of the newest revolutions in medicine. The American Health Information Community (AHIC) recommended in 2008 that the federal government should prepare for the storage and integration of genetic information into many facets of healthcare.[59] Their recommendations will initiate the necessary dialogue that must take place to prepare for bioinformatics to align with the practice of medicine. But, as pointed out by Dr. Varmus "the full potential of a DNA-based transformation of medicine will be realized only gradually, over the course of decades."[40]

References

1.   NCBI. A Science Primer. www.ncbi.nlm.nih.gov/About/primer/bioinformatics.html (Accessed July 1 2006)

2.   Biotech: Bioinformatics: Introduction www.biotech.icmb.utexas.edu/pages/bioinform/BIintro.html (Accessed July 10 2006)

3.   Bioinformatics Overview. Bioinformatics Web www.geocities.com/bioinformaticsweb/?200630/ (Accessed July 6 2006)

4. Butte AJ, Shah NH. Computationally translating molecular discoveries into tools for medicine: translational bioinformatics articles now featured in JAMIA. J Am Med Inform Assoc. 2011;18(4):352-353

5. Metagenomics. Human Microbiome Project. The NIH Common Fund. http://commonfund.nih.gov/hmp/overview.aspx (Accessed September 17 2011)

6. Genetics Home Reference. http://ghr.nlm.nih.gov/glossary=phenotype (Accessed September 17 2011)

7. Feero WG, Guttmacher AE, Collins FS. Genomic Medicine—An Updated Primer. NEJM 2010;362:2001-2011

8. National Institute of General Medical Sciences. The New Genetics. http://publications.nigms.nih.gov/thenewgenetics (Accessed June 24 2010)

9. Genome: The autobiography of a species in 23 chapters. Matt Ridley. Harper Perennial. 2006

10. Sarkar IN, Butte AJ, Lussier YA et al. Translational bioinformatics: linking knowledge across biological and clinical realms. J Am Med Inform Assoc 2011; 18:354-357

11. Altman RB, Kroemer HK, McCarty CA et a. Phamacogenomics: will the promise be fulfilled? Nat Rev Genet 2011;12:69-73

12. Buchan NS, Rajpal DK, Webster Y et al. The role of translational bioinformatics in drug discovery. Drug Discovery Today. 2011;16(9/10):426-434

13. Human Genome Project www.ornl.gov/sci/techsources/Human_Genome/project/info.shtml (Accessed July 5 2006)

14. Human Genome Project www.genome.gov (Accessed January 3 2008)

15. NCBI Human Genome Resources www.ncbi.nlm.nih.gov/genome/guide/human/ (Accessed July 19 2006)

16. DNA Arrays http://en.wikipedia.org/wiki/Dna_array (Accessed December 5 2006)

17. National Human Genome Research Institute http://www.genome.gov (Accessed September 17 2011)

18. Carlton JM et al. Draft genome sequence of the sexually transmitted pathogen Trichomonas vaginalis. Science 2007;315:207-212

19. Human Variome Project. www.humanvariomeproject.org (Accessed September 17 2011)

20. NCBI http://www.ncbi.nlm.nih.gov/ (Accessed September 17 2011)

21. GenBank www.ncbi.nlm.nih.gov/Genbank/ (Accessed September 17 2011)

22. The Online Mendelian Inheritance in Man. http://www.ncbi.nlm.nih.gov/entrez/query.fcgi?db=OMIM (Accessed September 17 2011)

23. World Community Grid www.worldcommunitygrid.org (Accessed September 17 2011)

24. Pharmacogenomics Knowledge Base http://www.pharmgkb.org/ (Accessed September 17 2011)

25. Framingham SNP Health Association Resource http://www.ncbi.nlm.nih.gov/projects/gap/cgi-bin/study.cgi?id=phs000007 (Accessed September 17 2011)

26. Mayo Clinic Bipolar Disorder Biobank http://mayoresearch.mayo.edu/mayo/research/bipolar-disorder-biobank/index.cfm (Accessed September 17 2011)

27. Informatics for Integrating Biology & the Bedside https://i2b2.org (Accessed September 17 2011)

28. Cincinnati's Children's Hospital i2b2 DataWarehouse https://i2b2.cchmc.org/ (Accessed September 18 2011)

29. CaBIG http://cabig.cancer.gov/ (Accessed September 18 2011)

30. Oster S et al. caGrid 1.0: An Enterprise Grid Infrastructure for Biomedical Research. JAMIA 2008;15:138-149

31. Shortliffe E and Cimino J (eds) Biomedical Informatics, Computer Applications in Health Care and Bioinformatics. 3rd edition. 2006. Springer Science and Media, LLC. New York, New York

32. Oracle and Thai Government to build medical and genetic database www.ihealthbeat.org July 13 2005 (Accessed August 1 2005)

33. DeCODE genetics http://www.decodeme.com (Accessed September 19 2011)

34. Celera Genomics http://www.celera.com (Accessed September 19 2011)

35. DNA Direct www.dnadirect.com (Accessed September 19 2011)

36. 23andMe www.23andme.com (September 19 2011)

37. Eriksson N, Macpherson JM, Tung JY et al. Web-based, Participant Driven Studies Yield Novel Genetic Associations for Common Traits. PLoS Gen 2010;6(6) e1000993

38. Google Co-Founder To Back DNA Database Study on Parkinsons. March 12 2009. www.ihealthbeat.org (Accessed March 12 2009)

39. Varmus H. Ten Years On—The Human Genome and Medicine. NEJM 2010;362:2028-2029

40. Hudson, KL, Holohan JD, Collins FS. Keeping Pace with the Times—the Genetic Information Nondiscrimination Act of 2008. NEJM 2008;358:26612663

41. Ioannidis JPA. Personalized Genetic Prediction: Too Limited, Too Expensive or Too Soon? Editorial. Annals of Internal Medicine 2009;150(2):139-141

42. Samani NJ, Tomaszewski M, Schunkert H. The personal genome—the future of personalized medicine? Lancet 2010;375:1497-1498

43. Paynter NP, Chasman DI, Buring JE et al. Cardiovascular Disease Risk Prediction With and Without Knowledge of Genetic Variation at Chromosome 9p21.3. Annals of Internal Medicine 2009;150(2):65-72

44. Meigs JB, Shrader P, Sullivan LM et al. Genotype Score in Addition to Common Risk Factors for Prediction of Type 2 Diabetes. NEJM 2008;359(21):2208-2219

45. Altman RB, Miller KS. 2010 Translational bioinformatics year in review. J Am Med Inform Assoc 2011;18:358-366

46. Pennisi, E. Breakthrough of the Year: Human Genetic Variation. Science 2007;318 (5858):1842-1843

47. Million Veteran Program http://www.research.va.gov/mvp/ (Accessed September 18 2011)

48. Kaiser Seeks Member's Genetic Info for Database. www.ihealthbeat.org February 15 2007 (Accessed February 16 2007)

49. Kaiser Permanente Research Program on Genes, Environment and Health (RPGEH) http://www.dor.kaiser.org/external/DORExternal/rpgeh/index.aspx (Accessed September 18 2011)

50. Electronic Medical Records and Genomics (eMERGE) Network https://www.mc.vanderbilt.edu/victr/dcc/projects/acc/index.php/About#About_the_eMERGE_Network (Accessed September 19 2011)

51. Ritchie MD, Denny JC, Crawford DC et al. Robust Replication of Genotype-Phenotype Associations across Multiple Diseases in an Electronic Health Record. Am J Hum Gen 2010;86:560-572

52. Kmiecik T, Sanders D. Integration of Genetic and Familial Data into Electronic Medical Records and Healthcare Processes. http://www.surgery.northwestern.edu/dos-contact/infosystems/Kmiecik%20Sanders%20Article.pdf (Accessed June 28 2009)

53. AHRQ Launches Project on Computer-Based Genetic Tools. September 23 2008 www.ihealthbeat.org (Accessed September 24 2008)

54. Ferro WG. New tool makes it easy to add crucial family history to EHRs. Perspectives. ACP Internist May 2009 p. 6

55. Family History http://familyhistory.hhs.gov (Accessed September 18 2011)

56. National Cancer Institute Gforge http://gforge.nci.nih.gov/projects/fhh (Accessed September 19 2011)

57. Kohane IS. Using EHRs to drive discovery in disease genomics. Nature reviews genetics. 2011;12:417-428

58. Ullman MH, Matthew JP. Emerging landscape of genomics in the EHR for personalized medicine. 2011;32(8):512-516

59. HHS considers adding genetic information to EHRs. HealthImagingNews June 12 2008. www.healthimaging.com (Accessed June 12 2008)

23

Public Health Informatics

ROBERT E. HOYT

JUSTICE MBIZO

NORA J. BAILEY

Learning Objectives

- Define public health informatics

- Define public health surveillance and how data is used in public health

- Explain the significance of information technology in the field of public health

- Explain the significance of syndromic surveillance for early detection of bioterrorism, emerging diseases and other health events

- Explain the significance and scope of global public health informatics

- Understand the workforce needs and competencies of a public health informatician

- List several of the current surveillance systems used in the field of public health

- Explain the function and purpose of the Public Health Information Network

Introduction

Public health is another medical sector that has been greatly influenced by advances in information technology over the past two decades. The overarching goal has been to monitor a variety of medical diseases and conditions rapidly and accurately so as to intervene as early as possible to detect, prevent, and mitigate the spread of epidemics, the effects of natural disasters, and bioterrorism. With the advent of the internet, ubiquitous computing, electronic health records and health information organizations this vision is now possible.

For much of the 20th Century, public health reporting and surveillance consisted of physicians, hospitals and clinics sending paper reports to local health departments, who in turn forwarded information to state health departments who sent the final data to the Centers for Disease Control and Prevention (CDC) via mail or fax and finally to the World Health Organization for certain diseases. Although paper reports are still used, the shift to electronic media and information technology has facilitated-more efficient methods of public health surveillance, community based outbreak detection and disease control.

The most critical component in any disease investigation is the availability of timely data and information to pinpoint the possible source of the outbreak. The proliferation of information technology into public health and medical fields have significantly improved disease surveillance and enhanced early detection of community or population based epidemics. Global events, ranging from the September 11, 2001 terrorist attacks, the emergence of severe acute respiratory syndrome (SARS) in 2002 in China, to the recent global H1N1 influenza outbreak reinforced the need for roburst interoperable surveillance systems. The terrorist events of September 11, 2001 in particular, the subsequent anthrax attacks across the United States elevated and reinforced public health to a national security issue increasing the need for biosurveillance and real-time data analysis to detect and respond to disease outbreaks and health events more rapidly.

In the following sections we will define public health informatics (PHI), discuss public health surveillance systems, discuss syndromic surveillance, geographic information systems and cover global public health informatics.

Definitions

- Public health: the science and art of preventing disease, prolonging life and promoting health through the organized efforts and informed choices of society, organizations, public and private, communities and individuals." [1]

- Public health informatics: "the systematic application of information and computer science and technology to public health practice, research and learning...." [2]

- Public health surveillance: "the ongoing systematic collection, analysis, and interpretation of health-related data essential to the planning, implementation and evaluation of public health practice, closely integrated with the timely dissemination of these data to those who need to know. The final link in the surveillance chain is the application of these data to prevention and control." [3]

- Syndromic surveillance: "surveillance using health-related data that precede diagnosis and signal a sufficient probability of a case or an outbreak to warrant further public health response." [4]

Public Health Surveillance

Public health surveillance is essential to understanding the health of a population. Until recent years, public health surveillance was primarily paper-based. However, with the increasing shift towards eHealth public health surveillance has embraced the field of public health informatics. [12] In order to study a large population we need interoperable technologies such as standards-based networks, databases and reporting software. Current electronic surveillance systems employ complex information technology and embedded statistical methods to gather and process large amounts of data and to display the information for networks of individuals and organizations at all levels of public health. Public health surveillance serves to:

- Estimate the significance of the problem

- Determine the distribution of illness

- Outline the natural history of a disease

- Detect epidemics

- Identify epidemiological and laboratory research needs

- Evaluate programs and control measures

- Detect changes in infectious diseases

- Monitor changes in health practices and behaviors

- Assess the quality and safety of health care, drugs, devices, diagnostics and procedures

- Support planning [13]

Types of Surveillance Systems

Public health surveillance systems can be classified based on data collection purpose and design. Table 23.1 demonstrates the more common categories. [14-18]

Table 23.1: Types of Surveillance Systems

Surveillance System	Definition/Description	Examples
Case surveillance systems	• Collect data on individual cases of a health event or disease with previously determined case definitions in respect to criteria for person, time, place, clinical & laboratory diagnosis • Analyze case counts and rates, trends over time and geographic clustering patterns • Historically, case surveillance has been the focus of most public health surveillance.	• National Notifiable Disease Surveillance System (NNDSS)
Syndromic surveillance systems	• Collect data on clusters of symptoms and clinical features of an undiagnosed disease or health event in near real time allowing for early detection, rapid response mobilization and reduced morbidity and mortality • Data can be obtained through specific surveillance systems as well as existing epidemiologic data such as insurance claims, school and work absenteeism reports, over the counter (OTC) medication sales, consumer driven health inquiries on the Internet, mortality reports and animal illnesses or deaths for syndromic surveillance. • Geographic and temporal aberration and geographic clustering analyses are performed with real-time syndromic surveillance data. • Syndromic surveillance systems can also be used to track longitudinal data and monitor disease trends.	• Real-time Outbreak Detection System (RODS) • Biosurveillance Common Operating Network (BCON) • BioSense 2.0
Sentinel surveillance systems	• Collect and analyze data from designated agencies selected for their geographic location, medical specialty, and ability to accurately diagnose and report high quality data. They include health facilities or laboratories in selected locations that report all cases of a certain health event or disease to analyze trends in the entire population. • Pros: Useful to monitor and identify suspected health events or diseases • Cons: Less reliable in assessing the magnitude of health events on a national level as well as rare events since data collection is limited to specific geographic locations.	• PulseNet • FoodNet • ILINet
Behavioral surveillance systems	• Collect data on health-risk behaviors, preventative health behaviors, and health care access in relation to chronic disease and injury. • Analyze the prevalence of behaviors as well as the trends in the prevalence of behaviors over time. • Information is most commonly collected by personal interview or examination • Inferential and descriptive analysis methods such as age-adjusted rates, linear regression, and weighted analyses are used. • Most acute when conducted regularly, every 3 to 5 years	• Behavioral Risk Factor Surveillance System (BRFSS) • Youth Risk Behavior Surveillance System (YRBSS) • National Health Interview Survey (NHIS) • Pregnancy Risk Assessment Monitoring System (PRAMS)

Surveillance System	Definition/Description	Examples
Integrated Disease Surveillance and Response (IDSR)	• Incorporates epidemiologic and laboratory data in systems designed to monitor communicable diseases at all levels of the public health jurisdiction, particularly in Africa. • Useful for: detecting, registering and confirming individual cases of disease; reporting, analysis, use, and feedback of data; and preparing for and responding to epidemics.	
Clinical Outcomes Surveillance	• Monitors clinical outcomes to study disease progression or regression in a population. • Analyzes the rates of and factors associated with clinical outcomes using descriptive and inferential methods such as incidence rates from probability samples	• Medical Monitoring Project that monitors and tracks HIV patients
Laboratory Based Surveillance	• Collects data from public health laboratories, which routinely conduct tests for viruses, bacteria, and other pathogens. • Used to detect and monitor infectious and food-borne diseases based on standard methods for identifying and reporting the genetic makeup of specific disease-causing agents. • Commonly used in case surveillance and sentinel surveillance	• PulseNet • National Case Surveillance for Enteric Bacterial Disease (CDC)

The CDC has a helpful web page dedicated to surveillance programs for state, tribal, local and territorial public health officials. [19]

Syndromic Surveillance

Syndromic surveillance is part of meaningful use; therefore a basic understanding is important. Syndromic surveillance means symptoms are monitored (like diarrhea or cough) before an actual diagnosis is made. If, for example, multiple individuals complain of stomach symptoms over a short period of time, one can assume there is an outbreak of gastroenteritis. The important thing to remember is that syndromic surveillance systems do not identify the cause of the outbreak, rather they provide data comparisons which allows public health official to initiate outbreak investigation techniques.

In addition to the obvious sources of health data, public health officials can also monitor and analyze: unexplained deaths, insurance claims, school absenteeism, work absenteeism, over the counter medication sales, Internet based health inquiries by the public and animal illnesses or deaths. [15]

Initially, public health officials were very interested in detecting trends or epidemics in infectious diseases, such as severe acute respiratory syndrome (SARS) and avian influenza. After the terrorist attacks and anthrax outbreak in 2001, they have had to improve biosurveillance to detect bioterrorism. The objective is to "identify illness clusters early, before diagnoses are confirmed and reported to public health agencies and to mobilize a rapid response, thereby reducing morbidity and mortality."[20] The challenge is to develop elaborate systems that can sort through the information and reduce the signal to noise ratio. The syndrome categories most commonly monitored are:

- Botulism-like illnesses
- Febrile (fever) illnesses (influenza-like illnesses)
- Gastrointestinal (stomach) symptoms
- Hemorrhagic (bleeding) illnesses
- Neurological syndromes
- Rash associated illnesses
- Respiratory syndromes
- Shock or coma

Ambulatory electronic health records (EHRs) are a potentially rich source of data that can be used to track disease trends and biosurveillance. EHRs contain both structured (e.g. ICD-9 coded) data as well as narrative free text. Hripcsak et al. assessed the value of outpatient EHR data for syndromic surveillance. Specifically, they developed systems to identify influenza-like illnesses and gastrointestinal infectious illnesses from Epic® EHR data from 13 community health centers. The first system analyzed structured EHR data and the second used natural language processing (MedLEE processor) of narrative data. The two systems were compared to influenza lab isolates and to a verified emergency room (ER) department surveillance system based on "chief complaint." The results showed that for influenza-like illnesses the structured and narrative data correlated well with proven cases of influenza and ER data. For gastrointestinal infectious diseases, the structured data correlated very well but the narrative data correlated less well. They concluded that EHR structured data was a reasonable source of biosurveillance data. [21]

Real-Time Outbreaks Detection System (RODS)

The RODS system was initially developed by researchers at the University of Pittsburg and was the first real-time detection system for outbreaks. RODS collected patient chief complaint data from eight hospitals in a single health-care system via Health Level 7 (HL7) messages in real time, categorized these data into syndrome categories by using a classifier based on International Classification of Diseases, Ninth Revision (ICD-9) codes, aggregated the data into daily syndrome counts and analyzed the data for anomalies possibly indicative of disease outbreaks. Much like the ESSENCE system, RODS system started with a set of mutually exclusive and exhaustive categories of eight syndromic categories. However, as the program has gone through revisions and refinement, the categories have been reduced to seven as follows: respiratory, gastrointestinal, botulinic, constitutional, neurologic, rash and hemorrhagic. Figure 23.1 shows the daily counts of respiratory cases for Washington County, PA in the period June-July 2003.

Figure 23.1: Daily counts of respiratory cases six month period, Washington County, PA 2003.

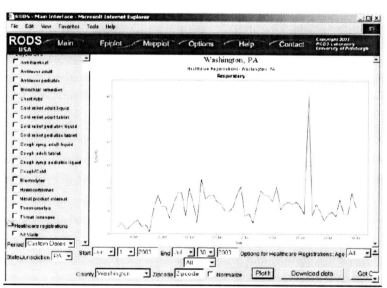

Source: Real-Time Outbreak and Disease Surveillance project.
⁺ The June 2003 increase corresponds to new hospitals being added to the system.
† The sudden increase on July 18, 2003, was caused by 60 persons reporting to one emergency department within 4 hours for carbon monoxide exposure.

In order to increase the adoption of the RODS system, the University of Pittsburg started offering software free of charge to public health departments. In 2003 the software was offered under an open source license and since then many more agencies have adopted the software for their use.[22]

Distribute

This project was created by the International Society for Disease Surveillance (www.syndromic.org), with the goal of supporting emergency department (ED) surveillance of influenza like illnesses (ILI). Figure 23.2 shows ILI reported over the last year in south eastern United States (region IV). [23]

Figure 23.2 Proportion of ED visits for ILI weekly 2011 (Courtesy Distribute)

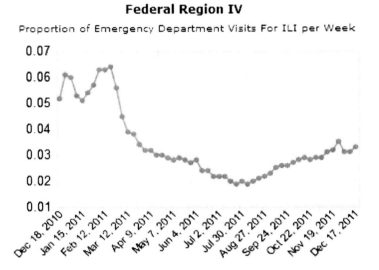

BioSense

This is a CDC national web-based program to improve disease detection, monitoring and situational awareness for healthcare organizations in the United States by reporting emergency room data. Participants include DOD, VA and civilian hospitals. The program addresses identification, tracking and management of naturally occurring events as well as bioterrorism. In 2010 BioSense was redesigned to integrate existing syndromic surveillance systems and allow for better regional sharing of information. The 2011 BioSense 2.0 allows state and local health departments to access data that will support syndromic surveillance systems under meaningful use. The goal is to provide a web based clearinghouse where data can be stored, searched and analyzed from and by multiple parties; decreasing the need for local health departments to purchase additional expensive information technologies.[24]

The Public Health Information Network

The Prevention and Public Health Fund, as part of the Affordable Healthcare Act of 2010, in conjunction with the Health Information Technology for Economic and Clinical Health (HITECH) Act has allowed the public health infrastructure to move into the eHealth era. Driven by the mission to prevent, reduce and treat disease, these initiatives focus on developing interoperable public health information systems that are beneficial to the healthcare of all Americans.[5-6]

The Public Health Information Network (PHIN) is a Centers for Disease Control and Prevention (CDC) initiative established to provide the framework for efficient public health information access, exchange, use, and collaboration among multi-level public health agencies and partners using a consensus of shared policies, standards, best practices, and services.[7]

Establishing messaging and vocabulary standards is a key strategy for PHIN, allowing for consistent interoperability between local, state and national public health entities as well as other agencies. The PHIN is currently working with the following Standard Development Organizations (SDOs): Systematic

Nomenclature for Medicine (SNOMED), Logical Observation Identifiers Names and Codes (LOINC), Health Level 7 (HL7), and Consolidated Health Informatics Initiative (CHI).[8] For more information about data standards, we refer readers to Chapter 6.

Electronic Health Records

Integral to this vision is interoperability with electronic health records (EHRs), as part of Meaningful Use of Health IT. As stated in the chapter on EHRs, Meaningful Use Stage 1 has several menu objectives with public health implications: the capability to transmit syndromic surveillance data to public health agencies, the capability to transmit data to immunization registries, and the capability for hospitals to transmit required disease Electronic Laboratory Reports (ELRs). [10]

As we see increased adoption of EHRs in the US and progression to Meaningful Use Stages 2 and 3, we should start to see new sources of data available for public health analysis. [10]

Public Health Information Network Update

On October 18, 2011, the "PHIN Messaging Guide for Syndromic Surveillance: Emergency Department and Urgent Care Data Version 1.0" was released, reaching a milestone for the public health objectives of Meaningful Use under the HITECH Act. This guide provides the HL7 2.5.1 messaging standard for the use of emergency department data as syndromic surveillance. [9]

Health Information Exchange (HIE)

We anticipate more public health reporting as a result of Meaningful Use for EHRs but a broader approach would be aggregating EHR/data shared with a health information organization. For further information about health information exchange we refer readers to Chapter 5.

A recent article outlined use cases that demonstrate the utility of HIE in public health:

- Mandated reporting of lab diagnoses: there is a predefined list of *notifiable diseases* (e.g. TB) that would benefit from electronic transmission to public health. In spite of that many states still rely on paper and results must be mapped to a standard vocabulary such as LOINC. A health information organization (HIO) could ensure proper identification, archiving and mapping. Mandated reporting could also trigger an alert of reportable diseases.

- Non-mandated reporting of lab data: There are several infectious diseases of interest that are not on the notifiable list but ideally tracked by public health. Additionally, antibiotic resistance patterns should be reported and shared with public health. A community wide antibiogram could be developed to educate local physicians about optimal prescribing patterns.

- Mandated reporting of physician-based diagnoses: physicians are separately required to report certain *notifiable diseases* but reporting is highly variable. This could be made easier with EHR reporting to the local HIO that in turn reports to public health. Data standards would be essential and alerts to appropriate public health staff, infection control officers, etc. would be possible.

- Non-mandated reporting of clinical data: syndromic surveillance will require symptom-related data from EHRs and emergency departments (EDs) to be sent and analyzed.

- Public health investigation: public health officials could query the HIO for additional clinical or demographic (age, gender, location, etc.) information about a case of interest.

- Clinical care in public health clinics: clinicians who treat patients in public health clinics could potentially benefit from access to a HIO.

- Population-level quality monitoring: HIE has the potential to give public health officials a glimpse of the quality of medical care in their area without chart reviews, across multiple health care systems.

- Mass-casualty events: HIOs might serve as a single point of contact for victims of a mass casualty. A record locator service might be able to keep track of admissions, discharges and transfer (ADT) data for the victims and their families.

- Disaster medical response: HIOs have the potential to make available patient data during a disaster when paper records might be destroyed or unavailable.

- Public health alerting--patient level: Theoretically, public health departments could alert all clinicians in a HIO about a case of TB where follow up is lost, for example. Public health officials could also warn hospitals about unique cases of highly resistant infectious organisms, particularly when patients tend to seek medical care at multiple institutions.

- Public health alerting - population level: Clinicians could be warned about trends in the community, for example viral culture results or antibiotic resistance trends.[11]

Geographic Information Systems (GIS)

As early as 1855 Dr. John Snow created a simple map to show where patients with cholera lived in London in relation to the drinking water source in the Soho District of London. Using his hand drawn map and basic epidemiological investigation techniques, much of which are still used today, he determined the source of the epidemic to be a common water pump. Epidemiology, public health surveillance and indeed the field of public health have improved significantly since the pioneer work of Snow and others after him. Much of this transformation has been the result of the emergence and proliferation of advanced computing technologies, the internet and other automated information systems.

Modern geographic information systems (GIS) use digitized maps from satellites or aerial photography. A Geographic Information System (GIS) is a system of hardware, software and data used for the mapping and analysis of geographic data. GIS provides access to large volumes of data; the ability to select, query, merge and spatially analyze data; and visually display data through maps. GIS can also provide geographic locations, trends, conditions and spatial patterns. Spatial data has a specific location such as longitude-latitude, whereas attribute data is the database that describes a feature on the map.

GIS maps are created by adding layers. Each layer on a GIS map has an attribute table that describes the layer. The data can be of two types: *Vector* or *Raster*. *Vector* data appears as points, lines or polycons (enclosed areas that have a perimeter like parcels of land). *Raster* data utilizes aerial photography and satellite imagery as a layer. Using GPS and mobile technology, field workers can enter epidemiologic data to populate a GIS. This geospatial visualization has been useful in tracking infectious diseases, public health disasters and bioterrorism. [25-26]

With the recent shift in public health focus to preventable chronic diseases, GIS has also been used to monitor chronic diseases and social and environmental determinants of health for public health policy. In early 2011, the Centers for Disease Control and Prevention launched a new project, Chronic Disease GIS Exchange. Designed for public health professionals and community leaders, GIS experts will use as an information exchange forum to network and collaborate with the goal of preventing heart disease, stroke and other chronic diseases. Data and information shared in this forum will be used in documenting the disease burden, informing policy decisions, enhancing partnerships and facilitating interventions from the use of GIS data.[27-29] Figure 23.3 shows a GIS display of diabetes incidence rates by State. [28]

Figure 23.3: GIS Map of diabetes diagnosis by county (Courtesy CDC Chronic Disease GIS Exchange)

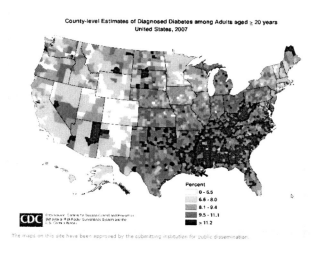

Virtually all of the biodetection systems mentioned have a GIS component that allows for the mapping of disease outbreak events giving public health practitioners the ability to timely deploy resources to control the outbreak and prevent further spread. Key variables can be inputted by zip code, latitude, longitude, that help public health disease investigators narrow down the source of the problem.

HealthMap is a global project to integrate infectious disease news and visualization using an Internet geographic map. This program classifies alerts by location and disease. For example, you can select "malaria" and "global" and see if there were any reported cases in the past 30 days. "Mouseover" an icon and you will see what is being reported in that area. A smartphone app "Outbreaks near me" details H1N1 (swine flu) outbreaks by locale, in near real time.[30] The program was developed by the Harvard-MIT Division of Health Sciences and Technology and a more detailed explanation of the system and architecture is provided at this reference.[31] Figure 23.4 shows a GIS display of global avian flu outbreaks

Public Health Informatics Workforce

As discussed, in order to most accurately and efficiently study the health of the population, information and communication technologies are essential to support the increasing demand for public health research and evidence based public health practice as a result of the aging US population. These technologies also require a diversity of human expertise for management, analysis, and communication of public health data. The Association of Schools of Public Health (ASPH) estimates that the field of public health will require 250,000 more workers by 2020 to avert a national public health crisis. [32] The transition to eHealth requires all public health workers to have some knowledge of IT depending on the demands of their position. In anticipation of this need, the CDC in collaboration with the University of Washington's School of Public Health and Community Medicine's Center for Public Health Informatics developed a list of informatics competencies for public health workers to meet the needs of the evolving public health field as well as for the Public Health Informatician. A Public Health Informatician is "a public health professional who works in practice, research, or academia and whose primary work function is to use informatics to improve population health."[33]

Figure 23.4: GIS display of global avian influenza outbreaks (Courtesy HealthMap)

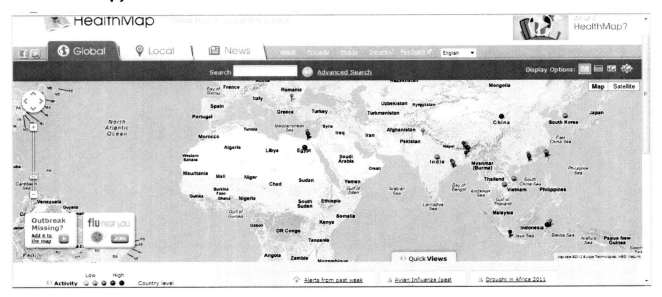

Global Public Health Informatics

Public health threats from chronic and infectious diseases, population health status, and health disparities within and across countries have gained global attention in part due to increasing personal mobility, economic globalization, and expansion of communication technologies. In fact, the global threat from chronic diseases was the focus of the 2011 UN General Assembly. Infectious diseases, such as influenza, polio, and HIV/AIDS, can quickly spread across national borders and are best curtailed through international cooperation and timely information sharing. New or re-purposed health information technologies provide critical support in the identification, monitoring, alerting, and responding to emerging diseases, pandemics, bioterrorism, and natural disasters. Simultaneously, health informatics has also emerged as an important tool in addressing population health goals and as a means to reduce health disparities between *developed* and *developing* nations.

World Health Organization

The leading international public health entity is the World Health Organization (WHO). Organized in 1948 as an agency of the United Nations (UN), WHO directs and coordinates public health efforts worldwide. WHO and its 195 Member States collaborate with other UN agencies, nongovernmental organizations, and the private sector to:

- Foster health security: Through its surveillance and disaster/epidemic response systems, WHO works to identify and curb outbreaks of emerging or epidemic-prone diseases. The revised 2007 International Health Regulations address the major forces contributing to epidemics including urbanization, environmental mismanagement, food preparation, and the overuse of antibiotics.

- Promote health development: Through this objective WHO works to increase access to life-saving and health-promoting interventions, particularly in poor, disadvantaged, or vulnerable groups. WHO's health development efforts focus on the treatment of chronic and infectious disease (e.g. diabetes), prevention and treatment of tropical diseases (e.g. malaria), women's health issues, and healthcare within African nations.

- Strengthen health systems: In poor and medically underserved areas, WHO endeavors to strengthen and supplement existing health systems. Activities include providing trained healthcare workers, access to essential drugs, and assistance in collecting vital health information.[34]

As discussed throughout this section, WHO increasingly relies on health information technology to carry out its objectives.

International Surveillance and Response Programs

The most visible role of WHO is to detect and respond to infectious disease outbreaks, pandemics, and disaster emergencies. Global surveillance of infectious disease, famines, and environmental disasters is implemented through a network of regional, national, and international institutes.[32] Government organizations (e.g. CDC), military networks (e.g. US Department of Defense's Global Emerging Infections Surveillance and Response System), and a host of public and private non-governmental organizations (NGOs) (e.g. Google, HealthMaps) monitor and report infectious diseases to WHO. Additionally, internet sites such as Epi-X or Pro-Med maintain discussions on current infectious diseases.

A 2007 review of 15 international surveillance and response programs (ISRPs) classified their activities into four basic components: surveillance, reporting, verification, and response.[35] The report found that the majority of these ISRPs focus on surveillance and reporting, while only six carry out all four activities. These six ISRP as well as other leading surveillance systems are described in the Appendix.

Regardless of the surveillance component performed by an ISRP, these organizations have benefited from the expansion of health information technology into the surveillance arena. Over the past decade, WHO and ISRPs have embraced web-based computing, mobile applications, GIS, and even text messaging. The role of health informatics within the major global surveillance organizations are discussed below.

Global Alert and Response (GAR): GAR is the integrated infectious disease surveillance program within WHO. A network of national, regional, and international agencies, governmental organizations (e.g. CDC) and military networks (e.g. US Department of Defense's Global Emerging Infectious Disease), GARs primary function is the facilitation of epidemic preparedness and response worldwide. This body is also responsible for maintaining and enhancing the global outbreak and bio-risk operational platforms.[36] Global monitoring and coordination are increasingly important in light of recent public health challenges such as outbreaks of severe acute respiratory syndrome (SARS) and influenza A (H1N1), the AIDS epidemic, and emerging new diseases and pathogens. Electronic surveillance capabilities have greatly enhanced the ability of GAR and its component functions to identify and respond to public health emergencies. Subsidiary functions under GAR include:

- International Health Regulations: 2005 revisions to WHO's International Heath Regulations (IHR) are aimed at improving global public health security and collaborative response to natural disasters, biological or chemical agents, and radioactive material release.[37] This legally-binding agreement provides a framework for the management of international public health emergencies, while also addressing the capacity of participating nations to detect, evaluate, alert, and respond to public health events. IHR specifies operational procedures for disease surveillance, notification and reporting of public health events and risks as well as for the coordination of international response to those events. The 2005 IHR allowed for the first time non-governmental sources to provide surveillance information to WHO. Participation by non-governmental contributors is as a positive step that pushes WHO to become more "dynamic, flexible, and forward-looking."[38]

- Early Warning Surveillance: GAR implemented an early warning surveillance response (EWARN) mechanism to effectively identify disease outbreaks and other health issues immediately following acute emergencies. An initial version of the system has been in use in

Haiti since the 2008 hurricane and expanded following the devastating Haitian earthquake in 2010. The system monitored public health issues such as injuries, mental health concerns, TB and HIV treatment programs, and disease trends. Inconsistency of data reporting, lack of trained personnel for data collection and technological errors among other problems interfered with the project from the start.[39] One solution that was developed in response to these challenges was a "virtual Google group" set up to improve communication. In remote, undeveloped areas of the world, WHO has encouraged Member States to develop early warning systems that use a variety of media including fax, telephone, the internet, and SMS to connect district or national surveillance officers with field collection efforts.

Global Public Health Intelligence Network (GPHIN): GPHIN was developed by the Public Health Agency of Canada to electronically monitor infectious disease outbreaks. Approximately 40 percent of the outbreaks investigated by WHO each year come from the GPHIN. This network "is a secure, internet-based 'early warning' system that gathers preliminary reports of public health significance in seven languages on a real-time, 24/7 basis."[40] GPHIN "continuously and systematically crawls web sites, news wires, local online newspapers, public health email services and electronic discussion groups for key words.[41] Although originally developed to detect infectious disease outbreaks, GPHIN now scans for food and water contamination, exposure to chemical and radioactive agents, bioterrorism, and natural disasters. It uses automated analysis to process the gathered data to alert human analysts to conduct additional review of any serious issues or trends. These data are then made available to WHO/GOARN and other subscribers through its web-based Microsoft/Java application and to the public through the WHO web site. GPHIN's automated data has significantly accelerated global outbreak detection.

Malaysia: Early Warning And Risk Navigation Systems

eWARNS is Malaysia's Early Warning And Risk Navigation Systems for natural disasters including rainfall, flash flood, soil erosion, landslide, tidal wave, and forest fire. Remote Sensing and Transmission Units (RSTU) placed throughout the country are used to predict floods and other natural disasters. Each RSTU collects rainfall data, senses the impact of the rain, and transmits the data via the internet to a receiving unit. The RSTU also acts as a web-server allowing the 'remote panel' to be viewed via the internet. The system alerts the public to real time risk levels and forecasts via SMS text messaging on their mobile phones.10 Information on daily rainfall, erosivity index, and erosion hazards are also available on the website.

http://www.ewarns.com.my/index.php?im=about

Global Outbreak Alert and Response Network (GOARN): The Global Outbreak Alert and Response Network was established by WHO in 1997. GOARNS has 420 global partners to collaboratively provide a rapid identification and response to outbreaks and alert the international community. Collaboration is provided by organizations like the Red Cross, the United Nations, humanitarian and scientific institutions, technical networks, laboratories, and surveillance and medical initiatives.[42]

Since 2000 GOARN has responded to more than 50 events worldwide, including SARS, Avian influenza and H1N1 influenza outbreak. Over one third of the surveillance information coming into GOARN is provided by GPHIN. Other surveillance information is provided by governmental agencies, universities, military agencies, and non-governmental organizations (NGO), such as the Red Cross and Médecins sans Frontières (Doctors without Borders). To facilitate global coordination, GOARN has established standardized operating procedures to be used by Member States and partnering organizations for identifying and responding to outbreaks. Features of the system include: alerts to the international community about outbreaks and technical collaboration on the rapid identification and response to outbreaks.[42] WHO's state of the art IT and communications systems ensure secure timely communications within GOARN and between GOARN and Member States and partnering entities thus facilitating the quick response and control of disease outbreaks.

Effective communication and collaboration between local and global responders to public health crises, hazards, and pandemics, is critical to successfully address the complex and diverse needs of the population after a disaster or during a public health emergency. GOARN and other responders recognize the benefit of integrating information and communication technology (ICT) into current operations.[43] Although sharing protocols of ICT appear to be a challenge, the emerging field of community informatics seems to provide the potential for inclusion of local health providers in emergency response efforts coordinated by global public health agencies.[44] Figure 23.5 depicts the complex and interdependent communication that must occur to ensure coordination of the local and global public health entities involved in disaster or public health emergency response.

Figure 23.5: Coordination between local and global public health organizations

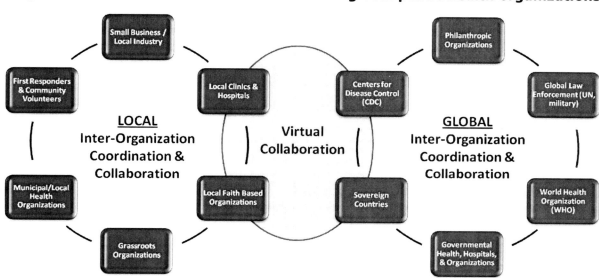

Other Global Public Health Activities

Surveillance and response to emergent health events maybe the most visible, but they are not the only functions of public health organizations. Public health is responsible for the prevention and control of disease, chronic and communicable diseases such as HIV/AIDS, TB, and polio and also plays a key role in health promotion and education. Historically, WHO and other public health organizations have struggled to provide even the most basic services to remote and poor areas around the globe. Health technology, particularly *mhealth*, has enabled public health agencies to reach out to isolated villages, connect with paraprofessional field workers, collect data, diagnosis disease, deliver disease management instructions, provide proficiency training to healthcare workers, and educate patients. Some of the organizations that deploy health technology in the fight to improve global health are identified in Table 23.2 on following page.

Global Health Information Technology Programs

Listed in alphabetical order below are a few of the premier organizations that facilitate the use of health information technology for public health:

- Center for Innovation in Global Health Technologies (CIGHT): A component of the Robert R. McCormick School of Engineering and Applied Science at Northwestern University, CIGHT collaborates with other universities, global healthcare companies, and non-profit organizations on the research and development of innovative and affordable healthcare technologies. The program focuses on three areas that are of concern in developing nations: HIV and associated diseases, saving lives at birth, and training healthcare workers to supplement physicians and nurses. http://www.cight.northwestern.edu/global-health-initiatives/index.html

- FHIi360-SATELLIFE: Created in 1987, SATELLIFE is a leader in using information technology to connect healthcare providers in developing nations to vital medical knowledge. Its GATHERdata™ project uses mobile devices to collect, report, and analyze real-time disease surveillance data. http://www.healthnet.org/

- Global Public Health Informatics Program (GPHIP): The Centers for Disease and Control (CDC) established a Global Public Health Informatics Program (GPHIP) in 2008 to collaborate with WHO and other international partners. "The Goal of GPHIP is to improve domestic and international public health informatics programs and advance the best informatics science, principles, strategies, standards, and practices."[46] GPHIP assists CDC-supported countries on developing and implementing innovative public health informatics solutions. Collaborative projects supported by GPHIP include a mobile-based information system for use in health emergencies and for surveys in China, an electronic integrated disease surveillance systems (EIDSS) in cooperation with Armenia, Azerbaijan, Georgia, Kazakhstan, Saudi Arabia, Ukraine, and Uzbekistan, and a national disease surveillance (NDS) and a health surveillance network (HSN) in Saudi Arabia. http://www.cdc.gov/globalhealth/programs/informatics.htm

- Information and Communication Technologies for Public Health Emergency Management (ICT4PHEM): Established by GAR in 2009, ICT4PHEM "is a technical collaboration of existing institutions and networks who pool human, technical and technological resources together to provide enhanced ICT solutions to predict, prevent and support Public Health Emergencies."[47] The objective of ICT4PHEM is to deploy ICT in the detection, assessment, verification and response to public health threats throughout the world. The initial meeting was held in April 2009 to discuss the need to develop, enhance and make available ICT tools to public health entities worldwide. http://www.who.int/csr/ict4phem/en/index.html

- WHO Global Observatory for eHealth (GOe): In 2005, the 58th World Health Assembly recognizing the need to incorporate emerging health information technologies into WHO and Member States adopted an eHealth strategy resolution. That same year WHO established GOe to study the impact of ehealth. The GOe conducted a survey of members in 2005 to establish a benchmark for each nation on its ehealth; a follow-up survey was conducted in 2009. Information on their findings relative to mobile technology, telemedicine, safety and security and other ehealth issues are available on their website: http://www.who.int/entity/goe/.

- Wireless Reach™: Through its Wireless Reach™ program, Qualcomm works with global partners to bring wireless technology to poor and remote areas around the world. Wireless Reach™ addresses education, entrepreneurship, public safety, and environment in addition to health. Its projects tend to be telemedicine related, although some have public health applicability. http://www.qualcomm.com/citizenship/wireless-reach

Table 23.2: Global Efforts to Improve Public Health through the use of Health Information Technology

Organization	Public Health Informatics Services
Cell-Life http://www.cell-life.org/	A not-for-profit organization that deploys mobile technology in the fight against HIV and other communicable diseases, primarily in South Africa. It has effectively used SMS to encourage HIV testing, to remind women to continue in prevention programs to curb mother-to-child transmission of HIV, increase antiretroviral therapy adherence, and provide family planning information.
Datadyne http://www.datadyne.org/	Datadyne offers applications for the use of cell phones to collect data, sending of mass SMS messages, and to provide continuing education to healthcare workers in remote areas through mobile devices.
Dimagi http://www.dimagi.com/	Dimagi in a for-profit company that builds custom mobile health and SMS solutions for resource-poor environments. It offers Windows Mobile 5 software devices to assist community health workers to screen HIV/AIDS patients, personalized SMS medication reminders to increase antiretroviral adherence in HIV patients, a mobile solution to improve home-based cancer care coordination, a portable web application for remote clinics to send cancer screening images to hospital-based physicians, SMS alerts for critical events, mobile applications for continuing education of remote healthcare workers, and a mobile application to increase compliance with WHO's Integrated Management of Childhood Illness program by remote health workers.
E Health Point http://ehealthpoint.com/?page_id=77	This project uses telemedicine to connect rural Indian villages to physicians and evidence based healthcare.
Mobile Alliance for Maternal Action (MAMA) http://www.mobilemamaalliance.org/	MAMA is a public-private partnership involving the US Agency for International Development, Johnson & Johnson, the United Nations Foundation, mHealth Alliance and BabyCenter. MAMA uses mobile phones to send audio and text health messages and reminders to new and expectant mothers.
mHealth Alliance http://www.mhealthalliance.org/	The mHealth Alliance is a public-private partnership between the UN Foundation, the Rockefeller Foundation, and The Vodafone Foundation. Its purpose is to harness the power of wireless technologies to improve health outcomes in low and middle income countries.
WHITIA-Essential Technologies for Safety Net Providers http://www.worldhealthimaging.org/index.html	WHITIA developed a low-cost, simple to use, self-contained digital x-ray unit. These units were initially deployed in Guatemala. WHITIA also provides telemedicine technology to connect village medical personnel with specialists and technology to enable high speed transmission of teleradiology images and healthcare data.

Future Trends

At the core of public health informatics is surveillance, a practice that relies on near-real time, high quality data. Largely because of the increased global use of the Internet, we are seeing an increase in analysis of aggregated data collected by both public and private organizations such as Google and various social media sites like Twitter and Facebook. Google.org recently launched three Internet-based projects utilizing revolutionary technology for public health research and policy development: *Google Flu Trends, Google Dengue Trends,* and *Google Crisis Response. Google Flu Trends* and *Dengue Trends* use aggregated data based on Google search queries to estimate disease activity in real-time.[48-49] Correlating strongly with data from the CDC, *Google Flu Trends* data is estimated to precede CDC results by about one week. [50-51] Ultimately, this methodology may be shown to be the most effective and fastest way to identify pandemic flu. Another venue for data aggregation analysis is social media. By examining data aggregated by user posts, researchers are gaining insight into health perceptions and behaviors as well as early detection of potential disease trends. Though criticized early on for the possibility of false reports and lack of specificity and sensitivity, social media's freely available, "real time" and statistically significant data is becoming as an essential tool for disease surveillance. [52-53]

Case Study

Mobiles in Malawi was initiated in the summer of 2007, by Josh Nesbitt who was working with a "rural Malawian hospital that serves 250,000 patients spread 100 miles in every direction. To reach remote patients, the hospital trained volunteer community health workers (CHWs) like Dickson Mtanga, a subsistence farmer. Dickson had to walk 35 miles to submit hand-written reports on 25 HIV-positive patients in his community. The hospital needed a simple means of communication."[42] Seeing the need Josh returned to the hospital the following year with mobile phones and a laptop running *FrontlineSMS*. In late 2008, *Mobiles in Malawi* merged with *MobilizeMRS,* an electronic medical records initiative that trained CHWs in structured data collection. The coming together of these efforts resulted in the creation of *FrontlineSMS:Medic* whose "mission was to help health workers communicate, coordinate patient care, and provide diagnostics using low-cost, appropriate technology....

"In six months, the pilot in Malawi using *FrontlineSMS* saved hospital staff 1200 hours of follow-up time and over $3,000 in motorbike fuel. Over 100 patients started tuberculosis treatment after their symptoms were noticed by CHWs and reported by text message. The SMS network brought the Home-Based Care unit to the homes of 130 patients who would not have otherwise received care, and texting saved 21 antiretroviral therapy (ART) monitors 900 hours of travel time, eliminating the need to hand deliver paper reports."[45]

Frontline SMS:Medic has since been deployed in Haiti after the 2010 earthquake where it was used by frontline disaster relief workers to text message urgent needs. "Using crowd-sourced translation, categorization, and geo-tagging, reports were created for first responders within 5 minutes of receiving an SMS. Over 80,000 messages were received in the first five weeks of operation, focusing relief efforts for thousands of Haitians." [45]

"In less than one year, *FrontlineSMS:Medic* expanded from 75 to 1,500 end users linked to clinics serving approximately 3.5 million patients. Growing from the first pilot at a single hospital in Malawi, they established programs in 40% of Malawi's district hospitals and implemented projects in nine other countries, including Honduras, Haiti, Uganda, Mali, Kenya, South Africa, Cameroon, India and Bangladesh." [45]

Frontline SMS has developed other mobile tools including: *PatientView*, a lightweight patient records system, *TextForms*, a text-based information collection module, and a messaging module for *OpenMRS. FrontlineSMS:Medic* recently changed its name to *Medic Mobile.*

Key Points

- Public health informatics is an important sub-category of health informatics
- Public health reporting will be part of meaningful use stages 1-3
- Public health surveillance is very broad and covers infectious diseases, epidemics, natural disasters and bioterrorism
- Geographic information systems provide a convenient display of medical information overlaid on geographical interface
- A myriad of new national and global public health informatics-related initiatives have been established

Conclusion

Public health is concerned with the health of populations, instead of individuals. In order to study large populations and track trends in health and other public health activities, paper-based reporting is no longer tenable. A robust public health network will require data standards, electronic health records and health information exchange. As a result of the HITECH Act and Affordable Care Act we are moving closer to the ideal goal of almost real time public health surveillance and reporting.

Acknowledgements

We would like to thank our graduate students for their assistance in preparing this chapter: Sara Beard BS, MPHc and Georgina Palombo MBA.

APPENDIX 23.1

International Surveillance Systems and Platforms (Adapted from Castillo-Salgado, 2010)

Name	Institution	Website Address	Components	Description and Activities
BTRP (Biological Threat Reduction Program)	Defense Threat Reduction Agency (US DoD)	http://www.dtra.mil/missions/NunLugar/BiologicalThreatReductionProgram.aspx	✓ Surveillance ✓ Reporting ✓ Verification ✓ Response	Working with partner nations, BTRP's focus is to prevent the proliferation of expertise, materials, equipment and technologies that could contribute to the development of biological weapons.
EPR (Epidemic and Pandemic Alert and Response)	WHO	http://www.afro.who.int/en/clusters-a-programmes/dpc/epidemic-a-pandemic-alert-and-response.html	✓ Surveillance ✓ Reporting ✓ Verification ✓ Response	EPR supports WHO Member States in the African Region to establish and implement functional integrated early warning and epidemic preparedness and response systems.
EUROFLU	WHO/European centers	http://www.euroflu.org/index.php	✓ Surveillance ✓ Reporting	Network of influenza morbidity and mortality surveillance reporting from health professionals in 53 countries and a laboratory network of European national influenza centers and two WHO influenza A/H5 reference laboratories.
GAINS (Global Animal Information System)	Wildlife Conservation Society with the support of USDA, USAID, FAO, and other agencies	http://www.gains.org	✓ Surveillance ✓ Reporting	Global initiative providing surveillance for influenza in wild birds. Collaborators in the GAINS network collect and analyze biologic samples from wild birds (which are caught and released), to identify locations of the avian influenza viral strain. The program disseminates information on avian influenza to governments, international agencies, and the public.
GDD (Global Disease Detection)	CDC	http://www.cdc.gov/globalhealth/gdder/gdd/	✓ Surveillance ✓ Reporting ✓ Verification ✓ Response	GDD is CDC's principal program for developing and strengthening global capacity to rapidly detect, accurately identify, and promptly contain emerging infectious disease and bioterrorist threats that occur internationally.
GEIS	US DoD	http://wrair-	✓ Surveillance ✓ Reporting	Designed to strengthen the prevention of, surveillance of, and

Name	Institution	Website Address	Components	Description and Activities
(Global Emerging Infections Surveillance and Response System)		ww.army.mil/index.p hp?view=preventive MedicineGeis	✓ Verification ✓ Response	response to infectious diseases that a) are a threat to military personnel and families, b) reduce medical readiness, or c) present a risk to U.S. national security. The DOD-GEIS mission is to increase DoD's emphasis on prevention of infectious diseases, strengthen and coordinate its surveillance and response efforts, and create a centralized coordination and communication hub to help organize DoD resources and link with U.S. and international efforts.
GOARN (Global Outbreak Alert and Response Network)	WHO	http://www.who.int/ csr/outbreaknetowrk/ en/	✓ Verification & Response	The main surveillance network of the WHO with the collaboration of more than 140 institutions. Receives surveillance information from the GPHIN and official country sources. Its mission is the rapid identification/confirmation and effective response to disease outbreaks of international public health importance.
GPEI (Global Polio Eradication Initiative)	Public-private partnership; includes WHO, CDC, UNICEF, Rotary International, and national governments	http://www.polioerad ication.org/	✓ Surveillance ✓ Reporting ✓ Verification ✓ Response	The four cornerstones of GPEI's efforts to eradicate polio worldwide are: (1) routine immunization, (2) supplementary immunization, (3) surveillance, and (4) targeted "mop-up" campaigns.
GPHIN (Global Public Health Information Network)	Public Health Agency, Government of Canada	http://www.cdc.gov/ globalhealth/GDD/gd doperation.htm	✓ Surveillance ✓ Reporting	Leading global web-based network providing surveillance information to WHO/GOARN and subscriber agencies. One of the first global monitoring systems using real time data from internet media sources to detect and report potential disease outbreaks.
HealthMap	Open-access GIS network supported by Google.org	http://www.healthma p.org/en	✓ Surveillance ✓ Reporting	Free internet GIS network collecting, organizing, and displaying infectious disease outbreaks. Integrates outbreak data of varying reliability, ranging from news sources to curated personal accounts (e.g. ProMED) to validated official alerts (e.g. WHO).
MedSys	European Commission	http://ec.europa.eu/h ealth/ph_threats/co m/preparedness/med ical_intelligence_en. htm	✓ Surveillance ✓ Reporting	Surveillance system available only to European Union member countries. The system includes an information scanning tool to support the surveillance of communicable diseases and early detection of bioterrorism activities in Europe.
ProMED-mail	International	http://www.promed	✓ Surveillance ✓ Reporting	Nonprofit, free email list network serving over 40,000 subscribers in

Name	Institution	Website Address	Components	Description and Activities
(Program for Monitoring Emerging Diseases)	Society for Infectious Diseases	mail.org/pls/apex/f?p=2400:1000		more than 150 countries. Global electronic reporting system since 1993. One of the leading email surveillance-reporting systems.
Regional Immunization Program of Americas			✓ Surveillance ✓ Reporting ✓ Verification ✓ Response	
Veratect Corporation	private biosurveillance firm (Kirkland, WA)	http://www.veratect.com/	✓ Surveillance ✓ Reporting	Leading private biosurveillance platform serving as an early warning system. It collects information from open-source reports and a global network of contacts. Tracks and locates global disease outbreaks and warns governments of any disease pattern compatible with an initial pandemic.
Voxiva System	Private company	http://www.voxiva.com/casestudies.php?caseid=30	✓ Surveillance ✓ Reporting	Electronic surveillance applications based on cell phones placed in remote places to report disease outbreaks. For use in low-resource environments.

References

1. Winslow, Charles-Edward Amory (1920 Jan 9). "The Untilled Fields of Public Health". Science 51 (1306): 23–33. doi:10.1126/science.51.1306.23. PMID 17838891.(Accessed October 4 2011)

2. Yasnoff, W., O'Carroll, P., Koo, D., Linkins, R., & Kilbourne, E. (2000). Public health informatics: Improving and transforming public health in the information age. Journal of Public Health Management Practice, 6(6), 67-75.

3. Centers for Disease Control. Comprehensive Plan for Epidemiologic Surveillance. Atlanta: US Department of Health and Human Services, Public Health Service; 1986.

4. Centers for Disease Control and Prevention http://www.cdc.gov/EPO/dphsi/syndromic.htm (Accessed September 20 2009)

5. Public Health Informatics Institute. (2010, June 30). Finding common ground: Collaborative requirements development for public health information systems. http://www.phii.org/resources/doc/CommonGround_Trifold_web.pdf (Accessed November 10 2011)

6. Foldy, S. (2011, February 11). Public health and the Health IT for Economic & Clinical Improvement (HITECH) Act: CDC's roles. www.cdc.gov/phin/library/about/IRGC_on_HITECH_20110211.pdf (Accessed December 10 2011)

7. Centers for Disease Control and Prevention, Public Health Information Network. (2011, June 8). Frequently Asked Questions (FAQs). http://www.cdc.gov/phin/about/faq1.html (Accessed December 10 2011)

8. Centers for Disease Control and Prevention, Public Health Information Network. (2011, May 20). Vocabulary. www.cdc.gov/phin/activities/vocabulary.html (Accessed December 11 2011)

9. PHIN Messaging Guide for Syndromic Surveillance: Emergency Department and Urgent Care Data Version 1.0 http://www.cdc.gov/ehrmeaningfuluse/docs/PHIN%20MSG%20Guide%20for%20SS%20ED%20and%20UC%20Data%20Release%201.pdf (Accessed December 11 2011)

10. Kass-Hout, T. (2011, March 7). Update on the BioSense Program Redesign, Meaningful Use, and syndromic surveillance [webinar]. Retrieved from http://198.246.98.21/osels/ph_informatics_technology/DOCS/PDF/Kass-Hout_CoE-PHI_03-07-2011.pdf

11. Shapiro JS, Mostashari F, Hripcsak G et al. Using Health Information Exchange to Improve Public Health. Am J Pub Health 2011;101:616-623

12. Krishnamurthy, R. & St. Louis, M. (2010). "Informatics and the management of surveillance data." In Principles and Practice of Public Health Surveillance. Third edition, Oxford Press. Eds. Lee LM, Teutsch SM, Thacker SB, St. Louis ME.

13. Teutsch, S. (2010). "Considerations in planning a surveillance system." In Principles and Practice of Public Health Surveillance. Third edition, Oxford Press. Eds. Lee LM, Teutsch SM, Thacker SB, St. Louis ME.

14. Sullivan, P., McKenna, M., Waller, L., Williamson, G. & Lee, L. (2010). "Analyzing and interpreting public health surveillance data." In Principles and Practice of Public Health Surveillance. Third edition, Oxford Press. Eds. Lee LM, Teutsch SM, Thacker SB, St. Louis ME.

15. Henning, K. (2004). Overview of syndromic surveillance: What is syndromic surveillance? MMWR, 53(Suppl), 5-11.

16. Nsubuga, P., White, M., Thacker, S....Trostle, M. (2006) Disease control priorities in developing countries. 2nd ed In Jamison DT, Breman JG, Measham AR, et al., editors.

17. McNaghten AD, Wolfe MI, Onorato I, Nakashima AK, Valdiserri RO, et al. (2007) Improving the Representativeness of Behavioral and Clinical Surveillance for Persons with HIV in the United States: The Rationale for Developing a Population-Based Approach. PLoS ONE 2(6): e550.

18. Anderson, M. (2008). Focus on Field Epidemiology. Volume 5, Issue 6. Chapel Hill, NC: UNC-Chapel Hill

19. CDC Surveillance information for state, tribal, local and territorial public health officials http://www.cdc.gov/stltpublichealth/Surveillance/index.html (Accessed December 20 2011)

20. Bioterrorism Preparedness and Response: Use of Information Technologies and Decision Support Systems www.ahrq.gov/clinic/epcsums/bioitsum.htm (Accessed September 21 2006)

21. Hripcsak G, Soulakis, ND, Li L et al. Syndromic Surveillance Using Ambulatory Electronic Health Records. JAMIA 2009;16:354-361

22. Espino J. et al. (2004) Removing a Barrier to Computer-Based Outbreak and Disease Surveillance – The RODS Open Source Project. MMWR, Supplement 53: 32-39

23. Distribute http://www.isdsdistribute.org/ (Accessed December 15 2011)

24. BioSense http://www.cdc.gov/biosense/ (Accessed December 15 2011)

25. ESRI. www.esri.com (Accessed December 10 2011)

26. Geographic Information Systems www.gis.com (December 10 2011)

27. GIS Wiki: The GIS Encyclopedia. Retrieved from http://wiki.gis.com/wiki/index.php (Accessed December 10 2011)

28. Ghirardelli, A., Quinn, V., & Foerster, S. (2010). Using Geographic Information Systems and Local Food Store Data in California's Low-Income Neighborhoods to Inform Community Initiatives and Resources. American Journal of Public Health, 100(11), p. 2156 -2162.

29. Centers for Disease Control and Prevention, National Center for Chronic Disease Prevention and Health Promotion, Division for Heart Disease and Stroke Prevention. (2011). Chronic Disease GIS Exchange. http://www.cdc.gov/DHDSP/maps/GISX/index.html (Accessed December 10 2011)

30. HealthMap. http://www.healthmap.org/en (Accessed March 2 2008)

31. Freifeld CC. Health Map: Global Infectious Disease Monitoring through Automated Classification and Visualization of Internet Media Reports. JAMIA 2008;15:150-157

32. Association of Schools of Public Health. (2008, February 27). ASPH Policy Brief Confronting the Public Health Workforce Crisis: Executive Summary. http://www.asph.org/UserFiles/WorkforceShortage2008Final.pdf (Accessed December 10 2011)

33. Centers for Disease Control and Prevention and The University of Washington School of Public Health and Community Medicine's Center for Public Health Informatics. (2009). Competencies for Public Health Informaticians 2009. http://www.cdc.gov/InformaticsCompetencies (Accessed December 10 2011)

34. World Health Organization. http://www.who.int/about/agenda/en/index.html (Accessed December 7, 2011)

35. Hitchcock P, Chamerlain A, Van Wagoner M, Inglesby T, and O'Toole T. Challenges to Global Surveillance and Response to Infectious Disease Outbreaks of International Importance. Biosecurity and Bioterrorism: Biodefense Strategy, Practice, and Science. Vol 5, Number 3, 2007. 206-227. DOI: 10.1089/bsp.2007.0041

36. Global Alert and Response (GAR). World Health Organization. http://www.who.int/csr/outbreaknetwork/en/ (Accessed November 21, 2011)

37. Castillo-Salgado C. Trends and Directions of Global Public Health Surveillance. Oxford Journals 2010; 32 (1): 93-109.

38. Fidler DP, "From International Sanitary Conventions to Global Health Security: The New International Health Regulations" Chinese Journal of International Law vol 4, issue 2, p 325-392. http://chinesejil.oxfordjournals.org/content/4/2/325.full (Accessed December 8, 2011)

39. Early Warning Surveillance and Response in Emergencies: Report of the second WHO technical workshop. World Health Organization. May 2011, Geneva, Switzerland http://whqlibdoc.who.int/hq/2011/WHO_HSE_GAR_DCE_2011.2_eng.pdf (Accessed December 9, 2011)

40. Global Public Health Intelligence Network (GPHIN). http://www.phac-aspc.gc.ca/media/nr-rp/2004/2004_gphin-rmispbk-eng.php (Accessed December 8, 2011)

41. Heymann, DL, "Dealing with Global Infectious Disease Emergencies" in Gunn SWA, et.al. Understanding the Global Dimensions of Health. 2005, 169-180

42. The Global Outbreak Alert and Response Network (GOARN). World Health Organization. http://www.who.int/csr/outbreaknetwork/en/ (Accessed November 21, 2011)

43. Gomez EA, Passerini K. Information and Communication Technologies (ICT) Options for Local and Global Communities in Health-Related Crisis Management 2010; 6 (2)

44. Stillman L, Linger H. Community Informatics and Information Systems: how can they be better connected?" The Information Society 25 (4): 1-1

45. FrontlineSMS:Medic. http://medic.frontlinesms.com/ (Accessed December 12, 2011)

46. Global Health Programs: Global Public Health Informatics. Centers for Disease Control and Prevention. http://www.cdc.gov/globalhealth/programs/informatics.htm (Accessed November 22, 2011)

47. Information and Communication Technologies for Public Health Emergency Management (ICT4PHEM). World Health Organization. http://www.who.int/csr/ict4phem/en/index.html (Accessed December 7, 2011)

48. Centers for Disease Control. Comprehensive Plan for Epidemiologic Surveillance. Atlanta: US Department of Health and Human Services, Public Health Service; 1986.

49. Friis, R., & Sellers, T. (2009). Epidemiology for Public Health Practice, 4th Ed. Subury, MA: Jones and Bartlett

50. Google.org. (2009). Flu trends: How does this work? Retrieved from http://www.google.org/flutrends/ (Accessed December 10 2011)

51. Ginsberg, J., Mohebbi, M., Patel, R., Brammer, L., Smolinski, M., & Brilliant, L. (2009). Detecting influenza epidemics using search engine query data. Nature, 457, 1012-1014.

52. Paul, M. & Dredze, M. (2011). You are what you tweet: Analyzing Twitter for public health. Proceedings of the Fifth International AAAI Conference on Weblogs and Social Media. Menlo Park, CA: The AAAI Press.

53. Brownstein, J. (2011, August 18). Using social media for disease surveillance. CNN.

24

---

e-Research

JOHN SHARP

Learning Objectives

After reading this chapter the reader should be able to:

- Understand the scope of eResearch and Clinical Research Informatics within the clinical research workflow

- Describe the use of EHR data in various phases of research including research originating from EHR data

- Conceptualize how informatics tools can be utilized in recruiting subjects for clinical research

- Detail how informatics supports the ongoing management of clinical trials

- Review the new trends in big data, real-time analytics and data mining

Introduction

Within the past ten years, there has been a dramatic shift from paper-based records in research to almost completely electronic. Paper case report forms being transposed into spreadsheets or early database programs are rapidly disappearing. Now every aspect of clinical research is supported by informatics tools. Several factors enabled this rapid change: availability of open source programming, major support from the National Center for Research Resources of the National Institutes of Health for informatics, consolidation of field of clinical research informatics with the American Medical Informatics Association, and academic medical centers' move toward securing patient data as a result of HIPAA and HITECH. These forces accelerated the move toward informatics permeating clinical research. But the most significant change is the adoption of electronic medical records.

In a perspective from the New England Journal of Medicine titled "Evidence Based Medicine in the EHR Era" the authors give examples of how an electronic cohort of patient data in the electronic medical record (EHR) can be used in clinical decision support. They conclude: "the growing presence of EHRs along with the development of sophisticated tools for real-time analysis of de-identified data sets will no doubt advance the use of this data driven approach to health care delivery."[1] So there is no doubt that health informatics and specifically eResearch will have a major contribution to evidence based medicine in the future. In fact, there are now informatics solutions for every phase of the research process. In this chapter we will explore the current state of these tools and their usefulness in promoting clinical research.

Preparatory to Research

The first step for a researcher with a question is to research the literature. It has been well documented that the medical literature is growing at a rate which overwhelms the practicing physician and the clinical researcher. Informatics tools are increasingly needed to assist with sorting through the literature and creating a reasonable background for any study. Fortunately, PubMed offers an array of tools which can be utilized on the site or integrated into a website or application using web services and RSS feeds. Entrez Programming Utilities provide a catalog of XML scripts as well as Perl scripts and other tools for custom extraction of medical journal data. A mobile version is also available.[2] The National Library of Medicine sponsors App contests to improve searches and create visualizations to improve data analysis. Google Scholar provides a broader database search which includes PubMed but also other scientific and academic publications (scholar.google.com). Google Books provides access to excerpts of books and allows searches through published works as well (books.google.com). For more details on online medical resources and search engine, please see additional chapters.

ClinicalTrials.gov provides the researcher with a search of all registered clinical trials within the U.S. As with PubMed, ClinicalTrials.gov provides an open API (Application Programming Interface) for linking and XML for connecting through web services.[3] For a wider search, the World Health Organization (WHO) provides a search tool which incorporates international trials.[4] Both PubMed and WHO now have mobile versions of the clinical trial search tools.

Research collaboration networks have seen significant growth in recent years. Research networks are typically web-based applications which include features similar to other social networks, such as a personal profile, opportunities to connect with others with similar interests and the ability to post status updates. Often research networks have personal profiles of researchers pre-populated with publications (thanks to integration with PubMed) and clinical trials (integration with ClinicalTrials.gov) and grants through the NIH Exporter.[5] With these rich data sources, some research networks have created semantic connections between researchers (vivoweb.org). However, most research networks provide search tools to enable finding connections between those with common interests. Three tools stand out, although many have been developed:

- Vivo. An open source tool developed at Cornell University, Vivo is a semantic web application (common framework that allows data to be shared and reused across application)[6]

- Harvard Profiles Catalyst. This is an open source community of over 130 member institutions with built-in network analysis and data visualization tool[7]

- SciVal Experts. This commercial solution also has modules to find research funding and measure benchmarks[8]

Other available tools generate National Institutes of Health (NIH) biosketches and add publications and grants dynamically (see Figure 24.1). Since research networks are relatively new, there is not substantial evidence of their effectiveness beyond anecdotal examples.

In addition to finding collaborators, clinical researchers would like to know the feasibility of their studies before they initiate them. One approach enabled by electronic medical record data is doing queries to evaluate adequate pools of patients to be recruited into the study. This requires a clinical data repository from EHR data with a query tool to search de-identified clinical information. By modifying inclusion and exclusion criteria, a researcher can find the appropriate cohort for recruitment based on a reasonable recruitment rate. There are already successful examples of this that have saved years of unsuccessful or under recruited studies.

Figure 24.1: Example of a biosketch from Harvard Profiles Catalyst

Electronic grant submission is now common for government agencies. Through the Office of Extramural Research at the NIH, grant submission and award management are all web-based. Forms are completed online and uploaded to the site, email alerts are available about the posting of new grants, and grant awards are posted online at the site. In addition, some Clinical Trial Management Systems (CTMS) discussed later have integration into the NIH electronic Research Administration Commons (eRA), which allows institutions to centrally manage grant submissions to the NIH.

Study Initiation

Informatics has a role in the initiation of studies as well. Volunteer recruitment can be enabled over the internet. Two approaches to volunteer recruitment are ResearchMatch and TrialX.

ResearchMatch provides a way to connect patients seeking clinical trials and researchers seeking volunteers.[9] (see Figure 24.2) Volunteers can create an account and indicate what their health issues are that may match with clinical trials. Researchers from institutions affiliated with the network can enter the clinical trials and contact information by completing an online form. Then the researcher can search volunteers and email them an invitation to participate. The volunteer can accept or decline to receive more information[10]. TrialX is a commercial venture which allows the volunteer to search clinical trials from ClinicalTrials.gov. Based on search terms, the user can see how closely their search matches available trials and then select a trial and email the investigator by registering on the site. Researchers can also register to list their trials and organizations can partner with TrialX to create custom listings of their studies. [11] Yet another model is a social network built around volunteering for clinical trials. ArmyOfWomen provides that platform and has provided thousands of volunteers for dozens of trials, initially for breast cancer but now for a variety of conditions.[12]

Figure 24.2: ResearchMatch program (Courtesy ResearchMatch)

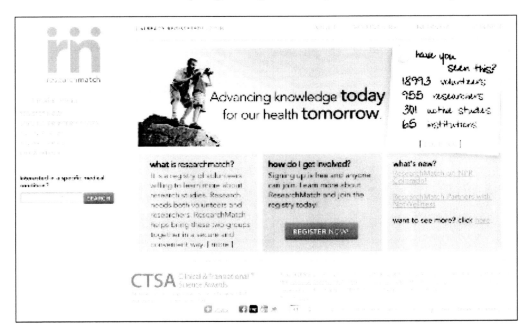

Online recruitment of subjects using social media is an emerging trend. Information on clinical trials using major social media outlets like Facebook and Twitter are new ventures. A transition from traditional advertising to online promotion of clinical trials is growing but many Institutional Review Boards are unfamiliar with this approach and need education to promote acceptance and establish standards for appropriate use. Another promising use of social media is provider groups and researchers developing relationships with online patient networks. These groups of ePatients are receptive to clinical trials and partnerships with researchers. A successful partnership was documented between women who have a rare cardiac condition and the Mayo Clinic. The women, whom already had an online community, were eager to participate in trials.[13] Patient social networks are already collecting data on their treatments and so the word about new clinical trials travels quickly.[14] Many healthcare organizations still caution patients and employees from using social media; in this context, patients should be cautioned that information on clinical trials communicated through social media must be evaluated like other online health content, with a critical mind.

Recruitment of subjects through the capabilities of the electronic medical record has two possible modes. First, the EHR can be used to find cohorts of eligible patients and create patient contact lists for recruitment.[15] Second, clinical trial alerts can be embedded within the EHR based on diagnoses, lab tests or other patient characteristics. The alert would typically remind the provider that their patient may be eligible for a clinical trial and who to contact.[16, 17]

Study Management and Data Management

There are several informatics tools which support study management and particularly managing research data. Clinical trial management systems (CTMS) are now common in academic medical centers. The purpose of these tools is to manage the planning, preparation, performance, and reporting of clinical trials. A CTMS has multiple functions in study management including: budget management, study calendar of patient visits, and creating electronic case report forms (eCRFs).[18] These tools can be open source or commercially available products.[19] The Cancer Biomedical Informatics Grid (CaBIG) has developed a CTMS Workspace with "modular, interoperable and standards-based software tools designed to meet diverse clinical trials management needs."[20] There is also a CTMS knowledge center which describes the NCI Clinical Trial Suite & Products.[21]

Some applications provide eForms or eCRFs with a focus on study data management. These tools enable the building of web-based forms for research without the support of programmers. Probably the most widely distributed tool is Research Electronic Data Capture (REDCap), which was developed at Vanderbilt University. REDCap provides a secure, web-based application based on PHP and MySQL, which can be installed locally and provides an online designer for creating data collection instruments. REDCap also provides a method for controlling user rights and user access groups as well as maintaining an audit trail[22, 23] (see Figure 24.3).

Figure 24.3: REDCap program (Courtesy Vanderbilt University)

OpenClinica is another example of a data management system. It is an open source tool which provides the ability to submit and extract data, manage protocols and other study administration tools. It enables compliance with Good Clinical Practice (GCP) and regulatory guidelines such as FDA regulations for electronic databases. OpenClinica provides a free community edition and a licensed enterprise addition. [24] CAISIS Cancer Data Management System, developed at Memorial Sloan Kettering Cancer Center, is an open source .NET application which provides eForms for study data collection. CAISIS has an active open source community supporting and enhancing the application. There are also some tools within CAISIS to import data from clinical systems.[25]

Integration of EHR data into clinical trials provides an efficient method to add routine data into the study database. While this feature is rarely available within commercial EHRs, the data from EHRs or clinical data warehouses can exported on study patients and then imported into study data management systems. The challenge is selecting the appropriate data, such as lab results from study visits, and exporting only that data. Some commercial data management systems have tools to automate this process .[26]

EHR data can be used exclusively to produce a variety of study types. For instance, epidemiologic research, studying population characteristics or trends, can be extracted from EHRs containing large groups, such as, from regional or national health systems. Biosurveillance studies are also enabled by EHR data. With daily or near real-time data on large populations, outbreaks of new infections or other disease trends can be tracked. Biosurveillance using EHR data has also been shown as a method of diagnosing Strep in real time.[27]

Identification of risk factors has been demonstrated through the use of EHR data. For instance a study from Harvard demonstrated the ability to rapidly identify risk of stroke associated with diabetes medication using signal detection analysis.[28] Another study, from Cleveland Clinic, used EHR data to predict six-year mortality risk in type II diabetes.[29] The Archimedes Model developed by David Eddy, provides predictive modeling for diabetes.[30] In addition to predictive studies, EHR data has been used in identifying post-operative complications, medication adherence and triggered adverse event reporting.[31, 32] From these and other uses, we can see that decision support is increasingly being supported by EHR data.

Comparative effectiveness research (CER) is of increasing interest related to healthcare reform and research sponsored by the Agency for Healthcare Research and Quality (AHRQ).[33] EHR data can answer some questions that clinical trials cannot and can often do so more quickly. Hoffman and Podgurski propose using EHR data to develop personalized comparisons of treatment effectiveness, applying the rich clinical data to decision support in a personalized medicine approach.[34] Observational studies, which infer causation from EHR data, can examine large cohorts who received different treatments and then evaluate the outcomes and costs associated with each. A study of diabetes management of 27,207 patients demonstrated the comparative effectiveness of using EHRs as opposed to paper records showing greater improvement in disease outcomes for those managed with EHRs.[35] The Institute of Medicine has developed a substantial workshop summary on the "Infrastructure Required for Comparative Effectiveness Research" which includes not only better research design, but a move from "siloed" evidence based medicine to "semantically integrated, information-based medicine" which requires "a substantial informatics platform to interpret, query and explore clinical data."[36]

What to do about data that is not routinely collected in EHRs? For instance, what about disease specific information which may be helpful in populating a disease registry? The solution is the use of smart forms within the EHR which are specific to a specialty clinic or treatment protocol. These forms must be designed with care to gather discrete clinical observations and judgments while being easy to complete in a busy clinical environment. Back-end integration with EHR data structure is essential.[37]

Collection of research data using medical devices is another informatics challenge. With more medical devices being integrated with the EHR or generating their own data bases, a significant amount of new clinical monitoring data is available for research. Whether these are EKG monitors, automated anesthesia records, implanted devices [38] or activity sensors [39] data collection from medical devices provides a method to quickly acquire research data for analysis.

Patient Reported Outcomes (PROs) is another area of growing emphasis in clinical research with the National Institutes of Health developing a program called PROMIS to focus attention on it. PROs "is the term used to denote health data that is provided by the patient through a system of reporting."[40] In the context of PROs, the use of tablet devices is gaining popularity as a method for collecting patient reported data at the point of care, such as, in the study of pain[41] or cognitive impairment.[42] Tablets also have broader uses, including social networking and cataloging relevant articles for research.[43]

Data Management Systems for FDA Regulated Studies

The unique requirements of the Food and Drug Administration (FDA) for data management for studies of new drugs and devices present challenges for informatics. The regulation 21 CFR Part 11: Electronic Records, Electronic Signatures[43] sets a high bar for implementing data management systems and their validation. In addition to selecting a system which is compatible with the regulatory requirements, significant validation test cases must be developed and executed. While this area is typically the purview of drug/device companies or contract research organizations, academic medical centers often require this capability to support early stage, investigator-initiated studies. Commercial systems such as PhaseForward[44] and Oracle Clinical [45] dominate this market, but open source tools like OpenClinca can also be validated in compliance with these regulations. Remote Data Capture (RDC) is a term often used for these systems which enable secure data collection over multiple study sites for large clinical trials.

Interfaces and Query Tools

In recent years, Clinical Data Repositories and Registries using EHR data have been developed at many academic medical centers. A review by Weiner et al.[46], discusses four such systems with a variety of features. One more broadly adopted tool, supported through the National Center for Research Resources (NCRR), is i2b2 (Informatics for Integrating Biology and the Bedside) which enables the secure storage and query of EHR and other data.[47]

Stanford University is creating their own clinical data repository called STRIDE, Stanford Translational Research Integrated Database Environment.[48] This repository has five functions: "Anonymized Patient Research Cohort Discovery, Electronic Chart Review for Research, IRB-Approved Clinical Data Extraction, Biospecimen Data Management, Data Management and Research Registries." Registries will become an even more important tool to track patients with chronic and rare diseases. A white paper by RemedyMD points out the essential elements of electronic registries, including robust reporting for non-technical staff, flexibility to accommodate adaptive studies, data visualization and interoperability.[49]

To support these large clinical data repositories, tools which support data mapping, semantic ontologies, and natural language processing have been developed. The National Center for Biomedical Ontology provides a repository of tools through its Bioportal for medical ontology standards and mapping. (see Figure 24.4) [50]

Figure 24.4: Bioportal Ontology Search (Courtesy NCBO)

Wynden et al. note that the two main challenges in maintaining an integrated data repository for research are, "the ability to gain regular access to source clinical systems and the preservation of semantics across systems during the aggregation process"[51]. Natural Language Process (NLP) is required when one seeks to mine clinical text notes, such as encounter notes, operative notes, radiology reports and discharge summaries. Many centers are developing such systems, such as cTAKES from the Mayo Clinic [52] and eNotes from Columbia.[53] Both examine notes and extract data elements based on structured vocabularies, such as LOINC® [Logical Observation Identifiers Names and Codes] for laboratory values.[54]

Health information exchange (HIE) is another technology which has potential for clinical research. Although developed primarily to enable care across health systems and states with various EHR implementations, it can be used in a de-identified mode to mine data for state or national trends including public health research. Health Query is a project of the Office of the National Coordinator for Health IT which is working on standard to develop a nationwide query capability.[55] If successful, it will promote epidemiologic research over broad patient populations.

Web services continue to expand in their support of many of the technologies noted above. For instance, the Columbia NLP tool utilized web services with "XML database storing documents represented using the Clinical Document Architecture (CDA) of Health Level 7 (HL7)."[56] At the Cleveland Clinic, a data warehouse and registry management tool are under development, utilizing RESTful web services to update and map data into a standard format for queries.[57]

The category of big data is now being defined in healthcare, not just business. Big data is typically defined in the multiple terabyte or petabyte range and creates unique management problems in traditional relational databases. Often, this scale of data requires cloud computing solutions for storage and analysis. A new focus on NOSQL databases and a group of tools developed by the Apache foundation is called Hadoop. "The Apache Hadoop software library is a framework that allows for the distributed processing of large data sets across clusters of computers using a simple programming model."[58] While some of the initial applications of these NOSQL databases are in genomics, other research applications, such as, exploring PACS (radiology images)[59] and multisite clinical trials may be future applications.[60] New analytic tools for large sets of EHR data are enabling data exploration. Explorys, a new spinoff company from the Cleveland Clinic using a Hadoop/MapReduce platform, is partnering with several health systems to store de-identified data for clinical exploration.[61]

Data Analysis

While Clinical Research Informatics has traditionally left the statistical analysis tools to their Biostatistical partners, with the wealth and volume of clinical data now available, some role in data analysis is appropriate. With tools like The R Project for Statistical Computing, an open source statistical package[62], there is the potential for integration of the statistical package with the data repository.[63] Tools like REDCap provide access to their API (Application Programming Interface) to connect directly to statistical programs. SAS also provides for integration of patient data from a variety of sources with tools for data cleaning, standardization and exploration.[64]

Data visualization has progressed beyond simple charts and graphs to a part of informatics which enables the researcher to see data patterns as part of data exploration and planning for analysis. Data visualization in research is in its early stages, so new approaches for how to visualize data need to be created and standardized. But when done well, visualization can help detect errors in the data and explore relationships.[65] The selection of visualization tools is key, and informaticists can aid in the selection of these tools as they do with other software. Tools like Tableau[66], Acesis[67] and functions embedded in statistical packages like SAS should be considered.

Real time analytics are also helpful tools for dealing with large datasets and clinical decision support. Real time analytics is the provision of analyzed data relatively instantly to support decision making. While this approach is relatively new in medicine, IBM's Watson project is proposing to provide this kind of service. This is closely tied to predictive analytics based on clinical data including discrete data, text and unstructured data.[68]

Future Trends

The future of eResearch is leading toward the nationwide learning healthcare system as described by the Institute of Medicine.[69] With the number of tools in active use as described in the chapter, further use and enhancement of these informatics resources combined with the broad adoption of EHRs, make huge amounts of clinical data available for analysis and further discovery. Research networks will enable collaboration that was not possible a decade ago. Research volunteer recruitment, which has been chronically low, can see new opportunities through web-based tools and social media. Study and data management, tied to paper records for so long, are now freed in a digital form for secondary use. Biosurveillance can detect new outbreaks in hours instead of weeks. Data poor registries now have the opposite challenge – large data and how to store and manage it. E-Research will enable researchers to reduce the time from "the creation and validations of new biomedical knowledge and translation of that knowledge into practice."

Key Points

- eResearch and Clinical Research Informatics have a role within every aspect of the clinical research workflow
- EHR data can be effectively utilized in clinical trials, registries, public health studies and can include research originating from EHR data
- Informatics tools are effective in recruiting subjects for clinical research
- Informatics supports the ongoing management of clinical trials including study calendars, data management, grant management and subject recruitment and consent.
- New trends include: big data, real-time analytics and data mining

Conclusion

The emergence of clinical research informatics as a field within bioinformatics has been made possible by major advances in technology and institutional support. Every aspect of clinical research now has a set of tools to support its processes. A mix of commercial-off-the-shelf tools, software-as-a-service applications (SaaS) and open source tools developed at academic medical centers have enabled this transformation. The growing availability of EMRs nationally is just beginning to make a contribution to clinical research and is poised to become a standard method for comparative effectiveness and population-based research. New devices, such as, tablets and smart phones, and the ability to obtain data from medical devices, increase the amount of data available for research. Data analysis and visualization tools enable researchers to quickly turn the data into usable information. eResearch is now maturing as a field of informatics.

Acknowledgement

The authors wish to acknowledge Jessica Pollack, RN, MSN, for her editing assistance on this chapter.

References

1. Frankovich J., Longhurst C.A., Sutherland S.M. Evidence Based Medicine in the EHR Era. N Engl J Med 2011; 365:1758 – 1759

2. National Center for Biotechnology Information: Entrez Programming Utilities Help: http://www.ncbi.nlm.nih.gov/books/NBK25500/ (Accessed November 28, 2011)

3. ClinicalTrials.gov http://clinicaltrials.gov/ct2/info/linking (Accessed November 28, 2011)

4. World Health Organization: International Clinical Trials Registry Platform Search Portal http://apps.who.int/trialsearch/ (Accessed November 28, 2011)

5. National Institutes of Health: ExPORTER http://projectreporter.nih.gov/exporter/ (Accessed November 28, 2011)

6. VIVO http://www.vivoweb.org (Accessed November 28, 2011)

7. Profiles Research Networking Software http://profiles.catalyst.harvard.edu/ (Accessed November 28, 2011)

8. SciVal http://www.info.scival.com/ (Accessed November 28, 2011)

9. Research Match https://www.researchmatch.org (Accessed November 28, 2011)

10. Harris PA, Scott KW, Lebo L, Hassan N, Lighter C, Pulley J. ResearchMatch: A National Registry to Recruit Volunteers for Clinical Research. Academic Medicine, 87:1, 1-8, 2012.

11. TrialX http://trialx.com (Accessed November 28, 2011)

12. Army of Women http://www.armyofwomen.org/ (Accessed November 28, 2011)

13. Tweet MS, Gulati R, Aase LE, Haynes SN. Spontaneous Coronary Artery Dissection: A Disease-Specific, Social Networking Community–Initiated Study. Mayo Clinic Proceedings September 2011 vol. 86 no. 9 845-850.

14. Wicks P, Massagli M, Frost J, Brownstein C, Okun S, Vaughan T, Bradley R, Heywood J. Sharing Health Data for Better Outcomes on PatientsLikeMe. Med Internet Res. 2010 Apr-Jun; 12(2): e19.

15. Pickett M, Sharp JW. Research Recruitment in Anesthesia Using EMR Data. American Medical Informatics Association Clinical Research Informatics Summit, 2011.

16. Embi PJ, Jain A, Clark J, Bizjack S, Hornung R, HarrisCM. Effect of a clinical trial alert system on physician participation in trial recruitment. Arch Intern Med 2005 Oct 24; 165(19): 2272-7.

17. Baum S. Penn Medicine to expand use of EMR software that finds clinical trial candidates, MedCity News, Oct. 31, 2011, http://www.medcitynews.com/2011/10/penn-medicine-to-expand-pilot-using-emr-app-for-clinical-trial-candidates/ (Accessed November 28, 2011)

18. Leroux H, McBride S, Gibson S. On selecting a clinical trial management system for large scale, multi-centre, multi-modal clinical research study. Stud Health Technol Inform. 2011; 168: 89-95.

19. Geyer J, Myers, K, Vander Stoep A, McCarty C, Palmera N, DeSalvo A, Implementing a low-cost web-based clinical trial management system for community studies: a case study. Clin Trials October 2011 vol. 8 no. 5 634-644.

20. Cancer Biomedical Informatics Grid (caBIG): Clinical Trial Management Systems (CTMS) Workspace https://cabig.nci.nih.gov/workspaces/CTMS/ (Accessed November 28, 2011)

21. Cancer Biomedical Informatics Grid (caBIG) Knowledge Center https://cabig-kc.nci.nih.gov/CTMS/KC/index.php/Main_Page (Accessed November 28, 2011)

22. Research Electronic Data Capture (REDCap) http://project-redcap.org/ (Accessed November 28, 2011)

23. Harris P, Taylor R, Thielke R, Payne J, Gonzalez N, Conde J, Research electronic data capture (REDCap) - A metadata-driven methodology and workflow process for providing translational research informatics support, J Biomed Inform. 2009 Apr;42(2):377-81.

24. OpenClinica http://www.openclinica.com (Accessed November 28, 2011)

25. Caisis http://www.caisis.org/ (Accessed November 28, 2011)

26. PPD: Software Products: eLoader
    http://www.ppdi.com/services/phase_ii_iiib/technology_services/software/eloader.htm (Accessed November 28, 2011)

27. Fine AM, Nizet V, Mandl KD. Improved diagnostic accuracy of Group A Streptococcal Pharyngitis with use of real-time Biosurveillance. Ann Intern Med 2011; 155:345-352.

28. Brownstein JS, Murphy SN, Goldfine AB, Grant RW, Sordo M, Gainer V, Colecchi JA, Dubey A, Nathan DM, Glaser JP, Kohane IS. Diabetes Care 33:526-531, 2010.

29. Wells BJ, Jain A, Arrigain S, Yu C, Rosenkrans WA, Kattan MW. Predicting 6-year mortality risk in patients with Type 2 Diabetes. Diabetes Care 31:2301-2306, 2008.

30. Stern M, Williams K, Eddy D, Kahn R. Validation of prediction of diabetes by the Archimedes Model and comparison with other predicting models. Diabetes Care 31: 1670-1671, 2008.

31. Murff HJ, FitzHenry F, Matheny ME, Gentry N, Kotter, KL, Crimin K,, Dittus RS, Rosen AK, Elkin PL, Brown SH, Speroff T. Automated identification of postoperative complications within an electronic medical record using natural language processing. JAMA 2011;306(8):848-855.

32. Linder JA, Haas JS, Iyer A, Labuzetta MA, Ibara M, Celeste M, Getty G, Bates DW. Secondary use of electronic health record data: spontaneous triggered adverse drug event reporting. Pharmacoepidemiol Drug Saf. 2010 Dec;19(12):1211-5.

33. Agency for Healthcare Research and Quality (AHRQ): Comparative Effectiveness Research Grant and ARRA Awards http://effectivehealthcare.ahrq.gov/index.cfm/comparative-effectiveness-research-grant-and-arra-awards/ (Accessed November 28, 2011)

34. Hoffman S, Podgurski A. Improving health care outcomes through personalized comparisons of treatment effectiveness base on electronic health records. Journal of Law, Medicine & Ethics, Fall 2011, 425-436.

35. Cebul RD, Love TE, Jain AK, Herbert CJ. Electronic health records and the quality of diabetes care. N Engl J Med 365; 9, 2011:825-833.

36. Olsen LA, Grossmann C, McGinnis JM. Learning What Works: Infrastructure Required for Comparative Effectiveness Research. Institute of Medicine, The National Academies Press, 2011, p.35.

37. Olsha-Yehiav M, Palchuk MB, Chang FY, Taylor DP, Schnipper JL, Linder JA, Li Q, Middleton B. Smart forms: Building condition-specific documentation and decision support tools for ambulatory EHR. AMIA Annu Symp Proc. 2005: 1066.

38. Greenlee R, Magid D, Go A, Smith D, Reynolds K, Gurwitz J, Cassidy-Bushrow A, Jackson N, Glenn K, Hammill S, Kadish A, Varosy P, Suits M, Garcia-Montilla R, Vidaillet H, Masoudi F. PS2-15: Linking Disparate Data Sources to Evaluate Implantable Cardioverter Defibrillator Outcomes in the Cardiovascular Research Network: Initial Lessons. Clin Med Res. 2011 Nov;9(3-4):151.

39. Stanley KG, Osgood ND, The Potential of Sensor-Based Monitoring as a Tool for Health Care, Health Promotion, and Research, Ann Fam Med July 1, 2011 vol. 9, no. 4 296-298.

40. PROMIS: What Patient Related Outcomes (PROs) Are http://www.nihpromis.org/Patients/PROs (Accessed November 28, 2011)

41. Minton O, Strasser F, Radbruch L, Stone P. Identification of Factors Associated with Fatigue in Advanced Cancer: A Subset Analysis of the European Palliative Care Research Collaborative Computerized Symptom Assessment Data Set. Journal of Pain and Symptom Management, 2011.03.025.

42. Kim H. Exploring technological opportunities for cognitive impairment screening. ACM CHI Conference on Human Factors in Computing Systems, 2011. http://dl.acm.org/citation.cfm?id=1979512 (Accessed November 28, 2011)

43. FDA: Code of Federal Regulations Title 21 http://www.accessdata.fda.gov/scripts/cdrh/cfdocs/cfcfr/cfrsearch.cfm?cfrpart=11 (Accessed November 28, 2011)

44. Oracle: Phase Forward http://www.phaseforward.com/ (Accessed November 28, 2011)

45. Oracle Clinical http://www.oracle.com/us/industries/life-sciences/046720.html (Accessed November 28, 2011)

46. Weiner MG, Lyman JA, Murphy S, Weiner M. Electronic health records: high-quality electronic data for higher-quality clinical research. Informatics in Primary Care, Volume 15, Number 2, June 2007 , pp. 121-127(7).

47. Murphy SN, Gainer V, Mendis M, Churchill S, Kohane I. Strategies for maintaining patient privacy in i2b2. J Am Med Inform Assoc. 2011 Oct 7.

48. Stanford Translational Research Integrated Database Environment (STRIDE) https://clinicalinformatics.stanford.edu/research/stride.html (Accessed November 28, 2011)

49. RemedyMD, Registries: The Missing Link Between Researchers and Clinicians. http://remedymd.com/WP-Registries-Missing-Link.php (Accessed November 28, 2011)

50. BioPortal http://bioportal.bioontology.org/ (Accessed November 28, 2011)

51. Wynden R, Weiner MG, Sim I, Gabriel D, Casale M, Carini S, Hastings S, Ervin D, Tu S, Gennari JH, Anderson N, Mobed K, Lakshminarayanan P, Massary M, Cucina RJ. Ontology Mapping and Data Discovery for the Translational Investigator. AMIA Summits Transl Sci Proc 2010, 2010:66-70.

52. Cancer Biomedical Informatics Grid (caBIG) Knowledge Center: OHNLP Documentation and Downloads https://cabig-kc.nci.nih.gov/Vocab/KC/index.php/OHNLP_Documentation_and_Downloads (Accessed November 28, 2011)

53. eNote - Electronic Notes in Medicine http://lucid.cpmc.columbia.edu/enote/ (Accessed November 28, 2011)

54. Logical Observation Identifiers Names and Codes (LOINC) http://loinc.org/ (Accessed November 28, 2011)

55. S&I Framework: Query Health – Query and Data Model Analysis http://wiki.siframework.org/Query+Health+-+Query+and+Data+Model+Analysis (Accessed November 28, 2011)

56. eNote - Electronic Notes in Medicine http://lucid.cpmc.columbia.edu/enote/ (Accessed November 28, 2011)

57. Sharp J, Pickett M. Design of a Registry Management Tool for EMR Data, AMIA Summits Transl Sci Proc 2011, 2011:119.

58. Apache Hadoop http://hadoop.apache.org/ (Accessed November 28, 2011)

59. Emory University: Dr. Fusheng Wang's Homepage http://userwww.service.emory.edu/~fwang22/projects.html (Accessed November 28, 2011)

60. ACM Digital Library: An application architecture to facilitate multi-site clinical trial collaboration in the cloud http://dl.acm.org/citation.cfm?id=1985511 (Accessed November 28, 2011)

61. Explorys http://www.explorys.net (Accessed November 28, 2011)

62. The R Project for Statistical Computing http://www.r-project.org/ (Accessed November 28, 2011)

63. Hothorn T, James D A, Ripley BD. R/S Interfaces to Databases. DSC 2001 Proceedings of the 2nd International Workshop on Distributed Statistical Computing. http://www.ci.tuwien.ac.at/Conferences/DSC-2001/Proceedings/HothornJamesRipley.pdf (Accessed November 28, 2011)

64. SAS: Health Outcomes and Patient Safety http://www.sas.com/industry/healthcare/provider/patient-safety.html (Accessed November 28, 2011)

65. Fox P, Hendler J. Changing the Equation on Scientific Data Visualization. Science 331:705-708, 11 February 2011.

66. Tableau http://www.tableausoftware.com/ (Accessed November 28, 2011)

67. Acesis http://www.acesis.com (Accessed November 28, 2011)

68. IBM: Content and Predictive Analytics for Healthcare http://www-01.ibm.com/software/ecm/content-analytics/predictive/healthcare.html (Accessed November 28, 2011)

69. Friedman CP, Wong AK, Blumenthal D. Achieving a Nationwide Learning Health System. Science Translational Medicine 2:57, 1-3, 10 November 2010.

Index

CPSIA information can be obtained at www.ICGtesting.com
Printed in the USA
LVOW09s0817270713

344677LV00001B/3/P

9 781105 437557